An Invincible Beast
Understanding the Hellenistic Pike-phalanx at War

An Invincible Beast

Understanding the Hellenistic Pike-phalanx at War

Christopher Anthony Matthew

Pen & Sword
MILITARY

First published in 2015 by
Pen and Sword Military

An imprint of
Pen & Sword Books Ltd
47 Church Street
Barnsley
South Yorkshire
S70 2AS

ISBN 978 1 78383 110 4

Printed and bound in England
By CPI Group (UK) Ltd, Croydon, CR0 4YY

Pen & Sword Books Ltd incorporates the Imprints of Pen & Sword Aviation, Pen & Sword Family History, Pen & Sword Maritime, Pen & Sword Military, Pen & Sword Discovery, Pen & Sword Politics, Pen & Sword Atlas, Pen & Sword Archaeology, Wharncliffe Local History, Wharncliffe True Crime, Wharncliffe Transport, Pen & Sword Select, Pen & Sword Military Classics, Leo Cooper, The Praetorian Press, Claymore Press, Remember When, Seaforth Publishing and Frontline Publishing

For a complete list of Pen & Sword titles please contact
PEN & SWORD BOOKS LIMITED
47 Church Street, Barnsley, South Yorkshire, S70 2AS, England
E-mail: enquiries@pen-and-sword.co.uk
Website: www.pen-and-sword.co.uk

Contents

Acknowledgements

A great number of people helped bring this project to life (in some cases literally). I wish to thank the following people without whose contributions the following work would not have been possible: Firstly, my wife Kate who has always expressed an interest in, and patience with, all aspects of my research; the following members of the Sydney Ancients re-enactment group for their invaluable support in putting many aspects of the Hellenistic pike-phalanx into action: David Armstrong, Krishna Armstrong, Peter Berecz, Paul 'Gross' Fisher, Craig Gascoigne, Mark Kelly, Peter MacKinnon, Robert Wheeler; Doug Nielsen and Anne Nielsen for their assistance and participation in the testing phases of this research; Wayne Robinson of the Routiers Pike & Musket Society for providing me with direction to information on sixteenth and seventeenth century pikes and military training; Andrew Parkin of the Great North Museum, Newcastle upon Tyne, for images and details of the Macedonian pike-butt held in their collection; Willem van Haarlem from the Allard Pierson Museum, Amsterdam for images and details of the Ptolemaic shield mould in their collection; Stavros Paspalas from the Australian Archaeological Institute in Athens for his help in obtaining images used in this publication; Dr. Bill Franzen of Australian Catholic University for his valuable input and assistance with the mathematical formulae used throughout this work; Dr. Ian Plant from Macquarie University for help with some of the ancient passages used in this research; the editors of *Antichthon* for allowing me to reproduce the information I published in the article *The Length of the Sarissa* in an earlier edition of the journal (No.46 [2012], pages 79-100); and all of the other friends and colleagues who offered their support and encouragement during the research and writing of this project.

CM
2014

List of Illustrations

List of Tables

List of Plates

Abbreviations

AAA	Ἀρχαιολογικὰ Ἀνάλεκτα ἐξ Ἀθηνῶν *(Athens Annals of Archaeology)*
ABSA	*Annual of the British School at Athens*
AHB	*Ancient History Bulletin*
AJA	*American Journal of Archaeology*
AncSoc	*Ancient Society*
AncW	*Ancient World*
AncWar	*Ancient Warfare*
AthMitt	*Mitteilungen des Deutschen Archäologischen Instituts, Athenische Abteilung*
BCH	*Bulletin de Correspondance Hellénique*
BCIS	*Bulletin of the Institute of Classical Studies*
CP	*Classical Philology*
CSCA	*California Studies in Classical Antiquity*
Historia	*Historia: Zeitschrift für Alte Geschichte*
JAAS	*Journal of the Arms and Armour Society*
JBT	*Journal of Battlefield Technology*
JHS	*Journal of Hellenic Studies*
JMH	*Journal of Military History*
JNAA	*Journal of the Numismatic Association of Australia*
JRMES	*Journal of Roman Military Equipment Studies*
OpRom	*Opuscula Romana*
PAPS	*Proceedings of the American Philosophical Society*
RÉMA	*Revue des Études Miltaires Anciennes*

Foreword

Under dynamic leaders such as Philip II and Alexander the Great, the ancient Macedonian war machine achieved a series of spectacular victories that have enshrined their names among the ranks of the most renowned generals of all time, and have ensured that the army they forged continues to be admired and studied thousands of years later. The backbone of the Macedonian army was the pike-phalanx, an intimidating mass of infantry who wielded distinctive and unusually long spears or pikes, known as *sarissas*. However, as this book points out, despite the fame and achievements of the ancient Macedonian military, profound gaps exist in our knowledge concerning almost every aspect of how this phalanx was equipped and functioned.

These uncertainties have spawned a particularly lively and long-standing set of debates regarding the weapons, organization, and tactics of the pike-phalanx, and various rival interpretations have been put forward. As usual, the root cause of the mysteries surrounding this topic is the scarcity of the ancient sources, which consist of a handful of descriptions of the Macedonian phalanx by ancient authors, some contested archaeological finds of pieces of equipment, and a scattering of often vague images in ancient art. Since all the participants in these disputes have been drawing upon this same set of ambiguous primary source material, the arguments have tended to go around and around over the same territory for generations, with the fervor of the disagreements matched only by the improvability of the assertions. The existing ancient evidence may suggest a range of potential answers to basic questions such as how long a *sarissa* was or how it was held, but within that range, it is impossible to settle upon a definitive conclusion, however vigorously proponents of the various theories might present their arguments. Thus, barring a fresh archaeological discovery, or the finding of a new text, these debates have been stuck in something of a rut whose parameters are delineated by the available evidence. The most significant accomplishment of this innovative book is that it offers a means of breaking free of this scholarly impasse by bringing a new methodology to bear upon these perennial mysteries.

One major way in which Christopher Matthew accomplishes this welcome task is by employing the techniques of what is sometimes termed 'experimental archaeology': a method in which the scholar reconstructs objects or practices from the ancient world – in this particular case, ancient arms and armour – and carries out a series of experiments in order to assess specific aspects of their characteristics and use. Vital components of such an approach are a close attention to the information provided by the full array of primary source materials as well as an adherence to vigorous standards of experimental protocol, including the meticulous recording of materials and procedures and the careful, measured, observation of results.

When properly done such a study employs a scientific methodology akin to that found in a laboratory, and one that is equally valid. Even if such an approach cannot definitively resolve a particular debate, it can often eliminate one or more suggested interpretations by revealing them to be untenable: for example, by demonstrating their incompatibility with the limitations imposed by human physiology. Therefore, experimental archaeology offers a new way to access information about the ancient world and an alternative approach to what previously seemed to be intractable problems. What gives validity to such studies is that they are based on unchanging laws of physics, geometry, chemistry, and biology. The flight of an arrow today is subject to precisely the same forces of gravity, drag, velocity, wind, and mass as one that was shot 2,000 years ago. Thus, if one can accurately replicate the physical characteristics of an ancient arrow, one can measure and discuss what it may or may not have been capable of doing, and this is true whether or not any ancient literary source explicitly discusses the details of its performance.

Such reconstructive or experimental archaeology is already well-established as a beneficial and valued tool in certain fields of study – for example, among academics that focus on the Neolithic or other prehistoric eras. In such circles, publications which describe experiments in flint-knapping or ceramic production are routine. This methodology has been a bit slower to find widespread application among archaeologists and historians who work on the classical cultures of the ancient Mediterranean. The area that has probably most embraced this approach is the study of classical warfare, a field which perhaps somewhat naturally invites attempts at reconstructing weapons and armour. This process has been facilitated by the efforts and interests of amateur military re-enactors and hobbyists, who often couple considerable scholarly erudition with practical experience in traditional artisanal crafts such as metal- or wood-working. Academics investigating the Roman legions or gladiator combat have done much useful work utilizing experimental archaeology that has greatly enhanced our understanding of these warriors. In recent years, however, a growing number of scholars have applied these methods to an exciting range of topics beyond the field of military history, producing insights into numerous aspects of ancient life and culture, from the masks worn in Greek theatrical performances to the hairstyles that adorned upper-class Roman women.

This book constitutes an ideal model of how to deploy the techniques of experimental archaeology in combination with the traditional scrutiny of textual and iconographic evidence. Christopher Matthew methodically takes us through every aspect of the Macedonian pike-phalanx, starting with its possible origins and then proceeding to analyse each individual piece of equipment that was employed by Macedonian phalangites, how it was held or worn, what its physical properties would have been, the implications of these characteristics for how it would have been held and used, how this in turn would have determined how the men were arranged in the phalanx, how groups of men would have stood and moved together, the effects the *sarissa* would have had upon an opponent, and how all of this information taken together can help us to understand how the pike-phalanx operated in actual combat. At each step of this investigation, Matthew applies the knowledge and insights gained at one stage to illuminate questions concerning the next, resulting in a cohesive and holistic analysis that resolves many long-standing debates.

To give but one example of how his practical, science-based, approach elucidates vexing issues, Matthew offers a careful examination of how the balance point of a *sarissa* would be variously affected by different-sized spearheads, the presence or absence of a buttspike, and the attachment of a metal shaft guard. With a weapon as unwieldy and ponderous as a 5m *sarissa*, the point of balance is an absolutely crucial matter that dictates where it could have been held and how it could have been used. A surprising number of earlier studies, however, have failed to fully take this simple but decisive point into account. By doing so, Matthew's research convincingly clarifies or resolves some of the perennial theories and disputes about these basic features of the weapon. Admittedly, this is not an entirely new approach to this topic; in a series of articles published in the 1970s, Minor M. Markle reported impressions derived from fashioning a replica *sarissa* and conducting a few basic tests (including placing a man holding it on a small horse). However, the experiments conducted by Matthew possess a level of scientific rigour and thoroughness previously neither attempted nor attained.

While in this foreword I have chosen to emphasize the experimental archaeological aspects of this book, I should stress that it also contains lengthy and detailed passages of more traditional forms of inquiry. It is by combining all of these analytical tools that Matthew amply achieves his goal of offering a novel and illuminating exploration of how the imposing Macedonian pike-phalanx operated and, for well over a century, dominated the battlefields of the ancient world.

Gregory S. Aldrete
Frankenthal Professor of History and Humanistic Studies,
University of Wisconsin-Green Bay.

Preface

In the fourth century BC a new style of warfare came onto the world stage which was to become the dominant form of fighting in the eastern Mediterranean for two centuries. That new system of combat was the Hellenistic pike-phalanx – a block of ranks and files made up of men wielding a long pike known as the *sarissa*.[1] The sight of such a serried array of lengthy pikes, wielded in the hands of heavily-armoured phalangites, struck terror into all who witnessed it. Polyaenus calls the formation an 'invincible beast'.[2] The Roman historian Livy states that 'the phalanx is irresistible when it is closely packed and bristling with extended pikes'.[3] At the battle of Gaugamela in 331BC, the phalanx of Alexander the Great is said to have 'rolled forward like a flood'.[4] Plutarch tells us that Aemilius Paulus, the Roman commander facing a Macedonian phalanx at the battle of Pydna in 168BC, 'had never seen a more fearful sight'.[5] Diodorus, in one of the greatest understatements found anywhere in ancient literature, states that the sight of the Macedonian phalanx merely 'causes concern'.[6]

The pike-phalanx has been an object of fierce study and endless debate almost from the moment that the first phalangite stepped forward onto the field of battle. Indeed, almost every facet of phalangite warfare is contested in some regard. Scholars have argued for ages over the origins of the pike-phalanx, its composition and deployment, the arms and armour of the phalangites therein, the formation's tactics, strategies, logistics and function. In fact, in regards to the campaigns of Alexander the Great, the only things that can be stated without touching upon one or more contested topics is that Alexander went on a long campaign against the Persians, made it all the way to India, came back to Babylon, and then died. Almost every other facet of Alexander's campaign is debated at some level or to some extent. Due to the limited source material, examinations of the conflicts of the later Hellenistic Age are equally contentious.

This lack of a total understanding of the Macedonian way of war seems quite strange when the influences it had on the development of military technology and the arts of war are considered. As well as dominating the conflicts of the eastern Mediterranean for two centuries, the principles of the Hellenistic pike-phalanx would influence later armies during the Byzantine periods, the

Muslim armies of the crusades, and even the large pike and musket armies of sixteenth and seventeenth century Europe. Yet, as noted, despite the long and extensive influence of the Hellenistic way of war, the specifics of this style of combat have been much debated. Markle states that, 'to understand these developments [i.e. the changes in military tactics brought about by the adoption of the *sarissa*] it is necessary to be clear about the limitations of the size and weight of the *sarissa*, since these factors determined how the weapon could be wielded in battle'.[7] Thus, at least according to Markle, one cannot expect to understand the broader construct of warfare in the Hellenistic Age without first understanding the very nature of the lengthy pike which changed the face of that warfare. Unfortunately, the investigation into Hellenistic warfare has not entirely held to this fundamental underlying principle. A comprehensive analysis of the behaviour of the individual phalangite, and the mechanics involved in wielding his weapons and armour correctly, on the battlefields of the Hellenistic world has really only begun with the advent of investigative techniques which have allowed scholars to move beyond the examination of literary accounts of battles and place themselves, quite literally in many regards, into the sandals of the warriors of the past.

What many of these newer techniques have demonstrated is that, in order to fully appreciate and understand what warfare in the Hellenistic world was like, things such as the sights, sounds, physical limitations and the weaponry of the time need to be carefully examined as a combined whole if modern scholarship is to have any hope of working out what it was like to take part in one of the large-scale battles of the Hellenistic Age. As Pietrykowski notes:

> Understanding the sights, sounds and emotions of the ancient battlefield is an often ignored prerequisite to understanding the temperament of the armies, the decisions of the commanders and the overall course of the action.[8]

This is where processes such as physical re-creation, experimental archaeology and ballistic science come to the fore. By re-creating the panoply of the Hellenistic phalangite to the best of modern ability, and then putting the elements of this equipment to the test in the physical world, many of the passages of the ancient texts are literally brought to life. This then allows for the validity of such passages, and of prior theories which may have been solely based upon an interpretation of them, to be seen in a way that is not possible using any other means of analysis.

The value of practical experience in historical investigation is not a new concept. Indeed, it is older than the pike-phalanx itself. Writing at the end of the fifth century BC, the Greek military writer Xenophon stated that, 'how to use a weapon can often be determined by simply holding it'.[9] At the end of the Hellenistic Age, the writer Polybius commented on the worth of prior military writings of his time by declaring that, 'any history, written solely on the review of memoirs and prior historical writings, is completely without value for its readers' – by which Polybius was declaring that, without an appreciation of the physical aspects of the topic being examined, any historical analysis or narrative is somewhat limited.[10] Such are the principles which underpin the use of physical re-creation as a means of examining history. Despite the early acknowledgement of the value of these techniques, the use of such methodologies has only been recently taken up by those investigating the past.

This, however, is not a negative reflection on the nature of previous scholarship as a whole. Rather it is the result of the focus of much of this earlier work. A great extent of the previous work done on the Hellenistic Period has focused more on the personalities and strategies of the time rather than attempting to understand the functionality of the men who fought the battles which helped shape the age. As Snodgrass notes, for the times of Philip II and Alexander the Great we know a good deal on the human and tactical side, but very little in term of arms and armour.[11]

Sekunda notes that, 'the literature dealing with all aspects of the military conquests of Alexander [the Great] is vast and growing constantly'.[12] In the opening of his book *Conquest and Empire*, Bosworth notes how, in 1976, books on Alexander the Great were being released at a rate of more than one per year.[13] Thirty-one years later, in 2007, Thomas noted that the number of books that had been published on Alexander the Great (as listed on *Questia*) numbered 4,897 and that the inclusion of articles on the same subject would more than quadruple that number.[14] If books and articles on Alexander's father Philip, earlier Macedonian history, and the time of the Successors following Alexander's death in 323BC were also counted, not to mention the vast corpus of extant ancient literature, and publications on art, inscriptions and archaeological reports, the sheer number of published works that touch on aspects of the warfare of the Hellenistic Age would be staggering. As Bosworth states, reading all of this literature would be a Herculean, and somewhat redundant, effort.[15] This is because, by and large, the majority of these works fall into one or more of a number of sub-categories (regardless of whether they are ancient or modern, book or article):

- Works which concentrate on the major personalities of the Hellenistic Age and their character.
- Works that examine specific elements of warfare in the Hellenistic Age, such as tactics, weaponry, organization or logistics, with a view to better understanding them.
- Narrative overviews and examinations of the battles and campaigns of the time.
- Examinations of topography, particularly of battlefields, to understand the part that terrain played in some of the major events of the age.
- General works on the time period, or on ancient warfare, and even on specific armies of the Hellenistic Age.
- Works that are a combination of all or part thereof.

Thus, many of these works often cover similar material – often accepting and reiterating what a previous work has stated as part of their broader narrative – and only some of them attempt to examine any part of Hellenistic warfare in a new way. Consequently, across this vast library of literary resources, all of the elements needed to compile a comprehensive understanding of the functionality of the foot-soldier of the Hellenistic Age are present – although the finer details required for such an analysis may be hidden within broader narratives, scholarly debates and/or investigations of other areas not specifically relating to that topic. However, a comprehensive understanding of the Hellenistic man-at-arms needs to be undertaken in order for a lens to be held up to other works, both ancient and modern, to see how they compare and correlate with each other. A detailed understanding of how the individual combatant of the time operated on the field of battle helps put things such as strategy, tactics, logistics, deployment, topography and operational functionality into their correct context within the larger construct of Hellenistic warfare. What this means is that, similar to the principles set out by Polybius centuries beforehand, without understanding the role of the individual (which is only really possible through physical re-creation), you cannot fully understand the role and function of the army and, therefore, it is more difficult to understand the true nature of the broader aspects of war in this time period. Consequently, a thorough understanding of how each member of the pike-phalanx performed on the battlefield is of the utmost importance to any investigation into an aspect of warfare in the Hellenistic Age.

However, unlike Polybius, it must be acknowledged that the review of prior historical memoirs and other writings is an integral part of the examination of history, and of the role of the individual in particular, and must, in fact, be the place in which any investigation has to begin. Polybius was fortunate enough to be able to compose his own analysis of the pike-phalanx in action at a time when, although in its waning years, such formations were still used on the battlefield. For the modern scholar this is no longer possible (at least not on a large scale) and, consequently, the ancient literary accounts of the pike-phalanx must always be where any investigation into this style of combat must commence. However, the review of the ancient literary record is only one element of that investigation. Along with ancient texts, any examination of warfare in the ancient world must also come to grips with other sources of evidence such as numismatics, epigraphy, archaeological interpretation, topographical analysis, art history, an understanding of logistics, physical dynamics, and every other piece of information and evidence that the researcher can access – usually through many of the modern works that have come beforehand. In essence, all of the things that are examined in the different categories of analysis must be combined to create a holistic image of Hellenistic combat.

But for all of their value, many ancient sources still only provide a limited amount of information that can be used when reconstructing an ancient battle or examining a style of ancient combat. Artistic depictions of massed combat are rare, for example, and are additionally burdened by the varying interpretation of the viewer as well as the limitations imposed by the skill of the artist and the medium they have used. Similarly, the archaeological record may provide the researcher with examples of weapons or armour from the time period in question, but provides no details about how they were used without careful and critical analysis.[16] Modern scholars derive their conclusions from the interpretation of a variety of sources and mediums found within the available evidence; yet they commonly reach greatly differing conclusions. This uncertainty shows that the precise nature of warfare in the Hellenistic world has been far from fully understood.

The use of purely theoretical reconstructions of a style of fighting does not allow for physical experimentation which is controlled, measurable and repeatable in order to test any conclusion based solely upon an interpretation of the source material. The ability to create tactical reconstructions and simulations of ancient battles, whether they are computer based or on paper, does provide a somewhat controlled means of replicating ancient warfare. However, such avenues of investigation are still limited by the amount of variables that are placed within the model and the rules under which these variables will operate.[17] It is unlikely that every possible variable can be considered and included in such an exercise.

Thus this examination follows a simple, yet still rather complex, premise: in order to fully understand the functionality of the phalangite on the battlefields of the Hellenistic world, the traditional sources of evidence, and the methodologies used to interpret them, cannot be solely relied upon. Rather, the best way to understand what the phalangite within the pike-phalanx was faced with and, in turn, to understand Hellenistic warfare even in its most basic context, the phalangite himself has to be re-created, along with his environment, by combining the information available in the traditional sources of evidence with the investigative techniques of physical re-creation, experimental archaeology and ballistic science.[18]

To aid this process, a complete set of phalangite equipment was created by skilled craftsmen; this was then worn and tested in a variety of experiments which analysed different aspects of combat in the Hellenistic world. It is only by wearing the phalangite's equipment, experiencing all of the limitations to movement and the senses that this panoply creates, extrapolating information from the imagery and descriptions of Hellenistic warfare given in the ancient source material, and then trying to replicate, via physical re-creation, the functions of the phalangite as is suggested by these sources, that the true nature of Hellenistic warfare can be understood. This process then allows for the accuracy of these sources and previous scholarship on the topic to be determined.

The design of this examination follows a series of progressive investigations with each one building on, and expanding upon, the findings of those that came before it. It begins with an examination of the scholarly debate over who created the pike-phalanx. It then moves on to examine the phalangite's primary weapon, the *sarissa*; its constituent parts, the length, weight and balance of the assembled weapon and how it could be wielded in combat. This understanding of the physical properties of the *sarissa*, and how this dictated the way the weapon could be used, allows for many theories on Hellenistic warfare to be assessed. The results of this examination demonstrate that many of the models forwarded by previous scholars for how the individual phalangite wielded his primary weapon have been based upon an incorrect interpretation of the available evidence and this has then influenced subsequent investigations into the nature of Hellenistic warfare. This, in some cases, in not totally the fault of these models as these same results show that some of the ancient literary descriptions of both the *sarissa* and the phalanx do not correlate with what is physically possible. Thus the results of these investigations not only highlight the value of physical re-creation as a research tool, but also show where these ancient sources need to be reconsidered and modern models appropriately revised.

However decades, if not centuries, of investigation into the Hellenistic pike-phalanx cannot be simply dismissed based solely on a re-examination of the phalangite's primary weapon. As such, physical re-creation, experimental archaeology and ballistic science are subsequently used to examine the functionality of the *sarissa* in the combat environment of the time by analysing such things as: how the pike could be repositioned within the massed confines of the phalanx, the strength and angle of impact of attacks made with this weapon, how long such offensive actions could be maintained, how well protected the phalangite was by the armour of the time and, therefore, what both he and his opponent would have aimed at during the course of a battle. The results of these tests are then compared to the literary, artistic and archaeological evidence to correlate (if possible) the test results with the ancient sources and modern theories. The results of these tests demonstrate that phalangite warfare was conducted in a manner that was vastly different to the way it has been interpreted by some previous scholars.

This holistic understanding of the dynamics of the phalangite in action is then used as the basis for a greater interpretation of the mechanics of Hellenistic combat; including the creation and maintenance of formations on the battlefield, tactics and strategy. Many of these facets cannot be fully understood without first developing a comprehensive understanding of the role of the individual who took part in these engagements. The use of physical re-creation allows for many of the questions relating to these areas to be addressed in a level of detail that is not possible through any other means of research. The results of this final part of the examination into the Hellenistic pike-phalanx at war show that much of the earlier written work is incomplete, inconclusive or simply incorrect.

A NOTE ON METHODOLOGY

An Invincible Beast is a follow-up work to an earlier investigation into the mechanics of warfare in Classical Greece which was published by Pen & Sword in 2012 as *A Storm of Spears: Understanding the Greek Hoplite at War*. However, *An Invincible Beast* is also intended to be a stand-alone work which assumes no familiarity with this earlier project on the part of the reader. People who have read the earlier work may find certain parts of this book (particularly areas where examination processes are described) somewhat familiar and possibly even slightly repetitive. This is not an attempt to deceive the reader by simply rehashing old material. On the contrary, this is an unavoidable by-product of using techniques that were successfully employed in a prior examination of one type of warfare in order to better understand the mechanics of conflict in a different part of the ancient world. I apologize for any sense of repetitiveness encountered by readers of *A Storm of Spears* and hope that it does not detract from the examination that follows.

A NOTE ON TERMINOLOGY

One of the biggest problems facing anyone who attempts to examine the warfare of the Hellenistic world is the varied, and often ambiguous, terminology used by the ancient writers. The Hellenistic pikeman, for example, goes by a number of names in the ancient texts – even in works (and even paragraphs) written by the same author. Aelian calls them *'peltasts'* and 'hoplites'.[19] Asclepiodotus also refers to them as 'hoplites'.[20] Polybius calls them 'hoplites', 'phalangites', and *'peltasts'*, while he also calls them, along with Arrian and Livy, *sarissophoroi*, or *'sarissa* bearers'.[21] Diodorus calls them 'hoplites' and 'phalangites', as does Livy in other parts of his work.[22] Plutarch, along with using many of the other, more common terms, also calls the pikemen of the army of Antiochus (c.196BC) *logchophoroi* (λογχοφόροι).[23] Troops are also collectively referred to as 'infantry' (πεζοί) or 'fighters' (πυκνότητα) in more generalized terms – which may or may not be including pikemen.[24] Livy calls some of the units fighting at battles such as Magnesia in 190BC and Pydna in 168BC *caetrati*, which some interpret (somewhat controversially) as a reference to pike-wielding phalangites.[25] There are numerous other examples of the interchangeable use of terms to describe the Hellenistic pikeman, but these few suffice to illustrate the point.

These various and interchangeably used terms create several problems when reading the ancient texts as many of them have dual meanings. The term 'hoplite', for example, was also the name of the heavily armoured spearman of Classical Greece who was armed and equipped in a completely different manner to the Hellenistic pikeman. The term could also be used to simply mean 'equipped'. Similarly, the term *'peltast'* can be used to distinguish a Macedonian pikeman – Aelian uses the term as meaning 'one who carries the *peltē*, the small Macedonian shield – as well as being the term used to describe skirmishers, missile troops and light infantry (which Aelian called the *psiloi*) by other writers who also carried a small shield. The only term that cannot be misinterpreted in any way is *sarissophoroi*.

Similarly, parts of the Hellenistic pikeman's equipment go under a variety of names. The pike itself, for example, could be referred to specifically as a *sarissa* or generally as a spear (*doru*) – which is the same name for the weapon carried by the classical hoplite.[26] Additionally, the pikeman's shield could be referred to specifically as a *peltē* or generally as an *aspis* – which is, again, the term used for the shields carried by the classical hoplite. In an attempt to avoid similar confusion within this work, unless the specific terminology of a passage is being addressed and is therefore required to be reproduced verbatim, the following terms will be used to distinguish the different troop types of the Classical and Hellenistic Age:

hoplite	The spearman of Classical Greece – armed with a large shield (*aspis*) 90cm in diameter and wielding a spear (*doru*) approximately 2.5m long and wielded in the right hand.
phalangite/ pikeman	The pikeman of the Hellenistic Age – armed with a small shield (*peltē*) 64cm in diameter and wielding a long pike (*sarissa*) over 5m in length and wielded in both hands.
psiloi/light troop/ skirmisher	Lightly-armoured troops fighting as slingers, archers or javelineers – often functioning in the guise of a protective screen ahead of the main formations of infantry.

Chapter One

Who Invented the Pike Phalanx?

Any reappraisal of the warfare of the Hellenistic Age must begin with an examination of the main infantry combatant of the time – a heavy infantryman (the phalangite) who fought in a massed formation of ranks and files (the phalanx), and who was armed with a long pike known as the *sarissa*. An understanding of how the individual phalangite functioned on the battlefields of the ancient world, how he interacted with those around him while in formation, and how his actions dictated, and were dictated by, the actions of others, forms the foundation upon which every subsequent investigation into the broader aspects of the warfare of the Hellenistic Age must be based. Similarly, any enquiry into the roles and functions of the individual phalangite must begin with an examination of who created this new style of warfare and when this occurred.

The Hellenistic pike-phalanx was a true military innovation in every sense of the term: it literally changed the face of warfare in the ancient world. For almost 200 years, from the rise of Macedon as a military power in the mid-fourth century BC, to their defeat at the hands of the Romans at Pydna in 168BC, the pike-wielding phalangite formed the core of almost every Hellenistic army to deploy on battlefields stretching from Italy to India. And yet, despite the dominance of this form of fighting for nearly two centuries, and despite the vast amount of modern and ancient literature dedicated to detailing the history of the Hellenistic world, and the organization of the Hellenistic formations which shaped it, there remains great contention among scholars as to who actually created the pike-phalanx and when.

Prior to the rise of Macedon as a military power, the main offensive combatant in the ancient Greek world was the heavily-armoured hoplite. The fundamentals of hoplite warfare emerged at the end of the Greek Archaic Age (c.750BC) and remained relatively unchanged for the next 400 years. The Greek hoplite was geared for hand-to-hand combat. Armed with a spear approximately 2.5m in length (the *doru*), carrying a large round shield (the *aspis*) almost 1m in diameter, and wearing a bronze helmet and breastplate, the equipment of the Greek hoplite was designed to both withstand and engage in the rigours of close-contact fighting where an opponent was never more than 2m away. Deployed in a large phalanx formation, and engaging an enemy front-on, hoplite armies dominated the battlefields of the Greek world for centuries.

Against more lightly armoured opponents such as the Persians, Greek hoplites greatly outclassed enemies who were not prepared for the same close-contact style of fighting. Testaments to the effectiveness of the hoplite in combat can be seen in the accounts of the battles of Marathon (490BC), Thermopylae (480BC) and Plataea (479BC) where the Greeks were able to inflict considerable casualties among invading armies which greatly outnumbered them, while suffering comparatively small losses of their own.[1] These successes made the Greek hoplite one of the most sought after mercenaries of the late fifth and early fourth centuries BC.

This dichotomy of war between the advantages of the hoplite compared to more lightly-armed troops only changed when Greeks fought against other Greeks. The battles of the long, bloody and costly Peloponnesian War (431-404BC) and its aftermath highlight how even-handed a hoplite versus hoplite engagement could be if both sides possessed strong enough morale to hold their ground.[2] The conflicts of this time also demonstrated where the weaknesses in the hoplite phalanx lay. When engaged by skirmishers, who fought the hoplite from a distance, who were mobile enough to attack the vulnerable flanks of the hoplite phalanx, and were quick enough to be able to withdraw while the encumbered hoplite was unable to pursue, armies composed mainly of hoplites suffered significant losses.[3] Thus, as the age of Classical Greece began to wane, the nature of warfare had fundamentally changed to a mode of fighting in which the effective use of troops armed in a variety of manners – from heavily armed hoplites, to skirmishers, to cavalry – was what now carried the day. An appreciation of tactics and generalship were now on the rise as a desired martial skill and many individuals who possessed these qualities and/or the innovative ability to come up with unconventional tactics – men such as Epaminondas and Pelopidas of Thebes, and able commanders such as Chabrias and Iphicrates of Athens – saw their stars, and those of their home states, well and truly shine.

However, following the end of the Peloponnesian War, many of the city-states of Greece were exhausted physically, politically and economically and very few states were able to hold sway over the others for any considerable period of time despite the ability of their military leaders. The ancient writer Justin surmises that, 'the states of Greece, while each sought the sovereignty of the country for itself, lost it as a body'.[4] In other words, the Greeks had fought themselves into a state of military impotence, where no one state was able to unite Greece under its hegemony, until a point was reached where the whole of Greece was in a position to be easily conquered by an outsider who had been spared the ravages of the previous decades of conflict. Into this vacuum marched the armies of the new emerging power in the Greek world – Macedon. Spared from much of the Peloponnesian War, Macedon was in a far better position than many of the Greek states to the south that it now sought to dominate.

Yet despite the internal weaknesses of many of the Greek city-states, the hoplite was still a formidable warrior on the field of battle. One of the tools which allowed Macedon to conquer Greece was the use of a style of fighting that had not yet been seen (for the most part) by many of the armies of the Greek city-states. This new style of fighting was the pike-phalanx. The employment of men armed with long pikes, rather than spears, allowed the Macedonians to fight the Greek hoplites in a conventional, hand-to-hand, manner. However, much like the skirmishers of the Peloponnesian War, the length of the *sarissa* allowed the Macedonians to engage the Greeks at a distance where the Greek hoplites were not able to respond in kind (see The Reach and Trajectory of Attacks made with the *sarissa* and [from page 167] The Anvil in Action [from page 375]).

The creation of the pike-phalanx, and particularly who created it, has been a topic of considerable scholarly debate for decades. Numerous theories have been forwarded, examining a variety of (albeit often ambiguous) ancient passages, and which offer a number of possible candidates for the creator of this new system of warfare. Adding to the scholarly controversy is the fact that the theories that have been forwarded cover a 170 year period of Macedonian history and four distinct rulers: Alexander I (495-450BC), Alexander II (370-368BC), Philip II (359-336BC), and Alexander III 'The Great' (336-323BC), with several others occasionally remembered. A critical examination of the ancient source material and the various modern hypotheses can now finally answer the question of 'who invented the pike-phalanx?'

DID AN ALEXANDER INVENT THE PIKE PHALANX?

The writer Harpocration relates how the historian Anaximenes, a contemporary of Alexander the Great who wrote the *Philippika*, a history of both Philip II and Alexander's reigns, mentions that the organization of the Macedonian military into distinct units was implemented by someone by the name of Alexander:

> Anaximenes, in the first book of the *Philippika*, speaking about Alexander, states: Next, after he accustomed those of the highest honour to ride on horseback, he called them 'Companions' [*hetairoi*], and, after he had divided the majority of the infantry into companies [*lochoi*] and files [*dekads*] and other commands, he named them 'Infantry Companions' [*pezhetairoi*], so that each of the two classes, by participating in the royal companionship, might continue to be very loyal.[5]

Aelian states that the first thing that must be done with a new levy of troops is to arrange them into files and larger units – in other words, to accustom them from the very beginning to the military structure required to create an effective fighting force.[6] This finds similarities with the passage of Anaximenes and shows that such practices were a clear part of a professional military institution under 'Alexander'. Unfortunately, which Alexander is being alluded to is not clear as there are three possible rulers that this text may be referring to: Alexander I, Alexander II and Alexander III.

THE CASE FOR ALEXANDER I
(king of Macedon 495-450BC)

The important element of Anaximenes' passage for the understanding of the development of the pike-phalanx, is his reference to the creation of the 'Foot Companions' (*pezhetairoi*) under the mysterious Alexander. Demosthenes, writing in the 340s BC, states that the *pezhetairoi* had a 'reputation for being remarkably well trained in military matters' and formed a part of the armed forces of Philip II.[7] Theopompus, a contemporary of Alexander the Great, states that the *pezhetairoi* had acted as the king's bodyguard – a definition later recalled by the lexicographer Photius.[8]

Thus the written sources concerning the role of the *pezhetairoi* seem to conflict. For Anaximenes and Demosthenes, the *pezhetairoi* seem to be a part of the infantry phalanx, whereas for Theopompus and Photius they only seem to be the guard of the king. However, Theopompus uses the imperfect tense (i.e. the *pezhetairoi* had formed a bodyguard) which suggests that Theopompus was aware that the *pezhetairoi* had once acted as the bodyguard for the king but that, at the time he was writing, they no longer performed this function. Thus, even if the *pezhetairoi* had once performed the role of a bodyguard unit, by the time of Alexander the Great they must have functioned in some other capacity – most likely as a unit or units within the main infantry of the Macedonian army. This suggests that the sources may not conflict as much as they first appear.

The nature of these ancient passages has led to many differing modern interpretations of who the *pezhetairoi* were. Connolly, for example, refers to the whole pike-wielding infantry as the 'Foot Companions' – a view shared by some other scholars.[9] Warry suggests that the term *pezhetairoi* could encompass not only the pikemen of the Macedonian army, but also the elite *hypaspists* as well.[10] However, this conclusion seems unlikely as the *pezhetairoi* and the *hypaspists* are often treated separately by the ancient writers (see following). Heckel suggests that Philip gave the term *pezhetairoi* to an elite unit of infantry and that this term was later transferred, in the time of Alexander the Great, to mean the whole pike-infantry.[11] Gabriel, on the other hand, states that Philip gave the name *pezhetairoi* to the whole pike-phalanx.[12]

Contrary to all of these suggestions, Erskine argues that, if the term *pezhetairoi* is to be applied to the whole pike-wielding infantry by the time of Alexander the Great, it is odd that the term appears only three times in the works of Arrian (see following) which, he suggests, indicates that it is a term for a select unit within the Macedonian infantry.[13] Erskine then argues that the *pezhetairoi* under Philip were a bodyguard (as per Theopompus and Photius) while, under Alexander the Great, they were a unit of infantry (as per Anaximenes and Demosthenes) with the guard role being taken up by the *hypaspists*.[14] Regardless of who the term *pezhetairoi* should be applied to, the evidence seems clear that, at least by the time of Alexander the Great, the 'Foot Companions' were operating on the battlefield as fully armed, pike-wielding, phalangites.

As noted by Erskine, in the writings of Arrian the *pezhetairoi* are specifically mentioned only three times. These passages all suggest that the *pezhetairoi* were part of the main body of pike-wielding infantry. For example, when the Macedonians deployed against the Pisidians in 334BC:

> On the right of the attacking force, Alexander had the *hypaspists* under his personal command. Next to them were the *pezhetairoi* – forming the whole centre of the line and commanded by the various officers whose turn of duty happened to fall upon that day...In advance of the right wing he stationed the archers and Agrianes, while his left was screened by the Thracian javelin units under Sitacles.[15]

Thus in this passage the *pezhetairoi* appear to be located within (or even constitute) the main infantry of the Macedonian line – quite distinct from the *hypaspists* and light troops which are listed separately. Following Erskine, the *pezhetairoi* in this passage do seem to be part of the main body of heavy infantry but the term may not necessarily be synonymous with the pike-phalanx as a whole. However, Arrian suggests that the *pezhetairoi* were under the command of various officers. The use of the plural suggests that there was more than one unit of 'Foot Companions'. This would then correlate with the idea that the term *pezhetairoi* should be applied to the main body of pike-infantry as is stated by Demosthenes. However, the important thing to note is that, even if the *pezhetairoi* are an individual unit or units within the pike-phalanx as some scholars suggest, they would still need to be armed in the manner of phalangites in order to perform within this larger formation on the battlefield. Consequently, in terms of armament at least, the members of *pezhetairoi* in the time of Alexander the Great and the pike-wielding phalangite are one and the same. Arrian recounts how Alexander the Great's heavy infantry who were facing the Pisidians were not eager to pursue their defeated foes due to their heavy equipment and the unknown nature of the local terrain.[16] The difficulty that these men had with their equipment and the terrain further suggests that the *pezhetairoi* were armed as phalangites.

In another passage, Arrian describes how Alexander was marching, 'with a unit of *hypaspists* and the *pezhetairoi*'.[17] Again, the *pezhetairoi* are separated from the *hypaspists* showing that they are not only two separately distinguishable units of troops, but the *pezhetairoi* were also no longer the guards of the king as they seem to have been under Philip. However, in this instance nothing is given to suggest the size of the unit (or units) of *pezhetairoi* so it cannot be automatically concluded that the term is indicative of the main infantry force as a whole or, conversely, an individual unit.

Later, during the mutiny of the army at Opis, Alexander reprimanded many insubordinate Macedonian units by giving their command to Persian officers, incorporating Persians into Macedonian units, and by creating 'a Persian unit of guards, Persian *pezhetairoi*, Persian 'infantry companions' (*asthetairoi*), and a unit of Persian Silver Shields'.[18] The separation of the Guards and Silver Shields from the *pezhetairoi* and *asthetairoi* here suggests that both units of 'Companions' were separate bodies of troops – but most likely still part of the heavy infantry. The reference to the coveted title of *pezhetairoi* may indicate that they were a special unit (or units) within the main infantry line.[19] These units were most likely those under the command of Alexander the Great's most reliable and experienced commanders and may have been his 'frontline' infantry

units. Alexander's cavalry was organized along similar lines with the Companion
Cavalry, of which overall command was held by Philotas at Gaugamela, while
individual units within the Macedonian cavalry were commanded by separate
officers.[20] Thus it seems that the *pezhetairoi* were a specially honoured or
designated group of units within the main line of infantry. Importantly, despite
such titles, the *pezhetairoi* still had to be equipped as pikemen. This then sheds
light on both Anaximenes' passage and on who created the pike-phalanx.

Brunt argues that, as the *pezhetairoi* already seem to exist in the time of
Philip as per Demosthenes, their creation (as per Anaximenes) has to have been
implemented by an Alexander who ruled prior to Philip.[21] Brunt goes on to state
that, in his view, 'Anaximenes undoubtedly referred to Alexander I'.[22] Brunt
concedes that he finds it impossible to believe that any one king could have
initiated all of the reforms listed by Anaximenes, and suggests that the first
reform that is listed (i.e. the creation of the Companion Cavalry) was initiated
by Alexander I, while the later reforms to the infantry, in a conflated passage,
were initiated by Archelaus who ruled from 413-399BC.[23]

Thucydides states that Archelaus 'reorganized the [Macedonian] cavalry,
the arms of the infantry, and equipment in general, so as to make the country
stronger for war than it had been under the previous eight kings'.[24] This
passage bears many similarities for the reforms of 'Alexander' detailed by
Anaximenes and may be where the theory of conflation has come from.
Hammond suggests that Archelaus had probably trained his men in the
manner of the classical hoplite.[25] English, on the other hand, dismisses the
statement of Thucydides by referring to passages of Polyaenus (2.1.17) and
Xenophon (*Hell.* 5.7.40) which suggest that Macedonia was not militarily
strong following the reign of Archelaus.[26]

During the Persian Wars (c.480BC), for example, Macedon had submitted to
Persian rule with little resistance.[27] In 430BC, Perdiccas II of Macedon was only
able to send a force of 200 cavalry to aid the city of Potidaea.[28] In 429BC, the
Macedonians did not have enough troops to withstand an invasion by Thrace,
Perdiccas 'never considered meeting [them] with his infantry', and Macedon
had to rely on cavalry and reinforcements called upon from elsewhere.[29] The
Spartans, who were fighting as allies of Macedon in 382BC, also encouraged
the Macedonians to hire mercenaries for their campaign against the Thracians.[30]

Demetrius of Phaleron, a late fourth century writer whose work is cited by
Polybius, has this to say about the military state of early Macedon:

...only these last fifty years immediately preceding our generation, you
will be able to understand the cruelty of Fortune. For can you suppose, if

some god had warned the Persians or their king, or the Macedonians or
their king, that in fifty years the very name of the Persians, who once were
masters of the world, would have been lost, and that the Macedonians,
whose name was before scarcely known, would become masters of it all,
that they would have believed it?[31]

In this passage Demetrius refers to a militarily weak Macedon without the
capacity for conquest at some time in its past. Demtrius was born in 350BC.
As such, 'these last fifty years immediately preceding our generation' would
mean that Demetrius was referring to a militarily impotent Macedonian
state around 400BC. This would correlate with the similar statements found
in the works of other writers.

All of these passages indicate that Macedon was not militarily strong
into the opening years of the fourth century BC – although according to
Thucydides it was still stronger than it had been. Thucydides specifically
states that the Macedonians under Archelaus 'were unable to stand up to
their enemies in battle' in a clear description of the military weakness of the
state despite Archelaus' reforms.[32] Justin states that Alexander I expanded
his kingdom through both diplomacy and acts of valour.[33] However,
such expansion is not indicative of the military capacities of Macedonia
as a whole or of their armament and fighting style – although references
to the small size of the Macedonian military prior to this time certainly
suggests a militarily weak state. Consequently, if Archelaus had reformed
the Macedonian army as Thucydides states, this can only have been an
improvement on a relatively minor scale which still made Macedon reliant
upon the support of allies and mercenaries for its military actions until at
least 382BC.

Several scholars have noted how the ancient Greeks and Romans
commonly ascribe blocks of reforms (either social, political or military)
to a single person from the distant past – citing the 'reforms' ascribed to
Lycurgus of Sparta, Solon of Athens and Servius Tullius of early Rome
as examples.[34] Develin states that, in a similar vein, Alexander I was,
'a reasonable candidate for mythologizing'.[35] Thus, according to this
argument, Alexander I, as a legendary king from Macedon's past, was used
by Anaximenes as an all-encompassing provider of the new Macedonian
army. Brunt, on the other hand, argues against a conflated text regarding
the establishment of the *pezhetairoi* by noting that Harpocration employs a
stylistic formula which he uses for verbatim quotations – 'Anaximenes, in the
first book of the *Philippika*, speaking about Alexander, says...' (Ἀναξιμένες

ἐν ᾷ Φιλιππικῶν περὶ Ἀλεξανδρου λέγων φησίν). As such, Brunt argues for the reforms of the Macedonian military under a single Alexander – whom he identifies as Alexander I.[36]

However, the lack of Macedonian military power into the fourth century BC is strong evidence to indicate that Alexander I is not the mysterious Alexander referred to by Anaximenes. Yet by the time that Perdiccas III was defeated by the Illyrians in 359BC, Macedon may have been able to place 10,000 men in the field.[37] Clearly something had happened prior to 359BC which had greatly bolstered Macedon's military manpower. However, a militarily weak Macedonia in the late fifth and early fourth centuries BC is not what one would expect if a series of reforms aimed at implementing a new and superior method of fighting had been introduced by Alexander I (or Archelaus for that matter) more than a century earlier. This then suggests that the reforms outlined by Anaximenes had occurred at some time after 382BC when Macedon had been encouraged to rely on mercenaries by the Spartans, and prior to Perdiccas' defeat by the Illyrians in 359BC. As a consequence, Alexander I seems unlikely to be the major reformer of the Macedonian military that is mentioned by Anaximenes, nor does Archelaus as is mentioned by Thucydides.

THE CASE FOR ALEXANDER II
(King of Macedon 370-368BC)

Some scholars dismiss the idea that the reformer mentioned by Anaximenes is Alexander II due to the short length of his reign.[38] Hammond suggests, based upon the passage of Thucydides, that Archelaus trained and equipped the Macedonian infantry along the lines of traditional Greek hoplites sometime between 413BC and 399BC while Alexander II first introduced the title of *pezhetairoi* as is stated by Anaximenes.[39] Gabriel offers that Alexander II 'made some tentative efforts to create infantry units that were armed and trained similarly to the hoplites of the Greek states' but concludes that these reforms 'did not amount to much'.[40]

Erskine suggests that Alexander II's reign was too short for a comprehensive reorganization of the military, but does offer that the insecurity of the time may have given Alexander II a particular interest in securing the loyalty of his men – which is the crux of the passage of Anaximenes – out of a need for personal safety.[41] Plutarch relates how Alexander II was involved in an internal conflict with Ptolemy of Aloros.[42] This conflict may have given the monarch the grounds to revamp his military and secure their loyalty that Erskine suggests. Develin suggests that Alexander II cannot be ruled out solely on the basis of the length of his reign (although he actually argues for Alexander III as Anaximenes' reformer) while Hammond openly favours Alexander II as the reformer of Anaximenes.[43]

Indeed, Alexander II cannot be so summarily dismissed because of the brevity of his reign. Throughout the history of ancient warfare, there are cases where a comprehensive series of military reforms have been implemented by a single commander within the space of a year. In 104BC, for example, the general Gaius Marius, implemented a series of three inter-dependent reforms aimed at altering and strengthening the structure, tactics and organization of the Roman legions – each of which would have required considerable time and resources to implement.[44] Consequently, and in agreement with Develin, Alexander II cannot be ruled out based solely upon the short length of his rule. In fact, the evidence seems to confirm that Alexander II is both Anaximenes' reformer and the monarch who incorporated the pike-phalanx into the Macedonian military.

Other fragments of Anaximenes' *Philippika* (F5-6, F27) indicate that the beginning of the work covered events prior to the time of Philip II – probably as part of an overview of Macedonian history and military affairs that acted as an introduction to the more focused timescale of the rest of the work. Bosworth, based upon the events detailed in these fragments, suggests that the material covered in the introductory chapter of Anaximenes cannot have gone beyond 359BC.[45] This seems to confirm that the reformer mentioned by Anaximenes has to be an Alexander who preceded Philip II. However, as noted, this reformer is unlikely to have been Alexander I due to the military weakness of the Macedonian state during and immediately after his reign. It then must be concluded that Anaximenes' reformer is none other than Alexander II.

What is of importance to the understanding of the development of Hellenistic warfare is the examination of the question that even if Alexander II did give part of the Macedonian infantry the name of *pezhetairoi* as detailed by Anaximenes, can it automatically be concluded that in doing so he also altered their equipment to create the pike-phalanx? Furthermore, if Alexander II did not create the pike-phalanx, but merely incorporated it into the Macedonian army, who did create it, and how did it become part of the Macedonian military? The evidence indicates that the pike-phalanx, which is commonly attributed to the Macedonians, was, in fact, invented by an Athenian who had close ties to Alexander II.

THE REFORMS OF IPHICRATES

In 374BC, the Athenian military commander Iphicrates reformed the armaments of the Classical Greek hoplite.[46] According to Cornelius Nepos:

> ...he introduced many novelties in military equipment as well as many improvements. For example, he changed the arms of the infantry. Whereas, before he became commander, they had used large shields, short spears and small swords, he, on the contrary, exchanged the *peltae* (or Thracian shields) for the round ones – for which reason the infantry have since been called '*peltasts*' – in order that the solders might be able to move and charge more easily when less burdened. He doubled the length of the spear and increased that of the sword. He changed the character of their breastplates; giving them linen ones in place of bronze...[47]

Diodorus provides similar details of the Iphicratean reforms:

> ...he devised many improvements to the tools of war, committing himself especially to the matter of weaponry. For example, the Greeks were using shields that were large and difficult to handle. These he discarded and replaced with small round ones of moderate size – thus successfully achieving two objectives: to provide the body with adequate cover, and to allow the bearer of the smaller shield, due to its lightness, to be completely free in his movements...and the infantry, who had been called 'hoplites' because of their heavy shield, had their name changed to '*peltasts*' because of the light *peltē* that they carried. In regards to the spear and sword, he made changes in the opposite direction: namely he doubled the length of the spear and made the sword twice as long as well...He made boots for the soldiers that were easy to untie and light and they continue to be called 'Iphicratids' to this day.[48]

Ueda-Sarson suggests that Iphicrates' reform was initially only an alteration to the equipment carried by the marines (ἐπιβάται) of the Athenian fleet.[49] Fuller, on the other hand, suggests that the reform was made to Thracian javelineers.[50] However, Diodorus specifically uses the term hoplite, rather than marine or skirmisher, to describe who it was that had their equipment altered and so it is uncertain where such conclusions have come from.[51]

The spear used by the classical hoplite had an average length of 255cm or about eight Greek feet.[52] By doubling the length of such a weapon, Iphicrates created a new spear sixteen Greek feet (around 512cm) in length.[53] A weapon of this length, in another reference to the result of the Iphicratean reforms, is mentioned by Arrian.[54] Markle suggests that Arrian's reference to a weapon in feet is merely an average rather than an exact figure.[55] However, Markle has not associated Arrian's passage with the reforms made by Iphicrates to the equipment of the classical hoplite and the creation of the new Iphicratean *peltast*. Rather Markle applies Arrian's passage to the equipment of the fully-formed Hellenistic phalangite who carried a longer weapon.

The word '*peltast*' has, more often than not, been used interchangeably with terms like 'skirmisher' or 'light infantry' by numerous scholars over the years. Thracian *peltasts*, for example, commonly operated in a skirmishing capacity. This is evidenced by Alexander the Great's use of them to attack elevated positions and his use of them as a screen ahead of his infantry line.[56] However, the *peltast* resulting from the reforms of Iphicrates was a completely different type of warrior and was distinctly different from the Thracian skirmisher.[57]

Interestingly, there are references to peoples in the Balkan and Black Sea regions using long spears and/or pikes prior to the adoption of the *sarissa* by the Macedonians – with some scholars suggesting that the *sarissa* had Balkan origins.[58] Didymus recounts how Philip II was wounded through the thigh by a Triballian weapon which he calls a *sarissa*.[59] The poet Lucian also describes a Thracian, armed in the manner of a phalangite, who uses his *sarissa* to slay a rider and his mount:

> ...the Thracian, standing his ground and crouching beneath his shield (*peltē*), parried his [i.e. Arsaces'] lance, and, planting his pike (*sarissa*) beneath him, pierced both man and horse with it.[60]

Best suggests that the Thracian *peltast*, rather than operating exclusively as a skirmisher, also regularly carried a long thrusting spear.[61] If true, then the Thracian *peltast* was much closer, in terms of their armament, to the Iphicratean *peltast* and the later Macedonian phalangite, and may have been a source of inspiration for the development of both of these troops.

Xenophon also describes the Armenian Chalybians as being armed in a manner similar to the Iphicratean *peltast*/Macedonian phalangite at the turn of the fourth century. He states that the accoutrements of the Chalybians were:

> ...corslets of linen reaching down to the groin, with a thick fringe of plaited cords instead of flaps. They also had greaves and helmets, and

on the belt a sword about as long as a Lacedaemonian dagger...They
also carried a spear about 15 cubits (686cm) long with a point only
at one end.[62]

Dillery calls this weapon 'impossibly long'.[63] Warner calls it 'astonishing,
[and] when not used in the phalanx unmanageably long'.[64] However, Polyaenus
describes pikes 16 cubits in length being used in 300BC which would indicate
that a 15 cubit weapon was not 'impossibly long' as Dillery would suggest.
Additionally, Xenophon makes no reference to the Chalybians using a shield.
This would have allowed both hands to be used to carry the lengthy spear in the
same manner as the later phalangite.

In a variant of the Balkan origin of the *sarissa* theory, Ueda-Sarson suggests
that Iphicrates was inspired to lengthen the spear of the hoplite based upon
observations of the use of longer weapons by both the Thracians and the
Egyptians.[65] He points to many similarities between troops in Thrace and
Egypt, both places where Iphicrates had campaigned extensively, and the later
Iphicratean *peltast*, such as the use of long spears by both and the wearing of
linen armour by the Egyptians.[66] While it is possible that Iphicrates was inspired
by what he saw in either, or both, of these places, and had possibly designed the
Iphicratean pike in order to out-reach these enemy weapons, the only thing that
can be said with any level of certainty is that, according to Diodorus and Nepos,
he seems to have been the first person to introduce such measures to Greece
through the creation of the Iphicratean *peltast*.

In his examination of the organization and tactics of Hellenistic armies, the
ancient military writer Aelian clearly distinguishes between the hoplite, the
Iphicratean *peltast* (whom he treats synonymously with the phalangite), and
the light infantry/skirmisher by stating that:

> ...infantry can come in one of three forms: the hoplites (ὁπλῖται), the
> *peltasts* (πελταστάι) and, thirdly, the *psiloi* (ψῖλοί) or 'naked' troops
> [i.e. the light infantry and skirmishers]. The hoplites carry the most
> equipment of all of the different types of foot soldier – using, according
> to the Macedonian manner, round shields and long spears (δόρατα). The
> *psiloi*, conversely, carry little or no equipment. They have neither body
> armour nor greaves, nor do they use long or round shields of any weight,
> but fight from a distance by casting missiles such as arrows, javelins and
> stones (either by hand or by using a sling). The *peltasts* wear the style of
> armour known as '*argilos*' – which is similar to Macedonian armour only
> lighter. This type of soldier carries a small shield (πέλτης μικρόν), and his

spear (δὸρατα) is much shorter than the Macedonian pike (σαρισσῶν). As such, his armour is in between that of the 'heavy infantry' and that of the 'light infantry' (being lighter than that of the hoplite but heavier than that of the *psiloi*) and this has often caused the *peltasts* to be confused with the 'light infantry.[67]

Similarly, Polybius describes the troops acting as a reserve in the army of Philip V as heavy-armed *peltasts* and describes Seleucid *peltasts*, which are elsewhere numbered as 10,000 strong, leading an attack through a breach in a city's defences.[68] These men could only be heavily armoured and are most likely phalangites.

Aelian states that the *peltast* carried a small shield called the *peltē*, wore a style of armour known as '*argilos*', and carried a spear smaller than the Macedonian pike.[69] The *peltē* is the name given to the shield carried by both the Iphicratean *peltast* and the Hellenistic phalangite – with the term *peltast* literally translating as 'one who carries the *peltē*'. As such, the shield borne by the Iphicratean *peltast* and the Hellenistic phalangite are one and the same. The word *argilos* (ἀργῖλος), used in reference to the armour worn by the Iphicratean *peltast*, means 'white clay'. This may be a reference by Nepos to the linen cuirass, or '*linothorax*': composite armour made from gluing several layers of cloth and/or hide together to create a material not unlike modern Kevlar. The finished cuirass provided similar protective properties to the bronze-plate cuirass worn by classical hoplites but weighed slightly less (as per Nepos' and Diodorus' account of the Iphicratean reforms and Aelian's description of the different armour worn by both hoplites and *peltasts*).

Representations of the Hellenistic linen cuirass in tomb paintings show many of them to have had a base colour of white, while others were brightly coloured. It may be that the cuirass was finished with a thin layer of a clay-based 'white-wash' or were painted with a chalk-based white gesso (which would account for the name *argilos*), which some soldiers appear to have left plain while others covered theirs with bright colours and intricate designs.[70] It is also possible that this 'white clay' was worked into the material itself to strengthen the fabric of the armour.[71] Regardless, of how the 'white clay' was used, the *argilos* cuirass worn by the *peltast* appears to be the same as the linothorax worn by the Hellenistic phalangite. Interestingly, many of the warriors depicted in these Hellenistic tomb paintings also wear high boots similar to the new 'Iphicratid' style of footwear mentioned by Diodorus. Pausanias, the assassin of Philip II, is also said to have been wearing high boots.[72] These provide further correlations between the equipment of the Iphicratean *peltast* and the Hellenistic man-at-arms.

The most important thing to note in regard to the reforms of Iphicrates is that a weapon doubled in length from eight to sixteen Greek feet could not have been wielded one-handed in the same manner as the traditional hoplite spear.[73] Such a weapon could only have been wielded using both hands to support its length and weight.[74] However, both hands could not have been employed to wield a lengthened spear while carrying other elements of the classical hoplite panoply at the same time – in particular the large hoplite shield. The *aspis* was a cumbersome piece of defensive equipment. Carved on a lathe to create a bowl-shaped wooden core almost 1m across, and occasionally faced with a thin layer of bronze, the *aspis* could weigh up to 7kg.[75] Such a hefty shield was carried by inserting the left forearm through a central armband (*porpax*) while the left hand grasped a cord (*antilabe*) which ran around the inner rim.[76] The bowl-like shape of the *aspis* allowed it to be positioned on the left shoulder when carried into combat which, in turn, allowed for much of its weight to be supported.[77] However, despite the fact that the grip of the left hand on the *antilabe* could be fully released, and the shield simply carried on the forearm and shoulder, the large diameter and rigid *porpax* of the *aspis* prevented the left hand from extending beyond the rim of the shield as would be necessary if it was to be used to help carry a longer and heavier weapon. The only way that such a weapon could be carried was through the use of a shield with a smaller diameter than the *aspis* which would then free the left hand to allow the longer weapon to be carried using both hands.

This was accomplished, as part of the Iphicratean reforms, through the replacement of the large *aspis* with the smaller *peltē* – the shield that continued to be used by later Hellenistic armies – as the reduced diameter of the *peltē* allowed the left hand to extend beyond the rim of the shield and so be free to help carry the weapon (see Bearing the Phalangite Panoply from page 133).[78] Due to its weight, the enlarged Iphicratean pike would have been wielded at waist level – in a variant of the hoplite's 'low position' – where less muscular stress was placed on the arms (Plate 1).[79] This, in turn, allowed for its use in battle for much longer periods of time (see Accuracy and Endurance when Fighting with the *Sarissa* from page 207). Featherstone and Ueda-Sarson suggest that a reduction in the size of the shield was one of the reforms of Philip II when he created the pike-phalanx.[80] This is clearly incorrect as a smaller shield had already been adopted to carry a pike as part of the Iphicratean reforms.

According to some ancient writers, by the time of Alexander the Great (the time Aelian says he is writing about) the pike had increased to 549cm (12 cubits) in length.[81] Thus, as Aelian correctly states, the weapon carried by the Iphicratean *peltast* (512cm) was much shorter than the pike carried by the

Macedonian phalangite. However, it is important to note that while the weapon of the Iphicratean *peltas*t was shorter than the Macedonian pike, it was still twice the length of the spear carried by the classical hoplite and had to be used in the same manner as the phalangite's *sarissa*.

The consequences of these reforms to military equipment were that the soldier of the Iphicratean reforms, as described by Diodorus and Nepos, bore all of the hallmarks of the Hellenistic phalangite: a long pike wielded low and in both hands, a small shield (called a *peltē* as per Asclepiodotus), and linen-composite armour and high boots as is evidenced by the tomb decorations of the Hellenistic period (see Plate 1).[82]

It can subsequently be concluded that the pike-phalanx common to the Macedonian armies of the Hellenistic period was not actually invented by a Macedonian at all but was, in fact, created by an Athenian. Tarn goes as far as to suggest that the Macedonians made no technical improvements to the art of war – an argument that can only assume that the pike-phalanx was invented by someone else.[83] Based on the available evidence, it is therefore possible to state that the pike (even if it was not actually called a *sarissa* at the time), and subsequently the pike-phalanx, was invented in 374BC, prior to the advent of the Hellenistic Period. What needs to be considered is how the product of such a reform ended up creating the pike-phalanx of the Macedonian military?

According to the ancient writer Aeschines, Iphicrates had close ties with the Macedonian royal house, had been adopted by Amyntas (king of Macedon 393-370BC) – the father of Alexander II and Philip II – and was a close friend of Alexander II.[84] Alexander II then ascended to the throne when Amyntas died in 370BC. Two years later in 368BC, after Alexander II had been assassinated, Iphicrates drove Pausanias, a claimant to the throne, from Macedon which set in motion a train of events which would eventually allow Philip to later become king.[85] This familial tie between Iphicrates and Alexander II would have placed Alexander in a prime position to be privy to the finer details of the Iphicratean reforms of 374BC – a reform which created the forerunner of the later Macedonian phalangite – four years prior to his ascendency to the throne. Markle states that, 'the close friendship between his [i.e. Alexander II's] family and the Athenian Iphicrates must have given him many ideas for the improvement of the military forces of Macedon.'[86] Bosworth suggests that Alexander II may have been influenced to restructure the Macedonian army into *dekads* by observations of the Theban military when they entered Macedon in 369BC and Pelopidas struck up a friendship with the young king.[87]

This seems unlikely as the Thebans did not base their military upon files of ten.

Aldrete, Bartell and Aldrete suggest that the young Philip's (he was around 8-years-old at the time of the Iphicratean reforms in 374BC) association with Iphicrates may have also influenced the Macedonian adoption of linen armour for their troops.[88] Bennett and Roberts also suggest that the young Philip may have found inspiration in seeing the Iphicratean reforms firsthand.[89] However, what such claims fail to consider is that both Alexander II and Perdiccas III would have also known of Iphicrates' reforms. As such, the adoption of the *linothorax* or any other element of the phalangite way of war by Macedon could have occurred under any of these rulers rather than Philip. Indeed, it must also be noted that Amyntas himself would have still been on the throne at the time of the Iphicratean reforms, and may have been privy to them as well. However, Anaximenes' reference to an Alexander as the reformer of the Macedonian military, rather than Amyntas, suggests that, even if Amyntas did witness the invention of the pike-phalanx by Iphicrates, he did not undertake a similar reform to his own military.

It is alternatively possible that the reforms to the Macedonian military, which included the adoption of the pike-phalanx, were begun under Amyntas following the reforms of Iphicrates in 374BC. If this was the case, they seem to have been completed by Alexander II after Amyntas had died in 370BC – an action for which Alexander is then given the full credit by Anaximenes. Subsequently, the brevity of Alexander II's reign poses no problem to understanding the development of the pike-phalanx as some scholars would argue, as the completion of a reform that had already begun years beforehand could have easily been accomplished in the space of a single year. English offers that the dates of the Iphicratean reforms are irrelevant.[90] However, due to the nature of the soldier resultant from these reforms, the close correlation between the Iphicratean *peltast* and the Hellenistic phalangite, and the close ties between Iphicrates and the Macedonian royal house, an understanding of these reforms within the broader chronological context of the military developments of Greece and Macedonia at the time is of the utmost importance.

The close association between Alexander II and Iphicrates provides something in favour of Alexander II being the reformer who incorporated the pike-phalanx into the Macedonian military that does not occur in any of the other theories that offer an alternative identity for its creator – a clear, contemporary, military precedent as to where the concept of the pike-phalanx had come from.[91] All of the other theories are reliant upon a different reading of history. If, for example, Alexander I is taken as Anaximenes' reformer, and the pike-phalanx was therefore created sometime between 495BC and 450BC, this does not explain why the Macedonian military was apparently so weak prior to 382BC. Furthermore, if Alexander I had invented the pike-phalanx, when Iphicrates created his own version of the pike-phalanx more than seven decades later in 374BC, there would have been little reason for it to be considered so noteworthy as to be recorded as such by Diodorus and Nepos. The fact that the Iphicratean reforms are treated as such groundbreaking developments in the technology of warfare by these writers suggests that nothing like it had been seen before in Greece. This in itself would dismiss any Macedonian monarch ruling prior to 374BC from contention as the creator of the Macedonian pike-phalanx.

Similarly, if a later Macedonian king such as Philip II or Alexander III is taken as the great reformer of the Macedonian military (see following), then it becomes curious as to why, when the pike-phalanx had been developed by someone with close ties to the Macedonian royal house in 374BC, such developments are assumed to have not been adopted by the Macedonians until sometime after 359BC when Philip came into power, or after 336BC when Alexander had ascended the throne. Sekunda suggests that Philip used the concepts of the Iphicratean reforms to alter the structure of the Macedonian army in 358BC and that the Macedonian word for the Iphicratean pike was 'sarisa'.[92] Rahe also suggests that it was Philip who was inspired to create the pike-phalanx, based upon the reforms of Iphicrates, during the time when he and his mother were under Iphicrates' protection.[93] Despite the errors in his interpretation of the Iphicratean reforms, Champion still calls the Iphicratean *peltast* the 'prototype for the introduction of the Macedonian pike-armed infantry' and, although he acknowledges an association between Iphicrates and Philip, does not specifically state that Philip was inspired by his reforms.[94]

It is uncertain why such models offer that Philip was inspired by the reforms of Iphicrates (no ancient text says that he was) when they had occurred sixteen years earlier and the incorporation of the details of the reform could have been implemented by any Macedonian king in the intervening period. Diodorus states that the use of the new arms and armour resultant from the Iphicratean reforms 'confirmed the initial test and from the success of the experiment won great fame for the inventive genius of the general [i.e. Iphicrates]'.[95] This raises the question: who saw these reforms as a success? Clearly, this cannot have been any of the

Greek city-states as they did not adopt a lengthy pike as the basis for their military at this time. Iphicrates' reported success could only have been with a state that adopted the results of the reforms. This is most likely Macedon during the reign of Amyntas and/or Alexander II.

THE CASE FOR ALEXANDER III – 'THE GREAT' (king of Macedon 336-323BC)

Despite this evidence, some scholars argue for a creation of the pike-phalanx after the time of Philip II. Markle suggests that the Macedonian army did not use the *sarissa* until Alexander the Great's battle at Gaugamela in 331BC.[96] Heckel and Jones interpret the passage of Anaximenes as referring to Alexander III.[97] Develin argues that what is being described by Anaximenes are changes to the structure of the Macedonian army under Alexander III rather than the invention of something new. Develin suggests that both the *hetairoi* and the *pezhetairoi* were already in existence by the time of Alexander III – as much of the available evidence would suggest – but that what Anaximenes is referring to is that Alexander III had implemented a series of reforms which altered the structure of the military.[98] Similarly, English states that the Alexander mentioned by Anaximenes is Alexander III, but that the reform being referred to was not a major one – merely a widening of the use of the term *pezhetairoi* to mean the whole infantry.[99] The initial problem with such conclusions is that, as noted earlier, the passage of Anaximenes refers to events in Macedonian history that occur prior to the reign of Philip – thus the Alexander being referred to cannot be Alexander the Great. Additionally, it should also be noted that, even if Anaximenes is referring to Alexander III, changes to military organization and terminology as some scholars suggest are not necessarily a direct correlation with the development of a new type of weaponry and style of fighting.

Markle's theory for a later creation of the *sarissa* in 331BC holds little merit. Polybius, for example, commenting on the eye-witness account of the battle of Issus by Callisthenes, specifically states that Alexander's men were 'carrying the *sarissa*' (σαρισοφόρον).[100] As Callisthenes was present at the engagement he is unlikely to have got this wrong. Consequently, the *sarissa* had to have been in use at least by the time of the battle in 333BC. However, does this then mean that Alexander may have invented the pike-phalanx at some time prior to his departure for Asia in 334BC?

Markle, arguing against the popular concept of Philip II as the reformer of the Macedonian military (see following) in favour of Alexander the Great, states that there was no real precedent for *sarissa*-armed infantry that Philip may have copied.[101] Griffith similarly states that there is no reference to a pike-phalanx prior to 360BC.[102] Such statements are clearly incorrect as the pike-wielding *peltast* of the Iphicratean reforms had been invented before Philip became king of Macedon. It also seems odd that, if Philip can be dismissed as the creator of the pike-phalanx due to a lack of prior precedent as these scholars claim, Alexander III can then be automatically assumed to be the creator of this formation when no prior precedent is ascribed to him either.

As part of his argument against earlier possible reformers (and in particular Philip II) introducing the pike-phalanx, Markle cites a passage of Polyaenus which recounts how a force of Macedonians was defeated by the Phocian Onomarchus in 353BC:

> When Onomarchus was deploying against the Macedonians, he placed a crescent-shaped ridge to his rear, concealed men on the peaks at both ends with rocks and rock-throwing engines, and led [the rest of] his forces onto the plain below. When the Macedonians came out against them and hurled javelins (ἠκροβολίσαντο), the Phocians pretended to flee into the hollow centre of the ridge. As the Macedonians, pursuing with an eager rush, pressed them, the men on the peaks threw rocks and crushed the Macedonian phalanx. Then indeed Onomarchus signaled for the Phocians to turn and attack the enemy. The Macedonians, under attack from behind while those above continued to throw rocks, retreated rapidly in great distress...[103]

Markle suggests that the Macedonians being referred to in this passage were not armed with *sarissae* otherwise their losses would have been less.[104] It is uncertain how such a generalised conclusion has been reached. Some of the forces opposing the Macedonians in this encounter were arranged on top of a crescent shaped ridge, were armed with stone throwing catapults, and hurled rocks and other missiles down upon the Macedonian line as it advanced. Regardless of what type of weapon they were armed with, the Macedonians would have clearly suffered losses in such an engagement.[105] At the Persian Gates in 330BC, the forces of Alexander the Great (some of whom would have been armed with the *sarissa* even following Markle's hypothesis) likewise suffered losses, from defenders who hurled and catapulted missiles down from the ridges above, and were similarly forced to withdraw.[106] Years earlier in Thrace, Alexander had been forced to come up with inventive counter-measures to protect his troops against wagons that were rolled down from heights onto his infantry.[107] Clearly, suffering casualties due to missile attacks delivered from high ground, or not, was not singularly dependent upon the type of weapon the troops were carrying.

Markle may be basing his conclusion that the Macedonians were not armed with pikes on the way that Polyaenus describes the Macedonians as skirmishing with missiles (ἠκροβολίσαντο). The term ἠκροβολίσαντο means something like 'hurled javelins', 'threw from afar' or 'fought with missiles'. One way of interpreting this statement is as a reference to the presence of light troops and skirmishers within the force under Philip's command. The phalanx itself (τὴν Μακεδονικὴν φάλαγγα) is only referred to later in the passage in reference to the troops that were drawn into the trap and attacked with stones from above.[108] Furthermore, both pikemen and skirmishers could be part of a 'phalanx' if the term is used generically to describe a deployed army as a whole. Skirmishers regularly formed a screen of advance troops ahead of the main infantry line, or protected the wings as part of their deployment within the 'phalanx'. The two terms are not mutually exclusive and the possibility that Philip had both skirmishers and phalangites in the encounter with Onomarchus cannot be discounted.

That the skirmishers appear to be the first to engage the Phocians in Polyaenus' description of the encounter follows the role played by such troops in many engagements that contain a pike-phalanx. When Alexander engaged the Pisidians in 334BC, archers and javelineers formed a screen across the front of his line.[109] Similarly, at Gaugamela in 331BC, missile troops formed a screen in advance of the right wing.[110] Thus the skirmishers of Philip in Polyaenus' account of the battle were simply part of a larger force – initially deployed ahead of the main line where they were the first to engage. Based upon the evidence, it seems likely that the army under Philip contained both a heavy infantry phalanx and contingents of skirmishers. What is unfortunately lacking is any description of how this heavy infantry was armed. Polyaenus' other reference to Philip's men training with the *sarissa* (4.2.10) would suggest that at least part of his army was using pikes. If it is assumed that phalangites did engage the Phocians of Onomarchus, it can only be concluded that, contrary to Markle's hypothesis, such weapons, and their required modes of deployment, had been incorporated into the Macedonian military sometime prior to 353BC.

Markle goes on to argue that it would have been unlikely for Philip to use a revolutionary tactical formation in the first major battles of his reign.[111] Again, it is uncertain where such a conclusion has come from. The whole basis of the technological advances of warfare (in any age) is to use something revolutionary which provides a tactical, strategic and/or technological advantage on the battlefield. There is little reason to conclude that Philip would not have used a formation such as a pike-phalanx, with all of the benefits that it gave, regardless of whether it was new or not. Polyaenus' reference to Philip's troops training with *sarissae* suggests that it was a standard piece of equipment at this time. This is consistent with the pike-phalanx being adopted some years earlier under Amyntas and/or Alexander II. Furthermore, as Iphicrates had already created such a formation twenty-five years earlier, it would have hardly been revolutionary by the time of Philip as Markle suggests.

The one thing that counters Markle's hypothesis is the archaeological record. Parts thought to have come from a *sarissa* were excavated from outside of a tumulus grave at Vergina in Macedonia which Andronicos dated to 340-320BC.[112] This would then place the weapon late into the reign of Philip at the earliest and the reign of Alexander III at the latest. If the earlier date is taken as accurate, then any theory suggesting that Alexander the Great was the creator of the pike-phalanx cannot be correct. The interment of this weapon in a tomb of this time period suggests that the pike already existed, and was probably in regular use, prior to the time of Alexander the Great. This then provides a physical correlation with Polyaenus' reference to the army of Philip training with pikes.

THE CASE FOR PHILIP II
(king of Macedon 359-336BC)

By far the most popular candidate among scholars for the creator of the pike-phalanx is Philip II of Macedon – the father of Alexander the Great.[113] Many of these theories cite various ancient passages as evidence to support their claims, and offer that the invention of the pike-phalanx was part of a series of reforms to the Macedonian military made by Philip which greatly increased its battlefield effectiveness. However, a critical review of the ancient literary material demonstrates that, not only does no ancient text actually state that Philip created the pike-phalanx, but that it is also unlikely that he even adopted this system of warfare from another source.

Connolly states that before the reign of Philip, the Macedonian infantry was lacking in discipline, training and organization and that Philip 'immediately set about reorganizing his army [to] bring it into the modern world' – dating these reforms to some point in time between 359BC and 345BC.[114] This sentiment is echoed by Cawkwell who states that Philip 'inherited a disorganized and defeated army, and he bequeathed to Alexander a superb instrument of war'.[115] Such claims seem to be based upon a passage of Diodorus which states that

Philip 'left armies so numerous and powerful that his son Alexander had no need to apply for allies in his attempt to overthrow Persian supremacy'.[116] English suggests that the proof that the Macedonian army was not well organized or trained prior to the reign of Philip is that they were defeated by the Illyrians in 359BC. English therefore concludes that any advancement in their effectiveness overall had to begin after this time.[117] Griffith suggests that the creation of the new Macedonian army was not done in one fell swoop but over all of the years of Philip's reign, but that the *sarissa* was used 'for the first time' in Philip's counter-offensive against the Illyrians in 358BC.[118] All of these theories imply that Philip was the creator of the pike-phalanx – although many of them do not specifically state so. Bennett and Roberts, on the other hand, claim that the pike-phalanx 'sprung fully formed from the head of that extraordinary monarch Philip II'.[119]

It is interesting to note the very non-committal nature of many of the claims forwarded by modern scholars in favour of Philip as creator of the pike-phalanx. Cawkwell, as noted, declares that Philip 'inherited a disorganized and defeated army, and he bequeathed to Alexander a superb instrument of war'. Both aspects of this statement are true. Perdiccas III had been defeated by the Illyrians in 359BC with the loss of nearly half of Macedon's military might.[120] Thus Philip did 'inherit a disorganized and defeated army' as Cawkwell claims. Furthermore, it can hardly be doubted that the army used by Alexander the Great in his conquest of Persia was, as Cawkwell calls it, 'a superb instrument of war'. Clearly, under Philip the position of the Macedonian military was greatly improved between the time that he came to the throne in 358BC and his assassination in 336BC. It seems to be the case that some scholars accept that the Hellenistic pike-phalanx was a fully formed and effective military institution by the time of Alexander the Great. These scholars also recognize that the Macedonian army had been defeated in 359BC – just prior to the reign of Philip II. Consequently, it seems that some scholars extrapolate backwards from the time of Alexander to conclude that such a strong formation as was used by Alexander cannot be the same kind of one that was defeated in 359BC, and therefore it had to have been created at some time in between these two events during the reign of Philip. However, can it be automatically assumed that this 'improvement' involved the creation of the pike-phalanx?

Using the defeat of 359BC to date the creation of the pike-phalanx actually contradicts itself. If it is assumed that the pike-phalanx had to be created by Philip because the Macedonians would have otherwise not been defeated in 359BC if they were armed with the *sarissa*, it then has to also be assumed that both the army of Philip and that of Alexander the Great would have been armed with pikes. However, both Philip and Alexander suffered defeats and military setbacks as well – such as Philip's defeat by the Phocians in 358BC, and Alexander being forced to initially withdraw from the Persian Gates in 330BC.[121] Yet a reference to a single defeat of a Macedonian army, possibly 10,000 strong, in 359BC is the very basis for the conclusion that it was not well organized, effective and/or armed with the *sarissa* prior to the reign of Philip. Does this then mean, following this same line of argument, that because both Philip and Alexander also suffered defeats their armies were not well organized, effective and/or armed with the *sarissa*? Clearly, this cannot be the case. Additionally, such things as the military capacity of the Illyrians, the command ability of Perdiccas III, the tactics employed on the day and the variable fortunes of war, seem to have never been considered as contributing factors as to why the Macedonians were defeated in 359BC. It is simply assumed that because they were defeated, they could not have been well armed. This can only be regarded as extreme supposition. Consequently, and importantly for the understanding of the development of the pike-phalanx, any reference to the defeat of the Macedonians in 359BC can in no way be taken as indicative of their armament at this time.

The use of backward extrapolation to support Philip as the creator of the pike-phalanx, and the inherent problems contained within such an approach, is an aspect of the examination of warfare in the Hellenistic Age that has not been lost on some other scholars. Markle, for example, declares that there is no evidence for the 'orthodox view' that Philip devised the *sarissa*-armed infantry phalanx.[122] Dickinson states that the creation of the pike phalanx 'has been extrapolated with some controversy to the beginning of the reign of Philip II'.[123] Such extrapolation can be seen in the work of Cawkwell who states that, because of the lack of detailed evidence relating to Philip, 'the military career of Philip may be admired but not discussed'.[124] However, Cawkwell goes on to suggest that, because Alexander the Great inherited Philip's army, and because he concludes that Alexander would have had little time to reorganize the army that he had inherited into the professional institution that it was, it is possible to extrapolate backwards from Alexander's campaign in Illyria in 335BC to determine what Philip had done to the Macedonian military beforehand – which, Cawkwell says, was to invent the pike-phalanx.[125] A more extreme form of extrapolation can be seen in the work of Everson who, as he believes that Philip invented the pike-phalanx, and because he believes that Philip may have been inspired to do so by what he saw when he was held hostage as a boy in the city of Thebes, concludes that one of the reasons why the Thebans were able to defeat the Spartans at Leuctra in 371BC was because they may have been using pikes rather than spears.[126] However, despite the fact that there is no reference to the Thebans employing pikes at Leuctra, other evidence additionally indicates that the Thebans continued to use the equipment of the standard Greek hoplite until 335BC at least (see following).

Gabriel states that 'the Macedonian army was born in crisis and Philip invented its structure, weapons and tactics in response to that crisis.'[127] In the year prior to Philip's ascendancy to the throne (i.e. 359BC), the Macedonian army may have been 10,000 strong. In that same year the Macedonian king, Perdiccas III, was defeated by the Illyrians with the loss of 4,000 men (and the king himself).[128] Gabriel suggests that Philip had to rebuild the army after this event and that part of this reform was the creation of the pike-phalanx.[129] Diodorus tells us that in the following year (358BC), Philip was able to place 10,000 men back in the field.[130] Thus Philip seems to have been able to rapidly replace almost half of the entire military might of Macedon. By 338BC Macedon, with the aid of allies, was able to place 30,000 infantry in the field for the battle of Chaeronea.[131] By 334BC, Macedon could field around 30,000 men just on its own.[132] While there is clearly a major act of recruitment being undertaken by Philip from the very beginning of his reign, can it also be assumed that in this time the 'structure, weapons and tactics' of the Macedonian military, as Gabriel puts it, were reformed as well?

The main problem with assigning the creation of the pike-phalanx to Philip is that no ancient source actually states that he did so.[133] Nonetheless some elements of claims made in favour of Philip cannot be totally dismissed. There is evidence which indicates that at least some parts of the Macedonian army were disorganized and lacking discipline when Philip came to the throne in 358BC. According to Arrian, when dealing with mutinous troops at Opis in 324BC, Alexander the Great harangued his men by declaring that:

> Philip found you a tribe of impoverished vagrants, most of you dressed in skins, feeding a few sheep on the hills and fighting feebly enough, to keep them from your neighbours – the Thracians, Triballians and Illyrians. He gave you cloaks to wear instead of skins. He brought you down from the hills onto the plains. He taught you to fight on equal terms with the enemy on your borders until you knew that your safety lay not, as once had been, in your mountain strongholds, but in your own valour.[134]

However, while Arrian does infer that Philip had gained control of an army which was, at least in part, undisciplined and ineffective, Arrian does not state that Philip, as part of his 'improvement' of the Macedonians, created a new style of warfare, nor does he specifically refer to the formation of the pike-phalanx. It is more likely that this passage refers to Philip's annexation of the regions of Upper Macedonia (the areas bordering the Thracians, Triballians and Illyrians as Arrian stated), which were regions that had enjoyed a semi-nomadic pastoralist existence, and the incorporation of the men from these regions into the new Macedonian military early in his reign.[135] As if in confirmation of this, Arrian has Alexander go on to state that Philip took these newly subject peoples and 'made you city-dwellers; he brought you laws; he civilized you'.[136]

Unfortunately, what Arrian fails to elaborate upon is the exact extent to which Philip reformed the Macedonian army that now incorporated these men from Upper Macedonia and what, if anything, these reforms were based upon or resulted in.[137] The only thing that Arrian clearly states is that the Macedonians (and possibly just those from Upper Macedonia at that) were taught to fight effectively against their neighbours. This cannot be interpreted as the invention of the pike-phalanx. None of the cultures neighbouring Macedonia that Arrian lists fought in the style of the pike-phalanx (with the possible exception of the Thracians), therefore improving the Macedonian way of war so that it was only 'on equal terms' with such opponents can hardly be a reference to the adoption of a new and dominant form of fighting. Furthermore, does 'on equal terms' mean the adoption of a new or similar style of fighting, or is it a reference to the Macedonians simply being more effective on the battlefield? Arrian's passage only states that Philip made the Macedonians more effective in their conduct of war and does not attribute any changes in the actual weaponry, and the subsequent required changes to deployment and style of fighting, used by the Macedonians to Philip. Subsequently, while Philip may have improved the structure and effectiveness of the Macedonian military, he may not have concurrently altered its weaponry as well.

Hammond interprets Arrian as saying that what Philip was doing was 'beginning to train an army of infantry on professional lines and equip them at his expense'.[138] While the gist of Arrian's passage suggests the improvement of elements of the Macedonian military, it does not indicate a move towards the professionalisation of the army as a whole, or even part of it. No doubt such claims have again sprung from the foreknowledge that the army of Alexander the Great was professional in its nature, and the generalised assumption that it was not previously. However, Arrian only outlines an improvement to the combat effectiveness of the Macedonian army – there is no reference to professionalisation. Furthermore, while the provision of cloaks mentioned by Arrian could be interpreted as the equipping of troops at state expense in some capacity, the passage does not elaborate on a reform to supply arms, armour and other equipment at state expense as a move towards professionalisation would warrant. Alternatively, Arrian's reference to the exchange of skins for cloaks may simply be a metaphoric expression to further illustrate how Philip had 'civilized' the nomadic pastoralists from northern Macedonia.

This acknowledgement that the Macedonian military became professional, but that the time of its evolution is unknown, is best summed up by Snodgrass who states:

> The poor and scattered peasant population of Macedonia could not have been expected to arm itself to the standard of the Greek hoplite, and it did not attempt to. Instead, it was turned, certainly by Philip II's time and conceivably even earlier, into an instrument that was altogether new, although its name – the phalanx – was a word as old as Homer.[139]

However, there are several passages within the ancient literature which indicate that the Macedonian military had already moved towards a certain level of professionalism prior to the reign of Philip II. Anaximenes' reference to troops under Alexander II being formed in to files and other units, for example, coupled with Aelian's reference to this being the first thing that needed to be done with new recruits, would suggest a military institution with a fairly set and rigid structure.[140] Aelian also states that the Macedonian style of countermarching (i.e. changing the facing of a unit by 180°) was abandoned by Philip II and Alexander the Great in favour of the Lacedaemonian method of countermarching.[141] This shows that the Macedonian method had been developed and used in the past and, as a result, a semi-professional military system, at least, had to have been in place prior to Philip II which would have undertaken such drill movements. Furthermore, Frontinus states that Philip reduced the number of attendants and servants attached to each *dekad* of the infantry.[142] This shows that the terminology and structure established by Alexander II (as detailed by Anaximenes) was still in effect by the reign of Philip II. Finally, when Alexander recruited troops for his campaign in Asia in 334BC, he enrolled men who had previously served with Philip II and with both Alexander II and Perdiccas III.[143] This demonstrates that at least some of the troops serving under Alexander the Great had been part of a military institution going back to at least 370BC – which would then correlate with Alexander II being the creator of the 'professional' Macedonian army with the adoption of the pike-phalanx.

Despite such references, some scholars suggest that Philip was inspired to create a new Macedonian army, based upon the pike-phalanx, by his study of the military conduct of the Thebans, and in particular their leader Epaminondas – a man whom Adcock calls the equal of Frederick the Great – when he was held hostage in the city of Thebes as a boy.[144] Plutarch tells us that the king of Macedonia at the time, Alexander II, had been involved in an internal power struggle with Ptolemy of Aloros which the Theban, Pelopidas, had been invited to arbitrate. Following the conclusion of the negotiations, Pelopidas 'took Philip, the king's brother, and thirty other sons of leading men in the state and brought them to Thebes as hostages'.[145] This period of detention may have had an impact on the young Philip. Yet the question remains: was he inspired to create the pike-phalanx during this time?

Cawkwell states that Philip 'derived from and developed the art of war as he saw it practiced in Greece, especially by the Thebans under Epaminondas'.[146] Warry suggests that, while a hostage in Thebes, 'Philip acquired an admiration for...the use of massed infantry as developed by Epaminondas'.[147] Connolly says that, as a hostage, 'the great Theban's [i.e. Epaminondas'] views were not wasted on Philip' and goes on to say that Philip reformed his heavy infantry into the phalanx 'no doubt based upon the Theban model, and later changed its organization from *dekads* (base units of ten) to multiples of eight sometime between 359BC and 345BC'.[148] Fuller suggests that Philip 'became grounded in the Theban art of war' while in Thebes due to his 'acquaintance with Epaminondas and Pelopidas'.[149]

There are several issues with such conclusions. Firstly, the ancient sources do not suggest that this occurred in any way shape or form. Plutarch only states that young Philip 'was believed to have become an ardent follower of Epaminondas, possibly because he comprehended his efficiency in wars and command'.[150] Thus Plutarch presents us with a passage of multiple conjectures: it is possible that Philip became an admirer of Epaminondas, although this is not stated as definite, and that, if it is true, then one possible reason for this admiration is that Philip may have recognized Epaminondas' abilities as a commander. Undoubtedly the fame of Epaminondas following the Theban victory over the Spartans at Leuctra in 371BC would have been considerable and it is perfectly reasonable to assume that a young man like Philip would have held the great general in high regard – just as Plutarch states. However, Philip being an admirer of Epaminondas is one thing, having him regularly tutored in the arts of war by Epaminondas (and by Pelopidas as well according to some) is something else entirely and completely foreign to what is stated in the ancient texts. Furthermore, Appian states that Epaminondas was away on campaign at the time when the young Philip was taken to Thebes.[151] Thus it is uncertain how Epaminondas could have even acted as a tutor to Philip. However, it does explain why Pelopidas, and not Epaminondas, had gone to Macedon to arbitrate the dispute between Alexander II and Ptolemy.

Indeed, Plutarch does not mention whether Philip even met Epaminondas while he was in Thebes, much less underwent regular instruction. Plutarch states that Philip's host and benefactor while he was in Thebes was Pammanes, the father of Epaminondas, and not Epaminondas himself.[152] It is highly likely, given this and the time at which Philip was in Thebes, that he would have heard a great deal about the Theban victory at Leuctra a few years earlier. In this engagement, the novel deployments and tactics executed by both Epaminondas and Pelopidas were crucial elements in securing their victory.[153] However, an appreciation for what these two commanders had accomplished on the field at Leuctra is completely different from having one or both of these men instruct the young Philip on the finer points of Theban military doctrine. Plutarch's statement actually provides no basis for the concept of a military education for Philip in Thebes at all. Nevertheless this is the foundation for many of the hypotheses suggesting that Philip was inspired to create the pike-phalanx. These theories can only be considered an extreme extrapolation of the passage at hand.

Additionally, even if it is accepted that Philip did gain some instruction in the arts of war while he was in Thebes, what exactly would he have been taught? Clearly any lesson on Theban warfare given to Philip (regardless of whom the lesson may have been given by) would have had nothing at all to do with the pike-phalanx as the Thebans, unlike the Macedonians, neither employed the pike as their main offensive weapon, nor were their formations of hoplites based upon a multiple of ten (as the early Macedonian pike-phalanx was) or sixteen (as per the later Macedonian phalanx).

At the battle of Delium in 424BC, for example, the Thebans deployed in a formation twenty-five ranks deep.[154] At the battle of Leuctra in 371BC they deployed 'at least fifty ranks deep'.[155] While both of these could be seen as multiples of ten (2½ files of ten at Delium, and five files of ten at Leuctra), Xenophon's reference to the Thebans being deployed 'at least fifty ranks deep' at Leuctra would suggest something larger which would then dismiss the use of a phalanx based upon a multiple of ten. The conclusion for the Thebans using files of ten men seems to be merely based on the numerical expediency that both twenty-five and fifty are divisible by ten (although not to whole numbers in the case of twenty-five). But why does modern scholarship conclude that the Theban military was based upon a multiple of ten? Why not five, or even twenty-five? Both of these numbers would at least divide into the figures given for Theban deployments at Delium and Leuctra by whole numbers. It seems that such conclusions are only based upon an extrapolation which assumes that Philip was inspired to create a pike-phalanx based upon files of ten while detained in Thebes and therefore this must have been how the Thebans organized their military. Yet even here such an assumption can only be based upon a conclusion that the Theban deployments at Delium and Leuctra were not improvisations to cater to the tactical requirements of the specific engagements, and the strategies of the individual commanders at the time, but were standard Theban practices – a conclusion for which there is no confirmatory evidence.

Units of ten ranks deep are only referred to as improvised units for classical hoplite formations – and even here only in general terms. For example, the Athenians at the Piraeus in 404BC were deployed 'not more than ten deep', and the Spartans at Mantinea in 370BC were deployed 'nine or ten shields deep'.[156] importantly, both of these formations are not references to the Thebans. *Dekads* (specific units of ten men) are only referred to in relation to later Hellenistic formations from the time of Alexander II onwards. Frontinus states that in the time of Philip II each unit of ten men was given one pack animal to carry their heavy equipment.[157] This suggests that standard deployments to such a depth were a Macedonian military characteristic, from the very beginning of Macedon's rise to dominance, rather than a Greek one.[158]

Furthermore, in all of the battles involving the Thebans into the time of Alexander the Great, their infantry are armed as hoplites with shields and spears, rather than pikes. Diodorus, for example, tells us that when Alexander the Great attacked the city of Thebes in 335BC, three years after they were 'introduced' to the Macedonian pike at Chaeronea, the Thebans were still fighting with swords, javelins and spears rather than the *sarissa*.[159] Consequently, even assuming that Philip had been in a position to observe Theban military training and deployment while he was detained there as a boy, it is uncertain how such practices could have provided the inspiration for the later Macedonian phalanx of phalangites armed with a lengthy pike and arrayed ten or sixteen ranks deep.

Featherstone goes as far as to state that, upon returning to Macedon, Philip 'improved upon the Theban phalanx, lengthening the pikes (*sarissa*) to sixteen, eighteen and even twenty-four feet'.[160] However, no ancient text indicates that Philip lengthened the pike to create the Macedonian phalanx, and *sarissae* over 12 cubits (576cm) long were not used until much later in the Hellenistic Period. Similarly, Skarmintzos suggests that, while a hostage in Thebes, Philip noticed that only the first two ranks of the traditional hoplite phalanx engaged in combat. Skarmintzos then offers that, upon his return to Macedon, Philip doubled the length of the hoplite spear, called it a '*sarissa*' and, due to the need for this new weapon to be carried in both hands, reduced the size of the shield as well.[161] All of these were the results of the reforms of Iphicrates and not inventions of Philip. As previously noted, the ties between Iphicrates and Amyntas/Alexander II would suggest that the adoption of a long pike and small shield had already taken place in Macedon several years prior to Philip's detention in Thebes.

It is also interesting to note that Philip himself may have been a provincial governor of eastern Macedonia shortly after his return from political detention in Thebes. Carystius of Pergamum claims that Euphraeus, an associate of Philip's brother Perdiccas III, encouraged the king to assign part of the kingdom to Philip to rule.[162] Anson suggests that this region was eastern Macedonia, as control of the troubled west would likely have been retained by the king himself.[163] If this is the case, then Philip may have seen the long spears used by the Thracians. Worthington suggests that Philip may have been encouraged to experiment with the military, based upon what he had seen in Thebes, while he was a governor by the ruling monarch of the time, Perdiccas III.[164] However, possible exposure to long Thracian spears does not necessarily mean that Philip was inspired to create the pike-phalanx based upon what he may have seen. Justin says that the Macedonians were perpetually in conflict with both the Thracians and the Illyrians.[165] Consequently, any ruler who came into contact with the Thracians would have recognized the value of using a longer pike to outreach the lengthy Thracian spears. Indeed, Amyntas had fought against the Thracians over the region of Olynthus in 383BC.[166] However, the militarily weak Macedonians at this time had to rely on assistance from the Spartans and other mercenaries in order to put an effective force into the field. Amyntas' experiences during this campaign, and his witness of the Thracian long spears, may have been the motivation behind the adoption of the Iphicratean *peltast* around 374BC – the use of a lengthy pike to outreach the spear used by a long standing enemy being one practical application for the adoption of a new style of fighting.

This would then correlate the beginnings of an adoption of the pike-phalanx under Amyntas, and the completion of such a reform under Alexander II as is suggested by Anaximenes.

Furthermore, the Illyrians appear to have been using shields very similar to the round Macedonian *peltē* by the end of the fifth century BC. A plate from Gradiste, Croatia, dated to c.400BC, depicts both cavalry and infantry carrying shields which, both in relative size and decoration, are almost identical to their later Macedonian counterparts.[167] A bronze shield now in the National History Museum in Tirana, Albania, although only 50cm in diameter, also possesses decorations which are identical to the Macedonian *peltē*.[168] Ueda-Sarson suggests that the use of this type of shield was introduced into Macedonia by Archelaus (king 413-399BC) as part of the military reforms that are recounted by Thucydides (2.100).[169] However, the small Macedonian military under Archelaus seems to have been based upon troops armed as traditional Greek hoplites. Thucydides states that in 423BC these troops were raised from native Greeks residing in Macedon.[170] Subsequently, it seems unlikely that Archelaus introduced a smaller Illyrian shield to the Macedonians as it would have been unsuitable (and to the Greeks quite foreign) for the style of hoplite warfare they were accustomed to.

Gabriel, on the other hand, suggests that the inspiration for the *sarissa* came from Homeric times. In the *Iliad*, the Trojan prince Hektor is described as carrying a spear 11 cubits (528cm) long which would have to have been wielded in both hands due to its length.[171] Gabriel theorises that this weapon, which bore many similarities to the common Macedonian hunting spear (αἰγανέη), was what the *sarissa* may have evolved from under the reforms of Philip.[172] Yet again, such conclusions are problematic. Even if a long, two-handed, hunting spear was the inspiration behind the development of the *sarissa*, why does such an innovation need to be specifically attributed to Philip when any other previous ruler or reformer, who would have undoubtedly seen such a weapon in action over most of the period of Macedonian history, is just as likely? Based upon the common usage of the long hunting spear by the Macedonian nobility, the attribution of a reform to develop it into the *sarissa* can only be considered a selective acceptance of all possible scenarios.

Added to this, there is also uncertainty as to when Philip was actually held hostage in Thebes and whether he would have been old enough to even appreciate some of the finer nuances of Theban military strategy, organization and tactics. Aymard suggests that Philip was a hostage between 369BC and 367BC and that he would have therefore been too young (about thirteen years of age) to understand the complexities of generalship.[173]

This conclusion is a view held by several other scholars.[174] Diodorus clearly places the beginning of Philip's period of detention in Thebes in the year 369/368BC.[175] Justin says that it was while a hostage in Thebes that Philip gained 'the first rudiments of education'.[176] This too suggests that Philip was in his early teens at the time. Alexander the Great, for example, only began to be tutored by Aristotle when he was fourteen years old.[177] In his account of the arrival of Philip in Thebes, Plutarch says that 'this was the same Philip who was to later make war upon the Greeks and deprive them of their freedom, but at this time he was no more than a boy'.[178] The clear contrast between the Philip before and after his period of detention in Plutarch's account would again suggest that he was quite young when he was taken as a political hostage to Thebes.

Other scholars suggest a later period of detention for Philip, offering that he was taken hostage in 367BC and then returned in 365BC.[179] In such models Philip would have been seventeen years old by the time of his return and, it is suggested, that he would have been quite capable of understanding Theban military developments at this age. Such conclusions must also assume that part of the 'education' that was given to a foreign political hostage (regardless of their age) was training them in your own military arts – which, if it was something novel, could only have disastrous consequences should the homeland of a returned hostage later make war upon you and you become a victim of your own creation.[180]

Hammond and Griffith base their conclusions that Philip was taken hostage in 367BC on a Boeotian decree which outlines a peace deal between Thebes and Macedon in 365BC.[181] Hammond and Griffith argue that one of the causes for this treaty was that Thebes needed timber from Macedonia in 365BC, and that part of this deal was the release of Philip and other hostages.[182] Justin states that 'this thing [i.e. peace with Thebes] gave very great promotion to the outstanding natural ability of Philip'.[183] However, this passage does not clarify whether Philip and the other hostages were released as a part of this settlement, or whether they had already been returned some years earlier (which would have then placed Thebes on a favourable standing with Macedon and in a position to request the needed timber). Worthington, on the other hand, suggests that Philip may have been released soon after the ascension of Perdiccas III in 368BC as the new Macedonian monarch exhibited pro-Theban/anti-Athenian tendencies and that Philip may have even persuaded Perdiccas to grant the Thebans access to the timber.[184]

The main problem with suggesting a later period for Philip's detention in Thebes is that it does not fit with the evidence. As well as not correlating with the timeframe given by Diodorus and Justin, who clearly place the beginning of Philip's detention in 369/368BC, Plutarch recounts how Alexander II was assassinated 'in the year after Philip was taken to Thebes' by Pelopidas.[185] Justin also states that not long after Philip was taken to Thebes Alexander II was killed.[186] As Alexander II died in 368BC, it can only be concluded that Philip was taken back to Thebes as a hostage by Pelopidas in 369BC – the same time as is given by Diodorus.[187] Interestingly, the majority of those who propose a later period of detention for Philip are those who favour him as the creator of the pike-phalanx and it seems that in many cases the later period is offered merely to support the possibility of Philip being old enough to be the reformer of the Macedonian military despite any evidence to the contrary.

It is also important to note that Plutarch's account of how the young Philip came to be held in Thebes is only one of three accounts and is the only one that places the young detainee in a relatively stable and friendly environment where he may have received positive care and instruction on any subject from the Thebans. Diodorus' account (16.2.2), in which Philip is given to the Thebans by the Illyrians, or Justin's (*Epit.* 7.5.1), in which he is ransomed to the Thebans by Alexander II, both place Philip in a more negative train of events where his treatment in Thebes may have been less benevolent.[188] However, Diodorus does state that Philip received an education in Pythagorean philosophy and 'availing himself of the same initial training' as Epaminondas, eventually achieved no lesser fame than Epaminondas.[189] Justin states that Philip gained an education 'in a city renowned for strict discipline and in the house of Epaminondas, an eminent philosopher and commander'.[190] Unfortunately, what the exact nature of this training and education was is not outlined by either writer and the ambiguity of the passages can be read as either referring to military tutelage or something else entirely. Whether the younger Philip was of an age to, and even in a position to, observe and appreciate some of the finer points of the Theban arts of war, despite the fact that these would not have included any reference to the pike-phalanx, will no doubt be the topic of continued scholarly debate.

Other ancient passages are also erroneously used by scholars to support the theory that Philip created the pike-phalanx. Diodorus, for example, states that '...he [i.e. Philip] devised the compact order and the equipment of the phalanx – imitating the close-order fighting with overlapping shields (*synaspismos*) of the warriors (*puknoteta*) at Troy, and was the first to organize the Macedonian phalanx'.[191] Diodorus dates this reform to around 359BC – the time when Perdiccas III had been defeated by the Illyrians and

Philip ascended to the throne of Macedon. Some scholars see this passage as a reference to Philip creating the pike-based phalanx at the very beginning of his reign. Champion states that this passage is the key to understanding the reforms of Philip – however, he also notes that this passage does not specifically mention the creation of the pike-phalanx.[192] Markle uses this passage as an earliest possible date for the erection of a monument in Veria due to its depiction of the *peltē* – which he assumes was part of Philip's military reforms.[193] Thompson, citing the above passage of Diodorus, states that 'a significant part of the reorganization [of the Macedonian army] was the creation of a formidable infantry phalanx' under Philip.[194] Gabriel also cites the same passage of Diodorus as evidence for Philip being the creator of the pike-phalanx.[195] Similarly, McDonnell-Staff, presumably basing his hypothesis on the passage of Diodorus, suggests that Philip invented an even closer order formation of 50cm per man for defensive purposes.[196]

Diodorus' passage presents a number of problems. Firstly, as noted by Champion, it in no way states that Philip armed the organized Macedonian army with pikes and small shields as would be required for phalangites. Furthermore, the ancient military writer Asclepiodotus outlines how Hellenistic infantry could be deployed in one of three orders: a close-order, with interlocked shields (*synaspismos*), with each man 1 cubit (48cm) in the Hellenistic period, from those around him on all sides; an intermediate-order (*meson*) with each man separated by 2 cubits (96cm), on all sides; and an open-order (*araiotaton*) with each man separated by 4 cubits (192cm), by width and depth.[197] The intermediate-order spacing is also described by both Aelian and Polybius, and the close-order formation is noted by Arrian.[198] It is Asclepiodotus' close-order of 48cm per man with interlocked shields (*synaspismos*) that is referred to by Diodorus in relation to Philip, and which is used by some scholars as evidence for him being the creator of the pike-phalanx.[199]

The problem in using Diodorus' passage as evidence for the adoption of the pike-phalanx by Philip is that phalangites wielding a *sarissa* are incapable of forming a close-order deployment of 48cm per man with interlocked shields for combat (see: Bearing the Phalangite Panoply from page 133). According to Aelian, the close-order formation was too compressed for phalangites to wield their pikes effectively for battle and phalangites only adopted a close-order formation to undertake such battlefield manoeuvres as 'wheeling' (changing the direction that the formation was facing) as the small interval of the close-order formation meant that the unit(s) would have a lesser distance to travel.[200] Importantly, Aelian states that when undertaking such a manoeuvre, the phalangites were required to carry their pikes vertically – no doubt due to the fact that any formation with 'interlocked shields' prevented them from deploying their weapons for battle and to make the process of wheeling easier.[201] Polybius similarly outlines the inability of a formation with lowered pikes to wheel about – especially once engaged.[202] Connolly, who used a replicated unit of pikemen to test the process of wheeling (plus other aspects of the phalanx) found that, with pikes lowered, it was impossible to wheel the phalanx on anything less than a 14 cubit radius and that, as per Aelian, marching and changing direction could only be accomplished when the pikes were held vertically.[203] Importantly, if phalangites are incapable of adopting a close-order formation for battle, then Diodorus must be referring to combatants armed in a different manner to the phalangite.

The only troops that could use the close-order formation in either an offensive or defensive capacity were traditional Greek hoplites armed with a spear and large shield (*aspis*). Due to the large diameter of the *aspis* (90cm), any formation of hoplites adopting the 48cm close-order deployment mentioned by Asclepiodotus automatically created an interlocking 'shield-wall' (*synaspismos*) as parts of their shields extended into the interval occupied by the men to both their left and right.[204] Furthermore, the elevated way in which the hoplite could wield his spear allowed for it to still be deployed for use in combat while the shield-wall was maintained.[205] Phalangites on the other hand, with the *sarissa* held at waist level, could in no way deploy their weapons for combat and interlock the smaller shields that they carried at the same time.

Troops armed as hoplites played a significant role in the army of Alexander the Great by creating a more mobile 'hinge' of infantry between the cavalry on the wings and the slower pike-phalanx in the centre. Alexander's elite troops, the *hypaspists*, saw some of the fiercest action of any of Alexander's army, under all sorts of conditions, in all theatres of operation, and in every engagement Alexander ever fought. In the literary accounts we have of Alexander's campaigns, there are more specific references to the contingents of *hypaspists* than there are for any other infantry unit within the whole army. In many of these references the *hypaspists* are deployed as a 'hinge' between the cavalry and the pike-phalanx.[206] Based upon the name of this unit, with its root in the term for the hoplite shield (*aspis*) it seems that this mobile and active 'hinge' of soldiers were armed as traditional Greek hoplites.[207]

Due to the inability of pike-wielding phalangites to adopt a close-order formation for either offensive or defensive purposes, it is unlikely that what Diodorus is referring to in his passage about Philip emulating the Greek close-order formations is the creation of the pike-phalanx. It is also unlikely that Diodorus is referring to the adoption of the close-order interval to aid the process of 'wheeling' by Philip. Even if it is assumed as such, this would then suggest that either the *sarissa* was already in use by 358BC, or that the Macedonian army at this time was not using the *sarissa* at all but was still employing troops armed as hoplites. Either case cannot be interpreted as the creation of the pike-phalanx.

If *sarissa*-armed infantry had already been adopted by Amyntas/Alexander II, Diodorus must, therefore, be referring to Philip incorporating something (or someone) else into the Macedonian military who could use this type of deployment. Based upon the very nature of the formation being described, Diodorus can only be referring to the use of contingents of troops armed as traditional Greek hoplites. Ellis says that if Diodorus is correct, then the 'new' Macedonian formation was really no different from any of its [Classical Greek] predecessors.[208] If Diodorus is referring to the incorporation of hoplites into the Macedonian battle-order, this statement is more or less correct – at least in regards to the deployment of these troops. Brunt, on the other hand, says that the formations used by the Greeks at Troy were not the same as the Classical phalanx, but this does not discount the notion that Philip or any of his predecessors had heavy infantry.[209]

Philip also seems to have placed a greater emphasis on cavalry as a strike weapon operating from the wings of the phalanx. This was also a practice used by Alexander the Great in all of his major engagements. Unfortunately, the accounts of the engagements Philip fought during his reign do not provide enough detail to determine whether troops armed as hoplites were positioned as a hinge between the cavalry and the phalanx as they were later used to great effect by Alexander. It can only be concluded that, based upon passages like that of Diodorus, Philip initiated a number of military reforms to the Macedonian system – some in regards to the cavalry and others to the infantry. However, the most important point of this passage is that the physical dynamics of the formation being described by Diodorus preclude this from being seen as a means of confirming (or even to suggest) that the pike-phalanx itself was created by Philip.

As Diodorus refers to Philip being the first to use troops who could deploy in a close-order formation, and so is the first to 'organize the Macedonian phalanx' as he puts it, this suggests that the use of such troops in any sort of capacity, whether protecting the flanks of the pike-phalanx as they were used by Alexander or otherwise, had not been a regular part of the composition of the Macedonian military in the past. It is clear that the Macedonians had employed the use of hoplites in the past, but Philip seems to have been the first to consciously use men of multiple arms in concert with each other, both to protect the flanks of the pike-phalanx, and to create an overall effective fighting force. Thus Diodorus' passage would be better read as saying that Philip 'was the first to organize the Macedonian phalanx [with multiple supporting arms]'.

A lack of more mobile troops protecting the vulnerable flanks of the pike-phalanx in the past may in part account for the defeat of the Macedonians under Perdiccas III by the Illyrians in 359BC and why, according to Diodorus, Philip incorporated the use of troops in such a manner almost immediately afterwards.[210] Thucydides, describes the Illyrians in 423BC as being unable to fight in any sort of regular formation and incapable of withstanding an organized infantry attack.[211] Diodorus, on the other hand, describes the Illyrians as fighting as hoplites by 385BC which would suggest the adoption of massed formations.[212] Despite the failings outlined by Thucydides, a more mobile form of combat as the Illyrians seem to have employed would be very effective when engaged against a pike-phalanx as the greater manoeuverability of individual combatants would have allowed the Illyrians to attack Perdiccas' more rigid formation from multiple directions at once – and especially against the vulnerable flanks of the phalanx if these were not being protected by other, more mobile, troops like hoplites or cavalry. This further supports the idea that a lack of protection was, at least in part, responsible for Perdiccas' defeat in 359BC, and was the essence of Philip's subsequent enhancements to the pike-phalanx shortly thereafter as reported by Diodorus.

Polyaenus states how as part of his training of the Macedonian army, Philip made his men embark on lengthy cross-county marches bearing all of their equipment which included helmets (κράνη), shields (πέλτας), greaves (κνημίδας) and the *sarissa* (σαρίσας), as well as provisions and utensils for daily use (καὶ μετὰ τῶν ὅπλων ἐπισιτισμὸν καὶ ὅσα σκεύη καθημερινῆς διαίτης).[213] If taken as accurate, this passage would indicate that Philip's army was carrying a lengthy pike. Champion cites this passage as proof that 'it is almost certain that Philip was the instigator [of the reform which created the pike-phalanx]'.[214] However, just because Philip's men were carrying the *sarissa*, does not mean that Philip invented it. Markle suggests that Philip made his men carry the *sarissa* because it was a relatively new piece of equipment.[215] Again, such interpretations are problematic. Professional troops train with their weapons constantly – not only when they are new. Alexander, for example, is said to have kept his troops busy with tactical exercises and constant training in the use of their weapons.[216] While Champion correctly surmises that it is 'incomprehensible that Philip would train his infantry with a weapon they did not use', use is not synonymous with the act of invention. As such, this passage can also not be used to confirm or deny Philip's part in the creation of the pike-phalanx – it can only be used to support the concept that the army at the time of Philip was already employing the *sarissa*. This, in turn, provides a *terminus ante quem* by suggesting that the development of this piece of weaponry had to have occurred during the reign of Philip at the latest – but it does not indicate that the *sarissa* was invented at this time.

It has been suggested that because there is scant evidence for Philip's battles, we cannot be sure how his troops were armed with any certainty.[217] However, if the pike-phalanx had been invented by Iphicrates in 374BC, and was then introduced into the Macedonian military by Alexander II around 370BC, it can be assumed that heavily armed phalangites were part of Philip's army as Demosthenes' reference to the *pezhetairoi* would suggest. This also explains why Philip's troops were training with the *sarissa* according to Polyaenus and why the reforms of Philip that are mentioned by Diodorus are more likely to be the incorporation of men armed as Classical hoplites in the Macedonian battle line rather than the creation of the pike-phalanx. Interestingly, both Polyaenus and Nepos describe Iphicrates as subjecting his men to extensive training including digging trenches and making camps, repairing equipment, and practising drill.[218] This method of physical training finds close parallels with practices carried out under Philip.[219] It is possible that Iphicrates' concept of regular physical training for troops to condition them to the turmoils of war was another element of military reform that was taken up by the Macedonians along with the adoption of the pike-phalanx to create a standing, professional, and well trained military.

Another point to consider is that the wholesale adoption of a completely different style of fighting by a state costs a considerable amount of money. In the city-states of Greece, where each member of their hoplite militia had to supply their own equipment, the state did not have to consider such expenses. However the Athenians, for example, do not seem to have adopted the revamped infantry of the Iphicratean reforms – preferring instead to continue to use men armed as traditional hoplites for some time.[220] This was most likely because it would have been impractical for all members of the part-time citizen militia of a city-state like Athens to abandon the expensive arms and armour of a hoplite that they had already purchased only to buy a completely different set of equipment for use in a new style of combat. Sekunda suggests that the reforms of Iphicrates were an attempt to redress the 'pauperisation of Greece during the fourth century' which 'had reduced the number of citizens who could afford to supply themselves with hoplite equipment'.[221] However, it is unlikely that changing a required panoply would reduce the amount of cost incurred by the individual to any great extent. For a state like Macedon, on the other hand, whose professional military required standardized equipment in order for the pike-phalanx to operate effectively, weapons and armour could only have been supplied by the State. The provision of arms, armour and shields to the whole Macedonian military must have therefore required a considerable initial outlay of funds.

The equipping of the Macedonian military at state expense is something that is commonly attributed to Philip II as part of the creation of the pike-phalanx. Diodorus does state that Philip 'equipped the men...with weapons' early in his reign.[222] However, the use of such means by Philip does not necessarily mean that he invented this practice and it cannot be discounted that such a procedure was already in place when Philip came to the throne.[223] Furthermore, Justin states that the Macedonian state was broke when Philip came into power in 358BC due to 'a series of wars' that had occurred previously.[224] We know, however, that Philip was able to place at least 10,000 men in the field the following year against the Illyrians (who had defeated Perdiccas III in 359BC and destroyed half of the Macedonian army) and, not long afterwards, was able to embark on several campaigns to gain access to the mines of Thessaly and Thrace which were then used to bolster Macedonian state finances.[225] Diodorus states that Philip was in command of 'his best men' during the Illyrian campaign.[226] This may suggest that the Macedonian army was even bigger under Philip than the 10,000 strong army that had been defeated in 359BC. Sheppard offers that:

> Philip could not put a native hoplite army of any size into the field after his brother's Illyrian disaster, he would either have to rely on mercenaries, which in Macedon's current political and economic state he could ill afford, or he would have to provide his men with a distinct advantage, in weaponry, tactics or both...The only plausible explanation is that Philip experimented, right from the start, with a new weapon and tactic.[227]

How then are we to correlate such statements? Can we conclude that Philip was able to take a bankrupt state, recruit numerous replacements, implement a series of expensive reforms to the operational deployment of the army (which included adopting a totally new form of equipment), train the recruits in this new style of warfare, and then campaign for several years to defeat the Illyrians and successfully annex the mines of neighbouring regions in order to recoup the cost? This seems unlikely on a number of issues. Not only was Philip able to quickly replace the 4,000 men lost by his brother (and possibly recruit more), but the cost of implementing a sweeping series of reforms to the military equipment of a 10,000 strong army would have been more expensive than the hiring of 4,000 mercenaries.

Anson, on the other hand, suggests that the adoption of the phalangite panoply would have involved 'far less expense' than the issuance of hoplite equipment.[228] This would seem to be incorrect. While the phalangite *peltē* was smaller than the hoplite *aspis*, and the cost of the linen composite armour worn by the phalangite would have been comparable to the production of bronze plate armour, the manufacture of the *sarissa* would have incurred considerably more expense than the production (or procurement) of hoplite spears.[229] Not only did the *sarissa* contain more than four times the amount of wood in its shaft (4,070g for the *sarissa* compared to 850g for the hoplite *doru*), but there was substantially more metal in the butt of the *sarissa* (1.07kg) compared to the average butt of the hoplite spear (329g). Thus any amount saved through the use of smaller shields was offset by the use of a weapon which required much more material, and therefore cost, to construct. Gabriel suggests that, due to the length of the *sarissa*, armour was not really required unless you were fighting against another pike-phalanx (which hardly anyone had in the time of Philip) and this then made the adoption of the pike-phalanx a cheaper option for the Macedonian state.[230] However, there is no evidence that the Macedonian phalangite fought without wearing armour. Tomb paintings, for example, all depict phalangites wearing the *linothorax* that was part of the Iphicratean reforms and, as per the later military manuals, it can only be assumed that linen armour was the standard equipment for the Macedonian military from the time the pike-phalanx was adopted.

Due to the expense of a reform which would create the pike-phalanx, and the barren nature of the Macedonian coffers in 358BC, Philip cannot have undertaken such an alteration to the structure of the Macedonian military as soon as he ascended the throne – he could have only campaigned against the Illyrians, Thessalians and Thracians with the army as it existed at the time. This raises several possibilities. Firstly, that the creation of the pike-phalanx, if attributable to Philip at all, was made after gaining access to the mines of Thessaly and Thrace – the revenue from which could then be used to offset the expense. The other alternative is that the army had already been reformed prior to Philip coming into power, and an army 10,000 strong, which incorporated contingents of newly equipped phalangites, had then been used in the 'series of wars' of the preceding years which Justin tells us left the state devoid of funds – most likely from both operational costs and from the expense of creating the phalangites in the first place. Due to the events of the time, this second scenario seems the most likely. This would then explain why the passage of Arrian's which details Alexander the Great's harangue to his troops at Opis in 324BC only mentions Philip supplying cloaks to the men of Upper Macedonia (if this passage is taken as accurate and at face value) rather than weapons and equipment, as this practice would have begun under Amyntas and Alexander II with the adoption of the pike-phalanx and would have still been in place by the time Philip ascended the throne.[231]

If Macedonian troops under Archelaus were hoplites, and the pike-phalanx had been introduced by Amyntas/Alexander II following the reforms of Iphicrates, this would further account for why Philip went on to initiate further reforms to the Macedonian military to both protect the vulnerable flanks of the pike-phalanx and to rely more heavily on cavalry to overcome enemy formations.[232] This would then correlate with what Arrian has Alexander say in his harrangue to his men at Opis in 324BC when he describes how Philip had taken the semi-nomadic people of Upper Macedonia and improved their efficiency on the battlefield in order to place them on 'equal terms' (at least militarily speaking) with the Thracians and Illyrians. Thus we see an evolutionary path for the development of the Macedonian military whch aligns with what is stated in the ancient sources: from hoplites under Amyntas, to pikemen under Amyntas/Alexander II; to a military defeat due to the vulnerabilities of the pike-phalanx under Perdiccas III, and finally to a combined, and mutually protective, force of hoplites, skirmishers, cavalry and pikemen under Philip II.

Is this use of mutually supporting arms one of the lessons that Philip may have taken away from his time in Thebes as a youth? Certainly nothing specific is outlined in the ancient sources. Consequently, it is only possible to find parallels between the tactical arts of the Thebans and those later employed by Philip, and then surmise that these may have been inspired by what Philip could have witnessed as a child. Prior to the time of Philip's detention, the Thebans had lost every major engagement they had been involved in since the Persian Wars (where they had been on the losing side at the battle of Plataea in 479BC) and for nearly a century they had been continuously beaten by the Athenians except when they were fighting as allies of Sparta.[233] Consequently, the most talked about event in Thebes during the time of Philip's detention would have undoubtedly been the Theban victory over the Spartans at Leuctra a few years earlier in 371BC.[234]

However, the Theban actions at the battle of Leuctra do not provide a great number of direct parallels with the later actions of Philip, and so the impact of this battle on the formation of Macedonian battle tactics is hard to gauge. For example, one of the aspects of military innovation which some scholars claim Philip derived from his knowledge of Leuctra is the use of an obliquely deployed phalanx.[235] At Leuctra, Epaminondas had deployed with his heavily weighted left wing advanced and with his allied contingents arranged obliquely to the right-rear.[236] While this deployment no doubt contributed greatly to the Theban victory, and while there is also little doubt that the young Philip would have been told about it, in later actions both he and Alexander the Great do not completely emulate it.[237] In his battle against Bardylis in 358BC, for example, Philip advanced his right wing (as opposed to the Theban left at Leuctra), nor did Philip mass his extended wing to any great depth.[238] At Chaeronea in 338BC, Philip also advanced his right wing to engage first, creating a battle line extending obliquely to the left-rear, and then slowly withdrew his right behind the main frontage of the line to create a new oblique position extending to the right-rear.[239] At Gaugamela in 331BC, Alexander also deployed in an oblique line with both of his wings refused.[240] None of these encounters closely mirror the Theban deployment at Leuctra. It seems that, if Philip did take any lesson about oblique deployments away from the accounts of the battle of Leuctra that he heard, it was how to implement a complete variant of the Theban tactics and make it his own.

One of the key elements of Theban battle tactics at Leuctra was the concentration of force against what was perceived to be a key point in the enemy line. Gabriel suggests that this key point was the strongest part of the enemy line rather than the weakest part.[241] While this is true of the Theban action at Leuctra, it is not true of the actions of Philip. At Leuctra the Theban forces were deployed in a deep but narrow formation on their left wing with the aim of using the smaller frontage of their formation to concentrate their attack against the point in the opposing Spartan line where their king, Cleombrotus, was positioned.[242] This concentration of force was ultimately successful. Cleombrotus himself was injured and removed from the fighting and the Spartans suffered significant casualties among their front rank officers.[243]

Philip also employed the concentration of force against a key point in the enemy line as a major element of their battle tactics. However, this was against the weakest part of the enemy line rather than the strongest. In his confrontation against Bardylis in 358BC, Philip noticed that the front of the Illyrian formation was strong while its flanks were weak. It was against this weakened part of the enemy line that his attack was directed.[244] At Chaeronea, Philip used his advancing right wing to commence actions against the Athenian left while the strongest part of their line, the position of the Theban Sacred Band, was opposite the Macedonian left.[245] In 335BC Alexander the Great, when engaging the Taulantians, also directed his attacks against the enemy formation at the point 'where they were likely to make the greatest onslaught on the enemy at his weakest point'.[246] It is only later that Alexander's tactics more closely follow the objectives of the Thebans at Leuctra by having the main thrust of his attacks aimed directly at the enemy leader. In every one of his four major battles, Alexander intentionally directed his main assault against the location of the enemy king or commander. However, even here, the method of delivery of these attacks (i.e. a flanking attack using cavalry) is very different from those employed by the Theban infantry at Leuctra. Badian suggests that this was a tactic learnt by Alexander at Chaeronea in 338BC where he was positioned opposite the Theban Sacred Band.[247] However, despite such possibilities, no clear parallels between Theban tactics and those of Philip or Alexander can be seen which would indicate that the Macedonian way of war was based upon a Theban model.

It has also been suggested that Philip learnt of the value of professional units, like the Sacred Band, and of experienced officers, while he was in Thebes.[248] However if a pike-phalanx, organized, equipped and trained at state expense, had been adopted as the Macedonian system by Amyntas and/or Alexander II no later than 368BC, the Macedonians would have already been operating under a professional system prior to Philip being taken/sent to Thebes. It is possible, depending upon the exact date of the adoption of the pike-phalanx by Macedon, that the young Philip may not have had a lot of opportunity to see this reform in effect. As such, he may very well have gained an appreciation for the use of professional units and officers in Thebes as is suggested, and then gained an even greater appreciation for it (and possibly modified it) when he saw the Macedonian military in more detail upon his return.

It is also suggested that Philip learned to appreciate the value of a strong co-ordination of allied units, and particularly cavalry, while in Thebes.[249] At Leuctra, the Theban cavalry had initially repelled the opposing Lacedaemonian cavalry before retiring to the left wing to protect the flank of Epaminondas' deep formation.[250] Once repositioned into this location, the Theban cavalry played little role in the ensuing encounter other than preventing the Spartan right from enveloping the Theban line. For Philip and Alexander on the other hand, the cavalry, this time deployed more heavily on the right rather than the left, became the strike weapon *par excellence* of the Macedonian army. In 358BC, Philip used his right wing cavalry to sweep around and attack the weakened flanks of the Illyrian formation.[251] This same principle was followed by Alexander who commenced every major engagement with an advance of the right wing cavalry, usually in concert with other units such as light troops or mounted archers, and then drove head-long for his major objective, the enemy king or commander. The tactical use of cavalry varies so much between its use at Leuctra by the Thebans and the methods used by Philip and Alexander that it is uncertain what sort of lesson about it (if any) the young Philip may have gained. Aelian goes as far as to state that Philip adopted the use of the cavalry wedge as a tactical formation from how it was used by the Scythians and Thracians, not the Thebans.[252]

Whether Philip had learnt any such tactical methodologies from his time in Thebes as some scholars suggest, or whether able commanders simply recognized the value of such moves on their own volition, is unlikely to be proven one way or the other. However, it is interesting that many of the manoeuvres that become the signature trademarks of warfare under Philip and Alexander find only cursory historical precedents in the military innovations of the Thebans under Epaminondas. This then highlights Philip's true military genius. There is no need to conclude that the Macedonian military was altered to adopt the Theban model. Rather Philip, through his own brilliance and innovation, may have taken only the most basic elements of the Theban system, and then altered them in such a way that, when they were applied to the new Macedonian way of war, they created the most effective fighting force that the world had yet seen. Despite the possible transmission, and then adaptation, of military theory and practice from Thebes through Philip, the one thing that can be definitively stated is that Philip would not have learnt anything during his time in Thebes about the value of the lengthy *sarissa* – a weapon that was not used by the Thebans, but would have already been in use in Macedon, at that time.

Based upon a review of the available evidence, it seems that there was no Macedonian 'creation' of the pike-phalanx at all, only a Macedonian 'adoption' of it under Alexander II (and possibly begun previously under Amyntas) based upon the reforms of Iphicrates of 374BC. This would then clearly make Alexander II the reformer that is mentioned by Anaximenes. This formation was then augmented (or 'organized' as Diodorus puts it) by Philip II following the defeat of Perdiccas III in 359BC and Philip's ascension to the Macedonian throne. This 'organization' involved the incorporation of men armed as more mobile hoplites within the main battle line and through a greater, co-ordinated, use of cavalry as a strike weapon. Thus each of these rulers and reformers, in their own way, left their own mark on the Macedonian system – all of which came together to provide Alexander the Great with a professional, experienced and adaptive military institution with which he was able to conquer most of the known world.

Chapter Two

The
Sarissa

The *sarissa* was one of the tools that helped create the Hellenistic world. The employment of massed formations of men wielding this weapon allowed the Macedonians to subjugate (and then tenuously unite) most of Greece, and then conquer, and subsequently fight over, an empire that stretched from the Aegean to Egypt and India. Consequently, any understanding of the warfare of the Hellenistic Age must include a thorough examination of this influential piece of weaponry.

Unfortunately, this is no easy task due to two separate, yet inter-related, problems. The first of these problems is that the design of this weapon changed during the time of the *sarissa*'s common usage. The length of the *sarissa* was altered across the Hellenistic Period as army after army sought to gain an advantage over those employing similar methods and weaponry. Additionally, some of the passages which describe certain aspects of the *sarissa* were written at a time after it was no longer used and so cannot be describing a contemporary weapon. As a consequence, the exact configuration of the *sarissa* has become a matter of controversy for scholars who have argued over different characteristics for the weapon and how the ancient writers described it. However, a critical examination of the available evidence not only allows for the developmental sequence of the *sarissa* across the Hellenistic Period to be established, it also allows for the specifications of the weapon's constituent parts to be identified as well. This, in turn, allows for the scholarly debate on the subject to be addressed and many of the problematic ancient passages relating to the weapons and tactics of Hellenistic warfare to be seen in their correct context.

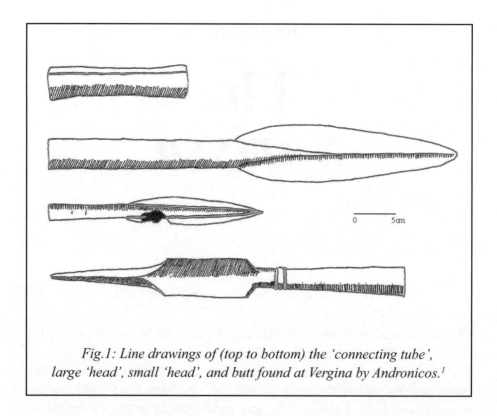

Fig.1: Line drawings of (top to bottom) the 'connecting tube', large 'head', small 'head', and butt found at Vergina by Andronicos.[1]

However, one of the main problems in determining the correct characteristics of the different constituent parts of the *sarissa* is the rather limited amount of available evidence. Like the earlier spear of the classical hoplite (*doru*) it appears that the *sarissa* was equipped with both a leaf shaped iron head on the forward tip of the shaft and a large heavy metallic spike (the *sauroter* of the hoplite spear) on the rearward end. Unlike the earlier hoplite spear, however, there is a distinct lack of examples of these metallic *sarissa* components that have survived within the archaeological record. As with many aspects of Hellenistic warfare, this dearth of clearly identifiable source material has led to much contention between scholars over the size of the different parts and the configuration of the *sarissa*. However, due to the interwoven nature of how the design of a weapon affects its performance, the various theories that have been forwarded regarding the characteristics of the *sarissa* affect subsequent models of Hellenistic warfare upon which they are based. This, in turn, has led to a vast number of interpretations of the mechanics of Hellenistic combat. However, by correlating the limited archaeological evidence with ancient descriptions and images of the *sarissa*, and then comparing this information to the physical properties of an assembled *sarissa*, a comprehensive outline of the configuration of the weapon as a whole, and broken into its constituent parts, can be compiled.

THE BUTT-SPIKE OF THE *SARISSA*

In 1970, Andronicos found what appears to be the butt-spike of a *sarissa*, along with pieces of various other weapons, outside of a tumulus grave at Vergina in Macedonia (Fig.1).

The finds were thought to have been items discarded by grave robbers who apparently viewed them as possessing no intrinsic value.[2] The butt that was found was described as:

> An iron spike. It is composed of three parts...the socket – the body, in the form of four wings – the point, with a pyramidal form...Total length: 445mm. Length of the socket: 180mm. Length of the point: 145mm. Maximum width (at wings): 40mm. Diameter of socket: 34mm. Thickness [of socket]: 2-2.5mm. Weight: 1,070g.[3]

Andronicos concluded that the butt had come from a Macedonian *sarissa* and dated the find to the late fourth century BC (c.340-320BC).[4]

The butt of the *sarissa* is rarely shown in Hellenistic art with such a configuration as the one found by Andronicos. In fact, there are no depictions of the *sarissa* at all which clearly distinguish the shape and size of all of the different parts of the weapon. Tomb paintings such as those found at Boscorale or Vergina regularly contain an individual holding a long shafted weapon of some kind which, unfortunately, extends beyond the margins of the mural in most cases, or is obscured by other features of the image. Consequently the exact size of the weapon being shown cannot be gauged and it is uncertain whether the weapon in such paintings is a spear or a *sarissa*. The paintings at both Boscorale and Vergina show weapons with butts more akin to the *sauroter* of the spear of the Classical hoplite than of weapons equipped with the butt found by Andronicos, but this is no indication of exactly which type of weapon they are.

Another butt-spike, now in the Shafton Collection in the Great North Museum in Newcastle upon Tyne (#111), is significantly different from the butt found at Vergina by Andronicos and is, in shape, closer to the *sauroter* of the hoplite spear. It has been suggested that the spike dates to the late fourth century BC – although dating it with any certainty is all but impossible (Plate 2).[5]

The distinctive wings of the Vergina find are absent from the Newcastle spike which is, similar to some versions of the hoplite *sauroter*, merely a tubular socket flaring out into an elongated spike – in this case conical in shape. What definitively identifies this spike as coming from a Macedonian weapon, rather than the spear of a Greek hoplite, are the painted letters MAK (for 'Makedonian') on the socket. Sekunda suggests that this is an indication of the weapon (or at least the butt-spike) being state issued equipment to the Macedonian military.[6] It may be representations of this style of butt-spike that feature more regularly in Hellenistic era tomb paintings.

The length of the Newcastle spike (380mm) is somewhat shorter than the butt found by Andronicos, but is larger than the average length of the *sauroter* that was affixed to the hoplite spear (which had an average length of 259mm).[7] The weight of the Newcastle spike (876g) is also less than that of the Vergina butt but, again, is distinctly greater than the average weight of the hoplite *sauroter* (329g). The weight of both the Vergina find and the Newcastle spike indicate that they have come from a weapon considerably heavier than a traditional hoplite spear. This suggests that they have both come from a *sarissa*. Additionally, the diameter of the socket of the Newcastle spike (33.5mm) is almost exactly the same as the diameter of the Vergina spike; both of which are larger than the average diameter of the socket of the hoplite *sauroter* (25mm). Even in the late Classical/early Hellenistic period when the Greeks more commonly used iron to make the *sauroter*, the average weight only increased to 545g while the diameter of the socket remained almost unchanged. This all but confirms that both the Newcastle spike and the Vergina find have come from a weapon with a shaft of a greater diameter than the average hoplite spear – most likely a *sarissa*. Traces of pitch inside the socket of the Newcastle spike indicates that it was fixed in place via the use of an adhesive rather than by having it riveted to the shaft. The absence of any hole from the use of rivets to secure it in place also indicates that the butt found at Vergina had to have been attached to the shaft of the weapon to which it belonged through the use of some form of adhesive.

Despite the lack of direct confirmatory evidence, Andronicos' find is generally assumed to be the butt of a *sarissa*. Everson claims that the Newcastle butt belongs to a Hellenistic cavalry lance rather than a pike.[8] Sekunda, on the other hand, attributes it to the *sarissa*.[9] If the dating of the Newcastle spike is correct, then it has clearly come from an infantry weapon as the butt of the Hellenistic cavalry lance during this time had a different configuration (see following).

Thus, like the *sauroter* of the hoplite spear, it appears that the spike on the rearward end of the *sarissa* could come in a variety of shapes with the only thing in common being the diameter of the socket (Table 1).

	Newcastle Spike	Vergina Find	Average Hoplite *Sauroter*
Metal	Iron	Iron	Bronze
Total length	380mm	445mm	259mm
Maximum width	36mm	40mm	21mm
Inner socket diameter	29mm	30mm	18mm
Total weight	876g	1070g	329g

Table 1: Comparison of the Newcastle spike,
the Vergina find and the average hoplite sauroter.[10]

Sekunda offers that it is uncertain whether the *sarissa* possessed a butt-spike or not.[11] However, what conclusively proves that it did is the fact that, without one, the *sarissa* could not be configured in a way that then correlates with the descriptions of this weapon that are found in the ancient texts. In other words, it was not the shape of the butt that was important to the functionality of the weapon it was attached to, but rather its weight as, when affixed to a shaft with a head at the opposite end, any difference in weight between the head and the butt would significantly influence the location of the weapon's point of balance, the performance of the weapon, and subsequently where it could be held in order to be used effectively in battle (see following).

Another important aspect of the butt-spike was its length. Both the Newcastle spike and the Vergina find are both smaller than 1 cubit (48cm in the Hellensitic Period) in size. This was important as a butt-spike could not be of a larger size without imposing on the ability of the person carrying it, and those around him, to conform to the intervals of the densely packed formation of the phalanx (see Bearing the Phalangite Panoply from page 133). The weights, lengths and socket diameters of both the Newcastle spike and the find from Vergina set them apart from the hoplite *sauroter* and indicate that they have come from two different weapons but with very similar configurations in terms of the size of their shaft. Again, these weapons are most likely the *sarissa*. Taking these two examples as the basis, the iron butt that was affixed to the rear end of the *sarissa* would have averaged 413mm in length, came in a variety of shapes, was attached to the shaft via a tubular socket with an inner diameter of 30mm, and had an average weight of around 973g.

THE HEAD OF THE *SARISSA*

Outside of the tumulus grave at Vergina, Andronicos also recovered two items which he labelled as 'lance heads' (Fig.1):

a) The point of a lance in iron...Total length: 51cm. Length of the socket: 23.5cm. Maximum width of the head: 6.7cm. Diameter of the socket: 3.6cm. Thickness of the wall of the socket: 2-3mm. Weight: 1,235g.[12]

b) The point of a lance in iron...Total length: 27.3cm. Length of the socket: 8.0cm. Maximum width of the head: 3.0cm. Diameter of the socket: 1.9cm. Thickness of the wall of the socket: 1-1.5mm. Weight: 97g.[13]

Andronicos claimed that the larger of the two 'heads' and the butt-spike recovered from the same site had come from the same weapon: a Macedonian *sarissa*. Andronicos reached this conclusion as the diameter of the socket for both of these items were of a similar size (34mm for the butt and 36mm for the head) and because there were many similarities in the metal used to make the two items.[14] Numerous scholars have accepted Andronicos' interpretation of the artefacts and have subsequently used a weapon configured with this large head in their own examinations of the *sarissa* and of Hellenistic warfare.[15]

Markle, for example, echoes Andronicos by stating that the larger of the two heads had come from a *sarissa* while he further suggests that the smaller had come from a regular hoplite spear.[16] Markle then uses this conclusion as the basis for identifying the different types of heads found elsewhere stating: 'a sufficient number of *sarissa* heads has been found to demonstrate that they can easily be distinguished from the traditional hoplite spear head' and that 'the striking difference is in their respective weights'.[17]

For example, a head found at Vergina by Petsas, which had an overall length of 55cm, was identified by Markle as another example of the head of a *sarissa* due to the similarities in size between it and the one found by Andronicos.[18] Using this conclusion as the basis, Markle then dismisses many interpretations of other finds. He states that not one example of the *sarissa* head can be identified from those recovered from excavations at Olynthus and that the ones identified as such by Robinson are too small to have come from the *sarissa*.[19] Markle continues, stating that the largest head found at Olynthus, which measured 29cm in length, would be more suited to a hoplite spear.[20] Another head found by Petsas, with a length of 34cm, was said to be too small for a *sarissa* by Markle, but too big for a hoplite spear.[21] Unfortunately, Markle did not elaborate on what weapon he thought the head had come from. Other heads found with lengths of 27cm and 31cm were interpreted by Markle as clearly coming from the hoplite spear.[22] Confusingly, the best preserved head found in the Macedonian *polyandrion* at Chaeronea measures 31cm.[23] Markle states that this is from a *sarissa*, but does not state why he believes that other heads of a similar size found elsewhere are not.[24] Champion similarly states that no *sarissa* heads have been found at Olynthus, but have been found at Chaeronea.[25]

Markle also published data on two more heads found at Vergina – one measuring 470mm and with a weight of 235g, and the other measuring 500mm with a weight of 297g.[26] Both of these finds were made from two halves hammer-welded together which made them much lighter than their respective sizes would suggest.[27] Connolly suggests that the smaller of the two appears closest to the type of head depicted in the representations of the *sarissa* in the Alexander Mosaic now in the Naples Museum (see Plate 5).[28]

There are numerous problems associated with such approaches to identifying the head of the *sarissa*. Firstly, the head of the Greek spear, whether that of the hoplite, the javelin or of the hunting spear, came in a variety of styles and sizes across the period of their common usage. Snodgrass, in his examination of Greek armour and weapons, classified Greek spearheads into fourteen different categories based upon their size, shape, the material they were constructed from and their period of use.[29] Among these classifications, Snodgrass refers to the 'J style' spearhead as the 'hoplite spear *par excellence*'.[30] The 'J style' spearhead is typified by its long socket, its long, narrow, blade, and its sloping shoulders. This type of spearhead saw service in warfare from the late geometric period (c.700BC) onwards.[31] The average dimensions of the 'J style' hoplite spear head are a length of 27.9cm, a width of 3.1cm, a socket with a diameter of 1.9cm and a total weight of 153g.[32] Despite being the 'hoplite spear *par excellence*' as Snodgrass describes it, the 'J style' is not the most common type found in the archaeological record – this is a distinction belonging to the smaller 'M style' which had a period of use from the early Protogeometric Age (c.1,000BC) to the fall of the Roman Republic (c.31BC). Thus even the more archaeologically common head of the hoplite spear, let alone that of the *sarissa*, cannot be easily distinguished by size alone. The identification of the *sarissa* head can only be made via comparison and correlation with other sources of evidence.

This is where Markle's conclusions encounter further problems. In his treatise *On Hunting*, the ancient writer Grattius outlines the unsuitability of the *sarissa* for such a pursuit due to its 'small teeth'.[33] This suggests that the *sarissa* was tipped with a head recognisably smaller than the large heads of regular hunting spears. This in itself would discount any correlation with the large head found by Andronicos and the head of the *sarissa*. Furthermore, the large 'head' as it was identified by Andronicos appears to actually be the butt of a Macedonian cavalry lance rather than the head of an infantry weapon.[34] Hammond notes that cavalrymen in the time of Philip II carried a lance with 'a blade at either end'.[35] This rearward 'blade' seems to have been larger than the forward tip of the lance – most likely to provide the weapon with a particular point of balance. A wall painting from 'Kinch's Tomb' at Lefkadia (c.310-290BC) clearly shows the rearward end of the cavalry lance equipped with a large secondary 'head' just like the one found at Vergina (Plate 3).

The depiction of Alexander the Great on the famous mosaic from Pompeii (#10020 in the Naples Museum) shows the rear end of the lance wielded by the young king. The detail of the rear end of this weapon, which is unfortunately fragmented and mostly missing, also seems to be tipped with a large, blade-like butt similar to that depicted in Kinch's Tomb (Plate 4).[36]

The image in the mosaic is thought to be a copy of a painting by Philoxenos of Eretria from the fourth century BC – possibly commissioned by Cassander after he had become ruler of Macedonia in 317BC.[37] Thus the image of the weapon carried by Alexander, although second-hand through the reproduction in the mosaic, is a contemporaneous depiction of a weapon used at the time, and the slightly later depiction of a similar weapon in Kinch's Tomb would suggest a fairly common usage of a cavalry lance with a large, blade-like, butt.[38]

Additionally, the Alexander mosaic also contains several representations of the *sarissa*. None of these possess an overly large head (see Plate 5). Nylander, who accepts the large 'head' found at Vergina as the head of a *sarissa*, states that 'it is interesting to note that the artist [who made the Alexander mosaic] never hints at such a conspicuous point on any of the lances depicted...'[39] Nylander's conundrum is easily explained. The reason why large heads are not depicted on the weapons shown in the Alexander mosaic is becuase the *sarissa* was not equipped with a large head like the one recovered from Vergina – just as Grattius states.

If the large 'head' recovered by Andronicos is indeed the butt from a cavalry lance as the pictorial evidence suggests, this not only makes Markle's considerations, and those similarly based upon Andronicos' conclusion, void, but also opens the question as to what size the head of the *sarissa* actually was. Some scholars, dismissing the large 'head' found by Andronicos (and of Markle's hypotheses), base their examinations of the configuration of the *sarissa* using the smaller 'head' that was discovered.[40] Connolly, for example, discounted the idea that the smaller head found at Vergina was that of a javelin without providing any particular reason for this conclusion.[41] However, all of the evidence points to this being the case.

The Greek javelin was equipped with a small leaf-shaped iron head and a small 'butt-cap' called a *styrakion* (στυράκιον).[42] The average weight of the *styrakion* was only 90g and had an average diameter of 1.9cm for its socket. At Vergina a *styrakion* measuring 6.3cm was recovered from Tumulus LXVIII Grave E along with a head (27.5cm, 97g) from the same weapon.[43] This was evidenced by traces of the wooden shaft that remained between the two items. A head that is only marginally heavier than the butt gave the Greek javelin a point of balance slightly forward of centre – the optimum place to grip a shafted missile weapon. The similarity between this clearly identifiable javelin head and the smaller one found by Andronicos, in terms of weight, length and socket size, clearly distinguish them as coming from the same type of weapon. Thus it seems that the smaller head found by Andronicos is that of a javelin and has not come from either a *sarissa* or a hoplite spear. What this means is that, while the butt of a *sarissa* was found, the head was regrettably absent from Andronicos' finds. This may be in part due to the motives that Andronicos assigned to them being discarded outside a tomb – a lack of immediate value for the tomb robbers who seem to have only taken the heads of the *sarissa* and cavalry lance found in the tomb while leaving the small head of the javelin, and the butts of the lance and *sarissa*, behind.

However, a lack of a clearly identifiable head for the *sarissa* within the archaeological record does not mean that examples of such items do not exist. The specifications of the head of the *sarissa* can be identified by other methods – in particular how the weight of the head affects the overall balance of the weapon (which is something that there is literary evidence for). Once the required characteristics of the *sarissa* head have been determined in this manner, this then allows for the possibility of whether any of the heads that have been recovered from Vergina, Olynthus and elsewhere, but which may have been dismissed as being *sarissa* heads by previous scholars, have possibly come from a *sarissa* or not (see the following section on The Balance of the *Sarissa*).

THE 'CONNECTING TUBE' OF THE *SARISSA*

The other item recovered from Vergina by Andronicos, along with the butt of the *sarissa* and the two 'heads', was a small, cylindrical, metal tube (Fig.1):

A socket in iron...with a concave profile...circular cross-section in the interior and polygonal on the exterior...length 17.0cm; Diameter 28mm (end 'a'), 32mm (end 'b'); Thickness: 2-3mm (end 'a'), 3-5mm (end 'b').[44]

The exact purpose of this item has caused considerable debate among scholars. Some scholars see the tube as the connector for a *sarissa* which came in two parts.[45] Others see the tube as something else. Dickinson states that whether the shaft of the *sarissa* is considered to have been a single piece of wood or two rests solely upon the interpretation of this iron tube and that the literary and archaeological sources are inadequate to resolve this issue.[46] However, by examining the physical properties of an assembled *sarissa*, the purpose of this item can be identified.

Those scholars who see the tube as a connector envisage a *sarissa* in which a forward section has the head attached, and a rearward section the butt, which could then be joined together through the use of the connecting tube to create a whole weapon.[47] A pike with such a configuration would certainly have its advantages in terms of logistics and operational adaptability. From a logistical perspective alone, being able to break a lengthy weapon down into two separate parts, possibly of the same length if it is assumed that the connecting tube was located in the middle, would make it much easier for the individual to carry.[48] Throughout his campaign, the troops under the command of Alexander the Great traversed all sorts of terrain including areas that were heavily wooded, steep and even covered in rainforest.[49] It seems almost incomprehensible that thousands of men could manoeuvre through such environments while carrying a pike over 5m in length without it becoming entangled in the foliage. Similarly, at Maleventum in 275BC, Pyrrhus led part of his army on a difficult night march, up the slopes of a heavily wooded hillside and along an overgrown goat track, in an attempt to outflank a Roman position.[50] The possibility of his men accomplishing such a feat while carrying pikes which, at this time, may have been over 7m in length is simply unthinkable.

Markle, in his argument for the *sarissa* only being used after 331BC, suggests that the account of the men of Perdiccas' contingent of phalangites marching across steep forested terrain for an attack on the rear of the Persian Gates in 330BC is indicative that they were not carrying lengthy weapons. Markle concludes that the Macedonians had to be carrying two different weapons – the longer *sarissa* for set-piece battles, and shorter spears for moving over rough and wooded country.[51] Similarly, English suggests that it would be difficult for troops to ferry lengthy weapons across rivers on makeshift rafts as Alexander had used to cross the Danube in 335BC (Arrian 1.3.5).[52] However, what Markle and English have not considered is that these men may have been carrying pikes that had been separated into two sections – both of which would be about the same size as a traditional hoplite spear – and therefore much easier to transport. Additionally, having the *sarissa* come in two different sections would also make repairing any damaged part easier as smaller lengths of timber could be used to replace broken shafts more easily than if the weapon came only in one piece.

It may be the case that, if the tube was used to join the two halves of a *sarissa* together, it was permanently affixed to the forward part of the weapon in order to create a shorter weapon that could be used on its own in a manner similar to the hoplite spear or javelin. Due to the difference in the weight between the head of the hoplite spear (153g) and its large metallic butt-spike (329g), the hoplite spear had a point of balance approximately 89cm from its rearward tip.[53] No weight for the tube found at Vergina is given by Andronicos. However, a replica of it weighs 200g. If this is attached to one end of a shaft 244cm in length, and a head 27cm in length and 174g in weight is attached to the other, creating a weapon 288cm, or half of a 12 cubit *sarissa*, in length, the resultant weapon would have a point of balance 132cm from its rearward tip or a little less than half-way up the shaft.[54] This point of balance would allow the weapon to be used as a thrusting spear to some degree, but its balance would also allow it to be thrown like a javelin as a more rearward point of balance makes flight almost impossible.[55] Thus the connecting tube would act as an improvised *styrakion* to provide a particular point of balance to the weapon.

Javelins seem to have been a fairly traditional Macedonian weapon. According to Arrian, Philotas was executed with a volley of javelins in 330BC (Φιλώταν μὲν κατακοντισθῆναι).[56] Curtius states that Philotas was killed with lances (*lanceis*) – by which he seems to mean javelins.[57] Curtius later describes how the Macedonian Horratas (called Coragus by Diodorus (17.100.2)) armed himself with 'regular' Macedonian weapons for a duel with the Athenian Dioxippus.[58] These arms included a small shield, a spear (which Curtius emphasizes was called a *sarissa*), a 'lance', and a sword (*Macedo iusta arma sumpserat, aereum clipeum hastamque – sarisam vocant – laeva tenens, dextera lanceam gladioque cinctus*). The correlation between the lance of Curtius and the javelin is confirmed by Diodorus who states that, during the ensuing contest, Coragus first cast a javelin at Dioxippus and then lowered his *sarissa* and advanced.[59] This would suggest that, while not recorded in the ancient texts, phalangites were capable of carrying both weapons at the same time. This would only be possible if the phalangite was initially stationary as this would then allow him to hold the *sarissa* upright in his left hand, cast the javelin with his right, deploy the pike, and then move forward (just as Diodorus recounts). However, if such an action was ever adopted for battle in a massed formation, it would seem unlikely that anyone other than the members of the front ranks would be able to cast javelins due to all of the vertically raised pikes.

Interestingly, if a javelin of some description was a common feature of the phalangite's panoply, this sheds light on the passage of Polyaenus which refers to the Macedonians 'skirmishing with missiles' (ἠκροβολίσαντο) during their encounter with Onomarchus in 353BC (see page 20).[60] It is possible that, at the beginning of this engagement, some of the phalangites first cast javelins at the enemy and then followed this up with an advance of levelled pikes. This would then correlate the actions of just a pike-phalanx with what Polyaenus states rather than the other option for reading the passage which has skirmishers engaging first followed by an advance of the pike-phalanx (see page 21). However, the fact that no such action is recorded anywhere else for a battle up to the end of the Hellenistic Age, other than the duel between two of Alexander's men, suggests that Horratus/Coragus had employed a 'heroic', and versatile, form of combat for his duel. Dioxippus is similarly described as fighting in a 'heroic' fashion – in the guise of Heracles, naked and armed with a club.[61] Unfortunately, what none of the sources provide is where the javelin that Horratus/Coragus was using had come from and it is interesting to note that Curtius' description of it as part of the 'regular' equipment of the Macedonians suggests that the portage of both a pike and a javelin may have been a relatively common occurance.

McDonnell-Staff suggests that phalangites carried not only their *sarissa*e but a short (1.5-1.8m long) throwing/stabbing weapon called a *longche* as well.[62] However, the term *longche* may not necessarily be a reference to a javelin. Plutarch, for example, describes the troops of Antiochus in 196BC as '*longche* bearers' (λογχοφόροι), but this is not necessarily indicative that they carried both this weapon and a *sarissa* – indeed the term λογχοφόροι is usually translated as 'pikemen' which would suggest that the *longche* and the *sarissa* were the same (or at least a similar) thing.[63] The term *longche* (λογχή) is generally only used by Classical Age writers to describe spears that are foreign in origin.[64] Plutarch, similarly using the term to describe a foreign weapon, states that Philip's thigh was pierced by a *longche* in a battle against the Triballians.[65] The use of the term *longche* would also be applicable to Greek descriptions of Macedonian weapons as the Greeks always viewed the Macedonians as slightly foreign and living on the fringes of the civilized world.[66]

The other possibility is that the term *longche* refers to the use of the front half of the *sarissa* as an improvised weapon. According to Arrian, one of the accounts of when Alexander the Great killed the general Cleitus in a drunken rage in 328BC, stated that the weapon that Alexander used was a *longche* which he had snatched away from one of his guards.[67] Hammond suggests that the guards were armed with *sarissae* - which would then make the terms *longche* and *sarissa* interchangeable.[68] However, a lengthy pike would be somewhat unwieldy indoors and not entirely suitable for someone acting as a guard protecting a king. This suggests that the guards were carrying a shorter weapon that they could use indoors and which was effective outside the massed formation of the phalanx. This again suggests that the *longche* may have only been the front half of a *sarissa*. Curtius called the killing weapon a lance (*lancea*) – the same name he uses for the javelins used to execute Philotas.[69] Plutarch also says Cleitus was killed with a lance (*aichmē* - αἰχμή).[70] This again would correlate with the use of only the front half of the *sarissa*.

Two inscriptions from Epidaurus, dated to the second half of the fourth century BC, the time when Philip and Alexander were campaigning against Greece with their 'foreign' weapons, detail how two 'walking wounded' visited the sanctuary of Asclepius to have wounds received from a *longche* cured (nos. 12 and 40).[71] One sufferer, Euhippus, had carried the head of the *longche* lodged in his jaw for six years! The other sufferer, Timon, had received a wound under the eye. What these inscriptions unfortunately do not tell us is whether these wounds were inflicted with a thrown or thrusted weapon.

It is also possible that the term *longche* is the correct name for a pike that comes in two parts, while the term *sarissa* could be the name for a single piece weapon – a name that later became a generic term for the Macedonian pike regardless of its configuration. If this is the case, then the Macedonian phalangite could have carried either a dual or single piece weapon depending upon the unit he was attached to and the duties that this contingent was expected to undertake. While the attribution of the term *longche* to a throwing/stabbing weapon used by Macedonian phalangites may not be accurate, McDonnell-Staff's conclusion that they carried such a weapon may be partially correct if it is assumed that the *longche* is the *sarissa*; either the front half or the whole pike. Either interpretation would then account for Plutarch describing pikemen as '*longche* bearers'.

When Hellenistic armies stormed cities that they were besieging, it is unlikely that any phalangites involved in the action were carrying their pikes while using such equipment as assault ladders or siege towers.[72] In the late third century BC, Philopoemen also used *caetrati*-pikemen in an amphibious night assault against the Spartans.[73] These troops are described as being armed with skirmishing weapons.[74] Williams suggests that these troops, while ordinarily pikemen, were using smaller weapons for this encounter as a lengthy pike would have been too cumbersome to carry on small boats, to carry over hilly terrain, and to use effectively at night.[75] However, it cannot be discounted that pikemen in such engagements may have only been using the forward halves of their *sarissae*. Similarly, Plutarch recounts how Antigonus wrote a message in the sand using the butt of his weapon.[76] It is unlikely that he was using a pike or cavalry lance over 5m in length to do this – either he was wielding a more traditional spear at the time, or was using one half of a *sarissa*.[77]

Arrian states that the shaft of the Macedonian cavalry lance was fashioned from the wood of the Cornelian Cherry (κράνεια).[78] Cornelian Cherry was also used to make the shafts of spears as well as bows, javelins and other weapons, predominantly used by non-Greeks.[79] However, there is no direct reference to the wood that the shaft of a *sarissa* was made from. Theophrastus uses the height of the Cornelian Cherry as a comparison for the length of the *sarissa* in the fourth century BC, but does not actually state that it was used to fashion its shaft – although many scholars interpret this passage as saying such.[80] According to Theophrastus, this tree had a height of about 12 cubits (576cm). Interestingly, Theophrastus states that the Cornelian Cherry has a relatively squat trunk and few long straight branches. As such, it is difficult to get a single piece of straight timber more than a few metres in length from this tree. Consequently, if the convention that the shaft of the *sarissa* was made from Cornelian Cherry is accepted, then it must also be accepted, by default, that the shaft came in at least two sections and thus required a connecting tube to hold it together – a consideration which seems to have escaped many scholars who argue for the use of this wood to fashion a single shaft for the *sarissa* (see also the following section on The Shaft of the *Sarissa* from page 63).

Other scholars assign a different purpose to the tube found by Andronicos – claiming that it is a foreshaft guard used to protect the shaft from being hacked through by an opponent.[81] Manti claims to see such foreshaft guards in the Alexander Mosaic from Pompeii.[82] However, what appears to be depicted in this mosaic are not separate foreshaft guards, but the socket of the small leaf-shaped head of the *sarissa* as there seems to only be a single item, rather than two separate pieces, mounted on the forward end of the weapon (Plate 5).

Manti also uses the size of the tubular 'guard' as proof for the existence of a smaller unit of measure in the Hellenistic world incorporating a cubit of 33cm (for the different models for the units of measure in the Hellenistic world see the following section on The Length of the *Sarissa* from page 66). Manti claims that the length of the 'guard', plus the head, equals a distance of 66cm or 2 cubits using this smaller standard.[83] This theory has to assume the use of a head similar to the larger of the two found at Vergina by Andronicos with a length of around 49cm. Manti concludes that, as the intermediate-order outlined by many of the ancient writers incorporates a spacing of 2 cubits per man, and because any weapon held by a more rearward rank would extend ahead of the line by an amount shorter than the weapon of the man in front by 2 cubits due to the interval he occupies, the first two cubits of the *sarissa* shaft had to be protected by the head and a foreshaft guard.[84]

The problem with this conclusion is that it is impossible for a phalangite to stand in a space only 66cm in size and deploy his weapon for combat while still conforming to the spatial requirements of the phalanx (see the chapter on Bearing the Phalangite Panoply from page 133). The minimum amount of space required is 96cm – based upon a system of measurements using a 48cm cubit rather than a smaller one of 33cm. Thus, 66cm does not actually equate to a measurement of 2 cubits, and any formation using the larger interval will present a serried wall of pikes with each successive weapon stepped back from the one held by the man in a more forward rank by 96cm rather than 66cm. As such, the 'exposed' parts of the weapon would not actually be wholly protected by the head and a 'foreshaft guard' only 17cm in length as Manti suggests.[85] This brings the interpretation of the small 'foreshaft guard' into question.

Another problem with Manti's interpretation is because the large 'head' found by Andronicos is used in the reconstruction. What conclusively illustrates that the metallic tube was not positioned immediately behind the large 'head' in the role of a foreshaft guard, is that the diameter of the socket for the large 'head' and the tube are different sizes. The socket of the head found by Androncos has a diameter of 36mm with the thickness of the wall of the socket ranging between 2-3mm. This would mean that the opening in the socket could accommodate part of the shaft that is between 30-32mm across.[86] However, the largest end of the metal tube found at Vergina has a diamter of 32mm, and a thickness ranging between 3-5mm, giving a size of the inner socket of somewhere between 22-26mm. It is interesting to note that the tube found at Vergina is not a solid cylinder. Rather it is a sheet of metal that has been wrapped around the shaft of a weapon. This is evidenced by a longitudinal split that runs the length of the tube.[87] Thus the tube could be fashioned to fit to a shaft of any diameter. It is possible that the tube has been compressed during the time prior to its recovery and that its diameter may have been slightly larger (which would then allow for it to be used as a foreshaft guard as some have suggested). However, the relative unifomity of the diameter of the tube as it currently exists would suggest otherwise and that its current diameter of 22-26mm is accurate. This then indicates that the socket of the large 'head' found at Vergina and the tube cannot have been abutted against each other, with the tube acting as a foreshaft guard, unless it is assumed that the thickness of the shaft was considerably reduced just behind the head in order to accommodate a guard of plate metal. This seems unlikely. However, if viewed as a connector, a split tube could be easily refashioned for use with replacement shafts which may not have been made to the exact specifications of the original weapon.

The final problem with the interpretation of the tube as a foreshaft guard is that, if positioned just behind the head, the addition of 200g of weight to the forward end of the weapon greatly affects the position of the balance unless the head is correspondingly reduced in size to a point where it would almost be offensively redundant (see the following section on The Balance of the *Sarissa* from page 81).

It is also possible that the tube was used to repair a split in a shafted weapon – although this may not have nescessarily been a *sarissa*. However, it would seem odd to include a broken weapon in the grave goods of the tomb of a Macedonian noble unless it is assumed that it was his personal weapon – perhaps one that he had been campaigning with for some time and had even died with. This would then account for the state of the weapon buried with him if the tube was indeed used as a repair. Alternatively, it is also possible that the tube was wrapped around the the shaft of a weapon in a particular location in order to adjust the distribution of weight so as to provide the weapon with its correct point of balance. The addition of extra weight to specific points of a shafted weapon was occasionally undertaken by the Classical Greek hoplite as evidenced by the *sauroter* of a hoplite spear now in the National Archaeological Museum in Athens (#6848) which has a ball of lead wrapped around it to add to the overall mass of the rearward end of the weapon which, in turn, would then alter the spear's point of balance (presumably to the correct one). Thus the use of a metal sheet, wrapped around the shaft, to correct a weapon's point of balance cannot be discounted as a possible function of the tube recovered from Vergina.

Finally, it must also be conceded that it is possible that the tube did not belong to a *sarissa* at all, but was part of the configuration of some other weapon or implement – such as the staff upon which a banner of some description hung. However, the logistical and tactical benefits of using a *sarissa* which came in two parts, and was then assembled for use in battle, suggest that this is the most viable interpretation for the configuration of the *sarissa*. Consequently, a mechanism had to be used to join the two halves of the weapon together. The most likely form of this mechanism, and the most likely purpose of the tube found at Vergina by Andronicos, is that it was a connecting tube for the *sarissa*.

THE SHAFT OF THE *SARISSA*

Importantly, it is not only the weight of the head and butt-spike which affects the dynamics and performance of the weapon, but also the configuration of the shaft as well. Here, like every other aspect of the design of the *sarissa*, scholarship does not agree. Some scholars state that the shaft of the *sarissa* was of a uniform diameter. Many of the proponents of this theory are those who accept the theory for the large 'head' found at Vergina being the tip of the *sarissa*. As both this large 'head' and the butt-spike found at the same time possess sockets with a similar size, some scholars have therefore concluded that the shaft had to have a uniform diameter of around 39mm along its entire length.[88]

Conversely, other scholars suggest that the shaft of the *sarissa* was tapered – with a smaller diameter at the front enlarging to a diameter of around 34mm at the rear.[89] Scholars who forward this theory are generally those who have rejected the identification of the large 'head' from Vergina as being that of the *sarissa* and favour the use of the smaller head with its narrower socket. In his examination of Greek armour and weapons, Snodgrass states that the shaft of the *sarissa* was obviously tapered as the weapon lacked a butt.[90] However, it must be noted that when Snodgrass first composed his work (1967) the butt at Vergina had not yet been discovered by Andronicos.

As has been demonstrated, there are immediate problems with both models. The large 'head' from Vergina is more likely to be the butt of a Hellenistic cavalry lance while the smaller head is most likely that of a javelin. As such neither model is using the correct constituent parts in their reconstruction of the configuration of the *sarissa*.

Unfortunately, while ancient sources like Arrian or Theophrastus provide one possibility for what wood the shaft of the *sarissa* was made from, such passages do not detail its exact shape. The Cornelian Cherry was relatively common throughout northern Greece, Asia Minor and Syria. However, despite these references, it cannot be automatically assumed that weapons were exclusively made from this type of wood. Both Tyrtaeus and Homer refer to ash being used to construct spears during earlier periods of Greek history.[91] Roman writer Statius specifically states that the Macedonians used *sarissa*e made of ash (*fraxineas Macetum vibrant de more sarisas*) and Theophrastus says that this type of tree was common in Macedon.[92] Pliny the Elder states that ash was a good wood for a spear shaft as it is lighter than the Cornelian Cherry and is more pliant than the 'service-tree'.[93] Thus Pliny's passage compares three different types of wood that had been, or were currently, used in the construction of weapons. In his third century BC treatise on siege warfare, Biton states how oak or ash should be used in the construction of the wheels and axles of war machines and artillery due to the strength of the wood. He further recommends that fir or pine be used to create the superstructure of these engines due to the strength of the wood.[94] Therophrastus also outlines the strength and lightness of oak.[95] It seems unlikely that woods that were known for both their lightness and strength would not be used in the manufacture of weaponry. It can therefore only be assumed that weapons in the Hellenistic period were fashioned from a variety of woods to create an optimum blend of lightness, strength and rigidity.[96]

It must also be considered that replacement shafts would have been made from whatever timber was available to an army in the field. It is unlikely that a campaigning army would have waited until they came across a grove of Cornelian Cherry before they replaced and/or repaired any damaged weaponry. English suggests that replacement shafts, made from Cornelian Cherry, were shipped to Alexander's army from Macedonia. This seems unlikely for several reasons. Firstly, and assuming that the shafts were made exclusively from Cornelian Cherry to begin with, if troops operating in Syria, for example, needed to repair or replace damaged weapons, they are unlikely to have waited for shafts to be sent all the way from Macedonia when the same timber was commonly available where they were. Secondly, if troops needed to repair their weapons quickly, they are unlikely to have waited for shafts made from a particular type of timber when something else that was closer at hand may have been both easier and quicker to fashion. Xenophon details how spoke-shaves were carried by Classical Greek armies when on campaign.[97] The later Macedonians are likely to have followed a similar practice. Dio Chrysostom (2.8-9) suggests that carpentry was one of the three most common occupations in Macedon prior to the time of Philip II, and Alexander's troops were able to manufacture all sorts of wooden items from furniture to defensive stakes (Arr. *Anab.* 4.2.1, 4.21.3, 4.29.7; Diod. Sic. 17.95.1-2). This suggests that, no matter which species of wood the shaft may have been initially manufactured from, replacement shafts could, and would, be fashioned from whatever timber was available to a Macedonian army in the field.

Regardless of which wood was used in the construction of the *sarissa*, the ancient texts omit details of both the diameter of the shaft and how its ends were shaped to accommodate the mounting of the head and the butt-spike. The artistic record is also of limited value in this regard as well. Depictions of the *sarissa*, such as those on the Alexander Mosaic, on funerary *stele*, or in Hellenistic tomb paintings generally show a shaft with a uniform thickness rather than one with a taper. However, this may simply be the result of the

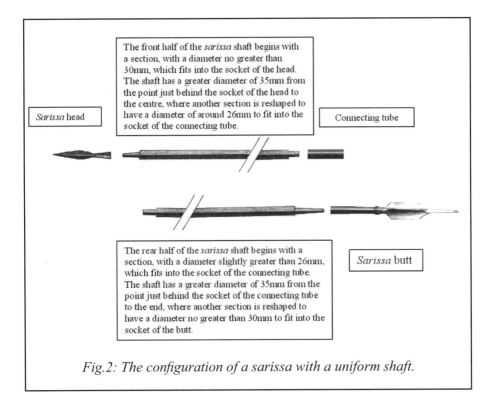

The front half of the *sarissa* shaft begins with a section, with a diameter no greater than 30mm, which fits into the socket of the head. The shaft has a greater diameter of 35mm from the point just behind the socket of the head to the centre, where another section is reshaped to have a diameter of around 26mm to fit into the socket of the connecting tube.

Sarissa head

Connecting tube

The rear half of the *sarissa* shaft begins with a section, with a diameter slightly greater than 26mm, which fits into the socket of the connecting tube. The shaft has a greater diameter of 35mm from the point just behind the socket of the connecting tube to the end, where another section is reshaped to have a diameter no greater than 30mm to fit into the socket of the butt.

Sarissa butt

Fig.2: The configuration of a sarissa with a uniform shaft.

ability of the artist and the medium used to create the image. One thing that should be considered is that the fashioning of a tapered shaft, particularly one over 5m in length, would require a high level of skill to manufacture accurately on the massed scale that would be required for the equipping of large Hellenistic armies.[98] Only experienced weapon makers could have manufactured such weapons. This leaves several possibilities. Firstly, that the *sarissa* was made with a tapered shaft by skilled craftsmen and that, if a phalangite's weapon broke while he was on campaign, he would have to wait for a replacement shaft to be made and/or delivered before his weapon could be repaired. On the other hand, if the shaft of the *sarissa* was made with a uniform width, and came in two pieces as the interpretation of the 'connecting tube' found at Vergina would suggest, then replacement shafts could be fashioned easily in the field using a spoke shave by troops themselves from whatever timber was available at the time.

The logistical expediency of having an army that was able to fashion and repair its own weaponry would suggest that this later scenario was the case and that the *sarissa* came with a uniform shaft. Consequently, if the butt of the *sarissa* possessed a socket with an inner diameter of about 30cm, and the walls of the socket are about 2-3mm thick, it can be concluded that the *sarissa* had a uniform shaft approximately 35mm in diameter – the ends of which would have been reshaped in order to accommodate the mounting of the head, butt and the connecting tube in the centre. The connecting tube, 32mm in diameter and with an inner socket of between 22-26mm, could easily be set onto a shaft with a larger diameter if the sections designed to insert into it were similarly reshaped to fit or if the rolled sheet of the tube was 'opened' slightly so that the elasticity of the metal would allow the smaller diameter tube to grip tightly onto a shaft with a slightly larger diameter (Fig.2).[99]

THE LENGTH OF THE *SARISSA*[100]

The overall length of the *sarissa* would include not only the shaft of the weapon, but the size of the butt and the head as well. The ancient military writer Asclepiodotus, writing at the turn of the first century BC, states that the *sarissa* (which he gives the same name as the hoplite spear (*doru*)) should be:

> no shorter than ten *pēcheis* (πήχεις), so that the part that projects forward
> of the line is no less than eight *pēcheis*. In no case, however, is the weapon
> longer than twleve *pēcheis* so as to project ten *pēcheis*.[101]

The *pēchus* was an ancient unit of measure otherwise known as the 'cubit'. Aelian, writing 200 years after Asclepiodotus in the early years of the second century AD, but basing his work on tactics upon much earlier treatises, recommends that the pike (which he also calls a *doru*) should be 'no shorter than 8 cubits and the longest no longer than a man could use and wield effectively' (a size which he, unfortunately, does not specify).[102] Thus it appears, upon initial reading, that these two passages are at odds with each other in terms of the size of the *sarissa* that they describe. However, Devine finds similarities between the terminology used by Asclepiodotus (δόρυ δὲ αὖ <u>οὐκ ἔλαττον δεκαπήχεος, ὥστε τὸ προπῖπτον αὐτοῦ εἶναι οὐκ</u> ἔλαττον ἢ ὀκτάπηκυ) and the terminology used by Aelian (δόρυ δὲ μὴ ἔλαττον ὀκταπήχους) and suggests that a central section (equivalent to the underlined part of Asclepiodotus' passage above) is missing from Aelian's text.[103] Devine concludes that both Asclepiodotus and Aelian are saying fundamentally the same thing: 'the spear [i.e *sarissa*] is not less than 10 cubits long and extends beyond the rank not less than 8 cubits'. If correct, both accounts outline the use of a weapon with a length of 10 cubits; with Asclepiodotus elaborating on a lengthier weapon of 12 cubits, which may be attributable to the weapon that Aelian refers to as the longest one which 'a man could use and wield effectively'.

Theophrastus, writing just after the death of Alexander the Great (c.322BC), states that the Cornelian Cherry tree grows to a height the same as the length of the longest *sarissa* – which he gives as 12 cubits.[104] The fact that Theophrastus refers to the 'longest' *sarissa* being 12 cubits in length implies that he was aware of shorter weapons as well – possibly the 10 cubit weapons mentioned by Asclepiodotus and Aelian. Thus there are four references, by three different authors from three different time periods, which are similar to each other; two definitively describing a weapon of 10 cubits and two a weapon of 12 cubits (plus a possible inference to the 12 cubit weapon by Aelian and a possible inference to the 10 cubit weapon by Theophrastus).

Polyaenus states that the garrison of Edessa was armed with *sarissae* 16 cubits in length when it fought against Cleonymus of Sparta at the end of the fourth century BC (c.300BC).[105] This length seems to be confirmed by Polybius who, writing just after the Roman defeat of the Macedonian phalanx at Pydna in 168BC, states that the *sarissae* used in earlier times were 16 cubits in length but, at the time he was writing, the weapon had been shortened to 14 cubits.[106] Somewhat confusingly, Aelian, two chapters after he possibly discussed a 10 cubit weapon, then echoes the writings of Polybius by stating that 'the length of his [i.e. the phalangite's] pike was initially 16 cubits, but in truth it ought to be 14 cubits'.[107]

These references to weapons of varying lengths have led to a number of different interpretations as to the size of the *sarissa* at different points throughout the Hellenistic Period. Sekunda, for example, states that because Theophrastus is writing after the time of Alexander the Great (albeit only by about a year), the weapons used in Hellenistic armies after Alexander died were 12 cubits in length.[108] Heckel and Jones, on the other hand, cite Theophrastus as evidence that the weapons used in the armies of Alexander and his father Philip were between 10 and 12 cubits in length.[109] Markle states that the best literary tradition says that the weapons used by the armies of Alexander and the Successors were between 10 and 12 cubits despite references to longer weapons being used in the later Successor Period.[110] Everson only states that the weapons carried by the armies of Philip and Alexander were 12 cubits in length without elaborating on when he thinks a 10 cubit weapon may have been used or whether the usage of 12 cubit weapons continued after Alexander's death.[111] Similarly, Snodgrass flatly states that the *sarissae* used throughout the fourth century BC were 12 cubits in length as per Theophrastus without any consideration of the evidence that implies the existence and/or usage of both shorter and longer weapons during this same period.[112]

Despite this somewhat confusing and contradictory corpus of scholarship on the length of the *sarissa*, it is possible to trace an evolutionary path for the size of the weapon from these ancient sources themselves. Two of the ancient accounts provide details of the length of the *sarissa* at a specific point in time during the Hellenistic Period. Polyaenus, for example, gives a length of 16 cubits for weapons used around the year 300BC (or 274BC according to Hammond and Walbank). Additionally, Polybius states that the weapons before his time (c.168BC) were 16 cubits in length but had subsequently been shortened to the 14 cubit weapons used in his day. The similarity between the passages of Polybius (18.29.2) and Aelian (*Tact.* 14) suggest that, at least in this latter passage, Aelian was commenting on the weapons of the mid-second century BC as well (if he was not just paraphrasing Polybius' passage without considering that it contradicts with what he had written earlier). In his introduction, Aelian advised the Emperor Hadrian (to whom the work was dedicated) that he may find a small pleasure in the fact that contained within the book were Alexander of Macedon's ways of marshalling an army.[113] This would suggest that much of the work (and also that of Asclepiodotus upon which it may have drawn) recounts the details of a Hellenistic army, including the size of their weapons, in the mid-late fourth century BC. However, the similarity between the passages of both Aelian and Polybius suggest that Aelian was, in fact, drawing upon sources which recounted the use of different length *sarrisae* throughout the whole Hellenistic Period.

Like Polybius, Theophrastus must also be referring to a weapon that existed in his own time (the late fourth century BC) and place of writing (Athens) as he uses it as a basis of comparison for the height of the Cornelian Cherry tree.[114] It would have been redundant for Theophrastus to have used the details of a weapon that would not have been commonly known to his intended audience as the basis for his comparison. This suggests that the *sarissae* at the time Theophrastus was writing were commonly 12 cubits in length. This suggests that the 10 cubit weapons detailed by Asclepiodotus and Aelian are references to pikes used in the early Hellenistic Period, and that the length of these weapons was increased to 12 cubits by the time of Theophrastus. From all this it seems that, in the mid-fourth century BC, the *sarissa* was between 10 and 12 cubits in length – the longest being deemed the greatest that a phalangite could wield effectively in battle. Despite this claim, by the end of the century, the length of the weapon had increased to 16 cubits only for the length to be subsequently reduced back to 14 cubits sometime before the end of the Hellenistic Period in the mid-second century BC (Table 2).

Reference	Given length of *Sarissa*	Time Period Referred To
Ascl. *Tact.* 5.1	10-12 cubits	Early Hellenistic Period
Ael. Tact. *Tact.* 12	Possibly 10 cubits	Early Hellenistic Period
Theoph. *Caus. pl.* 3.12	12 cubits	Early Hellenistic Period
Polyaenus, *Strat.* 2.29.2	16 cubits	Mid Hellenistic Period
Polyb. 18.29.2	16 cubits reduced to 14 cubits	Mid-Late Hellenistic Period
Ael. Tact. *Tact.* 14	16 cubits but really 14 cubits	Mid-Late Hellenistic Period

Table 2: References to the size of the sarissa for different time periods.

THE SIZE OF THE CUBIT

A further problem comes from converting the given lengths for the *sarissa* into modern equivalents. This is due to various interpretations of the size of the cubit. This problem is compounded by the fact that units of measure were not always standardized across the ancient Greek world but often varied from state to state. The smallest unit of measure was the *daktylos* (δάκτυλος) – a unit representing the thickness of a finger.[115] Four *daktyloi* made a 'palm' (παλαστή). Herodotus states that 'a foot (πούς) equaled four palms and a cubit equaled six palms'.[116]

Many scholars either argue for, or use in their examinations of Hellenistic warfare, an Attic *daktylos* of 1.8cm which results in an Attic cubit of 45cm.[117] However, a metrological relief, which has been dated to the time of Alexander the Great at the earliest, found on the island of Salamis in 1985, contains a measurement for a *daktylos* of 2.0cm, a foot of 32.2cm and a cubit of 48.7cm.[118] A *daktylos* of 2.0cm should, following the scale of increments outlined by Herodotus, result in an even foot of 32.0cm and a cubit of 48.0cm.[119] Dekoulakou-Sideris suggests that any discrepancy, which is only in millimetres, is attributable to a lack of precise stone cutting on the part of the mason who carved the metrological relief.[120] Thus it seems more likely that the Attic cubit at the time this relief was made was closer to 48.0cm than it was to 45.0cm. Other regions used a unit of measure similar to this larger Attic standard, with variances measured only in millimetres. The Olympic, or Peloponnesian, standard for example, was also based upon a foot of 32cm – which would equate to a *daktylos* of 2cm and a cubit of 48cm.[121] The cubit was also measured as the distance between the tip of the elbow and the tip of the middle finger:

On a person with an average height of 170cm (see below), this measures around 48cm. This further suggests that the cubit was in the vicinity of 48cm within most of the systems used across Greece.[122]

In an investigation of ancient construction techniques, Broneer identified what he called a 'Hellenistic foot' of 30.2cm – so named because it was believed to have come into use during the time of Alexander the Great.[123] This smaller foot would be based upon a *daktylos* of 1.9cm and a corresponding cubit of 45.4cm. The representation of what appears to be one of these 'Hellenistic feet' of 30.1cm also appears on the Salamis Metrological Relief and is what is used to date the engraving. This may be where modern references to a 45cm Attic cubit in the Hellenistic Period have originated. However, if so, such claims would ignore the presence of a 48.7cm cubit on the same relief.

Another metrological relief, now in the Ashmolean Museum, contains a representation of a 'foot' which measures 29.6cm. Michaelis calls this foot 'Samian' and believes that it was an Attic foot added later to the relief when Samos came under Athenian control in the fifth century BC.[124] Such a foot would, if accurately depicted, be part of a system incorporating a *daktylos* of 1.8cm and a cubit of 44.4cm; not far removed from the supposed 'Hellenistic' standard identified by Broneer. However, Dinsmoor's examination of the measurements of the buildings on the Athenian Acropolis, and other classical buildings in Attica, shows that the 'foot' that was in use in Attica in the mid-late fifth century BC measured 32.6cm (with a resultant cubit of 48.9cm).[125] If Michaelis is correct about the smaller cubit being an Attic measurement from the time when Samos came under Athenian control (early fifth century BC), it appears that this unit was replaced a few decades later with a larger unit of 48cm based upon the Olympic/Peloponnesian standard. Consequently, Broneer's 'Hellenistic foot' should actually be relabeled as an 'early-Attic foot' instead. This change from one standard system of measurements across the Greek world to another was most likely a result of the 'Coinage Decree' that was initiated by Athens to standardize all weights, measures and currency used in its allies, colonies and territories.[126] The dating of this decree has been controversial. Initially dated to around 430BC, stylistic considerations of a fragmentary epigraphic record of the decree from the island of Kos forced some scholars to revise this date to between 449BC and 445BC (which corresponds with the construction of the Acropolis under the building programme of Pericles). Further evidence in the form of another epigraphic copy of the decree from Hamaxitos, which became part of the Athenian empire in 427BC, has forced some to revise the date of this decree again to around 425BC.[127]

Regardless of the date of the actual decree, part of the standardising process must have been the imposition on the Athenian allies of a system of measurements based upon a *daktylos* of 2cm, a 'foot' of 32cm and a cubit of 48cm – a system of units that Athens itself appears to have adopted sometime before the construction of the Parthenon and the other buildings of the Acropolis in the mid-fifth century BC. Units of a similar measure were already in use in the Peloponnese and other parts of Greece as well as in Sicily and the Greek colonies in the west. By 'converting' all of its subjects and allies in the Aegean over to a unit of measure that was common throughout the rest of the Greek world, Athens not only facilitated trade among all the regions of Greece, and imposed another way of ensuring Athenian domination of its subject allies, but created a standardized system of measurements across the entire sphere of Hellenic influence that would continue to be used for centuries to come. Thus it is this unit of measure that is most likely the system that the ancient writers were using in their references to the length of the *sarissa*. Theophrastus for example, can only have been basing his comparison of the height of the Cornelian Cherry tree and the length of the *sarissa* on a contemporary local unit of measure that his audience would readily understand. This suggests that the weapons in service around the time that Theophrastus was writing were twelve 'late-Attic' cubits, or approximately 576cm, in length.

However, further controversy over the size of the *sarissa* arises from a number of theories that have been forwarded which suggest the cubit commonly used throughout Greece was of a different size again. In the 1880s, Hultsch suggested that the cubit measured 46.2cm while Dörpfeld suggested that it measured 44.3cm.[128] Then, in 1930, Tarn suggested the existence of a short 'Macedonian cubit' of only 33cm.[129] Tarn based his conclusions on two different factors. Firstly, he deduced that, as he believed that the Macedonian '*bematist's stade*' measured three-quarters of the Attic *stade*, all other Macedonian units of measure should follow suit when compared to the Attic. Secondly, Tarn claimed to have found the 'proof' of the existence of a short 'Macedonian cubit' in the writings of Arrian. Contained within his narrative history of the campaigns of Alexander the Great, Arrian describes the Indians as being the tallest race in Asia with most of them being over 5 cubits in height or not much less.[130] Later in the narrative, Arrian describes the Indian king Porus as 'a magnificent example of a man, over 5 cubits high'.[131] Arrian's description of Porus is echoed in the works of Diodorus while Plutarch gives the king a height of 4 cubits and one palm.[132]

Tarn, taking the ancient sources to be accurate descriptions of the king's stature, concluded that the authors (or at least their sources) had to have been giving measurements based upon a smaller Macedonian cubit of 33cm as this would give Porus a realistic height of 165cm. Tarn dismissed the possibility that Porus' size was given in larger Attic cubits as the resultant stature of 225cm (based upon the early-Attic cubit of 45cm) was unrealistically large. Tarn concedes that Arrian's use of the word ὑπέρ (meaning over, exceeding or beyond) to describe Porus' stature suggests that the given height is exaggerated, but concluded that the base of five cubits was an accurate measurement.[133]

Tarn's suggested small Macedonian cubit has raised more problems than it has solved in the study of ancient military history and has become an integral part of a debate that has raged over which unit of measure was used by which ancient writer and, subsequently, over the size of the *sarissa* mentioned in these ancient sources. Tarn himself, for example, suggests that Theophrastus was writing in terms of his small Macedonian cubits.[134] Alternatively, Mixter uses a cubit of 45cm for the basis of his examination into the *sarissa* in the fourth century BC.[135] Manti, on the other hand, accepts the existence of Tarn's 33cm *bematist's* cubit and suggests that this became the standard system of measurements in the wake of the Hellenisim spread by Alexander's advancing army.[136] Strangely, in an attempt to reconcile the sizes given by both Theophrastus and Asclepiodotus with those of the other authors (which Manti claims are all using the smaller cubit) Manti suggests that Theophrastus was using the Attic cubit as his standard and Asclepiodotus either copied Theophrastus or a contemporary source. Such a conclusion seems odd when it is considered that, even though Theophrastus was writing in Athens, he was writing just after the time of Alexander when Manti suggests that the smaller Macedonian cubit had already become the standard. Manti claims that both Theophrastus and Asclepiodotus are describing a weapon 5.4m in length (based upon 12 x 45cm cubits) that was in use in the time of Philip and Alexander and that this weapon was subsequently replaced by a smaller 4.8m weapon (based upon 14 smaller Macedonian cubits of 33cm) which he says is what is being described by Polybius and Aelian.[137]

Dickinson, in his examination of the *sarissa*, also claims that Polybius based his discussion of the Hellenistic phalanx on Tarn's smaller Macedonian cubit.[138] Such conclusions completely ignore statements like that made by Polybius himself which refer to the use of longer weapons during the late Hellenistic period of 16 cubits which, even in the smaller standard, equate to 5.3m in length – bigger than the supposed longest weapon used in the period according to Manti. Conversely, both Markle and Lane Fox flatly dismiss the existence of Tarn's proposed short Macedonian unit of measure and use a 45cm cubit in their examinations.[139]

By examining the evidence presented by Tarn, this mire of scholarly debate can be cleared. The evidence indicates that both sides of this debate are, in fact, incorrect. Not only does the review of this evidence show that a 45cm cubit was unlikely to have been that used by the ancient writers, it also shows that the claims for the existence of a 33cm Macedonian cubit, and subsequently all scholarship that is based upon it, are unfounded.

Firstly, there is little evidence that the '*bematist's stade*' was based upon a unit of measure incorporating a cubit of 33cm. The *bematists* were specialist surveyors – trained to walk with a regular, measured, step so that accurate distances could be recorded. Several such surveyors accompanied Alexander's expedition against Persia.[140] Many of the measurements taken by these *bematists* were later recounted in the works of Pliny and Strabo.[141] Engels suggests that the high level of accuracy in the recorded measurements taken by the *bematists* may be an indication that they used a specific measuring tool to make their calculations – such as the odometer described by Heron of Alexandria.[142]

What the texts of Pliny and Strabo demonstrate is that Alexander's surveyors were using a system of measurement based upon the larger 48cm cubit implemented by Athens in the mid-fifth century BC. Strabo, for example, recounts that the distance measured by the *bematists* between Alexandria Arieon (modern Herat in north-west Afghanistan) and Prophthasia (modern Juwain in south-west Afghanistan) was recorded as 1,600 *stades*.[143] The stade was an ancient unit of measurement equivalent to 600 Greek 'feet'. Thus in the system of Tarn's smaller cubit, which contains a foot of 22cm, the *stade* is equal to 132m.[144] In the system incorporating the 'early-Attic' cubit of 45cm, the *stade* equals 180m.[145] A *stade* based upon the larger 'late-Attic' system, with its cubit of 48cm, equals 192m.[146] Interestingly, the distance between the start and finish lines in the great stadium at Olympia (the length of the 'stadium' being what the term *stade* is based upon) measures 191m – further proof that the Olympic/Peloponnesian standard incorporated a cubit of around 48cm.

Based upon these different measurements, the distance from Alexandria Arieon to Prophthasia recorded by Strabo calculates to 211km, 288km and 307km when converted into the 33cm cubit system, the 45cm cubit system and the 48cm cubit system respectively. The actual distance between these two locations is 304km, which suggests the use of the larger 'late-Attic' cubit in the determination of distance.[147] Many of the other recorded measurements given by Strabo and Pliny, when converted, fall between distances in early and late Attic cubits. This suggests the use of either one or the other system for the recording of measurements with a margin of error of only a few per cent.[148]

However, the closeness of the converted and actual distance between Alexandria Arieon and Prophthasia suggests the use of the larger 'late-Attic' system. Importantly, even if the 'early-Attic' system was the one that was used with a lesser degree of precision, the recording of such distances clearly demonstrate that Alexander's *bematists* were not using a system of measurement that incorporated a small 33cm cubit as is suggested by Tarn. Even Tarn himself states that there is no doubt that Hellenistic explorers and navigators used a larger Attic cubit to measure distance, which seems to go directly against his own conclusions – although he does not state which Attic cubit he believes they used.[149] As noted, Manti suggested that the small Macedonian unit of measure became the standard across the Hellenistic world after being spread by Alexander's advancing army.[150] Such conclusions, however, go against the evidence which clearly indicates the use of a larger (48cm cubit) system by Alexander's surveyors and it is thus uncertain where the idea of a short '*bematist's* cubit' has come from. This in itself suggests that references to the length of the *sarissa* in the ancient sources are given in cubits larger than Tarn's 33cm 'Macedonian' standard.

Another reason to dismiss Tarn's small 33cm cubit is that there is no need to automatically assume that what Arrian and Diodorus are trying to relay in their description of the stature of the Indians or their rulers is an accurate portrayal of size as Tarn suggests. Ascribing 'larger than life' characteristics to both enemies and heroes was a relatively common literary motif in the ancient world. Herodotus, for example, claimed that Orestes, the son of king Agamemnon of Mycenae, was 7 cubits (ἑπταπήχεϊ) tall.[151] At the battle of Marathon, the Athenian Epizelus was suddenly struck blind by the sight of an opponent of 'great stature' (μέγαν).[152] Similarly, Plutarch describes the corpse of the hero Theseus as being of 'gigantic size' (μεγάλου σώματος).[153] Thus, there is no need to automatically assume that Arrian and Diodorus were using a previously unattested smaller Macedonian unit of measure to accurately describe the height of a race of people or their leader. It is more likely that they were basing their descriptions on the standard 'late-Attic' cubit of 48cm and using the great height of the Indians and their kings that this implies as a literary construct to portray them as more imposing and to make the succeeding victories of Alexander's army that occur later in their narratives appear all the more glorious.

Tarn suggested that the presence of 165cm tall Indian warriors (based upon his small 33cm cubit) would have made quite an impression on the ancient Greeks whom he describes as 'not tall'.[154] However, the evidence indicates that the ancient Greeks were around the same height, or taller, than 165cm on average. The remains of thirteen Spartan warriors unearthed in the Kerameikos in Athens are said to be the same height as that of a modern man – around 170cm.[155] Two other bodies, one of a young adult male and one of a mature male, excavated from near the Herian gates north-east of the Kerameikos (now T35 (case 111) and T37 (case 112) in the National Archaeological Museum in Athens) were dated to the mid-fifth century BC. Their bodies measured 180cm and 175cm respectively. Schwartz, in a review of the measurements of other ancient bodies found in Greece, suggests that the average size for an adult male was between 162cm and 165cm.[156] This all suggests that an average height for the ancient Greeks was around 170cm.[157]

Thus it is unclear what impression an Indian opponent of a smaller or similar size, based upon Tarn's smaller Macedonian cubit, would have had on such men to warrant its mention by Arrian and Diodorus. If Plutarch's stated stature for Porus of 4 cubits and one palm is taken as accurate, a conversion into Tarn's Macedonian cubits results in a height of only 137.5cm – much shorter than the average Greek. Clearly all of these ancient writers are trying to give an impression of imposing size. Using the larger 'late-Attic' system, Plutarch's stated stature for Porus equates to 200cm – a height taller than the average Greek and yet still realistic – suggesting that this may be an accurate description of the size of the Indian king. This further indicates that Tarn is in error with his suggestion of the existence of a small Macedonian cubit.

Another element that Tarn failed to consider is that there is an indication of the size of the cubit used by the Macedonians in the writings of Asclepiodotus. Asclepiodotus uses the 'palm' as a unit of measurement in his examination of the phalangite's shield (the *peltē*) – which he says was eight 'palms' (or 32 *daktyloi* or 2 'feet') in diameter.[158] Aelian gives the same measurements for the Macedonian shield in his treatise on tactics.[159] The diameter of the partial remains of two phalangite shields, found at Dodona and Vegora and dated to the third century BC, were estimated at around 66cm and 74cm respectively based upon reconstructions of their very fragmented bronze facings.[160] However, in the case of the Dodona find, some 80 per cent of the shield is missing and the remaining fragment is quite buckled. Similarly, in the case of the larger Vegora find, more of the centre of the shield has survived but the amount of remaining shield rim is almost the same as for the Dodona find. Thus, any calculation of shield diameter based upon these two fragmentary sections can only be regarded as estimates. Interestingly, Liampi stated that the estimated diameter of the Vegora shield was 66cm (the same diameter as the Dodona shield) in her 1990 article on the find, but then stated that the diameter was 73.6cm in her 1998 book on Macedonian shields.[161] Hammond states that 'the diameter of the

[Vegora] shield cannot be determined precisely' but agrees with Tarn that 'there can be no doubt that his [i.e. Asclepiodotus'] measurements were in terms of Macedonian 'feet' and 'cubits' '.[162] However, Hammond then goes on to re-examine the size of Macedonian units of measure, based upon the size of the various stylistic elements of the Vegora shield, to conclude that the Macedonian foot was 32cm (which would result in a Macedonian cubit of 48cm – a point Hammond failed to realise) and that the shield would have had a diameter of 66cm (as per Liampi's first estimate) when complete.[163]

A more accurate indication of the size of the phalangite's shield comes in the form of a complete limestone mould used to create the metallic coverings for the shields of phalangites in a Ptolemaic army (c.300BC) now in the Allard Pierson Museum, Amsterdam (Plate 6). The diameter of this mould measures 69-70cm.[164] However, it would create a covering large enough to have its outer edges (as indicated by the first ridgeline in the mould running around its circumference which is about 2cm, or one late-Attic *daktylos*, in from the outer edge) folded over the wooden core of a shield around 65cm in diameter. The covering of a Hellenistic shield found at Pergamon similarly measured around 65cm in diameter.[165]

The average size of the phalangite shield is given by Everson as between 65cm and 75cm.[166] Other scholars, basing their calculations on the early-Attic cubit of 45cm and Asclepiodotus' reference to the shield's diameter, state that the average size of the *peltē* was only 60cm.[167] However, the archaeological evidence clearly suggests that the shield was slightly larger than this. Using 65cm as the average, each of the two 'feet' of its diameter described by Asclepiodotus would therefore be around 32cm long and each *daktylos* about 2.0cm – the same as the Olympic/Peloponnesian/late-Attic standard that became the common unit of measurement throughout the Greek world in the last half of the fifth century BC.[168] Thus, as Hammond suggests, Asclepiodotus (and additionally Aelian) are clearly basing their writings on a larger 'Macedonian' cubit of 48cm and not one of 33cm as is suggested by Tarn – which is actually closer to the length of the Olympic/Peloponnesian/late-Attic 'foot'.[169]

Further proof that the works of many ancient writers are based upon a unit larger than Tarn's proposed Macedonian cubit comes from their description of the Hellenistic phalanx. Asclepiodotus, for example, states that formations could be deployed in a close-order, with interlocked shields, with each man 1 cubit from those around him on all sides, or in an intermediate-order (also known as a 'compact formation') with each man separated by 2 cubits on all sides.[170] The intermediate-order spacing is also described by both Aelian and Polybius and the close-order formation is noted by Arrian.[171] It is unlikely that these descriptions of interval are based upon Tarn's smaller Macedonian cubit as it is impossible for a phalangite to conform to an interval of only 33cm for the close-order formation while wearing armour, carrying a shield 65cm in diameter in a protective position across the front of his body and while wielding a *sarissa*. However, an interval based upon a cubit of 48cm can easily accommodate a phalangite (and even a classical hoplite with his much larger shield) with all of his equipment.[172] This indicates that Asclepiodotus, Arrian, Aelian and Polybius (*pace* Dickinson) were all using larger cubits as the standard unit of measure for their examinations of Hellenistic warfare.

Based upon this evidence, it appears that a) Tarn's proposed smaller Macedonian cubit of 33cm is unfounded, and b) that all of the ancient sources were using a larger 48cm cubit as the basis for their discussion on the length of the *sarissa* (and other elements of Hellenistic warfare). From this conclusion, the modern equivalent measurements for the length of the *sarissa*, at different times across the Hellenistic period and as given by the different ancient writers, can be determined (Table 3):

Reference and Period	Given length of *Sarissa*	Modern Equivalent (based on a cubit of 48cm)
Ascl. *Tact.* 5.1 (Early Hellenistic)	10 cubits	480cm
Ascl. *Tact.* 5.1 (Early Hellenistic)	12 cubits	576cm
Ael. Tact. *Tact.* 12 (Early Hellenistic)	Possibly 10 cubits	480cm
Theoph. *Caus. pl.* 3.12 (Early Hellenistic)	12 cubits	576cm
Polyaenus, *Strat.* 2.29.2 (Mid Hellenistic)	16 cubits	768cm
Polyb. 18.29.2 (Mid Hellenistic)	16 cubits	768cm
Polyb. 18.29.2 (Late Hellenistic)	16 cubits	672cm
Ael. Tact. *Tact.* 14 (Mid Hellenistic)	16 cubits	768cm
Ael. Tact. *Tact.* 14 (Late Hellenistic)	14 cubits	672cm

Table 3: References to the size of the sarissa
(in modern units) for different time periods.

THE PIKE OF THE IPHICRATEAN *PELTAST*

Additionally, there is a reference to the size of the forerunner to the *sarissa* in the works of Arrian – which also impacts on the establishment of a developmental chronology for the weapon. In his discussion on tactics, Arrian states that 'the *sarissa*'s size approached 16 feet'.[173] This equates to 512cm in the late-Attic standard – which is just under 11 cubits in length. This is shorter than most other references we have for the length of the *sarissa* and is the only one there is for a pike of this size. However, what Arrian is referring to in this passage is not the pike carried by the phalangite of the Hellenistic period, but the weapon carried by the predecessor of the Hellenistic phalangite – the Iphicratean *peltast* – whose weapon was a hoplite spear (255cm long) doubled in length.[174] Thus, it is also possible to state that in 374BC, the size of the pike, at the moment of its invention and prior to the actual advent of the Hellenistic Period, measured 16 Greek feet or 512cm in length (Table 4).

Period	Reference	Given length of *Sarissa*	Modern Equivalent (based on a cubit of 48cm)
Pre Hellenistic c.374-369 BC	Arr. *Tact*. 12.7 Diod. Sic. 15.44.3 Nepos, *Iphic*. 1.4	16 'feet'	512cm
Early Hellenistic (c.350-320 BC)	Ascl. *Tact*. 5.1 Ael. Tact. *Tact*. 12	10 cubits	480cm
	Ascl. *Tact*. 5.1 Theoph. *Caus. pl.* 3.12	12 cubits	576cm
Mid Hellenistic (c.300 BC)	Polyaenus, *Strat*. 2.29.2 Polyb. 18.29.2 Ael. Tact. *Tact*. 14	16 cubits	768cm
Late Hellenistic (c.168 BC)	Polyb. 18.29.2 Ael. Tact. *Tact*. 14	14 cubits	672cm

Table 4: The size of the sarissa at different times across the Hellenistic Period.

THE EVEN-FRONTED PHALANX

Another problematic passage that influences the debate over the length of the *sarissa* is found in the Byzantine-era abridgement of Polyaenus' *Stratagems of War* (sometimes referred to in translation as the 'Excerpts'). In this passage it is recommended that the weapons belonging to the members of the first three ranks of a pike phalanx should be of different lengths so that, when they are lowered, they will present a level frontage while the men in the subsequent ranks hold their weapons vertical until they are needed.[175] However, it is uncertain whether this passage should be attributed to a Greek, Macedonian or later formation of spearmen, and there is no reference to a historical individual or event to place the passage in its correct context. That this passage does not agree with other references to the spears of the first two ranks of a classical Greek phalanx reaching the enemy, or the weapons of the first five ranks projecting beyond the front of a Hellenistic phalanx, suggests that this passage should not be regarded as typical of either of these formations.[176] Regardless, as there is no historical precedent to verify that a pike formation with weapons of varying lengths was ever used, it can be suggested that the compiler of the *Excerpts* is merely providing a recommendation of one possible way of altering the phalanx rather than giving a description of an actual battlefield practice.

In fact, such a configuration as the *Excerpts* provide would pose all sorts of problems within the formation once it was engaged in battle – again suggesting that it was not an actual battlefield practice. Again the problems of this passage are identified through an understanding of the size of the cubit, the length of the *sarissa*, and the dynamics of phalanx based warfare. For example, if the men in the front rank were armed with 12 cubit *sarissae* which projected forward of the line by 10 cubits (as per Asclepiodotus), then the weapon held by the man in the second rank has to be long enough to cover both the 10 cubit projection of the first rank's weapon plus the space occupied by the man in the first rank as well. If each man occupies a space of 2 cubits (as per Asclepiodotus' description of an intermediate-order formation), then the weapons of the second rank would have to be 14 cubits long and those of the third rank 16 cubits long to make a level frontage of points as per Polyaenus' description of an even fronted phalanx.

A particular problem arises with a formation carrying weapons with these varying configurations if a man in the front rank is killed in action. If a man in the second rank (armed with a 14 cubit weapon) steps forward into the front rank, his weapon will automatically be 2 cubits longer than those carried by every other member of the front rank – and this assumes that he can actually advance forward and does not have the tip of his weapon pressed hard up against the shield of an opponent as the Macedonians did at Pydna.[177] If the man in the third rank then stepped forward to fill the spot vacated in the second, while his longer weapon would still reach the same distance as that of the man in front of him, the soldier who moved up to fill in the third rank (i.e. the man who had initially been in the fourth rank with his pike raised vertically) would have to be carrying a weapon 18 cubits in length to cover the same distance. This problem would compound as more and more people in the front ranks sporadically fell and the varying lengths of weapons across the frontage would compromise the integrity of the formation. If men in this file continued to fall, the man initially positioned in the fifth rank of the phalanx would have to be armed with a pike 20 cubits in length, the sixth rank with a 22 cubit weapon, and so on.

If the man in the second rank merely held his position in line when the man in the front rank fell, rather than moving forward to take his place in the line, while his weapon would still conform to the even frontage of the formation, there would now be a small gap in the frontage of the phalanx due to the missing weapon. While this is not a great disadvantage on its own, as more and more members of the first two ranks fell, dangerous holes would begin to form in the line which would compromise the integrity of the unit and allow opponents with swords or other short reach weapons to move inside the reach of the pike – the very thing that Polyaenus suggests the even frontage was designed to protect against.

Another problem is that the scenario outlined above only works to provide an even frontage if the formation always adopts, and then maintains, an intermediate interval of two cubits per man. If, as detailed above, the members of each succeeding rank carry weapons that are longer than those wielded by the man directly in front of them by two cubits, but the formation then adopts an open-order of four cubits per man, then the frontage of the formation would not be level as Polyaenus recommends. Nor would it be so if the phalangites adopted a close-order of one cubit per man and somehow managed to deploy their weapons for battle. That such problems exist within a formation wielding weapons of differing lengths, and that there is no reference to a *sarissa* of more than 16 cubits long in any of the ancient literature, suggests that Polyaenus' description of an evenly fronted formation was not a battlefield practice in any time period. Despite this, some scholars suggest that this was how the pike-phalanx may have functioned.[178] Yet, as Hammond points out, the close formation and the manoeuverability of the Macedonian phalanx depended upon the standardisation of the weapons and equipment used within it.[179]

Consequently, the passage found in the *Excerpts* of Polyaenus should not be considered factually in any examination of the length of the *sarissa* used during the Hellenistic period.[180]

It can thus be concluded that the system of measures that was used by all of the ancient commentators on Hellenistic warfare was one that had already been in use in some parts of Greece for some time, was adopted by Athens in the mid-fifth century BC, and was then later imposed upon all members of the Athenian Empire – a system incorporating a larger cubit of 48cm. This unit of measurement also seems to have been used in Macedon. The identification of the larger cubit used by the ancient sources allows for the long running scholarly debate over the length of the *sarissa* to be finally put to rest. Additionally, the identification of the size of the Hellenistic cubit allows for all previous scholarship, erroneously based upon one of the smaller units of measure, to be appropriately revised.

Furthermore, the identification of the larger cubit allows for the specifications of the *sarissa* to be determined. Using the 12 cubit (576cm) weapons of the early Hellenistic period as an example, if the butt had an average length of 44cm (as per the Vergina find), and the head had an average length of 27cm, then the shaft of the *sarissa* (not including the 'tongues' which would need to be reshaped to fit into the sockets of the head and butt) would have a length of around 505cm.

THE BALANCE OF THE *SARISSA*

The balance of any hand-held weapon greatly affects how it can be wielded. As such, for any hand-held weapon to work effectively in battle, it needs to be correctly balanced to suit the style of fighting that it was designed to be used in. The *sarissa* is no exception. It has been suggested that the large butt on the end of the *sarissa* acted as a counterweight.[181] However, if this conclusion is taken as simply meaning that the weight of the butt offsets the weight of the head, this then has serious implications for understanding the dynamics and configuration of the *sarissa* as a whole. If taken as being nothing other than an off-setting counterweight, then it has to be assumed that the weight of the butt was similar to the weight of the head. However, there is no evidence to suggest that this was the case.

As noted earlier, some scholars suggest that the larger of the 'heads' discovered at Vergina by Andronicos was the head of a *sarissa*. The weight of this head is 1,235g – considerably heavier than the butt that was discovered at the same time and which was thought to have come from the same weapon (1,070g). Even if this configuration was correct, the butt would clearly not act as a simple offset to the weight of the head. Additionally, a twelve cubit weapon with this configuration, and with a shaft of a uniform diameter, would possess a point of balance 305cm forward of the rearward tip of the weapon – more than half-way up the shaft – as calculated using the following formula (Fig.3).

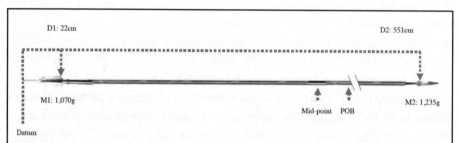

D1: 22cm   D2: 551cm

M1: 1,070g   Mid-point   POB   M2: 1,235g

Datum

Point of Balance (POB) = (M1 x D1) + (M2 x D2) / (M1 + M2)

Where:

M1 = the mass of the butt-spike

M2 = the mass of the *sarissa* head

D1 = the distance from the datum (the end of the weapon) to the point of balance of the butt-spike

D2 = the distance from the datum (the end of the weapon) to the point of balance of the *sarissa* head

Thus:

POB = (1,070 x 22) + (1,235 x 551) / (1,070 + 1,235)

     = 23,540 + 680,485 / 2,305

     = 704,025 / 2,305

     = 305.4 (cm from the datum or 53 per cent forward of the rearward end)

Fig.3: Calculation of the point of balance for a sarissa
with the butt and large 'head' found at Vergina.

Such a forward point of balance would make a *sarissa* with this configuration very difficult to wield and goes against all of the descriptions of the balance of the weapon found in the ancient literature. A depiction of what appears to be a *sarissa* in a tomb painting in Boscorale shows a grip, presumably of leather thonging, wrapped around the shaft. The location of this grip is at the rearward end of the weapon. This suggests that it had a rearward point of balance rather than one near the centre. Interestingly, depictions of the Macedonian cavalry lance, such as in the Alexander Mosaic and in the paintings from Kinch's Tomb, show this weapon to have a more central point of balance. This may account for some of the contention over the balance of the infantry *sarissa*.

Asclepiodotus says that the pike was held by the rearward 2 cubits (96cm); stating that 'the spear [i.e. the *sarissa*]...is not shorter than 10 cubits (480cm) [in length], so that the part that projects forward of the rank is no less than 8 cubits (384cm) – in no case, however, is it longer than 12 cubits (576cm), so as to project 10 cubits (480cm)'.[182] Aelian similarly states: 'the length of the *sarissae*, when the phalanx was first created, were 16 cubits [768cm] long but are, in fact, now 14 cubits [672cm] in length. Two cubits [96cm] of this length are taken up by the grip, being the distance between the hands, while the remaining 12 cubits [576cm] project ahead of the body'.[183]

However, Polybius says that the weapon was held by the last 4 cubits so that a 14 cubit weapon would project 10 cubits forward of the bearer.[184] Some scholars have accepted Polybius' description of the configuration of the *sarissa* over both Aelian and Ascelpiodotus.[185] Markle, for example, made his own replica *sarissa* 549cm in length (based upon twelve Attic cubits of 45cm each), with a uniform shaft 39mm in diameter and with copies of the large 'head' found by Andronicos attached to the forward end. The assembled weapon had a point of balance 51.5 per cent (or 283cm) forward of the rearward tip. This is almost exactly the same point of balance as for a *sarissa* with the same configuration as calculated above (see Fig.3). Connolly states that Markle had accepted Polybius' statement of where to hold the weapon (i.e. 4 cubits forward of its rearward end) but does not mention how his replica performed when wielded this way. When he constructed a replica *sarissa* with the same characteristics, Connolly found it almost impossible to lift the forward tip off the ground due to its weight and the weapon's forward point of balance.[186] This suggests that the identification of the large head found by Andronicos as the tip of a *sarissa* is incorrect and that the weapon possessed a much smaller, and lighter, head which would then shift the point of balance further to the rear.

Connolly, who favours a *sarissa* based upon a tapered shaft, suggests that a small head and a shaft with its thickness increasing towards the rear were designed to give the weapon a rearward point of balance.[187] Similarly, English also states that the large butt-spike on the end of the *sarissa* was designed to give it a rearward point of balance.[188] However, in a confused passage, English also states that the large head found by Andronicos had come from a *sarissa*.[189] Clearly both of these conclusions cannot be correct.[190]

Furthermore, Polybius seems to be incorrect in his description of the *sarissa* being held by the rear 4 cubits of the weapon. The main reason for this conclusion is that a weapon held at this point would not allow the person wielding it to conform to the intervals of the phalanx. Due to its weight, the *sarissa* has to have been wielded in both hands.[191] This is confirmed by both Polybius and Aelian who state that the distance between the hands equated to 2 cubits.[192] The left hand, the more forward of the two, would be placed at the weapon's point of balance and would act as a fulcrum while the right hand, to the rear, and most likely grasping the weapon close to the socket of the butt, provided the leverage to alter the pitch and angle of the weapon.[193] The forward (left) hand, would be positioned just behind the bearer's shield at the very front of the interval that he occupied. Thus the distance between the hands does not actually cover the whole interval of 2 cubits as Aelian and Asclepiodotus state, but the hands are within an interval of 2 cubits so that the large, winged, rear section of the butt and the thickness of the shield can be accommodated within the same interval as well (Fig.4).

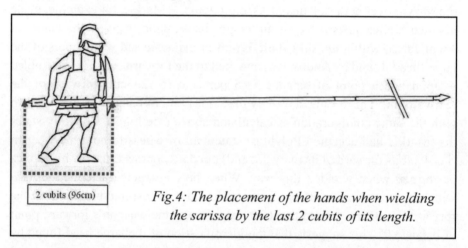

2 cubits (96cm)

Fig.4: The placement of the hands when wielding the sarissa by the last 2 cubits of its length.

Depictions of the phalangite in a combative posture are very rare in the artistic and archaeological record. However, one clear portrayal of the phalangite engaging in battle can be found on a bronze plaque recovered from Pergamon in Asia Minor. On this plaque two phalangites stand side-by-side with their *sarissae* leveled to engage with a mass of opposing cavalry and infantry (Plate 7).

   In the image, the foremost phalangite wields the *sarissa* with his right hand close to the rear end of the shaft – which slightly projects beyond the inner margin of the image but not all the way beyond the outer edge, thus clearly defining the rearward end of the *sarissa*. His forward left hand, which is concealed behind his shield, would grasp the shaft of the weapon a little forward of his body – almost in line with his forward left knee. Thus the phalangite (and it is assumed the more rearward one as well) is wielding his *sarissa* in a manner consistent with that described by Aelian, Arrian and Asclepiodotus – that is, about 2 cubits from the rear of the shaft. It has been suggested that the image on this plaque is a representation of the battle of Magnesia fought in 190BC between the Macedonians and the Romans (although this interpretation is somewhat contested).[194] If this is the case, this plaque would indicate that the Macedonians were wielding the *sarissa* in a more rearward fashion only years before Polybius would suggest otherwise.

Furthermore, if the position of the left hand was 4 cubits from the end, as per the description of wielding the *sarissa* given by Polybius, this affects how the phalangite wielding the weapon can deploy within a formation. For example, if a phalanx is deployed in an intermediate-order of 2 cubits per man, but the weapon is held 4 cubits from the rearward end, the 2 cubits of shaft required for the positioning of the hands would be accommodated within the interval occupied by the individual. However, the remaining 2 cubits (bringing it to Polybius' total of 4 cubits for the location of the grip) would extend backwards, beyond the space occupied by the person wielding the pike, and into the space occupied by the rank behind (Fig.5).[195]

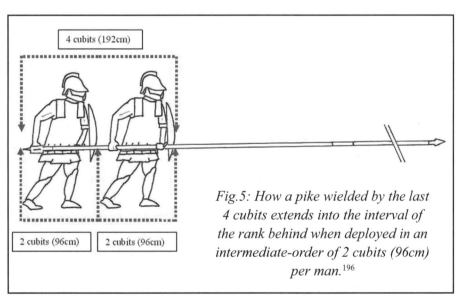

4 cubits (192cm)

2 cubits (96cm)   2 cubits (96cm)

Fig.5: How a pike wielded by the last 4 cubits extends into the interval of the rank behind when deployed in an intermediate-order of 2 cubits (96cm) per man.[196]

On the other hand, if the point of balance for the weapon is located only 2 cubits from its rearward end, as per Aelian and Asclepiodotus, then the grip, the butt and the shield are all accommodated within the interval occupied by the individual and do not extend into, nor impede, the man standing directly behind (Fig.6).

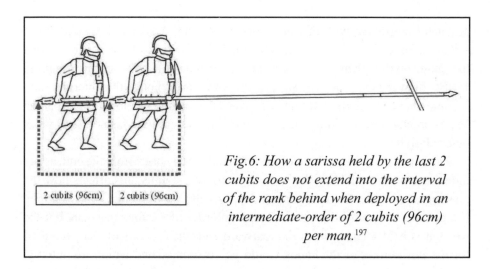

Fig.6: How a sarissa held by the last 2
cubits does not extend into the interval
of the rank behind when deployed in an
intermediate-order of 2 cubits (96cm)
per man.[197]

This is also a clear indication that phalangites could not have adopted the close-order formation of only 1 cubit per man unless they had their pikes raised vertically – as is stated by Aelian. If phalangites attempted to adopt a formation with this smaller interval for combat, their rearward hand and the butt of the pike would become entangled with the man standing directly behind. Consequently, any model of Hellenistic warfare that suggests that phalangites fought in a close-order formation with interlocked shields cannot be considered accurate.

Interestingly, the replica pikes made by Markle, which were devoid of a weighty butt on the rear end of the shaft, are somewhat more manageable when held at Polybius' 4 cubit mark. It is possible that the *sarissae* that Polybius is describing may have been similarly configured with no butt attached to the weapon. This suggests that it is possible that the *sarissa* came in two different forms: one with a large butt attached to the rear of the shaft, which gave the weapon a more rearward point of balance, and which allowed troops to adopt a more compact intermediate-order formation, and the other, devoid of the hefty butt and with a more forward point of balance, which would have been cheaper, faster and easier to manufacture, but which limited the deployment of troops to only the open order. Regardless of whether this is an accurate interpretation or not, the important thing to consider is that any formation containing phalangites wielding *sarissae* 4 cubits from the weapon's rearward end cannot have deployed in either a close-order of 48cm per man, or an intermediate-order of 96cm per man, without the section of the weapon that projects behind the bearer greatly impeding members of the rear ranks.

However, it must also be noted that Polybius' description of the phalanx cannot be based upon a formation deployed in an open-order of 4 cubits (192cm) per man either. Even though such a deployment would allow for weapons to be held at the 4 cubit mark and have the entire amount taken up by the grip and any rearward projection accommodated within the interval of the individual, Polybius specifically states at the beginning of his analysis that it is based upon an intermediate-order deployment of 2 cubits per man. Consequently, due to the spatial requirements of the phalangite within this type of phalanx, it can be concluded that the descriptions of the location of the grip on the *sarissa* as given by Aelian and Asclepiodotus are accurate, while the description given by Polybius needs to be appropriately revised. Interestingly, the location of a point of balance 96cm from the rear end of the weapon is not overly different from that of the spear that was carried by the Classical hoplite which had a point of balance 89cm from the rear. This may be due to the fact that the prototype of the *sarissa* was made by doubling the length of the hoplite spear as part of the reforms of Iphicrates in 374BC – all that was required was an increase in the mass of the butt on the reworked weapon to account for the weight of the longer shaft so that the same point of balance was retained.

It is also interesting to note that both the hoplite spear and the *sarissa* had a point of balance 2 cubits from the end of the weapon based upon the system of measurements that was in use at the time of their construction and common usage. The hoplite spear, for example, has a point of balance approximately 89cm from its rearward end. This equates to a distance of 2 cubits in the system of measurements used up until the mid-fifth century BC which, in Attica at least, incorporated a cubit of 45cm. The *sarissa*, on the other hand, has a point of balance 96cm from its rearward tip. This equates to a distance of 2 cubits using the larger 48cm cubit that was common in the Hellenistic period.

It thus seems likely that Polybius has incorrectly calculated the position of the grip of the *sarissa*. His description of a grip taking up the last 4 cubits of the *sarissa*'s length cannot be the result of a transcription error made when the text was copied as later in the passage Polybius goes into considerable detail to explain how a 14 cubit pike, carried at the 4 cubit mark, projects ahead of the bearer by 10 cubits and what effect this has for the weapons held by the members of the first five ranks of the phalanx.[198] This shows that Polybius (or his sources) firmly believed that the weapon was held by the rear four cubits. Polybius states that the weapons of his time were 14 cubits in length. However, if his analysis of the phalanx in action was in part based upon earlier source material whose descriptions were based upon earlier 12 cubit weapons, that source (á la Aelian and Asclepiodotus) would have most likely stated that, with a point of balance 2 cubits from the rearward end, a 12 cubit weapon projected 10 cubits ahead of the person carrying it. It is possible that Polybius has merely subtracted a 10 cubit projection from the overall length of his 14 cubit weapon, to incorrectly conclude that the *sarissa* was held by the last 4 cubits.

The identification of the correct location for the grip on the *sarissa* can now allow for its balance to be calculated which, in turn, allows the weight of the head required to accomplish this configuration to be determined. As noted, the forward (left) hand of the grip would be best positioned at the weapon's point of balance in order to act as a fulcrum while the controlling rearward (right) could be used to adjust its pitch and angle. Following the description of the grip given by Aelian and Asclepiodotus, the point of balance has to be 2 cubits from the rearward end of the weapon. With the location of the grip and point of balance identified, the required weight of the head can be calculated using the following formula (Fig.7):

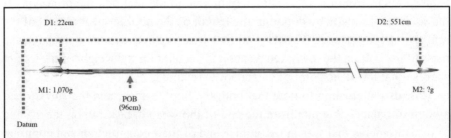

Point of Balance (POB) = (M1 x D1) + (M2 x D2) / (M1 + M2)
Where:
M1 = the mass of the butt-spike
M2 = the mass of the sarissa head
D1 = the distance from the datum (the end of the weapon) to the point of balance of the butt-spike
D2 = the distance from the datum (the end of the weapon) to the point of balance of the sarissa head
Thus:
POB (96cm) = (1,070 x 22) + (M2 x 551) / (1,070 + M2)

Therefore, M2 (the weight of the head) =
    96 x (1,070 +M2) = (1,070 x 22) + (551 x M2)
    (96 x 1,070) + (96 x M2) - (1,070 x 22) = (551 x M2)
    M2 = 1,070 x (96-22) / (551-96)
       = 1,070 x 74/455
       = 79,180/455
       = 174(g)

Fig.7: Calculation of the weight of the sarissa head based upon the weapon's point of balance.

Thus, for a weapon from the late fourth century BC, 576cm in overall length, with the butt found by Andronicos at Vergina attached to the rearward end of a uniform shaft, in order to obtain a point of balance 2 cubits (96cm) from the rearward end, the head of the weapon had to weigh around 174g. Consequently, it can be concluded that the large head found at Vergina cannot be the head from a *sarissa*. Additionally, the small head found at Vergina cannot have come from a *sarissa* either as it is too light, but is most likely to have come from a javelin. If a lighter version of the butt, such as the example from Newcastle upon Tyne, was affixed to the shaft, then the weight of the head would have to be less – 147g using the above method of calculation – but this is still heavier than the small head found at Vergina. If the shaft was tapered, rather than with a uniform diameter, so that the weight of the wood of the shaft was unevenly distributed towards the rear end of the weapon, then the head would have to be slightly heavier in order to create the correct point of balance. This indicates the complexities that would be involved for soldiers in the field if they were required to fashion their own replacement shafts if the *sarissa* was tapered. Even a slight variance in the taper would re-align the distribution of weight along the length of the assembled weapon and so alter its point of balance. This suggests that the shafts of the *sarissa* had a uniform diameter as it would make replacing broken ones easier.

The identification of the required weight of the head for a *sarissa* of this configuration allows for examples of what may be *sarissa* heads to be identified. The heads found at Olynthus, for example, which were identified by Robinson and Petsas as coming from a *sarissa*, but were dismissed as such by Markle, are, in fact, the perfect size and weight for a *sarissa*. Snodgrass states that, based upon finds at Olynthus, the *sarissa* had a head of 'moderate size'.[199] While not providing a specific weight, this statement is basically correct and allows for finds from other locations to be placed in a better context. The head found in the Macedonian *polyandrion* at Chaeronea, for example, is also well suited for a *sarissa*. Interestingly, the average head of the hoplite spear (27cm in length and with an average weight of 153g) is also perfectly suited for attachment to a *sarissa*. The weight of the head of the *sarissa* may be the result of the reforms of Iphicrates which altered the equipment of the Classical hoplite to create the forerunner to the Hellenistic phalangite in 374BC. The ancient sources state that the length of the hoplite's spear was doubled as part of this reform. Due to its length, a larger counter-weight would have needed to have been attached to the rear end of the weapon to offset the increased weight of the longer shaft. However, the head of the hoplite spear may have simply been retained to avoid having to alter all of the different components of the new weapon. This would explain why there are similarities between the size and weights of the head of both the *sarissa* and the hoplite spear, and the respective points of balance for both weapons. The use (or re-use) of hoplite spear heads on the *sarissa* also correlates with Grattius' description of the pike having 'small teeth' – at least smaller than those of traditional hunting spears of the time, and this would account for Grattius' conclusion that such a weapon was not well suited to such activities.[200]

Such determinations also demonstrate the unlikelihood of the metallic tube found at Vergina by Andronicos being a foreshaft guard as some scholars suggest. If an additional weight of 200g is added just behind the forward tip of the weapon, in order to maintain the same point of balance, the weight of the head would need to be correspondingly reduced to only 10g. If the weight of the head was not reduced, the addition of the weight of a foreshaft guard would shift the point of balance for the weapon forward of the 2 cubit position unless the weight of the butt was correspondingly increased. This indicates that the tube had to be used for some other purpose – most likely a connecting tube to join the two halves of the lengthy *sarissa* together which had less of an effect on the location of the point of balance.

THE WEIGHT OF THE *SARISSA*

How much did a weapon with such a configuration weigh? It has been suggested that the *sarissa* weighed between 6-7kg.[201] Markle, examined the characteristics of the *Cornus florida L*, a heavier member of the same family as the Cornelian Cherry, and determined that their densities were approximately the same.[202] During further investigations, Markle determined that oak possessed similar qualities to those of Cornel wood (although Markle did not specify which species of oak).[203] Markle concluded that the weight of Cornel wood was 13.6g per cubic inch (0.83g per cubic cm).[204] Thus the average shaft of a *sarissa*, 505cm long and with a uniform diameter of 35mm, would weigh approximately 4.07kg (not including the small amounts of shaft that would have to be refashioned to insert into the sockets of the head, butt and connecting tube).[205] If added to this is a weight of 1,070g for a butt with the same configuration as the one found at Vergina by Andronicos, 200g for a connecting tube, and 174g for a head appropriate to create the correct point of balance, a fully assembled, 12 cubit, *sarissa* of the early Hellenistic period, would weigh approximately 5.5kg – significantly lighter than the range offered by Sheppard.[206]

The overall weight of the weapon would vary if the lighter Newcastle butt, and the correspondingly required lighter head were attached to the weapon; bringing the overall weight of the assembled *sarissa* down to 5.25kg.

Additionally, the longer weapons of the later Hellenistic period would weigh more. In order to offset the increased weight of a longer shaft, either a heavier butt and/or a lighter head would have to have been attached to the weapon in order to provide it with the same point of balance 2 cubits from the end. This heavier butt and longer shaft would increase the weight of the overall weapon to just over 6kg.[207]

From a review of the available evidence, it is possible to determine that, in the early Hellenistic period at least, the *sarissa* had the following average characteristics (Table 5).

Component	Length	Width	Weight
Shaft	505cm	3.5cm diameter	4,070g
Head *	27cm	3.1cm	174g
Butt **	44.5cm	4cm	1,070g
Connecting Tube **	17cm ***	2.8cm diameter	200g****
TOTAL	**576.5cm**		**5,514g**
Point of Balance	96cm from rear tip		

Table 5: Tabulated data for the constituent parts of the sarissa.

* based upon the characteristics of the average hoplite spear
** based upon the finds from Vergina
*** this would have been located in the centre, joining the two halves of the *sarissa* together, and would therefore not contribute to the overall length of the assembled weapon.
**** based upon the weight of a replica.

Chapter Three

The Phalangite Panoply

As a warrior designed for hand-to-hand fighting, the Hellenistic phalangite bore a panoply of equipment designed to withstand the rigours of close combat and to make him effective on the field of battle. The specifics of some of the accoutrements of the phalangite are another area of the analysis of Hellenistic warfare which has prompted much scholarly debate. Snodgrass calls this lack of evidence 'a most defective picture' in regards to understanding the equipment of the phalangite in the time of Philip II and Alexander the Great.[1] This contention appears strange as there is solid evidence for the type of armour and equipment that the phalangite carried into battle across many differnt periods of the Hellenistic Age.

THE PHALANGITE SHIELD

Asclepiodotus provides a detailed description of the phalangite shield (the *peltē*): 'the best shield for use in the phalanx is the Macedonian: [made] of bronze, 8 palms in diameter, and not too concave.'[2] Similarly, Aelian states: 'the Macedonian shield, made of bronze, is best. It must not be too concave and should be 8 'palms' in diameter.'[3] In the unit of measure commonly used throughout the Hellenistic period, 8 palms (32 *daktyloi* or 2 Greek 'feet') equaled 64cm.[4] This size correlates with the recovered remains of phalangite shields which have a diameter of between 64-66cm (see pages 75-6).

Hammond suggests that the core of the shields was made of wicker.[5] Bennett and Roberts suggest that the shield was made from a leather and wood core.[6] Both seem unlikely. Perrin's translation of Plutarch's biography of the

Roman commander Aemilius Paulus interprets the term used to describe the *peltē, elaphrois de peltariois* (ἐλαφροῖς δὲ πελταρίοις), as meaning 'light wicker shield'.[7] However, the term only means 'light shield' – any terminology describing the use of wicker (such as πλέγμα (wickerwork) or ἀσπίς γέρρον (wicker shield)) is absent and there is no mention of a specific material that the shield is made from. The use of the term 'light shield' seems to be more of a direct literary contrast by Plutarch who, in the same passage, compares the Macedonian *peltē* to the Roman *scutum* – a shield which he describes as 'solid and reaching to the feet' (στερεοὺς καὶ ποδήρεις θυρεούς).

It is more likely that the *peltē* was constructed from a solid wooden core, rather than wicker, which was then faced with metal in much the same way that the hoplite *aspis* was made. The *aspis* was constructed from a blank of wooden beams glued together and turned on a lathe to create a bowl-shaped wooden core between 80cm and 122cm in diameter and 10cm deep with a 5-7cm offset rim (ἴτυς) running around its circumference. It was carried by inserting the left forearm through a central armband (the *porpax*), which the playwright Euripides states was custom made to suit the arm of each bearer, while the left hand grasped a corded hand-grip (the *antilabe*) which ran around the inner circumference of the shield.[8] Archaeological remains show that some hoplite shields had their rims faced with a reinforcing layer of bronze. Towards the end of the Classical Period, the entire outer surface of some shields was faced with a layer of bronze 0.5mm thick.[9] This method of construction is most likely how the core of the *peltē* was made.

Pliny the Elder states that the best woods for making shields were poplar and willow as the pliability of the wood meant that, if pierced by a weapon, the wood would contract and close around the impacting implement – thus making it harder for it to penetrate.[10] Krentz notes that a shield made from poplar or willow would weigh half as much as one made of oak, and two-thirds to three-quarters the weight of a shield made of pine (woods commonly used by modern re-enactors due to their availability).[11] If made from poplar or willow, a 64cm *peltē*, with attachments but minus a metallic facing, would weigh around 2.5kg.[12]

The bowl-like nature of the *aspis* allowed for some of its weight to be supported on the shoulder. In contrast, the Hellenistic *peltē* was too small and not curved enough to be carried in such a manner (nor did it need to be, as it was lighter than the *aspis*) but had to be of a configuration which allowed the bearer to extend his left hand beyond the rim of the shield in order to help wield the *sarissa*. This accounts for the recommendations of both Asclepiodotus and Aelian that the Macedonian shield should not be too concave. This requirement to allow the left hand to remain free also indicates that those scholars who suggest that the size of the shield was occasionally greater than 64cm are in error as any shield with a diameter larger than this would simply be too great in its dimensions to allow the pike to be carried for battle by the average person (see following).

Several scholars state that the *peltē* was carried by a simple neck strap (*ochane* – ὀχάνη) which went across the body and sat on the right shoulder.[13] A shoulder-strap, known as a *telamon*, had been a design element of shields used in the earlier Greek Archaic Age and may have been used with the hoplite *aspis* as well.[14] In the Hellenistic Period, Plutarch recounts how the Spartan king Cleomenes III trained his men to fight with a lengthy pike and to 'carry their shields by a shoulder-strap instead of by an armband' (καὶ τὴν ἀσπίδα φορεῖν δι'' ὀχάνης, μὴ διὰ πόρπακος).[15] Similarly, Plutarch also recounts how the Macedonians at Pydna in 168BC drew 'their shields from their shoulders round in front of them' (τῶν ἄλλων Μακεδόνων τάς τε πέλτας ἐξ ὤμου περισπασάντων).[16] This also suggests the use of a shoulder-strap to carry the *peltē* across the back when it was not in use. Markle says that a light weight was essential for any shield suspended by a neck strap – possibly following the concept that the *peltē* was made of wicker.[17]

Again, such conclusions seem incorrect. In order to function effectively as a piece of defensive armour, a shield, regardless of its shape or size, cannot simply be hung from the shoulder by a strap. A shield requires firm support so that it can be used to meet (and hopefully resist) incoming attacks, and a certain level of control to allow it to be moved in order to deflect blows. If the *peltē* was simply hung from the shoulder with no other means of support, any impact against the shield would just move it out of the way, leaving the bearer vulnerable to succeeding attacks and, due to the use the left hand to help carry the *sarissa*, with no available means of moving it back into a protective position. This indicates that the phalangite shield must have been more controlled.

There are two ways that this level of control can be achieved. The first is through the use of a rigid handle or grip set into the shield which is then grasped by the hand. The other is through the use of a central armband, through which the left forearm could be inserted, similar to the configuration of the hoplite *aspis*. The requirements of the phalangite *peltē* indicate that it must have had a similar configuration to the *aspis*. As the left hand needed to be free in order to help wield the *sarissa*, the way of carrying the *peltē* could not have employed a rigid handle. The only way that the shield could be controlled effectively, and yet still leave the left hand free to help carry the *sarissa*, would have been through the use of a central armband.[18]

The use of a central armband to carry the *peltē* is evidenced by representations of it being carried by phalangites on funerary *stelae*. On the *stele* of Nikolaos, son of Hadymos (inv. No. 2314: Kilkis Archaeological Museum) the *peltē* can clearly be seen carried in a way that the upper rim of the shield sits above the level of the shoulder (Plate 8). A phalangite in an almost identical posture can be seen on the *stele* of Zoilos (inv. No. 551: Museum of Macedonia, Skopje). The position of the shield in these representations suggests that it is being held in place by a means other than a shoulder strap extending across the body. The only way such a positioning of the shield could be accomplished is if it was mounted firmly on the left forearm, which was then held across the front of the body.

However, unlike the hoplite *aspis*, the weight of the phalangite *peltē* was not taken on the arm through the use of the *porpax* when the bearer positioned himself for battle. In order for the hands to both conform to a grip covering the last 2 cubits (96cm) of the weapon, and to provide a firm and stable grip on the shaft of the *sarissa*, they have to be separated. For the controlling right arm, the upper arm is angled out from the body and the forearm extends downward so that the hand can grasp the pike from above. This position allows the muscles of the arm to be employed to alter the angle and pitch of the weapon with relative ease. The left arm merely needs to support the weapon at its point of balance to act effectively in the roll of a fulcrum. The most natural position for the left arm is to have the upper arm tucked against the side of the body with the forearm extending away from the body and downwards so that the left hand can grasp the shaft of the pike at the correct location (see Plate 9). This reduces the amount of muscular stress placed upon an arm used to support a weapon weighing in excess of 5kg.

Due to the angle that the left forearm adopts while in this position, the *peltē* could not simply be carried by a central armband like the hoplite *aspis* as the force of gravity would draw the shield down the forearm and out of its protective position. This did not occur when wielding the *aspis* because the left hand gripped the corded *antilabe* which prevented the shield from sliding down the arm, and the left forearm was held across the body, parallel to the ground, when bearing the *aspis* for battle. Thus what was required to carry the *peltē* correctly was an alternative means to prevent it from sliding down the arm. This was achieved through the use of the *ochane*. The shoulder strap of the *peltē* not only took most of the weight of the shield, but it also kept it at a height where the left hand would still be able to extend beyond the rim to help wield the *sarissa*. The central armband of the *peltē* would have merely been used to give the bearer a level of control with his shield in combat rather than to support its weight – except in instances where the phalangite was standing at attention with the shield extended across his body as per the representations of phalangites on funerary *stelae*. The use of the *ochane* to take the weight of the *pelē* correlates with Plutarch's statement of how Cleomenes taught his men to fight with the shield supported by a shoulder strap rather than a *porpax* as the different posture required to carry the *sarissa* meant that a shield with a configuration similar to the *aspis* would not have worked. Importantly, due to variances in body shape and size, the *ochane* must have been custom made to suit each bearer so that it was the appropriate length to allow for the shield to be carried correctly.

Additionally, and also unlike the hoplite *aspis*, the *porpax* of the Macedonian *pelē* could not have been too wide. Surviving examples of the hoplite *porpax* have a width ranging between 8-14cm.[19] The armband on the *pelē* could not have been as wide as these largest examples. The armband would have still needed to have been mounted in the centre of the shield so that it was balanced. If the *porpax* was off-centre, the imbalanced distribution of weight would make the shield difficult to control and place considerable muscular stress on the left arm (Fig.8).

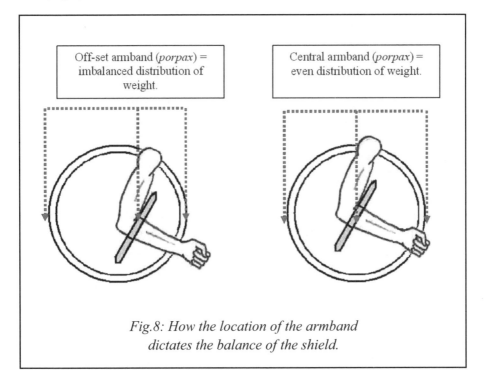

Off-set armband (*porpax*) = imbalanced distribution of weight.

Central armband (*porpax*) = even distribution of weight.

Fig.8: How the location of the armband dictates the balance of the shield.

The location of the *porpax* also provides clues to the further configuration of the *peltē* and how it had to be wielded. The average height of an ancient Greek soldier was around 170cm. On a person of this stature, the length of their forearm, from the crook of the elbow to centre of the palm, would measure around 34cm. The radius of the *peltē* was 32cm. This indicates that the central armband of the *peltē* could not have been wider than a few centimetres as the crook of the elbow had to sit almost in the exact centre of the shield so that enough of the hand would extend beyond the rim to help carry the *sarissa* (see Fig.8 (right) above). If the armband was wider, like the *porpax* of the hoplite *aspis*, the crook of the elbow would be set back by a few centimetres – enough to prevent the left hand from being used. The shield was also carried at a slight angle away from the axis of the forearm which provided more space for the left hand to be employed to carry the *sarissa* (see following).

It is also likely that the *peltē* had a small corded hand grip located near the outer edge of the shield, replacing the *antilabe* of the hoplite *aspis*, which the left hand could grasp. Heckel and Jones state that the *peltē* was carried by both an armband and a shoulder strap but that, due to the archaeological finds of the Hellenistic shield being only the bronze facings, it cannot be certain whether it possessed an *antilabe* or not – although they go on to suggest that it must have existed.[20] Some scholars suggest that this was more of a wrist strap rather than a hand grip, however this seems unlikely.[21] A hand grip would provide the bearer with even more control of his shield, and help prevent the *peltē* from sliding down the forearm, while still allowing the left hand to help carry the pike.

Importantly, the small difference between the radius of the *peltē* (32cm) and the length of the forearm and hand (34cm) does not leave sufficient room (only 2cm) for the hand to extend beyond the rim of the shield to grasp the shaft of the *sarissa* if the shield sits flush with the forearm. As such, the cord which made up the hand grip had to be longer. The extra length of the hand grip allows the forearm to be moved away from the shield – in effect placing the axis of the shield at a small angle to the axis of the forearm – which, in turn, provides the additional room required for the hand to fully grasp the weapon. The shield, positioned at this angle, is prevented from altering its position as its outer edge is in constant contact with the shaft of the *sarissa*, ahead of the left hand, and so the angle cannot be altered even by a direct impact. This not only made the positioning of the shield more secure, and provided enough room for the left hand to effectively grasp the shaft of the pike, but also meant that the majority of the hand was protected behind the stable outer edge of the shield – the only exposed parts being the fingers. This also meant that the shield, while still in its protective position across the front of the body, was set at a slight angle to any attack coming directly from the front which would make it easier for a relatively stationary shield to deflect the incoming blow (Plate 9).

Interestingly, a handgrip located towards the outer rim of the shield also provided support for the pike. By allowing the cord of the handgrip to be of a sufficient length so as to rest in the palm of the left hand, the hand can still extend beyond the rim of the shield in order to grasp the shaft of the *sarissa* without compromising the level of control of the shield itself. If the cord of the *ochane* also formed the cord of the handgrip by simply passing through the upper mounting, this creates a direct link between the shoulder strap, the handgrip and the left hand. Subsequently, due to this connection, the weight of the *sarissa* is partially borne by the *ochane* at the weapon's exact point of balance – alleviating some of the stresses placed on the muscles of the arm(s) caused by wielding a weapon weighing more than 5kg and allowing much of this weight to be borne by the shoulders (see Plate 9). This allows for the *sarissa* to be held in a combative position for a considerably longer period of time than is possible without the incorporation of the shoulder strap and handgrip to help wield it.[22]

Everson suggests that the use of a central armband and a wriststrap meant that, if the phalangite's pike broke and he had to resort to using his sword, the left arm could be pulled back slightly and the left hand used to grasp the wriststrap like a handle in order to use the *peltē* like an improvised hoplite *aspis*.[23] Everson cites the monument of Aemilius Paulus at Delphi as evidence for the use of the shield in this manner. However, the shields depicted on this monument are the larger *aspis* and not the *peltē*. Furthermore, while the possibility of using the *peltē* as an improvised *aspis* makes sense and is entirely possible, it is interesting to note that Everson did not consider the use of a handgrip rather than a wriststrap, and its effects on how the shield could be carried, when formulating it. Indeed, such alternative ways of wielding the *peltē* are easier to be made if a handgrip is employed on the shield rather than a wriststrap. That being said, there is nothing to indicate that Everson's theory for an alternate use of the *peltē* when employing a one-handed weapon is incorrect, and the versatility of such a shield would have had several advantages on the battlefield, or when storming a breach in a city's defences where one-handed weapons like swords or the front half or the *sarissa* were suitable. Importantly, for the *peltē* to be used as an improvised *aspis*, an armband would have to be both present (contrary to some scholarly opinions) and situated in the centre of the shield to provide it with a level of balance.

Thus it seems that the *peltē* was carried using a combination of a shoulder-strap to support its weight (and partially take the weight of the pike as well), a central armband and a handgrip to provide stability and control.[24] The accounts of Plutarch which merely outline the use of the *ochane* may be the basis for theories which suggest that the Macedonian shield was carried only by a shoulder strap. However, while this is fundamentally correct, the physical requirements for carrying the shield effectively in combat, suggest that in addition to a shoulder strap, both a *porpax* and a handgrip would have been used. In the case of the reference to the troops at Pydna, Plutarch's description of the use of the *ochane* seems to be more of a portrayal of how the shield was carried slung across the back by the shoulder strap while the phalangites were on the march and formed up for battle, rather than how it was carried for combat. It seems likely that, once moved into position in front of the body, the shield would have been held in position through the use of an armband and handgrip and the absence of a reference to them in this particular passage can be understood when it is considered that both the *porpax* and the wriststrap would not be used to carry a shield that is slung across the back.[25]

Using such a method of portage, the weight of the shield would be taken predominantly on the *ochane* when the phalangite adopted a combative posture. Contrary to Markle's hypothesis that the shield needed to be light in order for it to be carried solely by a shoulder-strap, a small *peltē* fashioned from a solid wooden core and weighing several kilograms can be supported in this manner for a considerable period of time. The similarities between the methods of portage for both the hoplite *aspis* and the Macedonian *peltē* (e.g. the similar use of a central armband and the replacement of the hoplite antilabe with the phalangite hand-grip) are likely to be further by-products of the reforms of Iphicrates which basically refashioned the *aspis* into a configuration that would allow the longer pike to be carried in combat.

Indeed, it seems that the *peltē* was merely a cut-down version of the *aspis* in most regards. The bowl-like nature of the *aspis* meant that the most angular part of the shield's 90cm diameter was near its outer edge. Additionally, the *aspis* possessed an offset rim several centimetres wide running around its circumference which, when positioned for battle, protected the lower half of the bearer's head. However, by trimming the outer-most 13cm off the entire circumference of the *aspis*, the offset rim and the pronounced angular sections of the 'bowl' are removed to leave a

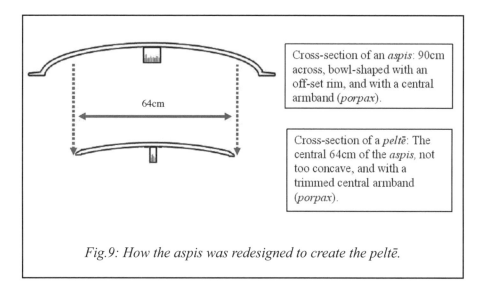

Cross-section of an *aspis*: 90cm across, bowl-shaped with an off-set rim, and with a central armband (*porpax*).

Cross-section of a *peltē*: The central 64cm of the *aspis*, not too concave, and with a trimmed central armband (*porpax*).

Fig.9: How the aspis was redesigned to create the peltē.

shield 64cm in diameter, with the *porpax* still mounted in the centre (the width of which would then be reduced), and with a shape 'not too concave' just as Asclepiodotus and Aelian describe the *peltē* (Fig.9).

Thus it seems that when the reforms of Iphicrates were implemented in 374BC, rather than create an entirely new type of shield, the hoplite *aspis* was merely cut-down, and the *porpax* reduced in width, to the size required for the wielding of a pike with both hands.[26] Thus, like the re-use of the hoplite spear head to give the correct point of balance to the new enlarged pike, the use of a smaller shield as part of the reforms of Iphicrates seems to have been more of a refashioning of hoplite equipment, just as Diodorus and Nepos state, to create a new style of fighting rather than the invention of something entirely original. The re-tooling of the hoplite *aspis* subsequently created the forerunner of the Hellenistic *peltē* at the same time that the doubling of the length of the hoplite spear created the forerunner to the *sarissa*.

Both Asclepiodotus and Aelian state that the *peltē* was bronze. However, the shield was most likely faced with metal rather than made entirely from it. The remains of shields recovered from both Vegora and Dodona, for example, are a thin bronze facing which would have been attached to a solid wood core. Similarly, the mould for the shields of a Ptolemaic army now housed in the Allard Pierson Museum (Plate 6) would have been made to create metallic facings for shields rather than the whole shield itself. A facing of bronze was an aspect of the design of the hoplite *aspis* – particularly in the last half of the Classical period – and the retention of a metallic facing for the phalangite *peltē* is most likely another hold-over from the reforms of Iphicrates which reconfigured hoplite equipment in the early fourth century BC. This facing would have provided structural stability to the finished shield. Should the inner wooden core split due to a weapon impact, for example, or if the glue holding the wooden core together did not hold, the metallic covering would keep the shield together and in its correct shape – albeit now with an internal weak spot.

Bronze was not the only metal that shields were faced with. Covering shields with different metals was a practice that was begun by Alexander the Great and continued across the Hellenistic period. Alexander covered the shields of his *hypaspists* with silver in recognition of their service – and the name of the unit changed to the 'Silver Shields' (*Argyraspides* – Ἀργυράσπιδες).[27] This unit later fought with Eumenes during the wars of the Successors.[28] A contingent of Silver Shields also fought at Magnesia in 190BC.[29] There are also references to units called the 'Bronze Shields' (*Chalkaspides* – Χαλκάσπιδες), the 'Gold Shields' (*Chrysaspides* – Χρυσάσπιδες) and the 'White Shields' (*Leukaspides* – Λευκάσπιδες) across the Successor period.[30] Similarly, Polybius recalls how the troops of Antiochus IV bore shields of gold, bronze and silver during a parade at Daphne in 168BC.[31] The Bible also describes Seleucid troops carrying shields of gold and bronze at the battle of Beth Zachariah in 162BC.[32] Everson suggests that the White Shields are an indication that some units had their shields painted rather than faced with metal.[33] The Amphipolis Decree, a set of military regulations set down in the second century BC, stipulates that a fine of one drachma would be incurred for the loss of a shield.[34] This is only half of the fine imposed for the loss of a set of greaves and suggests, following Everson, that not all of the shields used in the Hellenistic Period were faced with metals – especially precious ones.

The different facings on the shields were used, in part, to identify different units within an army. This is evidenced by units being referred to by the 'colour' of their shields. It has been further suggested that part of the symbolism of the different metals (or paint) used in the shield facing was to signify whether the troops were professional soldiers or part-time conscripts.[35] The army of Antigonus Doson in the late third century BC, for example, contained units of both Bronze Shields and White Shields. Plutarch says that his opponent, Cleomenes III of Sparta, 'set free those of the Helots who could pay five Attic minas... [and] armed 2,000 of them in Macedonian fashion as an offset to the White Shields of Antigonus'.[36] The 'offset' of the White Shields with a unit of raw recruits suggests that Antigonus' White Shields may have been something similar – making their white shields indicative of their status as McDonnell-Staff suggests. The eighth century poet Hesiod outlined five distinct 'Ages of Man' in his poem of creation, with four of them classified under the titles of different metals of decreasing value – gold, silver, bronze and iron.[37] It is possible that the metal facings used on Hellenistic shields were, in part, based upon this literary hierarchy or something similar. If this is the case, then the White Shields may have carried a *pelte* faced with a thin layer of polished iron.

However, it cannot be discounted that what Cleomenes was doing was simply countering a unit of enemy pikemen with one of his own, without any consideration of their potential or experience, which would then suggest that the colour of the shield was merely the identifying mark of the unit itself rather than an indication of their status or training. A unit of White Shields also fought fifty years later at Pydna in 168BC – although in this case their actions would indicate that they were more experienced than raw recruits which suggests that the colour of the shield was not indicative of status.[38] Regardless, even if the shield facings of different metals were used to indicate the status of the unit, there is no evidence for this occurring earlier in the Hellenistic Period than the time of the Successors and the majority of the shields in the time of Philip II and Alexander the Great seem to have been mostly faced with bronze.

What seems to have signified different units within one of these earlier armies, whose shields were all faced with bronze, was the design placed upon the shield itself. By far the most common design embossed into the metal facing of the *pelte* incorporated a series of concentric 'Cs' staggered around the outer edge of the shield and surrounding a central circle. This design could further incorporate the use of dots or stars, alternating with the concentric 'Cs', to create a fairly original pattern. The shield covering from Dodona carries this design, as does the Ptolemaic mould in the Allard Pierson Museum (Plate 6), as well as almost countless examples of the *pelte* from tomb paintings, sculptural reliefs and issues of 'shield coins' from right across the Hellenistic period (Plate 11).[39] The covering from Vegora carries a completely different design – that of the large Macedonian sunburst. This device can also been seen in the depiction of a shield in a wall painting in the Tomb of Lyson and Kallikles in Lefkadia. These two designs, the concentric 'C' pattern and the sunburst, are the only two devices that seem to have been used on phalangite shields – with the concentric 'C' pattern being by far the most common.[40]

However, while the outer designs on shields incorporating the concentric 'C' pattern seem to have been somewhat similar in many respects, there is a great variety of images that could be placed within the central circle of the design. These range from portrait busts, to geometric designs and patterns, to animals, to symbols like anchors, axes and lightning bolts. Yet there does not seem to be such a variety in these designs to suggest that each different one was used by an individual, similar to the personalised emblems on the Classical hoplite shield.[41] Indeed the issue of coinage, and the erection of monuments, bearing these designs suggests that they were indicative of more than the individual. Alternatively, the number of different designs is too great to suggest that they are indicative of an army as a whole (or its commander) either. Alexander the Great, for example, issued 'shield coins' with at least six different designs shown in the centre of the shield.[42] The most likely explanation for the variety in these designs is that the central motifs were indicative of particular units within an army.[43] The Vegora sunburst shield bears an inscription which translates as 'of Antigonus', while the Ptolemaic shield mould in the Allard Pierson Museum bears an inscription that reads 'of Ptolemy' and has a Gorgon head in the centre of the design (see Plate 6). Heckel and Jones suggest that these references to the ruler indicate that shields were pieces of state-issued equipment.[44] This would then suggest that it was the device, design or facing material (or a combination thereof) used in conjunction with such inscriptions which identified specific units within the army of the inscribed commander.

THE PHALANGITE HELMET

There is little doubt that the phalangite wore a helmet. Part of the equipment that Polyaenus states the troops of Philip were carrying was a helmet (κράνη).[45] Writing in the fourth century BC, Aeneas Tacticus states that helmets (περικεφαλαῖα) could be smuggled in to besieged cities.[46] The 'Amphipolis Decree' states that the fine for losing a helmet (*konos*) was 2 obols.[47] This suggests that helmets were standard issue to the Macedonian infantry and were an integral part of their equipment. In his biography of Aemilius Paulus, Plutarch states that Macedonian helmets were taken as spoils following the battle of Pydna in 168BC.[48] These passages indicate that helmets were a common part of the phalangite's accoutrements across the entire Hellenistic period. However, what type of helmet the Hellenistic phalangite wore has been something of a contentious scholarly issue.

There is evidence to suggest that there was a wide variety of helmet available to the Hellenistic infantryman. Dintsis separates the helmets of the Hellenistic Age into fourteen different categories (although not all of them were exclusively used just by the Macedonians), based upon their style, from the fourth century BC to the second (see Plate 10):[49]

Der Boiotische Helm (the Boeotian Helmet)[50] – in use between 475-50BC, the Boeotian helmet was made from hammering a flat sheet of plate metal onto a template to create the cranium of the helmet while any extra plating was hammered out relatively flat to create a broad brim.[51] The resultant helmet was open-faced but sections of the brim did provide some protection to the sides of the head. Archaeological finds suggest that, before the time of Alexander the Great, use of the Boeotian helmet was confined to eastern Greece. Following Alexander, the Boeotian helmet, based upon finds, was used in Italy, Greece, Macedon, Asia Minor, the Near East, Persia and Bactria.

Der Tiaraartige Helm (aka the Phrygian Helmet)[52] – in use between 450-100BC, the so called 'Phrygian' was a peaked helmet, with a backward sweeping cranium, and occasionally augmented with long separate cheek pieces which swept forward to protect the lower face. The distinguishing feature of the Phyrgian style was a large bulbous upward extension of the helmet in emulation of the cloth Phrygian cap upon which it may have been based. Examples of this style of helmet, dated to before the time of Alexander the Great, have been found in Italy, Greece, Crete, Thrace and Asia Minor. After the time of Alexander, examples are also found in the Crimea, the Near East, Egypt and Babylon.

Der Pilos/Konoshelm (the conical *pilos* helmet)[53] – possibly the most common form of helmet used by the Hellenistic infantryman. The *pilos* became popular in Greece during the time of the Peloponnesian War (431-404BC) and continued to be used until around 150BC. Its basic construction, fashioned from a single sheet of bronze hammered into a cone, allowed armies to quickly equip large numbers of troops and place them in the field. Early examples of the conical *pilos* have been found in southern Italy, Greece, Macedon, Asia Minor and Egypt. Following the reign of Alexander, examples of the *pilos* are also found in Thrace and the Levant.

Der Piloshelm mit runder kalotte (the rounded *pilos* helmet)[54] – a variant of the conical *pilos*, the rounded *pilos* was, as its name would suggest, more rounded than its conical counterpart. The rounded *pilos* was used across the same time as the conical version (430-150BC). Pre-Alexander examples of this type of helmet have been found in Taras in southern Italy and in Athens. The only post-Alexander finds of this helmet come from Macedon.

Der Konoshelm (the brimmed *pilos* helmet)[55] – in use from 175-100BC (at least according to Dintsis, but possibly earlier in date), the brimmed *pilos* was a hybrid of the basic *pilos* style helmet combined with the brim of the Boeotian style helmet. It seems that the manufacture of these helmets was achieved by using a slightly larger plate of metal than was required to construct a basic *pilos*, and then hammering the extra material (which would not have fitted in the mould or been trimmed off) to form the brim (the same method for the manufacture of the Boeotian style helmet). Examples of this helmet have been found in North Africa, Greece, Thrace, Asia Minor and Syria.

Der Korinthische Helm (the Corinthian helmet)[56] – in use from at least 670BC, and with variants, such as the Apulo-Corinthian helmets of southern Italy, still used until at least 100BC. The enclosed Corinthian style helmet common to the earlier Classical period of Greek history remained a relatively common feature of Hellenistic art which suggests that it may have had a continued, albeit limited, usage in the armies of Macedon. This style of helmet underwent developmental changes across its period of usage to provide greater protection via the incorporation of repousse ridges to reinforce the brow, the incorporation of openings for the ears to improve hearing and regularly larger apertures for the eyes. Due to its long and widespread usage, examples of the Corinthian helmet have been found from Italy to Greece, the Crimea and Egypt.

Der Chalkidische Helm (the Chalcidian helmet)[57] – the developmental predecessor of the Attic helmet, the Chalcidian differed by having much more rounded cheek guards (which were either fixed or hinged) and retained some form of nasal protection (albeit only small in some cases) across almost the entire period of usage for the helmet from 500-370BC. Common to Italy, Greece, Thrace and the Crimea in the period prior to the reign of Alexander the Great, the discontinued use of this helmet at the beginning of the Hellenistic Age has resulted in only limited finds (with somewhat ambiguous dating) of later examples in Macedon and Thrace.

Der Attische Helm (the Attic helmet)[58] – one of the most long lived of any of the Hellenistic helmets, early examples of this style have been dated to around 480BC, and variants continued to be used by the Romans up until AD410. This type of helmet was a development of the earlier Chalcidian style helmet used by some Classical Greek hoplites – although the Chalcidian style does still seem to have been used and depicted in Hellenistic art and on coins.[59] Where the Attic helmet differed from the earlier Chalcidian was that it was much more open faced, had a reinforcing brow ridge, either fixed or hinged angular cheek pieces, and a lack of nasal protection. Pre-Alexander examples of the Attic helmet have been found in Italy, Greece and Caria in southern Asia Minor. Examples dated to after the reign of Alexander have been found in Italy, Macedon, Thrace, Asia Minor, the Near East, Persia and India.

Der Pseudoattische Helm (aka the 'Thracian' helmet)[60] – a peaked helmet, with a backward sweeping cranium, and long separate cheek pieces which swept forward to protect the lower face. The distinguishing feature of the Thracian-style helmet was a pronounced ridge running from front to back over the top of the head. The 'Thracian' helmet, which Dintsis sees as a developmental offshoot of the Attic style, saw service from 430-100BC. The earliest finds of this type of helmet, predating the reign of Alexander III, come from Southern Italy, Attica, the Peloponnese and Caria in southern Asia Minor. Post-Alexander examples have been found in Southern Italy, Greece, Epirus, Thrace, the Crimea, Asia Minor and Egypt.

Der Pseudokorinthische Helm (the Pseudo-Corinthian helmet)[61] – a hybrid of almost every style of helmet merged together, the Pseudo-Corinthian helmet possessed the separate cheek pieces of the Attic style and the flanged brim and visor of the Boeotian and brimmed *pilos*. Unlike the Corinthian, the Pseudo-Corinthian helmet was not fully enclosed, but was much more open-faced like the Apulo-Corinthian style. This style seems to have had a relatively short, and much later, period of usage (160-0BC), and examples have been found in Sicily, Rome, Macedon, Asia Minor, the Near East and Egypt.

Der Glockenhelm (the 'Bell' helmet)[62] – seemingly another variant of the *pilos*, the bell helmet differed by having the somewhat conical, although rounded at the top, cranium of the helmet flare out at its lower rim to create a bell-like shape (hence the name). Bell helmets also differed from the *pilos* through the regular use of attached cheek guards, regular repousse ridging and, in one version, the large sweeping cranial ridge common to the Pseudo-Attic/Thracian style. Bell helmets have a very short period of use which covers the end of the Hellenistic Period (c.170-120BC), but examples have been found in Macedon, Illyria, Asia Minor, Cilicia, Armenia and the Levant.

Der Kappenhelm (the cap helmet)[63] – one of the earliest and most basic forms of helmet, examples belonging to this style date back as far as 1,400BC – but with a common period of usage between 270BC-AD100. The cap helmet was essentially an overturned metallic bowl, usually rounded but occasionally pointed, which sat on the head. Some later versions of this helmet, such as the Roman Montefortino style, incorporated the use of cheek guards, neck guards and crest holders. Cap helmets come in both plain and highly engraved variants and are generally found in Italy and central Europe – although a few examples have been found in Greece and Asia Minor.

Der Gesichtsmaskenhelm (the 'face-mask' helmet)[64] – this style of helmet refers to any type in which the cheek guards, or any other form of facial protection, are decorated in high relief in a depiction of the features of the face. The most common, Hellenistic Greek, example of this style of face-mask helmet is the so-called 'masked Phrygian' which has the bulbous upward extension of the cranium common to the standard Phrygian style helmet, but with the cheek/face guards carved in a representation of a beard and moustache. Masked Phrygian helmets seem to have been a phenomenon common to the age of Alexander the Great and the early Successors with found examples dated to 340-300BC.

Die Kausia (the Kausia)[65] – the *kausia* seems to have been a felt hat or cap similar to a beret or, as has been suggested, similar to the *chitrali* worn in regions of modern Afghanistan and Pakistan.[66] Artistic representations of what is thought to be the *kausia* (or at least something similar) have been found in central Italy, Illyria, Greece, Macedon, Thrace, Egypt and Bactria. The use of the *kausia* as a piece of military headwear has been much debated (see following).

Warry suggests that Greek helmets evolved during the fifth and fourth centuries BC with a view of improving vision, ventilation and hearing without sacrificing protection.[67] However, he is not entirely accurate as, while the development of many open-faced styles of helmet such as the *pilos*, the Boeotian, the Chalcidian and the Attic did offer improvements to hearing and ventilation compared to more enclosed types of headwear, such advantages came, in many cases, at the cost of a reduction in the level of protection given to the face, neck and throat. Julius Africanus states that the neck of a Greek wearing an open-faced *pilos* was particularly vulnerable to the swords of the Romans.[68] As such, each different style of helmet available to the Hellenistic man-at-arms possessed both benefits and drawbacks.

Many modern examinations of Hellenistic warfare often include illustrations of the phalangite wearing the *pilos* and/or either of the so-called Thracian or Phrygian styles of helmet.[69] The location of early finds of these helmets suggests that the names that are regularly ascribed to them are not indicative of their place of origin; rather they seem to be a reference to the similarities between them and the cloth hats worn in these areas.[70] Indeed, examples of the so-called 'Thracian' style of helmet that have actually been found in Thrace all date to the period after the reign of Alexander the Great, but are found in places like southern Italy, Athens, the Peloponnese and Caria in southern Asia Minor prior to this time.[71] This suggests that this style of helmet was developed in an area other than its namesake. Dintsis calls the 'Thracian' helmet a 'Pseudo-Attic' helmet – suggesting a possible Athenian place of origin.[72] Both the Thracian and Phrygian-style helmets were predominantly made of bronze.[73] The cranial ridge of the Thracian helmet, and the bulbous extension of the Phrygian, both of which might, or might not, have an additional crest of horsehair or feathers attached, reinforced the helmet and made the top of the cranium a more difficult target to penetrate.[74]

Depictions of the Thracian-style helmet can be seen in the wall paintings of the tomb of Lyson and Kallikles, on the Pergamum frieze now in Berlin, and on the Alexander Sarcophagus in Istanbul (although in this case the troops wearing them are *hypaspists* rather than phalangites).[75] A coin issued by Alexander the Great to commemorate his victory at the Hydaspes River in 326BC depicts a standing figure on the obverse, thought to be Alexander himself being crowned by the goddess Nike (Victory), wearing headgear with the bulbous crest of the Phrygian helmet.[76] A well preserved example of the Phrygian-style helmet housed in the collection of the Ioannina Museum has small metallic tubes mounted on either side of the bulbous crest which indicates that, at least in this instance, plumes could also be mounted on the helmet.[77] Both the Phrygian and Thracian-style helmets also possess reinforcing ridges across the brow. It has been suggested that these were developments incorporated into the design of the helmet in the third century BC following the Celtic invasions of the Greek world as a means of providing better defence against strikes delivered by a Celtic long slashing sword.[78]

An example of the Phrygian-style helmet was also found in the so-called 'Tomb of Philip' (otherwise known as Tomb II) at Vergina.[79] This helmet, however, is made of iron rather than bronze and is the earliest known example of this style of helmet made from this material. Everson suggests that expensive iron helmets may have only been worn by the wealthy aristocracy while the general rank-and-file soldier wore a helmet made of bronze.[80] The helmets worn by the *hypaspists* on the Alexander Sarcophagus were originally painted blue. Sekunda interprets this as meaning that the helmets were painted a blue colour in real life and were indicative of helmets worn only by the infantry.[81] Everson, on the other hand, offers the possibility that the blue pigment used on the sarcophagus was meant to represent an iron helmet – which would actually go against his own conclusion.[82] The wall paintings from the tomb of Lyson and Kallikles certainly suggest that some helmets were painted as Sekunda suggests. One helmet is depicted as yellow with a black and red stripe running around it, silver-bluish cheek pieces and a large orange plume on the top and smaller white plumes on either side. The other helmet is red with a black peak and yellow cheek pieces and crest-ridge. Everson suggests that the yellow in these depictions is meant to represent the original bronze while the other colours are painted decoration.[83] While this is one possible interpretation, it is also possible that the silver-blue colour was meant to represent iron, the yellow bronze and the other colours decoration. Indeed, there are almost countless ways in which these depictions can be read and no one theory is more likely than the others.

Depictions of Attic-style helmets from the Hellenistic period can also be found in the tomb of Lyson and Kallikles as well as in the decoration for the tomb of Antiochus II.[84] On coinage, Seleucus I is always depicted wearing an Attic-style helmet. An elaborate form of Attic helmet, fashioned to represent the head of a lion (contrary to the image that is supposedly of Alexander in a Phrygian helmet on his commemorative Hydaspes coins), is worn by the figure of Alexander on the frieze of the Alexander Sarcophagus of king Abdalonymus of Sidon in the Istanbul Archaeological Museum.

The basic design, and relative ease of manufacture, would have made helmets like the *pilos* the easiest for states to issue to troops when there was a requirement to quickly equip large numbers of soldiers. This explains why such helmets became more commonplace during the period of the Peloponnesian War. Similar considerations would have been made by the Macedonians who needed to equip large numbers of troops relatively quickly (and cheaply) at state expense. The widespread adoption of the *pilos* by the Macedonians appears to have been a measure implemented by Alexander the Great. Julius Africanus states:

'Some modifications were made to this equipment [i.e. that used by the Greeks] by the Macedonian *epigones* [The Successors of Alexander] who, due to the varying character of war, created unique equipment for the battles they waged among themselves and against the Barbarians [i.e. the Persians]. For instance, they unobstructed the combatant's vision through the introduction of the Lakonian *pilos*...This innovation has been attributed to the warrior-king [i.e. Alexander].'[85]

Hatzopoulos and Juhel suggest that this passage indicates that the common impression of the widespread use of the Thracian and Phrygian helmet by the Macedonian infantryman is incorrect.[86] However, while this may have been the case after the aforementioned reform of Alexander III, it is most likely that styles of helmet other than the *pilos* were more common during the time of Philip II and Alexander II. The short-lived period of the use for masked Phrygian helmets also falls under the reign of Alexander III. Thus it seems that, even after the widespread adoption of the *pilos*, all of the helmets common to the Greek world were still in use by members of the Macedonian phalanx.

That being said, the *pilos* is certainly the most commonly depicted helmet in both Hellenistic art and on coinage from the time of Alexander onwards. The funerary monument of Nikolaos depicts a phalangite wearing a basic *pilos* helmet (Plate 8). The bronze plaque from Pergamon (Plate 7) shows two phalangites poised for battle also wearing *pilos*-style helmets – the foremost one of which carries a crest. *Pilos* helmets are also a regular feature on the obverse side of 'shield coins' issued under Alexander and the Successors. The helmets in these later depictions are usually adorned with transverse crests and cheek pieces and may be the helmets worn by officers (see Plate 11). These crests seem to be less stiff than those that were mounted on helmets in Classical Greece. In his play *Perikeiromene* ('The Girl Who Gets her Head Shaved'), Menander describes a character as a 'feather-crested *chiliarchos*' (πτεροφόρᾳ χιλιάρξῳ).[87] This suggests that, other than horse-hair crests, officers of certain ranks had feather plumes on their helmets as the distinguishing insignia of their rank. This is confirmed by a well-preserved example of a Phrygian helmet in the Ioannina Museum which has four feather holders mounted on it (one at either temple and one on either side of the top of the helmet). This suggests that, not only were feather plumes a common form of identification for some ranks, but that the Phrygian style of helmet was also used by some officers of Hellenistic infantry units.

Snodgrass, on the other hand, suggests that the Boeotian style was the most popular helmet for use by both infantry and cavalry – although upon what basis this claim is made is not stated.[88] A clear depiction of this type of helmet can be seen on the Ephesus Relief in which the helmet is presented in a 'still life' image of different elements of a warrior's equipment. The top of one of these types of helmet, or possibly a brimmed *pilos* or bell helmet, adorned with a crest, can also be seen on the Alexander Mosaic from Pompeii. Slightly in front of, and below, the outstretched right hand of the Persian king Darius in this mosaic can be seen a number of marching phalangites. The top of a helmet, bearing many similarities to the brimmed-style *pilos*, is clearly visible.

The question remains as to how widespread the use of any of these types of helmet was within the Macedonian military at any particular point in time. Certainly, from a logistical and economic perspective, the use of more basic helmets like the conical or rounded *pilos* and the brimmed *pilos*, would make more sense in an army whose equipment was supplied at state expense and may have needed to have been manufactured both cheaply and rapidly. Yet this ability to produce helmets quickly was offset by the reduced level of protection given to the wearer. In some cases any disadvantage that this caused was offset by alternative means. For example, Alexander is said to have instructed his men to shave as he considered that, in battle, there was nothing more opportune to grab than an opponent's beard.[89] Such an act would only be possible against men wearing a more open-faced style of helmet like the *pilos*.

However, the evidence for the use of other types of helmet (some of which were more enclosed), such as the Thracian, Phrygian (masked or otherwise),

Chalcidian and Attic, would also indicate that these styles were also in use. It is possible that helmets like the Phrygian were more commonly used by units of the Macedonian army that had originated from the eastern parts of Macedonia – where they would have had more exposure to the cloth caps upon which their shape may be based. Similarly, the Attic and 'Thracian' style helmet, and the Boeotian-style helmet and rimmed *pilos*, may have been more commonly used by units from southern Macedonia who had more regular contact with the Greeks. Thus the helmet, as well as the shield, may have been a distinguishing characteristic of different contingents within the Macedonian military who were grouped according to their region of origin. However, the regular depiction of the conical *pilos* helmet, combined with the literary evidence, would suggest, despite any regional variances in the equipment of the Hellenistic infantryman, that the basic *pilos* was the common 'state issued' helmet of the Macedonian military.

It is also possible that, much like soldiers today, individual troops customized their equipment to suit their likes and needs (although in modern armies this is usually more to do with the layout of the webbing rather than the helmet). Certainly, if a phalangite issued with a basic *pilos* came across a helmet which provided more protection like a Thracian or a Phrygian on the battlefield or elsewhere, it is completely understandable that he would pick it up and use it for himself. The Amphipolis Decree specifically states that a helmet of some sort was a required piece of equipment. However, if there were no regulations in place which stated that only specific types of helmets could be used by the Macedonian phalangite, then it is possible that, while much of the rest of his equipment was standardized (such as the shield and *sarissa*) there could have been a great variety of helmets worn across an army containing hundreds and thousands of men.

Finally, there is also the *kausia* to consider. Antipater describes the *kausia* as 'from olden times the Macedonian's comfortable [piece of] equipment, shelter in a snow storm and a helmet in war, thirsting to drink your sweat'.[90] An inscription from the sanctuary of Leto in Xanthus lists the *kausia* among other pieces of military equipment that were forbidden to be worn within the sacred precinct.[91] Both Antipater's description and the inscription from Xanthus would suggest that the *kausia* had some form of military purpose. Philo of Byzantium even suggests it could be used to help convey secret, presumably military, messages.[92]

Strangely, for a piece of military headgear which seems to have been relatively common-place and had a variety of uses, depictions of the *kausia* are relatively rare in Hellenistic art – as are depictions of phalangites in general – other than in some tomb decorations which contain scenes that are noncombative in theme.[93] Saatsoglou-Paliadeli suggests that this indicates that the *kausia* was more of a replacement helmet rather than a true piece of traditional defensive armour.[94] Heckel and Jones, and Sheppard, on the other hand, suggest that the rear ranks of the phalanx went without metal helmets and only wore the *kausia* instead.[95] This seems unlikely. Both commanders and regular soldiers would need to be prepared for the men in the rear ranks to move forward during a battle if the men ahead of them in the file were killed or if the files of the formations were 'doubled' (i.e. moving the rear half of the file forward to double the number of men across the front of the line). If the rear ranks had little or no protective armour (some scholars suggest that the rear ranks did not wear body armour either – see following) then, if they had to move forward to replace a fallen comrade, their reduced defensive properties would significantly compromise the integrity of the entire formation. Metal helmets would additionally give everyone in the formation a level of protection against missile fire, and allow every rank to be prepared for the possibility that the entire formation might have to suddenly about-face to confront an unexpected enemy attack from the rear. Thus the very functionality of the phalanx relied upon every member of the formation being armoured in a similar manner as well as carrying a standardized shield and pike.

What then was the *kausia*? Clues to its identity and function come from the way it is described by Antipater and how he states that it could be used both in cases of variant temperature and in war where it soaked up sweat. Both of these functions were attributes of the felt or leather arming caps commonly worn by the Greeks under their helmets. These caps, usually a felt or leather rounded *pilos*-style cap, helped pad out the space between the head and the helmet to make it fit more securely on the head, provided a layer of padding to make the helmet more comfortable to wear and absorb the energy from any weapon impact, and helped absorb sweat and dissipate heat (these last points were vital when wearing a bronze helmet while fighting during a Greek summer). Julius Africanus states that the basic *pilos* helmet came in two parts; a headpiece and an outer layer of bronze.[96] Africanus also states that, if the wearer was struck on the head by a missile, 'while the outer layer becomes dented...and gives way, [a] projectile does not penetrate the inner covering of the head'.[97] This must be a reference to the metal helmet and the padded arming cap beneath.

The very name of the *kausia* is derived from the word *kausōn* (καύσων) which means 'extreme heat' and may be a further indication of its primary function – to dissipate heat within the helmet. Consequently, it is possible that the *kausia* was the Macedonian variant of the arming cap worn by the Greek hoplite.[98] This seems to be confirmed by Eustathius who describes the *kausia* as 'a Macedonian covering for the head coming from the *pilos*, like a [Persian] tiara, which protected from extreme heat and, to a certain extent, made a contribution to the helmet'.[99] It therefore seems that the *kausia* had a similar function to the felt *pilos* arming cap and would have therefore been worn under a metal helmet to help dissipate the heat that built up inside it. It would then also soak up sweat just as Anitpater states. This interpretation would then explain the lack of depictions of the *kausia* in artistic representations of the phalangite wearing his combative armour as it would be hidden under the helmet. Despite its important military functions, it is unlikely that vast numbers of phalangites would have forgone the protection of a metal helmet, even a basic bronze *pilos*, and only worn their cloth or leather arming caps on the battlefield.

The *kausia* also seems to have been easily recognisable as particularly Macedonian. Polyaenus recounts how Memnon, the commander of Greek mercenaries fighting for the Persians in 334BC, was almost able to deceive the inhabitants of Cyzicus into betraying their city to his forces by approaching the walls while he and his officers were wearing the *kausia*.[100] The people of Cyzicus, thinking that the approaching forces were Macedonian allies, opened their gates and only closed them again when they recognized that the troops approaching were not who they thought they were. This shows that the *kausia* was a piece of headwear that was individually Macedonian and easily recognisable as such.

THE PHALANGITE'S ARMOUR

Griffith and English suggest that the Hellenistic phalangite wore little or no armour to aid mobility.[101] Other scholars have reached similar conclusions.[102] Such statements seem unlikely as they go against all of the available evidence. The grave *stelae* of Zoilos and of Nikolaos both depict phalangites wearing armour. In the case of the image of Zoilos this is a *linothorax* (also known as a 'tube and yoke corslet') or linen cuirass (θώρακες λίνεοι). The Greeks seem to have had access to armour made from cloth as early as the Archaic Age (if not earlier). In the *Iliad*, Homer describes Ajax the Smaller as wearing a *linothorax* (λινοθώρηξ).[103] The first pictoral reference to this type of armour comes from the early sixth century BC (c.570BC) in the form of an illustration on the Francois Vase (Museo Archaeologico Etrusco inv. 4209). Linen cuirasses were commonly used throughout the Classical Period (based upon the regularity of their depiction in vase paintings) and continued well into the Hellenistic Period and, according to Diodorus and Nepos, became the standard armour of the pikeman following the reforms of Iphicrates in 374BC.[104]

The *linothorax* came in two or three sections. The torso was protected by a large panel which wrapped around the body to be secured on the left-hand side via a series of tie straps and rings.[105] The seam was most likely on the left-hand side as this was the side protected by the shield – however the panels may have also overlapped in order to remove the seam as a weak point and to provide a double layer of protection to the side of the body that was angled toward the enemy. The front part of the panel extended upwards to protect the wearer's upper chest, the sides had cut-away scallops to accomodate the arms, and the lower portion of the panel, from just above the hips, was cut into a series of flaps (*pteruges*) which both provided protection for the groin and upper thighs, but also facilitated movement and bending. A second layer of longer *pteruges*, staggered so as to cover the gaps in the front layer, was in some cases positioned behind the first. This second layer may have been a separately constructed panel which was then attached to the inside of the main body plate, or could have been part of the initial construction of the main panel. Attached to the upper back was another panel – the centre of which extended slightly upwards to protect the back of the neck while the sides were fashioned into two tabs (*epomides*) which bent forward over the shoulders, and then tied to rings affixed to the chest, to hold the cuirass in place while providing protection for the shoulders (see Plate 19).

In the fourth century BC, the *epomides* of some sets of armour seem to have been reduced in size compared to their Classical counterparts (based upon vase paintings), but sets of small *pteruges* could be attached to the outer rim of the *epomides*. Bardunias suggests that the use of these *pteruges* was an offset against the reduced protection for the arm given by the smaller *epomides*.[106] However, not all evidence for the *linothorax* in the Hellenistic period follows this trend. The phalangite depicted in the Tomb B of the Bella Brothers wears a *linothorax* with wide *epomides* and no *pteruges* protecting the arm. Thus it seems that there was a variety of styles of *linothorax* available during the Hellenistic Period (as there was in the Classical). Regardless of its configuration, a *linothorax* would weigh in the vacinity of 5.0kg.[107]

Classical Period vase paintings show the *epomides* of the *linothorax* standing upright when they were not tied down and this suggests that the material that the cuirass was constructed from was quite stiff. The need to have *pteruges* cut into the lower part of the armour to allow movement also indicates the use of a stiffened material. The actual material that such armour was constructed from has been a topic of much scholarly debate with theories suggesting layers of glued linen, hide covered with linen, quilted linen, or plain leather.[108] Bardunias points out that linen was a costly commodity in the ancient world and suggests that the massed production (and acquisition) of enough material to make hundreds, if not thousands, of sets of armour just from linen alone would have been financially prohibitive.[109] Aldrete, Bartell and Aldrete, on the other hand, have calculated that the cost of a set of linen armour was less than the cost of plate bronze, and was comparable to the cost of making armour solely from leather.[110] Representations of the *linothorax* in vase paintings are usually coloured white – a tone generally only used to to designate natural materials like cloth and skin. This suggests that the outer layer of the *linothorax* at least must have been cloth while the core could have been of a material stiff enough to account for things like *epomides* standing up when not tied down. It is therefore possible that the *linothorax* may have been made of a more readily available material such as hide, but faced with an outer layer of linen.

Linen armour, in whatever format, appears to have been quite common across the Hellenistic Period. Thompson goes as far as to suggest that all phalangites wore linen cuirasses.[111] Wall paintings in the tomb of Lyson and Kallikles, and also Tomb B of the Bella Brothers, clearly depict the *linothorax*. Coins issued by Alexander the Great also show a standing figure wearing what appears to be a *linothorax*.[112] This suggests a fairly widespread use of the *linothorax* even among monarchs and others who could afford lavish tombs – further suggesting that officers may have worn a *linothorax* instead of a muscled plate cuirass (see following). Alexander the Great also seems to have favoured linen armour. At the battle of Gaugamela, Alexander wore a set of Persian linen armour which had been captured as spoils following the battle of Issus.[113] A section of the frieze from the sanctuary of Athena Nikephoros in Pergamum (now in Berlin) similarly depicts a *linothorax*, however this set of armour is augmented with a sash that is tied around the chest and knotted at the front. McDonnell-Staff suggests that this sash was an indication of the rank of an officer.[114] This conclusion would then further correlate with the concept that the *linothorax* was worn by some wealthy nobles.

Another possibility is that such belts were used to help put the armour on and were therefore not indicative of rank. The stiffened *linothorax* had to have the edges of its left hand side pulled together so that the armour could be tied on by an assistant. Tying a belt or sash around the armour and then pulling it tight aids this process by forcing the two edges together and holding them in place. If a belt or sash is not used, then the person wearing the armour has to press it together (while they are wearing it) while an assistant ties the armour in place. This process can be somewhat awkward depending upon the rigidity of the material. It is also possible that such belts or sashes were not worn into battle. Alexander the Great had instructed his men to shave to remove the possibility of enemies grabbing onto a phalangite's beard. It would therefore seem strange for officers positioned in the front rank of the phalanx (or any other phalangite) to wear an accessory which would also provide an opponent an opportunity to grapple with them. This suggests that either the belt/sash may have simply been an implement to help put the armour on or, if the sash was a mark of distinction, then it may have only been parade decoration for an officer.

Another possible use for the sash – one not exclusive of being a mark of rank and/or an aid to putting the armour on – is that it could be used to help hold the sword in position. The sword was regularly suspended from a strap slung across the body. However, unless the scabbard is particularly loose, attempting to draw the sword (which in turn would require the pike not to be held) merely lifts up the sword and the scabbard. However, if a sash is tied around the body and over the strap from which the sword is hung, this effectively ties the sword in place, making it much easier to draw and preventing the weapon from moving around as the phalangite moves. Such a use of the sash would then require that it be worn in battle. This then leaves open the possibility that it was also worn as an insignia of rank as well as a pre-battle aid for putting on the *linothorax*.

The depiction of a phalangite on the *stele* of Nikolaos shows him wearing the bronze muscled cuirass common to the Classical hoplite rather than the *linothorax* (see Plate 8) – albeit an evolved version of the muscled cuirass that incorporates the shoulder pieces that were common to the design of the *linothorax*. The plate muscled cuirass was a common type of armour used by the Classical hoplite and variants of this style of protection continued to be used across the Hellenistic period.[115] The armour came in two parts – a plate for the front and a plate for the back. Both the front and rear plates were hand crafted from single sheets of plate bronze with an average thickness of 1mm. The front plate was highly detailed with representative muscles of the chest and abdomen. Pausanias describes how such armour was made in two halves and was secured by buckles and straps (most likely one at each of the shoulders and two on each side of the body).[116] The use of buckles to secure the two halves of this type of armour is also indicated by a section of cuirass from the Classical Period, now in the British Museum, which has a clear impression of the buckle pressed into the plate.[117] Another way the cuirass was secured in antiquity was through the use of an elaborate system of pins and/or hinges.[118]

The cuirass had openings for the arms and neck which were usually folded over to create a more rounded edge and to prevent the lip of the armour from digging into the flesh. By the Hellenistic Period certain enhancements had been made to the muscled cuirass. Rows of *pteruges* made from leather or the same composite linen material as the *linothorax*, were suspended from the lower rim to provide protection for the groin and upper thighs. *Pteruges* could also be suspended from the upper rim of the arm holes – most likely to provide better protection to the shoulders and upper arms which were left somewhat uncovered through the use of the smaller *peltē*. Finally, additional shoulder sections, reminiscent of those on the *linothorax* were attached to some muscled cuirasses to further protect the shoulders as is shown on the *stele* of Nikolaos. In its entirety, a plate bronze muscled cuirass would weigh in the region of 5-6kg.[119] This is not overly different to the weight of a *linothorax* although, in keeping with descriptions of this type of armour by the likes of Nepos, the linen armour is slightly lighter on average. However, the muscled cuirass differed from the *linothorax* in that its shape, wider at the shoulders and narrower at the waist, allowed it to sit on the hips. The barrel-like construction of the *linothorax*, on the other hand, meant that most of its weight was taken on the shoulders.

English suggests that it was only officers that wore armour in the Macedonian army as a mark of their status while the rank-and-file soldiery did not.[120] Connolly suggests that by Polybius' day, the front ranks were wearing metal cuirasses, the middle ranks were wearing linen cuirasses, and the rear ranks were not wearing any armour.[121] Similarly, Snodgrass suggests that by the time of Philip's siege of Olynthus in 348BC, neither the phalangites nor the *hypaspists* were wearing any armour and were thus not really 'heavy infantry'.[122] Alternatively, Warry suggests that the regular phalangite wore the muscled cuirass and the Thracian-style helmet.[123] However, such conclusions go against the pictorial and literary evidence of the general use of the *linothorax* and other styles of helmet like the conical *pilos*. Arrian describes Alexander's troops attacking the city of Sagalassos in Pisidia in 334BC as 'heavily armoured' (βαρύτητα τῶν ὅπλων).[124] During the siege of Tyre in 333BC, the Tyrians poured heated sand down on the attacking Macedonians which, according to Diodorus, 'sifted down beneath breastplates and tunics...scorching the skin with intense heat'.[125] While it may be possible that both of these passages are referring only to officers within Alexander's army, although this seems unlikely, other passages are more specific about the armour worn by the regular phalangite.[126]

Nepos states that the Iphicratean *peltast*, the forefunner to the Hellenistic phalangite, wore linen armour which was lighter than that worn by the Classical hoplite.[127] Plutarch states that Alexander's men waded across the swollen Hydaspes river 'in full armour'.[128] Diodorus and Curtius note how the reinforcements that came to Alexander in India in 326BC brought with them 'elegant suits of armour for 25,000 infantrymen'.[129] According to Curtius, these panoplies, which he also numbers at 25,000, were inlaid with silver and gold and that after receiving them Alexander ordered that the old sets of amour should be burned.[130] The sheer number of panoplies received suggests that they were for more than just the officers of Alexander's army despite any embellishments that may have been added to them with precious metals. Furthermore, the fact that the older sets of armour in the camp were able to be burned suggests that they were made from some form of combustible material. This indicates that the armour that was being worn was most likely the linen and/or hide *linothorax*.[131]

Aelian describes how Hellenistic pikemen wore either a style of armour known as the *argilos*, which may be a reference to the *linothorax*, or a corslet [*lorica* – λωρίκια] or an over-corslet [*epilorica* – ἐπιλωρίκια], often with an additional protective layer possibly of either hide or metal scales.[132] The Amphipolis decree also states that a soldier would incur a fine of 2 obols for losing his armour (*kotthybos*).[133] This cannot be a fine imposed upon an officer for such an offence as they are treated separately in the decree and are fined double the amount of the regular soldier. Thus the *kotthybos* must be a reference to something worn by the regular soldier.[134]

Interestingly, the Amphipolis decree uses different terminology for the armour worn by officers. Two types of armour are described; the breastplate (*thorax*), and the half-breastplate (*hemithorakion*).[135] It seems likely that the *kotthybos* ascribed to the regular soldier was either the *linothorax* or some other form of light armour, while the breast plate and half-breastplate ascribed to officers were plate metal. A clear indication that both the thorax and the *hemithorakion* are made from a more expensive material than the *kotthybos* is indicated by the level of the fine imposed on the offending officer. In the event of the loss of a shield, *sarissa* or greaves, the officer was fined double the amount that would be imposed upon a regular soldier. However, the loss of a thorax by an officer incurred a fine of 2 drachmae – six times the fine of 2 obols for the loss of a *kotthybos* by an infantryman, and the fine for the loss of a *hemithorakion* was 1 drachma or three times the fine for the *kotthybos*.

The fact that there is a half-breastplate indicates that the armour for officers came in two sections – most likely the front and back plates of the muscled cuirass. This is also evidenced by the fine for a full thorax being double that of the *hemithorakion* which would have only contained the front plate. If the attribution of the muscled cuirass with someone of rank is correct, then the armour worn by the figure on the funerary *stele* of Nikolaos would designate him as an officer. Hatzopoulos and Juhel, who identify the armour depicted on this relief as that mentioned in the Amphipolis decree, suggest that Nikolaos was a junior ranking officer – possibly a file leader (*lochargos*) or a file closer (*ouragos*).[136] Even here it is uncertain whether the metal half-breastplate was specifically used by officers of lower rank while full cuirasses, such as that depicted in the *stele* of Nikolaos, were used by more senior ones. Polyaenus states how Alexander the Great issued *hemithorakiae* to his men so that only their fronts would be protected, which then forced them to stand and fight as running away would expose their unarmoured backs.[137] If there is any truth to this statement at all, then it must either be assumed that Polyaenus is referring to an issue of metal half-breastplates to junior officers like the file leaders, or that it is possible that some members of the rank-and-file were also wearing metal armour. Similarly, Diodorus describes Demetrius' troops attacking Rhodes in 305BC as wearing 'gleaming armour' (ἀποστιλβόντων ὅπλων).[138] It is uncertain if this is a reference to a metal-faced shield (in which case the passage should be read as 'gleaming shields') or to the wearing of metal armour such as a muscled cuirass. Plutarch refers to Macedonian troops at the battle of Pydna in 168BC as wearing both gilded armour and scarlet cloaks and that the battlefield was later littered with the armour of the fallen.[139] This would then suggest that it was more than just the officers of the front ranks who were wearing metal cuirasses in the late Hellenistic Period.

It is unlikely that all pikemen wore a muscled cuirass. The ancient texts recount how in the reign of Philip II, and in the early years of the reign of Alexander the Great, the Macedonian state was nearly bankrupt. It is doubtful that a state in such a dire financial situation would have been in a position to requisition large numbers of plate-metal cuirasses. Additionally, the reference to Alexander ordering the old sets of armour worn by his men to be burned indicates that, by the time of his reign, they were wearing armour made of something other than metal.

It seems most likely that, in regards to the pike infantry, it was the officers of the infantry who wore the metal breastplates as a mark of distinction, as a means of identification and recognition, and because it provided good protection.[140] The cost of having a breastplate manufactured would also suggest that these items were only used by wealthy nobles – possibly only mid- to high-ranking officers – who may have paid for their own armour while the common soldier was provided with his equipment by the state. More senior officers undoubtedly wore armour as well, regardless of whether they were acting as a commander of infantry or cavalry. Alexander the Great, for example, received wounds which we are told pierced his elaborate breastplate, while later Demetrius is described as wearing both 'royal armour' (ὅπλοις βασιλικοῖς) and 'defensive armour' (σκεπαστηρίοις ὅπλοις).[141] On the Alexander Sarcophagus, Alexander is depicted wearing a purple cloak. Not only was the use of this color a mark of royalty, as per Demetrius' armour, but purple cloaks could also be handed out as awards to men of valour.[142]

In Tomb II at Vergina a set of armour was discovered that was constructed from iron plates, with gold decorations and fittings, fashioned in the same shape as a *linothorax* with a barrel shaped torso, shoulder pieces and most likely *pteruges* suspended from the lower rim (although these are now missing).[143] At 5mm thick, this iron cuirass is both thicker and undoubtedly heavier than the standard bronze plate cuirass which had a thickness of only 1-2mm – although no details of the weight of this cuirass have ever been published.[144] Traces of cloth both on the inside and outside of the armour have led Everson to suggest that the common *linothorax* may have been fashioned from similar metal plates faced on the front and back with linen or some other material.[145] However, due to the gold fittings on the front of the armour found in the tomb at Vergina, it is unlikely that the entire front would have been covered with cloth. Regardless, the materials used indicate that this armour undoubtedly belonged to someone very wealthy and influential – although whether this was Philip II or not is still a topic of considerable scholarly debate – as is whether or not this type of armour was a one-off set or not. Consequently, it is difficult to use this example of plate iron armour to suggest that the armour worn by the regular phalangite was fabricated in the same manner.

Officers of lower ranks, such as the junior file leaders and closers, may have still used a more basic *linothorax* made from hide and linen. This would then explain why depictions of this type of armour are found in some paintings in the tombs of the wealthy.[146] It then follows that the remainder of the infantry, the regular soldiery who made up the bulk of the Hellenistic pike-phalanx, also bore the cheaper *linothorax* which would have been issued at state expense.[147] Thus it seems that the type of armour a person wore within a Hellenistic army was very much dependent upon their status, their rank, and their position within the formation.

Regardless of the type of armour worn by the phalangite, beneath it would have been both tunics and additional padding. Xenophon outlines the benefits of having a well fitting cuirass and the drawbacks of having one which does not fit the body properly.[148] However, the wounds and turmoils of campaigning would have altered both the body shape of the individual and the armour over time. Additionally, Jarva suggests that ancient workshops may have possessed several wooden templates of different sizes upon which sets of armour were hammered to a shape which best fit the individual.[149] When Dionysius of Syracuse was equipping his troops for a war against Carthage, models of the different kinds of armour to be made were distributed amongst the craftsmen so that his mercenaries could wear the armour to which they were accustomed.[150] This also suggests the use of models or templates for the mass production of armour. Consequently, plate armour was generally not custom made to suit the individual. Rather individuals took possession of a set which fitted them best. The barrel-like torso of the *linothorax* also does not fit snuggly to the body but merely hangs from the shoulders and so, as long as the sides of the armour could be tied together with little discomfort to the wearer, a standard *linothorax* could be used even if it did not closely fit the wearer's body shape.

The result of this was that most armour in the ancient world required some form of padding underneath it to make it more comfortable to wear and to prevent chafing.[151] This is most likely to have been done when wearing the plate 'muscled' bronze cuirass as the seams where the front and back plates of the armour join would often pinch the flesh merely through the motion of walking, particularly along the shoulders and neck, unless some form of padding was also worn.[152] Furthermore, the manner in which the *linothorax* is borne mainly on the shoulders also requires additional padding underneath.

Tunics (*chiton, exomis*) were undoubtedly worn by the Hellenistic phalangite, as is evidenced by both tomb paintings and literary passages, and this garment would have been the first layer of protection against chafing.[153] The Classical hoplite commonly wore a felt or leather *pilos* cap under the helmet to provide an extra level of cushioning. There is also evidence from the artistic record that they placed padding underneath the greave to lessen chafing to the ankle. As these two areas were padded, there seems little reason to doubt that some form of padding was also placed under the armour, if a lining was not attached to the armour itself, at the places where it rubbed.[154] Additional padding may have come in the form of a bib of felt, leather or sheepskin (either option is just as likely as the other if all of the available means of providing padding beneath the armour were not used concurrently) which sat on the shoulders to cushion the armour.

Alternatively some form of padded tunic like the later Roman *subamarlis* may have been used.[155] Such a practice would have undoubtedly transferred to the professional armies of Macedon.

Another piece of defensive armour that the phalangite could have worn was a set of metal greaves (*knemides* – κνημίδες) to protect the lower legs. Polyaenus describes how the troops of Philip II were wearing greaves on their cross-country training exercises.[156] Additionally, the Amphipolis Decree outlines a fine of 2 obols for the loss of these items, and Aeneas Tacticus describes how greaves could be smuggled into a besieged city.[157] Greaves are also depicted in the decorations of the Tomb of Lyson and Kallikles, on the Ephesus relief, and the figure on the *stele* of Zoilos appears to also be wearing greaves. This suggests that greaves were a common part of the pikeman's equipment across the Hellenistic period.

The greaves worn by the Classical Greek hoplite were fashioned from single sheets of beaten bronze, were custom moulded to fit the shape of the lower leg, and wrapped around the shin and calf muscle with the opening in the guard running along the back of the leg. Most greaves also extended upwards to protect the knee. The common greave of the Classical Period was held in place via the elasticity of the metal rather than using any form of strap. Depictions of the greaves used by Hellenistic pikemen indicate that some of them were of a similar design.

Other greaves used in this period were of a more basic configuration. Four sets of greaves were found at Vergina. Some of these show less anatomical detailing than the common greave of the Classical Period, although most are of the same 'clip on' design, while some are more like metal plates tied to the front of the shin by two straps that wrapped around the lower leg. Traces of pigment on the representations of the *hypaspists* on the Alexander Sarcophagus show that many of the greaves being worn had a red lining and a red strap holding them in place that ran just below the knee.[158] This lining and strap was most likely leather. A similar type of shinguard is depicted on the Pergamum frieze which shows a basic greave with two straps – one just below the knee and the other just above the ankle.[159] This later type of greave, which is more basic in its design, would have been much faster and cheaper to manufacture due to it not being custom made to fit the legs of the individual wearer. If the armies of the Hellenistic period were issued with greaves by the state, this was most likely the type that would have been issued. Additionally, strap-on greaves could be worn over tall boots – such as those created by Iphicrates – whereas the clip-on variety could only be worn on bare legs and when sandals or similar footwear were worn.

The more expensive, and custom made, clip-on greaves with anatomical detailing would have most like been worn only by those who could have afforded such items – mainly officers.

Indeed, the greave may not have been required issue and may have depended upon the type of footwear that the phalangite was wearing. Plutarch states that it was customary for the Macedonians to wear high boots (*krepides*).[160] The depiction of a phalangite in Tomb B of the Bella Brothers similarly shows him wearing the high boots of the Iphicratid reforms but without greaves even though he is wearing most other elements of the phalangite's equipment. On the frieze from the Artemision at Magnesia, 15 per cent of the soldiers depicted are similarly wearing high boots.[161] Such boots would have provided a minimal level of protection against impacts to the lower legs. Tomb paintings from Alexandria, dating to the late third century BC, not only depict phalangites wearing high boots, but also some form of cloth covering or socks on their feet.[162] These would have provided not only a level of comfort to the wearer, but would have added to the level of protection for the feet as well – albeit only slightly.[163] It may have only been those in the front ranks who wore greaves as these men were not only the officers who may have been able to afford such items if they were not supplied by the state, but were also the ones whose legs were the most vulnerable to injury. Heckel and Jones, on the other hand, suggest that greaves were standard issue to all phalangites and were a particularly mandatory requirement for the men in the more rearward ranks as the butt-spikes on the pikes held by the men ahead of them were liable to cause injuries to the legs.[164] However, such a conclusion clearly ignores the fact that many phalangites are depicted without greaves. Furthermore, due to the way that the pike was carried for battle, it posed no threat at all to the man behind (see Bearing the Phalangite Panoply from page 133). It seems that the more common pikeman may have only worn the high Macedonian boot as outlined by Plutarch and possibly the tie-on form of greave, and accepted the amount of protection that they gave as no more was required in the more rearward ranks of the phalanx.

Everson suggests that the use of greaves began to decline during the reigns of Philip II and Alexander the Great possibly due to the cost of supplying such items at state expense. It cannot be ruled out that a decline in the use of the greave was also the result of a corresponding increase in the use of boots following the reforms of Iphicrates in 374BC. Despite evidence such as the Amphipolis Decree, which suggests the regular use of greaves by common soldiers up to the end of the third century BC, Everson further suggests that greaves may have only been used by officers in most armies of the Hellenistic period.[165] If this is the case then, much like the use (or not) of the metal muscled cuirass, the use (or not) of greaves by some officers may have been a matter of choice. This would further correlate with the idea that, while the regular soldier was issued with his equipment by the state, wealthy nobles serving as officers may have had to either supply their own or at least had the ability to augment their issued equipment with extra or alternative items.

The common use of footwear of some description by the pikemen of Alexander the Great's army seems to be referred to indirectly in the works of Plutarch. At the battle of Granicus in 334BC, the ground of the far bank of the river is described by Plutarch as 'wet and treacherous due to the clay'.[166] Hammond, who toured the site of the battle and studied the terrain, described the type of clay present on the banks of the Granicus: 'when wet it is very slippery for a man wearing smooth-soled shoes, but it gives reasonable footing for a horse or a man's bare foot if he uses his toes.'[167] Hammond also cites a reference in the works of Thucydides (3.22.2) in which hoplites from the city state of Plataea at Mytilene in 428BC only wore a shoe on their left foot to prevent them from slipping in the mud. This would suggest that, similar to Thucydides, the slipperiness of the terrain at the Granicus is on account of the men of the Macedonian phalanx being unable to gain a decent purchase on the clay-based soil of the battlefield due to their footwear. What is not stated is the exact nature of the footwear.

THE PHALANGITE'S SWORD

As a secondary weapon the phalangite also carried a sword. Polyaenus, in his description of the equipment carried by the troops of Philip, makes no mention of swords, but other evidence indicates that they were a common element of the phalangite's panoply. Yet like most other elements of the phalangite's panoply, the exact characteristics of this weapon are hard to distinguish across the Hellenistic Period.

Nepos states that Iphicrates increased the length of the hoplite's sword as part of his reforms of 374BC while Diodorus states that the length of the weapon was doubled.[168] Such statements are difficult to correlate with the archaeological record. The Classical hoplite usually carried a sword of one of two styles: one with a long, leaf-shaped, double-edged blade ideal for stabbing (the *xiphos*) or a long recurved blade, heavily weighted towards the forward tip and with a single edge in the inside of the recurve, which was ideal for hacking and slashing (variously known as the *kopis*, *macharia* or *falcata*) (Plate 10).[169] Both of these weapons were still in widespread use in the early Hellenistic

Period. Most surviving examples of the *macharia* range from between 35cm and 70cm in length – although an example found in the cuirass tomb at Thesprotia measured 77cm.[170] Snodgrass states that all Hellenistic infantry carried swords of moderate size for thrusting.[171] Conversely, Connolly states that the hacking *kopis* was the most popularly used sword.[172] However, the evidence suggests the use of not only the thrusting *xiphos*-style sword, and the hacking *macharia*, but also the use of small daggers by phalangites across the Hellenistic period (see following).

Everson suggests that the longer examples of the *macharia* are cavalry weapons.[173] However, it is also possible that the smaller weapons (of around 35cm in length) are of the style that was subsequently increased in size to 70cm (or doubled as Diodorus states) as part of the Iphicratean reforms to then become a standard infantry weapon of the time. It is also possible that the hoplites reformed by Iphicrates were carrying some kind of small dagger instead of a sword. The Spartans, for example, were famous, and occasionally the subject of ridicule, because of the short dagger-like swords, a small version of the *xiphos*, that they carried.[174] If it was this type of sword that was being carried by the pre-Iphicratean hoplite, then doubling its length would simply make a sword of normal size and provide the bearer with a weapon with more reach.

Sheppard suggests that the hacking *kopis* was a post-Alexander addition to the phalangite's panoply and that it was probably used when the phalanx broke up as this would provide the bearer with more room to swing it.[175] Another possibility to consider is that, if the hacking sword was truly an element of the equipment of the phalangite during the age of the Successors, much of the warfare in this time was pike-phalanx versus pike-phalanx. In such conditions, a heavy hacking weapon could have been utilized to try and sever the heads from opposing weapons by those whose own pikes may have broken, but who were required to maintain their position in order to maintain the integrity of the formation. Thus a heavy hacking blade would provide such individuals with both defensive and offensive options.

There is clear evidence that both the leaf-shaped *xiphos* and the hacking *macharia* were used across the Hellenistic period by phalangites. The frieze from the Sanctuary of Athena Nikephoros in Pergamum (now in Berlin) clearly depicts a *macharia*. The Amphipolis Decree also uses the term *macharia* to outline the fine incurred for losing a sword.[176] On the *stele* of Nikolaos, the bottom of a scabbard for a sword can clearly be seen hanging down below the shield (see Plate 8). Hatzopoulos and Juhel interpret this sword as a *macharia* as per the Amphipolis Decree.[177] However, the shape of the scabbard is closer to that used to contain the straighter *xiphos*. The Ephesus relief depicts a sword in a similar scabbard to that on the *stele* of Nikolaos. The shape of the scabbard identifies this sword as a straight-bladed *xiphos* rather than the curved *macharia*. Straight swords are also depicted in the wall painting in the Tomb of Lyson and Kallikles. This evidence suggests that both styles of sword were in common usage across the Hellenistic Period.

Interestingly, Plutarch states that Macedonians fighting at Pydna in 168BC were armed only with small daggers (μικροῖς μὲν ἐγχειριδίοις).[178] Plutarch also describes how Alexander defended himself at Malli with a dagger.[179] The word *macharia* has been translated by some as meaning 'dagger'. Heckel and Jones suggest that this may have been a small curved knife.[180] If this is the case, then this would suggest the use of only small stabbing swords similar to those that had been used by the Spartans centuries beforehand. Tarn suggests that the sword played only a minor role in Hellenistic fighting.[181] When it is considered how difficult it was for an opponent to get close enough to an opposing phalangite when they attacked front-on and the pike-phalanx retained its integrity this certainly makes sense. Plutarch recounts how at the battle of Pydna the Romans were only able to close with the Macedonians once gaps had formed in the line; prior to this the Romans had been kept at a distance by the length of the *sarissa*.[182] Under such combat conditions a shorter reach weapon like a sword was basically superfluous. However, if the phalanx began to break and the fighting changed to a close-quarters mode of combat, then it was the *sarissa* which became a redundant instrument and a short-range offensive weapon would have been necessary. Anderson suggests that the larger varieties of sword were only carried by officers while the general rank-and-file phalangite carried a smaller dagger as 'if the enemy did penetrate the spearline there would be no room to swing swords in the close press'.[183]

Certainly in the confines of a phalanx, a thrusting weapon like the *xiphos* or a dagger would have been much better suited than the hacking *macharia* which required considerably more room to wield and swing.[184] Heckel and Jones suggest that a slashing *macharia* would have only been used when the phalanx had broken and further suggest that such swords may have only been carried after the Macedonians had begun to face opponents who could exploit gaps in their formations like the Romans.[185] However, the depictions and literary references to the *macharia* indicate that this weapon was used prior to this time. Additionally, Polybius states that the Romans had to reinforce the rims of their shields to withstand the blows delivered by hacking swords.[186] While no direct connection to the Macedonians is made by Polybius, the need to reinforce the upper rims of the Roman *scutum* against hacking blows would suggest the use of a heavier hacking weapon like the *macharia* by some of the Macedonians (and many of the other people fought by the Romans such as the Celts). Consequently it is almost impossible to determine what type of swords were carried by the phalangite at different points across the Hellenistic Period. It cannot be discounted that, along with both the *xiphos* and the *macharia*, smaller daggers were carried by some Macedonian phalangites and that all three types of weapon saw concurrent service. It can only be concluded that some form of sword was a mandatory part of the phalangite's equipment, as is indicated by the Amphipolis Decree, regardless of what type was carried. Whether the preference for one type of sword/dagger over another was based upon individual preference, unit origins, or some other influencing factor is far from certain.

THE WEIGHT OF THE PHALANGITE'S PANOPLY

Due to the variety of different elements that may have been worn, it is very difficult to estimate an overall weight for a complete phalangite panoply. Determining an estimated weight for the panoply would need to consider such things as the length of the pike that was being carried at which specific point across the Hellenistic Period that was being examined, the type of helmet and armour being worn, whether greaves were worn or not, and so on. The numerous possible combinations of arms and armour that the phalangite may have carried may be a reason why so few modern scholars have attempted to estimate the weight of the full assemblage. An estimated weight for the phalangite panoply can be determined so long as the characteristics of any pieces of equipment specific to a particular time-period are taken into consideration. An Iphicratean *peltast* for example, at the very advent of the Hellenistic Age when the pike-phalanx was first created, bearing the full equipment described by Diodorus and Nepos, would carry an array weighing just over 19kg (Table 6).[187]

Using the above information as a starting point, the weight carried by phalangites at other time periods can also be calculated. For example, in calculating the weight of the panoply carried by a phalangite during the time of Alexander the Great, the weight of the Iphicratean pike (4.92kg) needs only to be replaced with the weight of the larger 12 cubit *sarissa* of that time (5.50kg) to arrive at a total weight for the panoply of 19.7kg. If greaves were to be added, then an additional 1.2kg would need to be factored in to arrive

Item	Description	Weight
Shield	A small *peltē:* 64cm in diameter, fashioned from a solid wood core, with a metal facing, central armband, shoulder-strap and hand-grip.	4.6kg*
Pike	16 Greek feet (512cm) with a uniform shaft 35mm in diameter and with a butt** and small head***.	4.92kg
Sword	A *xiphos*: iron, 72cm in length with metal fittings and leather wrapped grip, with a leather scabbard with metal fittings.	1.5kg
Armour	A *linothorax* made from 5mm ox-hide covered with a layer of linen on the inside and outside, painted, decorated and sealed, with leather tie straps.	5.1kg
Helmet	A basic conical *pilos* helmet of beaten bronze 1.5mm thick with an accompanying padded cap.	1.8kg
Tunic, Footwear & Padding	A basic knee-length tunic, high leather 'Iphicratid' boots and a sheepskin bib for padding beneath the armour.	1.2kg
TOTAL		**19.12kg**

* based upon the replica made from pine used in this examination. As noted on page 94, a shield made from willow or poplar would weigh less than this.
** based upon the one found at Vergina by Andronicos.
*** based upon a head of around 175g needed to create a weapon with a point of balance 2 cubits from the end of the shaft.

Table 6: The weight of the panoply
carried by the Iphicratean peltast c.374BC.

at a total weight for the panoply of 20.9kg. If a different style of helmet is considered, such as a Thracian, Phrygian or Boeotian, then the weight of this alternative item would also need to be incorporated into the calculation along with the weight of any crest or plume that the helmet may have had. If the phalangite in question was an officer, who may have been wearing the muscled plate cuirass, then the 5.1kg weight for the *linothorax* would need to be replaced with a figure of 5.6kg for the bronze breastplate – giving a total weight of 21.4kg. For phalangites of the later period, once the specific type of armour and helmet was factored into the calculation, the only other thing that would need to be considered is the weight of the longer pikes which were used during the time of the Successors.[188]

Aelian states that the *peltast* bore a panoply lighter than that of the hoplite.[189] The hoplite of the Classical Age, bearing a panoply incorporating the large *aspis* shield, a 2.5m spear, crested Corinthian-style helmet, muscled cuirass, greaves, tunic, padding and footwear, carried around 21kg of equipment. Thus Aelian is fundamentally correct in that the Iphicratean *peltast*, carrying 19.12kg of equipment, did carry less. Furthermore, Aelian seems to be merely discussing armour in his text and not considering weapons as well. If the weight for the sword, and the weight of the respective primary weapon for both the *peltast* and the hoplite (those being 4.92kg for the pike of the *peltast*, and 1.5kg for the spear of the hoplite) are not considered, then the *peltast*'s shield armour and helmet weigh 14.2kg – substantially less than the shield and armour of the hoplite which weighs around 19.5kg.

English suggests that phalangites were lighter than hoplites and that the common assumption that phalangites were 'heavy infantry' has led to a lot of confusion.[190] However, when the weight of just the arms and armour carried by the phalangite in the time of Alexander the Great is compared to that of the Classical hoplite, it can be seen that their respective equipment weighs about the same (around 21kg). While the *linothorax* common to the phalangite is slightly lighter than the muscled plate cuirass of the hoplite, and while the phalangite's *peltē* is smaller and lighter than the hoplite *aspis*, the phalangite's *sarissa* is considerably heavier than the hoplite *doru*. However, if the weapons themselves are not considered, then the phalangite is more lightly equipped than the hoplite as English suggests. Regardless of whether the *sarissa* is considered or not, the weight of the remainder of the panoply would still clearly class the phalangite as 'heavy infantry' compared to the light infantry which Aelian describes as wearing no armour at all. At Gaza in 312BC some of Demetrius' infantry abandoned their heavy arms and armour in order that they might flee more easily.[191] Polybius states that the

Macedonians 'have difficulty holding only their pikes when on the march and in supporting the fatigue caused by their weight'.[192] This clearly demonstrates the weight and encumberance of the lengthy *sarissa* and other elements of the phalangite's panoply.

Yet arms and armour was not all that the phalangite carried. Polyaenus states that the troops under Philip II also carried rations and utensils for daily use.[193] Frontinus states that Philip assigned one servant to every *dekad* of ten infantrymen to carry their grinding mill and ropes (presumably for their tent), while each man was required to carry flour for thirty days.[194] The amount of flour required for each man on a daily basis to provide him with enough nourishment is around 1.3kg. Around 2lt of water are also required daily.[195] This brings Frontinus' statement into question as, in flour alone, thirty days' rations equates to a load of 39kg for each phalangite to bear on top of his arms and armour. Porting the water as well would amount to an additional weight of around 60kg (not including the containers it was stored in). Frontinus additionally states that Philip prohibited the use of wagons, so almost everything the army required would have needed to have been carried – although if heavy items like mass volumes of water were to be transported, this would have undoubtedly been carried in carts or on pack animals while the phalangite may have only carried enough personally for a day or two. Even if only a fraction of the stated rations were actually carried, we can infer that the Macedonian phalangite carried a considerable weight in food, water, cooking utensils and other personal items while on campaign.

The self-portage of rations (and presumably other items of equipment as well) seems to have been a standard requirement of the organization of Macedonian armies across the entire Hellenistic Period. When marching against the Nabateans in 312BC, for example, the army of Demetrius were required to carry rations for several days that did not require cooking.[196] For his march into Egypt in 306BC, the army of Antigonus carried rations for ten days.[197] Added to this, many phalangites seem to have carried items of plunder on their person as well. Eumenes is said to have remarked that, if his army took too many spoils, they would be too weighed down for flight if needed.[198] All of this would have added considerable weight to the phalangite's overall panoply of equipment.

Chapter Four

Bearing The Phalangite Panoply

Both the lengthy *sarissa* and the small *peltē* shield were the hallmarks of the Hellenistic man-at-arms.[1] The characteristics and encumbrance of these pieces of offensive and defensive armament dictate how they could be wielded and, more than any other element of the panoply, how the phalangite had to position his body to engage in combat. The positioning of the body would, in turn, influence how effectively the *sarissa* could be wielded, and how well protected the phalangite was by his shield – thus determining the effectiveness of any combative posture used to employ them.

Markle states that in order to understand the changes in military tactics brought about by the creation of the pike, 'it is necessary to be clear about the limitations of the size and weight of the *sarissa*, since these factors determined how the weapon could be wielded in battle'.[2] However, while the design of the *sarissa* played a definite part in how the phalangite functioned on the battlefield, the pike was only one element of his panoply. In order to fully comprehend how the equipment carried by the phalangite influenced his performance on the field of battle, the limitations of all of the elements of the panoply combined must be considered.

Regardless of the time-period in question, and regardless of the method of fighting that is to be employed, there are certain fundamental criteria that need to be met in order for any combative posture to be effective. For the Hellenistic phalangite, armed with a lengthy pike and small shield, any combative posture must:

a) Allow the shield to be positioned in such a way so that it could be used defensively;

b) Allow the phalangite to move on the battlefield without any restrictions, particularly in a forward direction;

c) Provide stable footing to enable the phalangite to remain upright during the rigours of close combat;

d) Allow the phalangite to maintain his position within the confines of the phalanx and conform to any limitations of space dictated by that formation; and, most importantly

e) Allow the *sarissa* to be positioned in such a way so that the arms had a natural range of movement enabling the phalangite to engage an opponent offensively.

BODY POSTURE AND THE DEFENSIVE POSITIONING OF THE SHIELD

The natural location for a shield to be positioned when in a combat situation is across the front of the body – there is literally no point in carrying a heavy piece of defensive equipment like a shield into battle if it is not going to be used to help protect the person carrying it. Unfortunately the ancient sources provide few details of how the phalangite positioned his body in order to place the shield in its most effective position for battle.

Some scholars suggest that in order to carry the weighty *sarissa*, the phalangite had to stand in a side-on posture with both feet facing forward (Fig.10).[3]

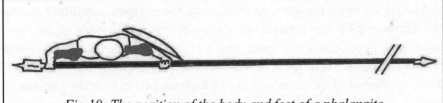

Fig.10: The position of the body and feet of a phalangite in a side-on position.

However, a side-on posture only satisfies some of the criteria necessary for an effective combative posture. Defensively, standing side-on in this manner does place the shield in a protective position – albeit a very limited one.

Standing side-on also creates the smallest target profile of the upper body, and places many of the vital organs at the furthest distance from any attack delivered from the front while the shield provides protection against any attack coming from the bearer's front-left and, to a limited extent, from directly ahead. However, the body is completely open to attacks delivered from the bearer's front-right which, due to the way in which the left hand is used to help wield the pike, leaves the phalangite with no means of altering the position of his shield in order to meet or deflect such attacks.

The diagrams accompanying Connolly's discussion of the phalanx also depict the shields with the lower rim extended.[4] For such a position to be achieved, the left elbow would have to be raised and the arm rotated outwards. However, as the left hand is used to help wield the *sarissa*, this contorts the arms and places a lot of stress on the muscles. Additionally, as the arm is raised, if the *ochane* was not appropriately shortened each time the arm was rotated and elevated in such a manner, the shoulder-strap would relax and no longer be supporting the weight of the shield or the pike. It is doubtful that anyone would be able to carry a shafted weapon weighing more than 5kg in this position for anything more than a few minutes. It is further unlikely that the shield would be positioned in this manner during combat as any strike made against the shield would be deflected dangerously upwards towards the bearer's head. This suggests that an angled position, with the lower rim extended outwards, was not the standard placement for the phalangite's shield.

In the suggested side-on stance, the left arm is extended outwards, akimbo to the body, and is relatively straight. Due to the use of a central armband and hand grip to help carry the shield, the axis of the shield must therefore follow the axis of the forearm which, in this case, results in the shield being positioned at an angle towards the left-rear (see Fig.10 above). A shield placed in this position does cover the bearer from just below the shoulder to just above the left knee; however, due to the angle of the shield and the near full extension of the left arm, it is difficult to brace the shield in position using outward pressure exerted using the left thigh.

A shield which is unsupported in this manner is liable to injure the person carrying it when in a combat situation. A shield supported solely by a shoulder strap yet mounted on the forearm held in such a way has nothing bracing it in place. Any strike made against the top of the shield would cause it to rotate on the arm, potentially forcing the rim of the shield to impact with the upper arm or, depending upon how high the shield was being carried, the bearer's face. Even if the shield did not impact with the bearer's head, the new angle that the shield had been pushed into would allow for the enemy weapon which caused it to move (for example the *sarissa* of an opposing phalangite) to simply slide upward into the bearer's face. The development and use of more open-face

styles of helmet like the conical *pilos* would suggest that the phalangite shield
was never used in a way which would endanger the bearer's head either directly
or indirectly and as such the shield cannot have been positioned in a way that
would result in such dangers.

Due to the presence of the *ochane*, any strike which hit the bottom of the
shield would not alter the angle of the shield too much, or push the rim into the
bearer's thigh to cause an impact injury, as the shoulder-strap would be pulled
taut by the slight rotation of the shield caused by the impact which would then
prevent it from being rotated any further. However, the positioning of the arm
at such an angle as is required to adopt a side-on posture cannot be considered
defensively stable and can therefore not be considered conformance with the
first necessary criteria for an effective combative posture. This alone would
suggest that it was unlikely that the phalangite used this posture to engage in
combat.

The most likely posture that a phalangite adopted for battle is an oblique
posture with the left foot forward, the feet well apart and the right foot back,
and with the upper torso rotated to the right by approximately 45° (Fig.11).

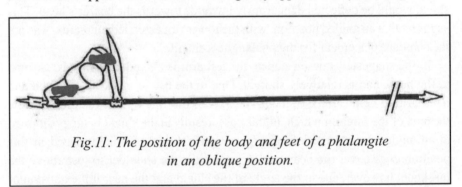

*Fig.11: The position of the body and feet of a phalangite
in an oblique position.*

This is clearly the posture of the phalangites depicted on the bronze plaque from
Pergamon – one of the few artistic representations of the phalangite in action
(Plate 7) and the use of an oblique posture forms the basis of other modern
models and illustrations of the phalanx in action.[5]

Defensively, positioning the left foot forward in the manner of the oblique
posture causes the torso to rotate, or allows it to be easily rotated, so that the
chest faces to the front-right. This presents the left shoulder forward which,
in turn, allows the left upper arm to be held beside the body while the left
forearm, supporting the shield, can be extended across the front of the body
relatively parallel with the front of the space that the bearer occupies. Due to
the need for the two hands that are used to wield the pike to be at least 70cm

apart, the left forearm is held at a slight forward angle which causes the shield to extend towards the left-rear. However, this angle is only marginal and does not negatively affect the combat effectiveness of the posture (see following).

By positioning the body in an oblique posture with the left leg forward, the left thigh can be used to support the lower rim of the shield while, at the same time, the left shoulder can be used to brace the upper rim in position. This is made easier if the bearer bends forward slightly at the waist, something that cannot be done in the side-on posture, which not only lowers the shield slightly so that it is more easily supported by the leg, but also places the shoulder and the rest of the torso in a protective position behind the shield. Leaning slightly forward to carry the *peltē* is exactly how the phalangites are depicted on the Pergamon plaque and the resulting posture places the shield in a position where it covers the bearer from shoulder to knee (see Plate 7). Thus, unlike the positioning of the shield in the side-on posture, a shield held with the body in an oblique posture creates a very stable and well supported defensive position which an incoming strike would find difficult to move; and so attacks would not be dangerously deflected towards the bearer's head. Furthermore, due to the slight concave shape of the *peltē*, even when the shield is held at a marginal angle as a result of the required spacing of the hands, the surface that is presented directly towards the front is still flat – which then creates the most effective surface for countering incoming attacks. This is an aspect of the shield's positioning that does not occur in the side-on posture which, due to the more acute angle of the shield, can only present an angled surface to any attack which will simply cause it to deflect.

Standing obliquely does present a larger target profile of the body than a side-on posture. However, the vital organs are still angled away from any attack delivered from the front while a shield 64cm in diameter carried in this manner protects the bearer from the shoulder to just above the knee, in effect covering his entire torso behind the shield, while the lower limbs were protected by either greaves or high 'Iphicratid' boots and the head was protected by the helmet. In an oblique position the body is also better protected from any attack coming from the bearer's right side. While not fully protected by the shield, the angle of the body, and the way that the left forearm is held across it, protects the bearer from most attacks coming in from anything less than a 45° angle from the front-right. The bearer would still be vulnerable to attacks coming from an angle greater than this – such as missile fire, infantry or cavalry attacks coming from the flank – but this is true of almost any combative posture. Based upon an examination of how the *peltē* could be positioned in both the side-on and oblique body postures, it is clear that the phalangite's shield can only be used in an effective defensive manner when the body is rotated to the right by about 45°. This means that, of the two possible combative postures for wielding the *sarissa*, it is only the oblique position that conforms to the first necessary requirement for an effective combative posture.

BODY POSTURE AND MOVEMENT WITHOUT RESTRICTION

As the pike-phalanx could be used both offensively and defensively, the ability to advance into battle was a vital aspect of the mechanics of phalangite warfare. This is clearly indicated by the battles of the wars of the Successor period. If the phalanx was completely immobile, and two opposing sides faced-off against each other but had no means of engaging, then the battle would simply not take place except through the use of other arms such as cavalry or light troops. However, not all of the possible ways for wielding the *sarissa* allow forward motion to be accomplished without impediment in conformance with the second necessary criteria for an effective combative posture.

Standing side-on, for example, greatly restricts the movement of the phalangite. While maintaining this position the ability to move forward is limited to a shuffling sidestep – basically the right (rearward) foot would need to be turned out and, with each step, brought up behind the forward left foot, before the left foot was then moved forward. If a more natural method of walking is employed, the body naturally rotates back to the left to provide the right leg with a freer range of movement. However, in doing so, the whole posture of the body changes from a side-on stance to an oblique position. Importantly, because the left arm is held outwards in the proposed side-on stance, moving the body into an oblique posture to facilitate movement also swings the left arm to the left – in turn causing the *sarissa* and the shield to swing to the left – which would then leave the phalangite vulnerable to a frontal attack. This indicates that such a method could not have been employed by the phalangite as, due to the massed confines of the phalanx, any weapon which moved in such a manner would swing into the space occupied by the man in front of him (except for those actually in the front ranks whose weapon would become entangled in those held by the file next to him).

While it is possible, although unlikely, that phalangites advanced into battle in this manner, it could not have been done rapidly. Nonetheless the ancient sources contain numerous references that the phalanx was capable of being more than a lumbering mass of extended pikes. Polyaenus states that Philip's infantry made 'a committed attack' against the Athenians at Chaeronea in

338BC.[6] At Gaugamela in 331BC, Alexander's infantry line 'rolled forward like a flood'.[7] While both of these passages are not necessarily indicative of speed, the commitment of these attacks suggests they were done at something more than a slow shuffling pace. Clearer evidence for the phalanx's use of speed comes from the description of the battle of Pydna in 168BC. Plutarch states that the Macedonian advance was so bold and swift that the Romans did not have time to move very far from their camp before the first man was slain.[8] Thus it seems that the pike-phalanx had the ability to advance at a different pace depending upon the intentions of the commander and the tactical situation at the time.

However, it is almost impossible to move quickly with the body turned side-on and with the feet facing forward as the illustration of the side-on stance suggests. This is particularly so if the left arm is elevated and rotated to extend the lower rim of the shield (which suggests it was not done), and even more so if armour is being worn. Even the lighter *linothorax* with its *pteruges* greatly inhibits movement when the body is placed in this position due to the stiffness of the fabric. Furthermore, certain styles of helmet also restrict the movement of the head when a side-on posture is adopted. For a side-on stance to be adopted, the head has to be turned entirely to the left so that the phalangite can look directly ahead over his left shoulder to see any potential targets and to identify any obstacles or terrain that he would have to traverse. However, the elongated cheek pieces of the Thracian and Phrygian type of helmet that was used by some combatants in the Hellenistic Age sweep down below the jaw to provide additional protection to the neck and throat. When wearing either of these styles of helmet, it is impossible to fully rotate the head to the side as the cheek flanges come into contact with the chest or cuirass and inhibit further rotation – preventing the phalangite from looking directly ahead at the tip of his weapon and/or any opponent (Plate 12). While it is possible that this means that the Thracian and/or Phrygian-style helmets were not worn by phalangites as some scholars suggest, their depiction in Hellenistic art does indicate that they were used in conjunction with other styles like the Boeotian and the *pilos*.[9] This, coupled with the fact that it is almost impossible to adopt a side-on position while wearing armour of any kind, suggests that the phalangite did not adopt a side-on posture to wield the *sarissa*.

Standing in an oblique posture, on the other hand, provides the phalangite with a full range of movement. As the body is only rotated by around 45°, any armour that is worn creates no impediment to forward motion or the position of the body. Forward movement, either at a walk, trot or even a run can be accomplished by simply walking/running forward in a natural motion.

Importantly, such an action does not affect the wielding of the *sarissa* in a leveled position to engage in combat nor necessitate the removal of the shield from a protective position across the front of the body. The rotation of the head is also natural when an oblique body posture is adopted. As with the use of a side-on stance, the cheek pieces of the Thracian or Phrygian helmet still connect with the chest/cuirass when adopting an oblique body posture. However this occurs at the point when the head is facing forward and the gaze is directly ahead (Plate 12). This is due to the angled lower edge of the cheek piece on these helmets. Had these helmets possessed cheek pieces which did not have angled lower edges, but rather ones that were squared off, the ability to turn the head would have been similarly impeded. This suggests that the angular nature of the cheek pieces of both the Thracian and Phrygian-style helmets were conscious design considerations made in order to facilitate the oblique body posture needed to wield both the *sarissa* and the *peltē* correctly.

BODY POSTURE AND FIRM FOOTING IN COMBAT

Firm and stable footing is a necessity in any form of hand-to-hand combat. This was fully understood within the context of the combat of the ancient world. The Archaic Greeks, for example, were advised to stand 'firm set astride the ground' (στηριχθεὶς ἐπὶ γῆς).[10] The Roman writer Vegetius refers to a positioning of the body with a preferred foot 'forward in combat' depending upon which sort of weapon was being used.[11] This demonstrates that the Archaic Greeks and the later Romans well understood the importance of balance and correct footwork in combat. The professional armies of Hellenistic Macedon are unlikely to have been any different.

The ability to transfer the body's centre of gravity and to provide stable footing is crucial to maintaining balance during the rigours of combat on terrain that might be soft, uneven or covered with the detritus of battle.[12] Arrian states that infantry 'push with their shoulders and sides' (τόυς ὤμους καὶ τὰς πλευρὰς αἱ ἐνερείσεις).[13] The term ἐνερείσεις (derived from the verb ἐπείδω) has several definitions including to push (with), thrust (with), lean in (with) or lay (upon). Regardless of which definition is used, it is clear that the terminology is meant to convey the idea that one side of the body is forward of the other. This could relate to a description of either a side-on or oblique body posture. However, an examination of how the body's centre of gravity, and hence its balance, can be altered in either of these combative stances demonstrates that the oblique body posture is the most effective for engaging in hand-to-hand combat.

Standing side-on would allow a phalangite to brace himself into position to resist any strong impact against his shield. The rearward right leg, with the foot turned outwards, provides solid support for the body to resist any force exerted against the shield or body from the front. The side-on posture also allows the phalangite to exert his own counter-pressure against this force, or to deliver it against an opponent, by simply leaning towards the enemy and pushing using the strength of the right leg. However, a person in a side-on position has difficulty altering the body's centre of gravity in every direction. By leaning either towards or away from an opponent, a phalangite could easily adapt his body posture in response to any pressure exerted against his shield. What a side-on position lacks is any ability to adjust the body's position and centre of gravity to react to any pressure coming from the sides – such as jostling by the men in the files beside him, an attack from the flank, or even topographical influences – in order to remain upright. Diodorus, for example, states that the current of the Tigris was so strong that some of Alexander's troops were swept to their deaths as they tried to cross it in 331BC because the force of the water 'deprived them of their footing' and the remaining troops were forced to link arms to avoid a similar fate.[14] Some battles were fought across watercourses which had similarly strong currents such as at Granicus and Issus.[15] The placement of the feet in the side-on position (which are basically in line with each other along the front-back axis of the posture) means that a phalangite could be easily knocked sideways due to the application of pressure from the side.[16] If, on the other hand, the feet are separated left-right to provide more stable footing while manoeuvres such as river crossings are undertaken, this naturally alters the posture of the body into an oblique position.

Standing obliquely with the left foot forward and the feet shoulder width apart is the most effective position for the transference of body weight and the body's centre of gravity. This is particularly so if the legs are slightly bent and the knee joints unlocked as this allows the phalangite to slightly squat and lean the body forward – which is required for him to gain the most protection from the small diameter of the *peltē*. Again, such aspects of combative posture are illustrated on the Pergamon plaque (Plate 7). Bending the knees both lowers and stabilizes the body's centre of gravity and makes the posture more stable and more likely to be able to adapt to changing conditions and pressures. With the knees bent, the joints unlocked and the left shoulder presented towards the front, a phalangite would be able to easily lean towards an opponent to his front (just as Arrian describes) while being braced in position by the rearward right leg. The unlocked knees allow the legs to flex or compress and so absorb any pressure regardless of the direction from which it had come. The stability of this stance is enhanced further if the right foot is turned outwards while standing still – making a static line of phalangites very stable and adaptive to battlefield conditions. This ability to strengthen the position is not available to the side-on stance unless significant alterations to the posture as a whole are made. On the other hand, the oblique posture, even with the smallest stride distance between the feet, allows the phalangite to use the strength of their rear leg to press forward, brace themselves in position if needed, and move with stable and adaptive footing all at the same time. Subsequently, Arrian's description of infantry pushing with their shoulders and sides is most likely a reference to the use of an oblique body posture with the left shoulder leading. Thus the physical dynamics of an efficient body posture correlates with the literary references and indicates that it was an oblique body posture that was used by the Hellenistic phalangite to bear his panoply in combat.

BODY POSTURE AND CONFORMATION TO THE LIMITATIONS OF THE PHALANX

Perhaps the most important aspect of the body posture of the phalangite was that it needed to allow the individual to conform to the massed style of fighting that it was used in. The Hellenistic phalangite did not fight as an individual. Rather he was a part of the densely packed phalanx. Consequently, the phalangite's body posture, and the way he wielded the *sarissa*, were not only dictated by how to most effectively carry this weapon, but also by the limitations placed upon the individual by all of those around him and the spacing of the formation that he was positioned within.

As noted earlier, the ancient writer Asclepiodotus outlines how infantry could be deployed in one of three orders: a close-order, with interlocked shields (*synaspismos*), with each man 1 cubit (48cm) from those around him on all sides (τὸ πυκνότατον, καθ' ὃ συνησπικὼς ἕκαστος ἀπὸ τῶν ἄλλων πανταχόθεν διέστηκεν πηχυαῖον διάστημα); an intermediate-order (*meson*), also known as a 'compact formation' (*puknosis*), with each man separated by 2 cubits (96cm), on all sides (τό τε μέσον, ὃ καὶ πύκνωσιν ἐπονομάζουσιν, ᾧ διεστήκασι πανταχόθεν δύο πήχεις ἀπ' ἀλλήλων); and an open-order (*araiotaton*) with each man separated by 4 cubits (192cm), by width and depth (τό τε ἀραιότατον, καθ' ὃ ἀλλήλων ἀπέχουσι κατά τε μῆκος καὶ βάθος ἕκαστοι πήχεις τέσσαρας).[17]

Yet not all of these were combative deployments. Asclepiodotus states that the open order was a 'natural' formation, while Aelian states that this deployment was only used when the situation called for it.[18] What situation is not specified, but the openness of the formation would suggest that it was used primarily for marching as the troops were less likely to become entangled with each other, there would be less external threats requiring the adoption of a closer order, and there would be little need for the troops to remain in step. The intermediate-order formation is described as the general offensive deployment by Arrian and Polybius, and Asclepiodotus and Aelian both say that it was used to advance upon an enemy.[19] The close-order formation with interlocked shields, on the other hand, is said to have possessed little offensive value for the pike-phalanx. Both Asclepiodotus and Aelian state that it was used to resist an enemy attack while Arian states that such a compressed formation limited the type of deployments that could be adopted (such as making an oblique line all but impossible).[20] Aelian also goes on to describe that one of the uses for the close-order formation was during the process of 'wheeling' the formation, as the more compressed nature of the close-order deployment meant that the troops had less distance to travel.[21] Importantly, both Asclepiodotus and Aelian state that when this manoeuvre was undertaken, the weapons of the men in the formation had to be held vertically so as to not impede each other.[22] This was also no doubt due to the fact that men carrying a shield with a diameter of 64cm, held across the body in a protective position in order to receive an enemy attack (as Asclepiodotus and Aelian put it) could create an interlocked shield wall, but due to the way in which the *sarissa* was wielded, made it impossible to position such a weapon for battle at the same time (see following).

Many modern models of Hellenistic warfare incorporate the use of phalangites deployed in an offensive close-order formation, in contradiction to the ancient manuals – although the terminology used in some of these models makes interpreting them somewhat confusing. McDonnell-Staff, for example, suggests that phalangites used a side-on stance so that they could stand in a 50cm space and that this resulted in a formation with interlocked shields (*synaspismos*).[23] Warry, who favours an oblique posture, also has phalangites creating a formation with 'interlocked shields' in close-order.[24] Snodgrass, who does not examine combative posture, offers that the *synaspismos* formation of 45cm per man (based upon the smaller Attic cubit of the early fifth century BC) was commonly used by Successor armies without detailing whether he thought it was (or was not) used by the armies of Philip II and Alexander the Great.[25] Sabin, in his modelling of ancient warfare, suggests that phalangites used various intervals in their formations (again based upon the smaller Attic cubit of 45cm and Asclepiodotus' descriptions of formations) of between 45cm and 90cm per man.[26]

The main problem with such models is that phalangites bearing a long pike and a shield 64cm in diameter are physically incapable of creating a combative formation with interlocked shields in an interval of only 45-50cm per man. Frontinus states that the phalanx did not like fighting in cramped conditions.[27] This suggests that phalangites required a certain amount of space in order to wield the *sarissa* effectively. Several ancient sources also detail how the pikes of the first five ranks of the phalanx projected between the files and ahead of the line.[28] As such, the minimum spacing between each file was not only limited by the amount of space that each individual man had to occupy, but it also had to leave enough room between the files themselves so that the weapons of the first five ranks could be lowered for battle.

An average person with a height of about 170cm has a width across the shoulders of approximately 45cm. When adopting an oblique body posture to wield the *sarissa*, the amount of lateral (i.e. side-to-side) space that this person would take up would be around 32cm.[29] The extension of the right upper arm to wield the *sarissa* increases the required width of this interval by around 16cm – bringing a total lateral width just to cater for the body of the individual alone to 48cm (the same as the close-order interval detailed in the ancient literature). The right-hand edge of this interval would be delineated by the shaft of the *sarissa* which projected directly ahead of the person wielding it. Added to this, up to 24cm of the shield would project to the bearer's left when he adopted an oblique body posture. Thus the minimum lateral interval occupied by a single phalangite adopting an oblique body posture, and with his weapon lowered for battle, was 72cm (Fig.12).

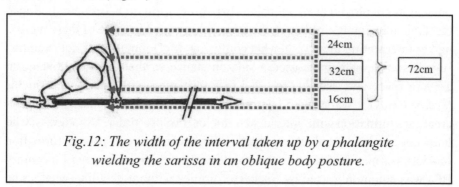

Fig.12: The width of the interval taken up by a phalangite wielding the sarissa in an oblique body posture.

Smith calculates the horizontal space occupied by a phalangite in an oblique body posture with his shield held across his front as being between 78cm and 92cm.[30] Smith further claims that this allowed the other pikes, which he suggests were held at shoulder height, projecting over the shoulder of the man in front, to be deployed.[31] However, it is unlikely that the weighty *sarissa* was carried at shoulder height. This means that there had to be enough room to the right side of the individual phalangite to cater for the presented pikes of the more rearward ranks (see following).

The projection of a section of the *peltē* to the left of the bearer would, if a weapon other than a *sarissa* held at waist level was being carried, allow a man to the left to move in, behind the bearer's shield so as to create a close-order interlocking 'shield wall', just as Asclepiodotus describes it, similar to the way that the larger *aspis* allowed the Classical hoplite to create such a formation.[32] Similarly, if the pike was raised vertically, phalangites would also be able to create an interlocking shield wall. However, the presence of this shield wall would then prevent the *sarissa* from being lowered for combat. This explains why both Asclepiodotus and Aelian state that the pikes had to be held vertically when this order of formation was adopted for manoeuvres such as 'wheeling' and demonstrates that such a deployment was not offensive or defensive (using the *sarissa* to hold an enemy back, for example) in nature.

The difference between the shield-wall of the Classical hoplite and the shield wall of the phalangite is that Classical hoplites wielded their shorter spears with one hand and in an elevated manner which allowed the weapons to be positioned above the line of interlocking shields when poised for combat. The phalangite, on the other hand, wielded the *sarissa* with both hands at waist height which prevents a shield wall from being created at the same time. Due to the position of the shield on the left arm, and how the left hand is used to help wield the pike, the bulk of the shield sits at the same elevation as the *sarissa* when it is lowered for combat. Thus even two phalangites standing side-by-side would not be able to interlock their shields due to the presence of the weapon held by the man on the left. Consequently, the minimum distance between two phalangites positioned adjacently for battle is the distance from the *sarissa* carried by the man on the right, to the left hand edge of his *peltē* if his shield is abutted up against the shaft of the weapon carried by the man to the left – or a distance of some 72cm (Fig.13).

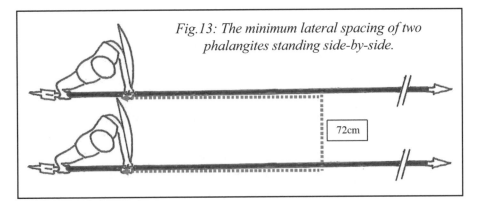

Fig.13: The minimum lateral spacing of two phalangites standing side-by-side.

72cm

In the image on the Pergamon plaque, the shields of the two adjacent phalangites are not overlapping and the shaft of the *sarissa* held by the rearward phalangite can be clearly seen extending between the two individual figures (see Plate 7).[33] This shows that these two individuals could not have been deployed with their shields interlocked and are most likely positioned in a formation no less than 72cm per man. This interval in itself is greater than the close-order formation outlined by Asclepiodotus and demonstrates that the pike-phalanx cannot have adopted a close-order formation for combat as some theories suggest.

Further to this, the additional four pikes belonging to the ranks behind had to also be accommodated within the space between each file. Subsequently, the phalangites of each file had to be even further apart. Due to the weight of the *sarissa*, the weapon held by each subsequent rank could not have been elevated higher so that that it was positioned above the one held by the man in front (or over the shoulder of the man in front as is suggested by Smith or Hammond). The weapons could have only been deployed level and relatively parallel with each other and maintained in this position for the required (and undeterminable) amount of time needed for any engagement to play out. Plutarch states that the presented pikes of the Macedonian phalanx all sat at the same level and presented 'a dense barricade' to any opponent.[34] Such a reference can only mean that each pike was held at the same level as all the others. This means that, between each file, there had to be enough space to accommodate the physical characteristics of a total of five lowered pikes.

The *sarissa* had a shaft of around 3.5cm in diameter. However, it is unlikely that the weapons of each rank were hard pressed up against each other as this would simply jam the hands of the men carrying them between the weapons. A small amount of space had to have been afforded for the placement and dimensions of the hands in order to avoid any injury. Allowing 6cm of space per man for the thickness of the shaft of the *sarissa*, the fingers wrapped around it, and a small amount of room to move, four additional pikes would add another 24cm to the required interval between each file.[35] If this is added to the 72cm required for the man in the front rank, the total lateral space required between two files of phalangites, with the weapons of five ranks of men projecting between each file, is 96cm – the intermediate-order outlined by Asclepiodotus (Fig.14).

Undoubtedly, the phalanx would not conform precisely to this interval all of the time. Gaps between the weapons of each man, for example, would open and compress through the simple movements of engaging in combat.[36] Additionally, it is unlikely that all five pikes belonging to each file were leveled exactly side-by-side. Due to variances in the height of each person, and movement of the

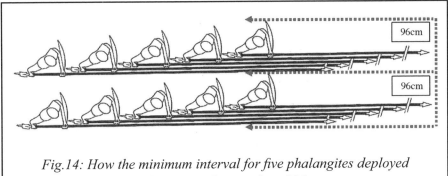

*Fig.14: How the minimum interval for five phalangites deployed
for battle is the intermediate-order of 96cm per man.*

formation over open ground, and the natural flex in the length of the *sarissa*, the pikes may have been positioned at slightly different elevations which varied over time. This would provide extra room between each weapon which would further reduce the risk of the hands becoming jammed between two adjacent weapons. Furthermore, as the men in ranks two to five were incapable of thrusting their weapon forward in an offensive manner due to the presence of the man in front of them, but would have merely held them in place, the men in these ranks could have stood slightly more erect than the man in the front rank (see: The Reach and Trajectory of Attacks made with the *Sarissa* from page 167). This would not only further reduce the risk of the hands of the man in the first rank becoming jammed between the other weapons of the file as he engaged in combat, but would also provide him with a small amount of additional room to undertake offensive actions without compromising the integrity of the formation. Yet despite any differences in the positioning of the weapons, the five presented pikes of each file would still only occupy a lateral space of about 24cm. What this understanding of the spatial requirements of the phalanx shows is that the minimum amount of room required for the formation to deploy for battle had to be the 96cm spacing of the intermediate-order at the very least.

Some ancient writers employ terminology relating to a close-order formation in their descriptions of the phalanx in action. Arrian, for example, states that Alexander's men adopted a *synaspismos* formation at the Hydaspes River in 326BC.[37] Plutarch also uses the term *synaspismos* in his descriptions of some Macedonian formations.[38] Some scholars have accepted these passages at face value – suggesting that the pike-phalanx used the close-order interval with interlocked shields.[39] English, on the other hand, recognizing that such a deployment is impossible for men armed with a *sarissa* and a *pelte*, suggests that the use of such terminology merely means the adoption of 'as compact an order as possible' and that such a formation required an interval of around 90cm per man.[40] This is undoubtedly correct due to the limitations of the phalanx. This is not a literary motif exclusive to Arrian and Plutarch, but the use of similar terms is found in the works of other ancient writers as well.

Curtius, for example, uses the term *testudo* three times in his account of the campaigns of Alexander to describe closely packed Greek and Macedonian formations. Tarn suggests that Curtius is using a piece of contemporary Roman terminology and placing it in an anachronistic context by ascribing it to the Greeks.[41] However, the true context of the terminology is dependent upon an understanding of the passages it is being used in. The *testudo* was a Roman defensive formation created by each legionary moving close to the man beside him so that the rim of each shield (*scutum*) touched 'edge-to-edge' rather than overlapped. This presented a solid wall of shields towards any attack, particularly to thrown missiles.[42] Members of the rear ranks of a *testudo* could also hold their shields above their heads and those of the men ahead them, in effect placing a roof over the formation to provide further protection. However, the important thing to note is that it was only the shields placed on the top of the formation which overlapped (usually just with those behind them which partially sat on top of the one in front) while those on the sides were abutted edge-to-edge. Thus the *testudo* was not entirely a formation with interlocked shields.

At 5.3.21, Curtius states that Alexander's men, who were being showered with large stones and other missiles, found no protection in the adoption of a 'tortoise formation' (*Nec stare ergo poterant nec niti, ne testudine quidam protegi, cum tantae molis onera porpellerent barbari*). The terminology here is somewhat ambiguous. When attacked by large numbers of missiles, troops will bunch together for protection regardless of the configuration of shield they are carrying. As previously noted, it is unlikely that troops bearing the small Macedonian *peltē* would be able to create a formation with inter-locked shields unless their pikes were raised vertically – in which case they could either deploy in a close-order with interlocked shields, or in a bunched formation with the rims of their shields touching. Either case bears many similarities to different aspects of the Roman *testudo* formation with its abutted and over-lapping shields. At 5.3.23 the same troops withdraw with their shields 'held above their heads' (*scutis super capita consertis*). Any such formation, whether in close or intermediate order, would justify the use of the term *testudo* by Curtius to describe it for a contemporary Roman audience. At 7.9.3, Curtius states that while crossing the Jaxartes River, some of Alexander's men adopted a 'tortoise formation with their shields' (*scutorum testudine*) to protect the oarsmen who were otherwise unarmoured.

It is unlikely that this use of the word *testudo* is meant to be taken as a row of inter-locked shields in this instance. It is more likely that the term is here cognate with the use of a shield to provide protection, much in the way that the Roman legionary was protected behind his shield when in the *testudo* formation. The oarsmen on the vessel would have been seated some distance apart and it is further unlikely that shields held to cover them would overlap. It is possible that the spacing between each man was enough to allow the rims of each shield to just touch (as in the *testudo* formation of the Roman legionary) and this may account for Curtius' use of the term. Thus Curtius' use of the term *testudo* to describe Greek formations is not as anachronistic as Tarn suggests, but is the use of a contemporary simile to convey to an audience how the shields borne by the phalangites were positioned.[43]

The inability for phalangites to adopt a close-order formation also demonstrates that Diodorus' reference to Philip adopting the use of a close-order formation with inter-locked shields (*synaspismos*), in emulation of the Greeks at Troy, cannot be interpreted as the creation of the pike-phalanx (see page 34). Diodorus is clearly not using the term *synaspismos* in a more general manner to merely mean a closely ordered formation as he specifically states that the formation has inter-locked shields. He is clearly describing the 48cm close-order formation of the manuals. Consequently, Philip's reform to the Macedonian phalanx could only have been accomplished by troops who were armed in a way which would allow them to use such a formation – Greek hoplites.[44] Hammond suggests that the use of the smaller *peltē* allowed phalangites to create an even closer order than that used by Greek hoplites.[45] This is clearly incorrect from an offensive perspective based upon the spatial requirements of the pike-phalanx.

A comprehension of the minimum amount of space required for the phalanx to deploy also highlights the deficiencies in the proposed side-on stance for wielding the *sarissa*. If a man turns side-on to wield the pike, the width of his body from shoulder to shoulder, the extended right arm, and the akimbo left arm, take up an interval, front-to-back, of about 96cm – the same size as the intermediate-order interval and twice as large as the close-order interval.[46] Thus by adopting this posture, a phalangite would be incapable of adopting the very deployment which forms the basis for many examinations of phalangite warfare.

Connolly gets around this problem by suggesting that a close-order formation, with the phalangites using a side-on stance, was 96cm front-to-back and 45cm side-to-side.[47] Similarly English, citing passages of Aelian (11) and Diodorus (17.57.5) suggests that the 'compact order' was 1 cubit per man left-to-right and 2 cubits front-to-back to accommodate the grip, which must assume the adoption of a side-on posture as per Connolly, and that such a formation was used offensively.[48] This passage highlights the problematic nature of many examinations of the Hellenistic pike-phalanx. Firstly, the 'compact order', as English calls it, was another name for the intermediate-order of 96cm per man according to Asclepiodotus. It is unlikely such a term could be applied to any formation employing the close-order 48cm interval, even if only left-to-right, as English suggests. Additionally, Asclepiodotus and the other military writers clearly state that the intervals they are describing are the same size both left-to-right and front-to-back. Thus any model of phalangite warfare which suggests a difference between the lateral distance across the file to the distance front-to-back cannot be correct.

Hammond states that the pikes of the first five ranks were presented when the formation was arranged in a 'close order'.[49] Again there are issues with such claims. If by 'close order' Hammond means the interval of 45-50cm per man, then this statement cannot be considered correct as it is impossible for the pikes of the phalanx to be deployed in such a way while in such a compressed formation and any claim to the contrary goes against what is outlined in the ancient literature. Polybius himself states that it was the 'compact' (i.e. intermediate) order formation of 2 cubits per man which allowed the pikes of the first five ranks to be presented.[50] Additionally, even if it is assumed that phalangites could somehow adopt a 45-50cm interval and still lower their weapons, a pike 576cm in length would be able to project ahead of the line even from the ninth rank so long as sufficient lateral space to accommodate all of the levelled weapons was provided. In this interpretation, it can only be assumed that, while it would be possible for the first nine ranks of the phalanx to present their weapons for combat, for some reason only the first five actually did so. Alternatively, if by 'close order' Hammond is actually referring to the intermediate-order of 96cm per man outlined in the manuals, then this statement would be correct in regards to the number of pikes that could be presented per file, but the terminology used to describe the interval is incorrect.

Furthermore, while Asclepiodotus and Aelian state that the close-order formation was not used offensively but could be used for 'wheeling' with the pikes held vertically, according to both writers the close-order could also be used to receive an enemy attack.[51] Due to the physical requirements of the pike-phalanx, this passage can be interpreted in one of two ways. Firstly, if seen as a reference to a pike-phalanx, it must be automatically assumed that the phalangites would have been standing in an 'attention' (προσεχέτω) position with the *sarissa* raised vertically. This suggests that, if this is the case, the enemy attack that is being referred to is incoming missile fire against a static pike-phalanx, where the phalangite could use his left arm, which is not holding the pike while in this position, to raise the shield to protect himself, rather than an attack delivered by infantry or cavalry which would require the deployment of the pikes in order to repel it (which cannot be done in close-order).[52] This then finds parallels in Curtius' first use of the term *testudo* to describe a contingent of phalangites under missile fire. Alternatively, Aelian may be referring to a close-order shield wall of troops armed as Greek hoplites who can act offensively while maintaining an interlocking shield-wall.[53] Importantly, either interpretation cannot be taken as an offensive use of the close-order formation by phalangites.[54]

Other scholars base their examinations of Hellenistic warfare on the intermediate-order interval of around 96cm per man. Champion, for example, states that Pyrrhus' 6-7,000 man frontage at the battle of Asculum in 280BC covered a distance of around 7km.[55] This must assume an interval of around 1m per man deployed in an intermediate-order. However, in many of these examinations, the terminology used is often confused and/or erroneous and it is uncertain as to whether some of the proponents of these models have explicitly understood the consequences of such statements in relation to the broader understanding of the dynamics of phalangite combat.

Warry, for example, while using the oblique body posture in his examination, still has phalangites adopt what is labelled as a 'close-order' and then an even closer order with 'locked shields'.[56] There are several problems with how this possible formation is presented. Firstly, Warry's 'close-order' seems to be the intermediate-order of 96cm per man. English similarly says that phalangites occupied 2 cubits per man in 'close-order' which can also only be seen as a misinterpretation of the ancient literature.[57] Secondly, in the depiction of the formation with 'locked shields' given by Warry, the shields are neither 'interlocked', nor are they even 'brought together', which are the two ways in which the word *synaspismos* can be translated. This lack of inter-locking shields, and therefore a lack of conformance with the terminology used to describe the close-order formation, also occurs in any model which incorporates the use of a side-on posture to wield the *sarissa* due to the way that the shield is angled towards the left-rear – further indicating that this posture was probably never used. Finally, in both of Warry's depictions of the 'close-order' and the 'locked shields' formation, the pikes belonging to the more rearward ranks are passing either above or below the shields of the men in the forward ranks (it is difficult to tell which from the diagrams). Regardless, due to the position of the shield and the weight of the *sarissa*, a deployment of rearward pikes either above or below the shield of the man in front is almost physically impossible.

Gabriel states that the 'compact order' (*pyknosis*) is 2 cubits of 45cm each (or 90cm total) per man, while the close-order (*synaspismos*) is 45cm per man and was used to engage in combat.[58] While the terminology used in these statements correlates with that of Asclepiodotus, the size of the cubit is based upon the earlier Attic standard incorporating a 45cm cubit rather than the Hellenistic standard with its 48cm cubit and contradicts Aelian's statement that phalangites could not use a close-order formation offensively.[59] Despite such statements, Gabriel later states, similar to Champion, that a 20,000 man force, deployed to a depth of ten deep, would create a frontage covering 2km.[60] This again must assume the use of the intermediate-order of around 1m per man.

An understanding of the spatial requirements of the phalanx formation also demonstrates that Polybius is unlikely to be correct in his description of the *sarissa* being held by the rear 4 cubits of the shaft.[61] If the weapon is held by the last 2 cubits as per the descriptions of Arrian, Aelian and Asclepiodotus, when deployed in an intermediate-order, the butt-spike of any weapon held is automatically pressed into the shield of the man standing directly behind (see Figs.6 and 14). Thus a man in a rearward rank can get no closer to the man in front of him than the point where both his shield and the butt-spike of the man in front of him meet. Heckel and Jones suggest that greaves were a mandatory element of the phalangite's panoply, particularly for those in the middle and rear ranks, due to the potential danger of receiving wounds to the legs from the butts of the weapons held by the men in the front ranks.[62] However, as the butt-spike pointed toward the shield of the man behind, rather than towards the legs, there is little danger to the lower limbs. The level at which the *sarissa* was held would additionally allow the man in the rearward rank to press his shield into the spike of the weapon in front and so brace it into position. This is also another reason why the side-on posture is unlikely to have been used as, due to the acute angle in which the shield is carried when using such a posture, the butt of a weapon held by a man in a forward rank would point at the abdomen of the man behind – which would run a risk of inflicting accidental injuries during battle.

Importantly for the understanding of the deployment of the phalanx, when employing a 2 cubit grip on the weapon, having the man in a rearward rank press his shield into the butt of the weapon in front equates to each man occupying a space of 96cm – in accordance with the descriptions of the intermediate-order formation. However, if there was an additional 2 cubits of the weapon projecting rearward as per Polybius, then the distance between each man would have to be 4 cubits (i.e. the distance from the weapon bearer's leading left hand to the point behind him where the spike of the *sarissa* connected with the rearward man's shield). Thus, at least front-to-back, phalangites wielding weapons in such a manner could only deploy in an open-order formation of 4 cubits (192cm) per man front-to-back at best. Some scholars who favour the Polybian model get around this problem by having the extra 2 cubits of pike project under the shield of the man behind, or have the men slightly offset to accommodate the rear of the pike.[63] However, due to the position and elevation of the shield, this is a physical impossibility unless it is assumed that the men in the forward ranks have their weapons angled upwards so that the rearward sections can pass under the shields of the men behind. However, this would mean that they would then not be able to be lowered to engage in combat. Connolly notes how it has been suggested that the large butt on the *sarissa* could pose a threat to the men in the rank behind but then goes on to say that, by holding the weapon at the 4 cubit mark as per Polybius, the pike actually passes beyond the man directly behind.[64] What Connolly seems to have failed to consider is that this then means that the butt would be pointing directly at the man two ranks behind rather than the man immediately behind. Regardless, due to the position of the shield, this concept of a 4 cubit point of grip on the *sarissa* seems incorrect.

An understanding of the spatial requirements of both the phalanx as a whole, and that to wield the *sarissa*, also shows that Tarn's proposed smaller, 33cm, Macedonian cubit cannot be correct (see the previous section on the Length of the *Sarissa* from page 66). Regardless of the configuration of the weapon used, it is physically impossible for the men of the phalanx to occupy a space only 66cm square in an intermediate-order deployment based upon this unit of measure and still adopt an effective combative posture. In his reconstructive experiments, Connolly found it impossible to march a unit of phalangites forward when deployed in an interval less than 69cm per man.[65]

Thus we are left with two options for interpreting the description of holding the *sarissa* by Polybius. Either the passage itself is incorrect (whether this was the fault of Polybius himself, his sources, or later transcription is unknown), or that Polybius is correct, but in accepting this it must also be accepted that pike-phalanxes in the second century BC only fought using an open-order deployment of 4 cubits per man for which there is no evidence. Either way, any modern theory which examines Hellenistic warfare based solely on the description of Polybius needs to be appropriately revised.

By examining the spatial dynamics and requirements of the phalanx as a whole, it can be seen that an oblique body posture was required for the phalangite to effectively bear and wield his panoply. The side-on stance simply occupies too much room to conform with the descriptions of the phalanx found in the ancient literature and places the phalangite in a position where he may be accidentally injured by the man in front during the trials of combat. The oblique posture, on the other hand, not only conforms to the spatial requirements of the phalanx (when a weapon with a point of balance 2 cubits from the rearward end is used), but also allows the phalangite to retain his shield in a protective position across the front of his body which, by pressing it into the butt-spike on a weapon held by the man in front, not only braces that weapon in position, but actually indicates to the rearward man that he is positioned properly within an intermediate-order interval. All of these elements would come in to play simultaneously to create a formation that was stable, protected, and conformed to the needs of the phalanx.

BODY POSTURE AND THE EFFECTIVE OFFENSIVE USE OF THE *SARISSA*

One of the most important aspects of any combative posture is the ability to act offensively. Yet by its very nature, the offensive abilities of a phalangite within a phalanx were greatly restricted. Indeed, anyone other than the members of the front ranks would not have been able to thrust their weapon forward into an attack at all. This was due to how both hands were used to wield the *sarissa*, how the shield was strapped to the left forearm at the same time, and how that shield was pressed into the butt-spike on the pike held by the man in a more forward rank when conforming to an intermediate-order deployment. Thus not only did pressing the shield into the butt-spike ahead of you brace that weapon in position, but in a negative capacity, it prevented you from performing any offensive action with the *sarissa* at the same time. It was only the experienced officers of the front ranks of the phalanx who were not similarly impeded and were thus capable of undertaking offensive actions directly to their front. The members of the rear ranks could engage an opponent with their weapons, but this was only under certain battlefield conditions (see The Anvil in Action from page 374). Yet even in these instances, body posture dictated how effectively the *sarissa* could be used.

If a side-on body posture is adopted, the position of the body greatly restricts any movement of the weapon. Any attacking motion has to swing the arms from right to left as the *sarissa* is thrust forward. However, in doing so the right upper arm almost immediately impacts with the upper body and so limits the amount that the weapon can be moved. The construction of the shoulder joint itself also restricts the amount of movement across the body that the right arm can make. Additionally, the presence of rigid shoulder sections on any armour worn further impedes the movement of the arms in this position (see: The Reach and Trajectory of Attacks made with the *Sarissa* from page 167). Consequently, it seems unlikely that such a posture was used to wield the *sarissa* in combat.

By standing obliquely, on the other hand, the right shoulder is not restricted in its range of motion. The angled position of the body creates a gap between the torso and the upper arm (see Fig.14). The ball and socket construction of the shoulder joint is also not impeded as any thrusting motion does not have to pass across the body, but the weapon (and arms) can simply be pushed/swung forward. The angle of the body also negates any impediment on the motion of the arms which might be caused by the shoulder sections of the armour when using another posture. Rather the angle of the body, the gap between the right arm and the torso, and the lack of restriction by the cuirass, provide the phalangite with a full range of motion for both arms with which to engage an opponent effectively.

CONCLUSION

The following table (Table 7) summarises the required criteria that any stance a phalangite could adopt must satisfy in order to be considered an effective combative posture and how the two suggested techniques (the side-on and the oblique positions) either conform, or not, to these requirements:

Criteria	Side-on Stance	Oblique Stance
Allow the shield to be positioned in such a way that it could be used defensively.	No	Yes
Allow the phalangite to move on the battlefield without any restrictions, particularly in a forward direction.	No	Yes
Provide stable footing to enable the phalangite to remain upright during the rigours of close combat.	No	Yes
Allowed the phalangite to maintain his position within the confines of the phalanx and conform to any limitations of space dictated by that formation.	No	Yes
Allow the *sarissa* to be positioned in such a way that the arms had a natural range of movement so that the phalangite could engage an opponent offensively.	No	Yes

Table 7: A summary of the necessary criteria
for an effective phalangite combative posture per stance.

As can be seen, it is only by positioning the body at an oblique angle that a phalangite is able to satisfy all of the necessary criteria for an effective combative posture. The understanding of the use of such a posture to carry the phalangite's equipment into battle further influences all subsequent examinations into how the *sarissa* could be wielded and its effects – both in terms of the reach and trajectory of any attack made with it, how the weapon's position could be altered, and every other fundamental aspect of engaging in combat with a pike in the Hellenistic Age.

Chapter Five

Phalangite Drill

It is highly unlikely that the phalangite held the *sarissa* in only one position. Whether through the process of marching, deploying in formation on the battlefield, or through preparing to engage an enemy, the phalangite undoubtedly moved through a number of different positions during the course of a battle. However, no ancient text comprehensively details the movements employed by the individual phalangite to alter the positioning of his weapon in preparation for, and to engage in, combat.

Other than references to the location of the grip found in sources like Arrian, Aelian, Asclepiodotus and Polybius, there are very few references to specific movements which can be considered part of 'phalangite drill' in the ancient texts. Even in military manuals, the focus of the work tends to be much broader in its scope; detailing the organization and movement of formations rather than the individual combatant. This may account for why an examination of the elements of phalangite drill have not been a part of any modern exploration of the warfare of the Hellenistic Age other than to recognize that phalangites needed to be well trained in order to effectively use the *sarissa* within the massed formation of the phalanx.[1] Yet the details of phalangite drill are essential in order to better understand the functionality and limitations of the pike-phalanx. The rudiments of phalangite drill must therefore be gleaned from passages that are scattered throughout the ancient narratives and military manuals.

Aelian, for example, concludes his examination of the phalanx by providing a simple overview of some of the words of command used when deploying an army. Many of these relate to individual drill movements with the *sarissa*.

Among those listed by Aelian are instructions to 'stand at attention' (*prosechē*), to 'raise spears' (*anō ta doru*) – by which he means pikes, to 'face to the spear' (*epi doru klinai*) – in other words, to turn to the right, and to 'wheel to the spear' (*epi doru epistrophe*) – in other words, to have the whole formation swing like a gate to the right.[2] Asclepiodotus offers similar instructions in his account of phalangite commands, as does Arrian.[3]

Thus from these passages we begin to see the basic movements of the phalangite with his weapons. The first movement, 'stand at attention', for example, is not a combative posture. In such a position, the phalangite most likely stood facing towards the front – with the shield held across the front of his body and the *sarissa* raised vertically, held only by the right hand, and with the butt placed upon the ground. Such positions are common in funerary *stele* such as that of Nikolaos (see Plate 8).

In the second position, 'raise spears', the phalangite would have raised the pike vertically, lifting the butt off the ground. It seems most likely that when this occurred, the left hand would be brought across to grasp the *sarissa* as well, in order to help carry its weight. Interestingly, when in the former 'attention' position, the right hand would grip a weapon which had its butt on the ground at a point higher than the weapon's point of balance. Thus as the weapon is raised, the left hand would have to move into a position where it could grasp the shaft at the correct point of balance, and the grip of the right hand would then be momentarily released and the hand moved to its correct location as the weapon was raised.

The final two commands, turning and wheeling to the right, would require the pike to be held in this elevated vertical position due to the confined nature of the phalanx. Aelian specifically states that the process of wheeling could only be undertaken while the *sarissa* is held vertically.[4] This would also apply to the process of changing the facing of the individual due to the fact that it is impossible for someone in a massed formation to turn 90° to either their left or right while a weapon like a lengthy pike is leveled for combat.

Perhaps the best description of phalangite drill movements can be found in Arrian's *Anabasis*. Arrian describes how, during a demonstration in Thrace in 335BC, Alexander the Great arranged his men in formation:

> ...and then he gave the infantry the order to raise their pikes vertically then, at the command, to extend them for battle, and now, by inclining to the right, closed up the spears, and then back again to the left. The whole phalanx moved smartly forward and then turned (*paragoge*) this way and that.[5]

Some earlier translations of this passage do not seem to be entirely correct in regards to the drill movements that are described. For example, de Sélincourt's translation (Penguin, 1971) reads:

> Then he gave the order for the heavy infantry first to erect their spears, and afterwards, at the word of command, to lower their massed points as for attack, swinging them, again at the word of command, now to the right, now to the left. The whole phalanx then moved smartly forward, and, wheeling this way and that...

The term ἐγκλῖναι, translated by de Sélincourt as 'swinging', would be much better translated as 'inclining'. Clearly, phalangites deployed in a massed phalanx would not be able to swing their leveled pikes either left or right to any great extent due to the men positioned in the files beside them. However, de Sélincourt's interpretation suggests exactly that. What this passage seems to be describing, rather than any movement of the pike itself, is the closing up of the intervals between the files, in effect moving from an open-order to an intermediate one, by having the files side-step (or incline) to the right and then open the intervals back up by side-stepping back to the left. Furthermore, the term translated by de Sélincourt as 'wheeling' (*paragoge* - παρήγαγε) is actually the term for changing the facing of individual phalangites (i.e. turning to face to the right or left) while the term for wheeling, changing the facing of the formation as a whole by having it swing 90° like a gate, is *epistrophe* (ἐπιστροφή).[6]

Thus from Arrian we get a slightly more detailed image of how a phalangite in the forward ranks deployed his weapon for battle. Firstly, once in formation, the pike would be raised vertically. This would undoubtedly be into a position where the butt was off the ground. The pike was then lowered into a combative position. Polybius describes how, at the battle of Sellasia in 222BC, both sides lowered their pikes (καταβαλοῦσαι τὰς σαρίσας) and advanced upon each other.[7] Due to the difference in the placement of the feet when standing at attention with the pike held vertically and the oblique posture used to engage in combat, any movement whereby the pike was lowered into position would also have to include a corresponding movement of the feet with the left foot moving forward as the *sarissa* was lowered so that the body posture could help support the weapon and provide stable footing in combat. Such a combative posture is shown on the only artistic representation of the phalangite in action – the Pergamon plaque (see Plate 7).

Consequently each phalangite, even when standing at attention, would have to be spaced at least 2 cubits from both the man in front of him and behind him so that each member of the phalanx would be able to take the necessary step forward with their left foot to adopt an oblique body posture, have the correct 2-cubit grip on the weapon, and conform with the interval of an intermediate-order phalanx at the very least. However, Arrian's description of the files closing to their right suggests that, in this instance at least, the phalanx was initially deployed in an open-order interval of 4 cubits per man as, if the phalanx was already in a 2-cubit intermediate order with their pikes lowered, they would not be able to close their files any further.

Arrian goes on to state that, once they had frightened their enemy with this display of drill movements, Alexander's troops clashed their weapons against their shields (καὶ τοῖς δόρασι δουπῆσαι πρὸς τὰς ἀσπίδας).[8] Beating on the shield with the pike seems to have been a fairly common practice in the Hellenistic period. Diodorus, for example, states that, in order to frighten the horses pulling the Persian scythe-equipped chariots that they faced at Gaugamela in 331BC, Alexander ordered his men to 'beat their shields with their pikes' (ταῖς σαρίσαις τὰς ἀσπίδας τύπτειν).[9] In another example of such actions, the army of Eumenes was similarly instructed to 'pick up their shields and beat upon them with their pikes and raise the battle-cry' (ἀσπίδας ἀνείλοντο κὰι ταῖς σαρίσαις ἐπιδουπήσαντες ἠλάλαξαν).[10]

Despite such passages which clearly describe this action being performed with a pike (except for Arrian who uses the word for spear), Markle says that the *sarissa* could not be beaten against the shield and concludes that the phalangite was not armed with a pike but rather with a spear.[11] However, when standing in an attention position with the butt of the pike resting on the ground, the shaft of the *sarissa* can be easily beaten against a shield held across the front of the body. Markle's conclusion is all the more interesting as Arrian describes Alexander's troops as carrying *sarissae* only a few paragraphs earlier and holding them leveled in order to flatten a grain field.[12] Importantly, the ability to beat the *sarissa* against the shield shows that the *peltē* had to possess an armband at the very least as it is much more difficult to move a shield that is only secured by a shoulder-strap into a position where the shaft of the pike can be beaten against it.

In another reference to a phalangite drill movement, Polybius also refers to how the men of the rearward ranks of the phalanx marched into battle with their pikes held at an angle over the heads of the men in front of them to provide cover from missile fire.[13] When standing in an oblique posture, with the left elbow tucked into the side of the body, elevating the tip of the *sarissa* is accomplished by simply raising the left forearm while straightening the right arm downward at the same time. Importantly, raising the left forearm increases the elevation of the shield slightly, due to how the arm bends at the elbow which is also the location of the centre of a shield mounted onto the forearm through the use of a *porpax*. This places the shield in a slightly higher position to provide better protection to the bearer against incoming missile fire, the very thing Polybius states this type of posture was used for, without compromising the vision of the bearer (see Plate 13).

Pitching a lengthy pike forward in this manner can be facilitated by sliding the left hand forward of the weapon's point of balance (the point where the left hand would normally hold it) by a short way. Not only does this allow for more of the pike to extend behind the forward left hand, in turn allowing the right hand to maintain its position on a pike which is now held at an angle rather than levelled, but shifting the placement of the left hand forward alters the distribution of weight behind the leading hand. By holding the pike ahead of its point of balance, there is more weight behind the left hand than there is ahead of it. This results in gravity forcing the rearward end of the weapon down, and the forward tip correspondingly up, which makes the weapon much easier to angle over the head of the man in front and keep it there. Importantly, due to the way that the pike is angled rather than levelled, allowing more of the weapon to be behind the leading left hand still allows the weapon to be contained within the 96cm interval of the intermediate-order spacing.

Once battle had begun, the members of the rear ranks of the phalanx would have kept their pikes angled in this manner. This not only maintained the overhead protection given to the forward ranks but, should a man in the front rank fall, the remainder of the file would be able to move forward so that his position was re-occupied and a combative posture adopted. If, for example, a member of the first rank fell, the members of ranks two to five would be able to simply step forward to take his place and they would already be in a combat position as their pikes would have already been leveled. The man who had been in rank six, and who now moved into rank five as the file moved forward, would have originally had his weapon angled upward. However, from this position, the pike could simply be lowered into a combat posture as no change of the grip on the weapon is required in order to adjust its pitch (see Plate 13).

Importantly, as the pike is lowered into a leveled position, the left hand can simply be slid back into place as the pitch of the weapon is decreased. This means that at any stage during the alteration of the pitch of the weapon, either up or down, both hands would remain in contact with it, and the phalangite's hold on the weapon would always be secure. Additionally, it means that when the *sarissa* is presented for battle, it would be held again at its correct point of balance by the left hand, and the rearward end of the weapon would remain within the interval that the phalangite was occupying within the massed phalanx.

When simply marching, rather than deploying for battle, the Classical Greek hoplite marched with his spear sloped back over his shoulder. It is most likely that the Hellenistic phalangite operated in the same manner. Polyaenus states how the troops of Philip II undertook lengthy marches of up to 35km carrying a *sarissa* and other pieces of equipment.[14] It is unlikely that each phalangite held their weapon in some other position, such as carrying it vertically, for the entire duration of one of these marches. Consequently, the *sarissa* had to be carried in a manner which would allow it to be ported without causing the carrier undue fatigue. This suggests that it was not carried vertically as the arms would become tired long before the march was over. Instead, the pike must have been supported in some way and the only means of accomplishing this would have been to have the weapon supported on the shoulder.

Even here an understanding of this posture can shed light on the configuration of the weapon. Holding the pike near the butt allows for the weapon's point of balance to be sat on the shoulder with the weapon projecting upwards and slightly to the rear at an angle not too far beyond vertical. However, a weapon over 5.5m long is difficult to maintain in this position for a long time even in open country. Even a slight change in the elevation of the weapon will move the point of balance off the shoulder and make it very difficult to carry. If the weapon is angled back too far, the sheer leverage of the weight of the weapon, working in concert with the force of gravity, will cause the head to tip further and further back, running the risk of becoming entangled with the pikes carried by the men behind, unless an increased amount of counter-pressure is applied through the arms to pull the tip back up. This results in considerable muscular stress being placed on the arms. It is possible that, with a shield slung across the back through the use of the *ochane*, both hands were used to support a *sarissa* carried in this manner. However, this would then mean that no other pieces of equipment could be carried.

In wooded terrain the angle of such a lengthy weapon would have to undoubtedly be altered in order to allow the bearer to pass unhindered under trees and other foliage. On uneven terrain, the jostling of the weapon through natural movement would also undoubtedly alter the angle at which it was carried. And yet, for such a lengthy weapon, the range of angle at which it can be ported easily is only very small – much smaller than either of these two types of terrain would allow to be consistently maintained. English suggests that the use of the *sarissa* was abandoned in 327BC to make Alexander's army more mobile – possibly for the impending campaign in India. English cites a lack of direct references to the *sarissa* in the ancient sources as proof of its discontinued use.[15] However, at the battle of the Hydaspes in 326BC, Alexander is said to have noted how the advantage lay with the Macedonians due to their long spears – a statement which can only be a reference to the use of the *sarissa* by Alexander's infantry in India.[16] Later in the campaign, Alexander received reinforcements in the form of 30,000 local youths equipped and trained in what is described as 'the Macedonian manner'.[17] Following the mutiny at Opis in 324BC, Alexander also personally oversaw the training of thousands of youths 'so that they might be brought up in the Macedonian manner in other respects and especially in the arts of war'.[18] If, following English's line of argument, the Macedonians had abandoned the use of the *sarissa*, they must have been using some other weapon instead – such as a spear. However, the spear was such a common weapon across the ancient world that the use of it can hardly be described as specifically 'Macedonian'. The references to these youths and the manner of their training suggest that they were equipped with weapons that were regarded as a distinguishing feature of the Macedonian military. Such armament can only have been the continued use of the *sarissa*. Furthermore, coinage that Alexander had struck to commemorate the victory at the Hydaspes in 326BC, shows a standing infantryman, thought by some to be Alexander himself, carrying what appears to be a *sarissa*.[19] Consequently, if the *sarissa* was employed on Alexander's Indian campaign, then it must have been configured in a way that would allow for its easy portage through the terrain. This suggests that the *sarissa* did not come as a single piece.

With a connecting tube used to join two halves of a *sarissa* together, the lengthy pike could be disassembled for ease of transport while the phalangite was on the march. If the connecting tube was positioned in the centre of a weapon 5.76m in length as was carried by Alexander's phalangites, this would mean that each half was 288cm in length. This is not much longer than the traditional hoplite spear which had a length of 255cm.[20] The two halves of the *sarissa* could be laid side-by-side, and then tied together, so that the weapon as a whole could be carried much more easily sloped over the shoulder, and supported only by the right hand, while on the march. Importantly, not only would the shorter length of a disassembled *sarissa* make it much more easy to carry, but by having each half of the weapon strapped together, the weight of both the butt and the connecting tube (a total of 1,230g) could be situated at the bottom, while only a small head weighing approximately 174g would be at the top. This would greatly reduce the amount of counter force required to maintain the pike in a position sloped over the shoulder if the angle away from the vertical was increased due to movement, terrain, foliage or fatigue. Alternatively, both the butt and the head could be positioned at the bottom of a dis-assembled pike – leaving only the small weight of the connecting tube at the top (see Plate 14).

Once the army had arrived on the battlefield, the two halves of the pike could be untied and assembled by simply slotting the front half onto the rearward section before the phalangite took his place in the line. In all of the accounts of battles across the Hellenistic period there are very few references to spontaneous actions like a column being ambushed. Most engagements are set-piece battles where both sides arrive on a battlefield, sometimes the day, or even days, beforehand, and deploy before any fighting commenced. Under such conditions a phalangite would have ample time to prepare his weapon for combat. Yet set-piece battles with enough time for preparation could not have been guaranteed even though they were the norm. Thus the *sarissa* needed to be able to be used on short notice if the need arose. Even in a spontaneous action a pike that came in two parts could still be utilized. Should the column that a phalangite was marching in come under unexpected attack, if there was not time for the full *sarissa* to be assembled for battle, the front half could simply be used either as a javelin or a spear. If even less time was available, the two halves of the *sarissa* could still be presented even while they were tied together so that the head pointed at an approaching enemy.

Thus from the ancient literary sources we have references (or inferences) to five different positions for the *sarissa*: sloped over the shoulder, held vertically with the butt on the ground while standing at attention, with the pike raised vertically with the butt off the ground, with it angled forward to provide cover for the men in the forward ranks, and levelled for combat. Using these positions as the basis, the fundamentals of 'phalangite drill' can be determined by simply working out the specific movements that needed to be performed in order to move the *sarissa* from one of these positions to another.

An important aspect of any sequence of drill movements is that it has to be dual directional – with the ability for the phalangite to be able to change the position of his weapon from one stance to another and then back again. There would be no point, for example, in establishing a set of drill movements which allowed the phalangite to lower his weapon for combat which did not, in turn, allow him to raise it back up again once the battle was over. The basic drill movements outlined in the ancient texts provide just such a means. A weapon held vertically, or sloped over the shoulder, can be easily shifted forward to be pitched over the head of the man in front, or directly lowered into a combative posture. This process can also be performed in the reverse order – from lowered to pitched, vertical or sloped over the shoulder. Importantly, in both directions, the placement of the hands is not required to be altered in any significant way, good balance of the weapon is maintained, and both hands never let go of the weapon in the combative postures (pitched, held vertically and lowered). Thus the rudiments of hoplite drill are relatively simple – employing only five set positions – and would have been quite easy for soldiers to train with. The benefits of these simple motions and postures meant that the phalanx as a whole could function effectively on the battlefield and use adaptive combative motions while using the *sarissa* as both an offensive weapon and a means of supplying limited protection.

Another point to note is that the fundamentals of phalangite drill witnessed something of a resurgence in popularity during the sixteenth and seventeenth centuries. During this time, the large pike and musket formations which raged across Europe were heavily modelled on the Hellenistic pike-phalanx of two millennia earlier.[21] Military manuals such as those written by Arrian, Aelian and Asclepiodotus were re-released in a variety of European languages and dedicated to different ruling monarchs of the time as suggestions of how they should marshal their armies. Many of these works contained illustrations to help clarify the concepts that they described (as many of the ancient originals upon which they are based had done). In these later copies, the figures used to represent the phalangite were rendered as contemporary, sixteenth or seventeenth century, pikemen that readers of the time could more easily associate with. However, many of the pikemen of this time presented their weapon for battle in a very different way to the Hellenistic phalangite – holding the pike elevated at shoulder height, with the rearward end cupped in the right hand, rather than held low at waist level.[22]

Smith suggests that this was how the phalangite wielded the *sarissa*, stating that 'he fought not with a sword, but with an exceptionally long spear used over the shoulders of his comrades...'[23] Similarly, in a diagram of the pikeman in action, Hammond suggests that the *sarissa* was couched into the armpit – in the same way that a medieval knight carried his lance.[24] Both of these postures seem unlikely. Not only do they go against the depiction of phalangites in a combative posture on the Pergamon plaque, and references to the phalangite lowering their weapons for action, but the pike of the sixteenth century was configured differently to the *sarissa* with no large butt on the end of the shaft.[25] This made the pike much lighter (2.5-3kg) compared to the 5.5kg *sarissa* and, as such, the sixteenth century pike would have been much easier to carry aloft than the *sarissa*. Additionally, the sixteenth century pikeman did not have a shield strapped to his left forearm (a detail also missing from Hammond's diagram) and so this weight did not need to be held aloft either. Interestingly, in order to carry the pike in such an elevated position, the shield is also raised due to the way it is strapped to the left forearm. If the pike is raised to shoulder height as per Smith's suggestion, the shield obstructs the bearer's view to the extent that he cannot actually see to his front. All of these factors suggest that the pike was not held in any form of elevated position.

Another point to consider is how such a deployment of the *sarissa* would translate within the confines of the phalanx. If the man in the second rank, for example, was shorter than the man positioned ahead of him, it would be difficult, if not impossible depending upon the difference in height, for the man in the second rank to position his spear at shoulder height and yet still have it project over the shoulder of the man in front and still be considered levelled for combat. Such a configuration of the phalanx would only work if each file of the formation was arranged from the shortest man at the front to the tallest man at the back. And yet, we are told that the front rank of the phalanx was made up of officers.[26] It is unlikely that a short stature was the only qualifying criteria for promotion within the Macedonian army and, as such, the height of the men in the front rank could not be guaranteed. This then suggests that the pikes of each file were not deployed 'over the shoulders' as Smith suggests, but were positioned in a way that each of the weapons held by the first five ranks could be presented for battle regardless of the stature of the individual. The only way this is possible is by having the pikes lowered to waist height and set side-by-side so that the file then occupies the intermediate-order interval of 96cm per man.

Polybius states that pikes of the Macedonians were 'lowered' into position at the battle of Sellasia in 222BC (καταβαλοῦσαι τὰς σαρίσας) to create a hedge of *sarissae* (συμφράξαντες τὰς σαρίσας).[27] Variants of a reference to the 'lowering of pikes' into a combative posture can be found in the descriptions of many other phalangite engagements.[28] This suggests that these weapons were lowered to waist height rather than held at shoulder height otherwise another term, such as 'raised' (ανασχόμενος) rather than 'lowered', would have been a more appropriate description. Finally, what suggests that the *sarissa* was not held in the same manner as the later pike is the way that this posture, and all other postures, affected the way in which this weapon could be used in combat.

Chapter Six

The Reach and Trajectory of Attacks made with the *Sarissa*

Once it was moved into a position for combat, how (if at all) was the *sarissa* used to engage in offensive actions? As with any hand-held weapon, the reach and trajectory of any attack made with the *sarissa* will have an influence on how effective that attack is in a combative situation. This, in turn, impacts on our understanding of the warfare of the Hellenistic Age. As Dickinson states: 'the understanding of the degree of manoeuverability of the *sarissa* affects the tactical interpretation of the progress of Alexander's battles'.[1] Yet despite such reasonings, scholars are at odds on such an issue. As well as there being some debate over how the phalangite wielded his pike, scholars also offer various hypotheses about how the *sarissa* was employed.

Warry, for example, suggests that the *sarissa* 'must have given the formation greater thrusting power' and that the phalanx was equally prepared to thrust with their pikes or push with their shields.[2] Milns suggests that the *sarissa* gave the Macedonian soldier the advantage in a set-piece battle of being able to deliver the 'first strike'.[3] Pietrykowski states that phalangites would 'lunge forward at the moment of contact with their great two-handed pikes'.[4] Worthington suggests that strikes with the *sarissa* were not great thrusts but that the phalangite may have simply jabbed back and forth with his weapon when in combat.[5] All of these passages suggest the employment of some kind of thrusting action by the phalanx but do not elaborate on which ranks are thought to have undertaken such actions. Gabriel, on the other hand, suggests that when pikes were employed like 'stabbing bayonets', an opponent had to face five pikes coming at him.[6] This would imply that all members of the first five ranks of the phalanx were able to use their weapons offensively. Similarly Smith, who favours a method of wielding the *sarissa* at shoulder height in a manner akin to the sixteenth century pikeman, suggests that the rear ranks of the phalanx could thrust over the shoulders of the men ahead of them and that phalanx combat required little individual movement other than for thrusting actions.[7]

There are several problems with such conclusions. The ancient texts describe how the regular deployment for the pike-phalanx was in the intermediate-order of 96cm per man. Within such a confined mass of men, the shield of anyone in a rearward rank would be pressed up against the butt of the weapon held by the man in front (see Fig.14). This would occur regardless of the posture used to carry the pike (i.e. at waist level or at shoulder height). Due to the way that the left forearm is inserted through the central armband of the phalangite shield, coupled with the way that the left hand is employed to help carry the weapon in a two-handed fashion, any phalangite in anything other than the front rank would have the movement of their arms inhibited by the presence of the man in front of him. The result of this is that, regardless of how the pike was wielded, it was only members of the front rank who would be able to thrust their weapons forward in an offensive capacity. This partially accounts for why the front rank of the Hellenistic phalanx was made up entirely of experienced officers.[8] Both Aelian and Arrian state that the weapons held by the first two ranks of the phalanx could reach the enemy.[9] However, due to the confined nature of the pike-phalanx which meant that those in the more rearward ranks had very limited movement, coupled with the way that the pikes held by members of the second rank were staggered back from those held by the front rank by a distance of 2 cubits (96cm), it seems unlikely that a phalangite in the second rank of a formation could have thrust his weapon forward to engage an enemy. Consequently, the statements of Aelian and Arrian seem to be references to the capabilities of formations of troops armed as Classical hoplites, whose more elevated and one-handed method of wielding the short *doru* allowed them to thrust their weapons over that of the man in front, rather than phalangites.[10] Yet the question remains: how far could even a phalangite in the front rank of the pike-phalanx thrust the *sarissa* forward in order to deliver a strike against an opponent?

All weapons, regardless of whether they are hand-held like pikes, spears and swords, or missile weapons like bows, slings or javelins, have what are known as an *effective range* and a *combat range* (sometimes called a *battle range*). For a missile weapon, the *effective range* is the maximum distance that a projectile (e.g. an arrow, sling bullet or javelin) will travel when released, fired or cast. The *combat range*, on the other hand, is a measure of how far such projectiles can be used to effectively inflict casualties amongst a distant enemy. For example, a modern military rifle may be able to fire a bullet for more than a kilometre. This is the weapon's *effective range*. However, due to variables which affect the aim of the weapon such as the line of sight, the diminishing size of targets as the distance to the target increases, the effect of gravity on the trajectory of the projectile and so on, the same rifle may only be able to be used to effectively engage opponents on the field of battle up to a distance of several hundred metres (its *combat/battle range*).

Hand-held weapons are no different. For any hand-held weapon the *effective range* equates to the distance that the tip of the weapon can be thrust forward when the arm (or arms in the case of two-handed weapons) wielding it is at its full extension. The *combat range* of the same weapon equates to the distance that the tip of the weapon can be extended forward, connect with a target, and still have enough extension remaining in the arm for the weapon to be driven into that target.

Open hand martial arts and sports like boxing work on the same principles. In the martial arts, the *effective range* of a punch can be considered to be the length of the arm when it is at its full extension. The *combat range* of a punch can therefore be considered to be the amount of distance required for the punch to reach the target, but still have enough extension left in the arm to be able to deliver sufficient impact energy to the target in order to make the strike effective. In martial arts this distance equates to when the arm is roughly 75 per cent of its full extension.[11] This is because, as the arm extends further and further into the strike, the arm begins to slow. This decrease in speed results in a reduction of the amount of impact energy delivered by the strike.[12] If a punch or kick is delivered with the limb at full extension (i.e. at the very limits of the *effective range*), the limb cannot be extended further and 'penetrate' the target. This is a fundamental principle of open hand combat techniques and explains why practitioners of martial arts are taught to punch or kick *through* a target rather than *at* it. By aiming at a point beyond the intended target, the punch or kick will be delivered before the limb reaches its full extension, and the greatest amount of impact energy will be transferred. It is this principle which allows martial artists to punch or kick through such things as wooden boards, bricks or roof tiles.

As thrusting weapons like the *sarissa* function as an extension of the bearer's arms, the same principles apply. Thus the *combat range* for a weapon like the *sarissa* is when the arms of the person wielding it are at 75 per cent of their full extension. The actual 'reach' of the weapon is increased by the distance between the most forward of the wielding hands and the impacting tip, but the fundamentals of a strike delivered with the *sarissa* are no different to those of any other thrusting weapon, punch, or kick. While it is still possible for soft targets such as unarmoured areas of the body, or even harder targets like shields and armour, to be penetrated up to the limits of the *effective range* of the weapon (in effect making the *effective range* and the *combat range* one and the same), or for targets to be penetrated at distances closer than when the arms are at 75 per cent of their full extension (but not closer than the distance that the tip of the weapon extends ahead of the body before the strike is made), the 75 per cent of extension range provides the most efficient delivery of impact energy and can thus be considered the *optimum combat range* for the *sarissa* (or any other hand-held weapon) (Fig.15).

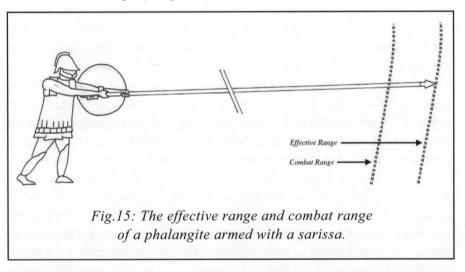

Fig.15: The effective range and combat range
of a phalangite armed with a sarissa.

The limits of the *effective range* cannot be altered beyond the full extension of the arm. While it is possible to hold a weapon further back to gain extra reach, this is usually contrary to the weapon's point of balance and, regardless, the tip of the weapon can still not be thrust any further forward than the distance that the arm can fully extend. Consequently, the limits of *effective range* for any hand-held weapon will always be when the arm is fully extended, and the *optimum combat range* for the same weapon will always be when the arm is at 75 per cent extension.

Similar to many aspects of the dynamics of phalangite combat, the ancient texts offer no help when determining the *effective range* and *combat range* of the *sarissa*. Such detail would have most likely been seen as unwarranted by most ancient writers who were more concerned with discussions of personalities, events, tactics and stratagems than they were with analyzing the fundamentals of certain pieces of weaponry. Similarly, modern scholarship has not fully examined the effectiveness of the *sarissa* as an offensive weapon other than offering generalized statements about the members of the pike-phalanx 'thrusting' during the turmoils of combat. This is an area where processes such as physical re-creation, experimental archaeology and practical testing can fill in gaps within the current knowledge base by putting replica weapons to trial.

In order to examine the *effective range* of the *sarissa*, a series of simple tests were conducted using interested volunteers and re-enactors. While none of the test participants possessed extensive experience with wielding the *sarissa*, the re-enactors owned numerous sets of replica armour (of both the *linothorax* and plate metal type), as well as helmets and shields appropriate for the Hellenistic Period. This allowed for all of the test participants to be kitted out as phalangites in various forms which, in turn, allowed for any influence that different types of armour and headwear had on the employment of the *sarissa* to be observed. Tests were conducted for the three proposed methods of wielding the *sarissa* found in modern scholarship: a low position with the pike wielded at waist level, a couched position with the pike tucked up into the right armpit, and a high position with the pike held above shoulder height as per the use of the pike by sixteenth century pikemen.

A large screen was erected to which a three metre long horizontal scale was attached at a height of 1m from the ground. The scale was calibrated into alternating black and white sections 20cm long to allow differentiations in the movement of the *sarissa* to be easily determined in photographs. Each of the alternating sections was additionally calibrated into centimetres so that accurate distance measurements could be taken by data collectors during the tests. A moveable vertical scale, 2m tall and with the same calibrations was used to measure the range and changes in elevation of thrusts made from each of the tested postures (see following). A line (hereafter referred to as the *zero line*) was also marked on the floor 3m from the beginning of the horizontal scale so that there was enough distance between the *zero line* and the furthest end of the horizontal scale to accommodate the range of the thrusts being examined.

The same weapon (a pike 12 cubits (576cm long) with a point of balance 96cm from the rearward end) was used by all test participants so that any influence created by the configuration of the weapon itself would be standardized across all of the tests. Each participant was tested individually by having them stand at the *zero line* with the *sarissa* held in its 'ready position' for one of the three proposed combative postures. The vertical scale was then moved forward until it touched the tip of the projecting pike. The height of the tip was then read off the vertical scale. The distance that the tip projected forward of the bearer was measured and recorded as the distance from the point where the vertical scale intersected the horizontal scale back to the *zero line*. The movable vertical scale was then removed and the participant extended the *sarissa* forward from its 'ready position' to the fullest possible extension of the arms for the given posture. When full-arm extension was reached, the participant held their position while the vertical scale was moved back in place and new measurements of range and elevation were taken (see Plates 15-17). This 'thrust' was not done at speed as would have occurred under real combat conditions. Rather it was done slowly so that the tests were conducted in a safe environment and so that measurements could be taken. Each step of the process (with the participant in their 'ready position' and with their weapon at full extension) was also recorded with a still camera. Any comments that the participant had regarding the tested posture were also recorded. When the data for the range of a particular technique was gathered, the participant then went through the same process for the two remaining postures.

It must be noted that there are several aspects of the tests which make them differ from actual combat conditions. The ground upon which the tests were conducted was flat whereas any of the major battlefields of the Hellenistic Age can be considered as anything but totally level. The participants were also not jostled by anyone around them as they would have been in the massed confines of a pike-phalanx. Nor were there any moving targets, nor a need to act defensively or commit a strong offensive action as would have been required in a real battle. All of these factors would have undoubtedly affected the overall reach of a phalangite with the *sarissa* under combat conditions. Additionally, the movement of the participant from one test to the next meant that each 'thrust' was not made from the exact same point as the previous one. However, by using the *zero line* as a guide, the overall impact of this last factor was minimized in regards to its effect on the test results.

In regards to the lack of real combat conditions, in a battle any consideration of the true nature of the *effective range* of the *sarissa* would not have occurred to a phalangite. Due to the extensive training of the Macedonian army, a phalangite would have understood what he could or could not do with this primary weapon quite well. The phalangite's main concern in battle would have been effectively engaging his opponent, not understanding the dynamics of how he did so. Yet such knowledge is absent from the modern researcher not physically familiar with the combative techniques of the Hellenistic Age. Subsequently, these tests were designed to provide only the most basic understanding of the *effective range* of the *sarissa* within a safe testing environment.

The results of these tests demonstrate the considerable number of variables which come into play when employing the *sarissa* offensively. If held at the correct point of balance, and with the forward left hand directly above the *zero line*, the amount of a 12 cubit weapon that projected ahead of the bearer should have been 10 cubits or 480cm. However, the results of the tests showed that a varying amount of the weapon projected forward of each participant (Table 8).

Posture										
Participant		A	B	C	D	E	F	G	H	**Avg**
Low (waist height)		500	478	490	497	486	485	500	503	**492**
Couched		520	495	485	487	452	450	515	516	**440**
High (shoulder height)		500	475	495	500	450	495	520	555	**498**

Table 8: The projected distance of the tip of the sarissa in the 'ready position' per posture and participant (centimetres).

These differences can be accounted for with the consideration of a number of factors:
1. Participants may have not been holding the weapon with their left hand at its precise point of balance.
2. Participants may have been standing back from, or slight forward of, the *zero line* or leaning towards or away from it.
3. Participants may have been standing in a way in which, despite holding the weapon correctly and having their left foot on the *zero line*, their leading hand was not above the *zero line* but was either behind it or slightly beyond it.

Despite such factors, it can be seen that in both the low and high positions the pike projected beyond the bearer by an average distance that was close to the estimated 10 cubits (480cm) with a variance of only a few per cent. The weapon projected to a lesser distance when the couched posture was used as when adopting this position the right hand is pulled back in line with the right shoulder rather than being positioned slightly forward of the shoulder as occurs in the low position. This means that the left hand, and the weapon itself, are also drawn further back, resulting in less of the weapon projecting ahead of the bearer. Importantly, as the weapon is drawn back in the couched posture, this also means that more of the weapon projects behind the bearer. However, in a packed phalanx adopting an intermediate-order of 96cm per man, the man behind would be pressed in close to the back of the man in front and, as a result, the pike could not be drawn back this far. This suggests that the couched posture could not have been used to wield the *sarissa*.

A similar variance in data was also seen in the measurements taken when the *sarissa* was moved to the full possible extension of the arm in each of the combative postures (Table 9).

Posture									
Participant	A	B	C	D	E	F	G	H	**Avg**
Low (waist height)	550	511	550	532	515	520	560	470	**540**
Couched	545	510	525	511	480	530	550	555	**527**
High (shoulder height)	555	506	530	525	505	550	560	570	**538**

Table 9: The projected distance of the tip of the sarissa at full arm extension per posture and participant (centimetres).

As well as the factors which affected the extension of the pike in the 'ready position' listed above, the varied results of the full extension test were the result of a number of additional factors:

1. Some participants remained upright throughout the test while others leaned into the delivery of their strike (compare the participant pictured in Plate 15 – who remained upright for the testing of the low position – to the participant leaning into the strike delivered from a couched position in Plate 16).
2. Due to the level of flex in the lengthy *sarissa* (regardless of whether it is a single piece weapon or in two parts), the tip would move even when it was attempted to hold it in place, and this affected the overall distance of the projection.[13]
3. Due to the length of the *sarissa* and its rearward point of balance, an alteration of the angle of the left wrist of only 5° would alter the pitch of the weapon by up to 42cm. This also affected the overall reach.
4. Some participants attempted to keep their weapon level (regardless of the strike being tested) while others allowed the weapon to follow the path dictated by the natural movement of the arms.
5. The different statures and arm lengths of the test participants meant that 'full extension' for some would be greater for some than for others.

Despite these additional variables, it can be seen that different postures allowed the pike to be advanced a different distance depending upon the posture employed: 48cm for the low posture, 87cm for the couched posture; and 40cm for the high posture. It was observed that the participants using the couched posture tended to lean into their strikes more than in others and this accounts for the greater distance the weapon travelled in these tests. Interestingly, the average amount of distance the pike advanced from the low posture was 48cm. A similar result was recorded by Connolly who, in his own experiments, found that the *sarissa* could not be thrust forward much more than a cubit.[14]

Any thrusting action delivered from either the low or couched positions follows a relatively natural movement of both arms as they extend. For the low position, both arms only need to swing forward to deliver the strike. There is no unnatural contortion of the limbs and the wrists can flex as required which allows the grip on the weapon to be maintained and the weapon itself can be kept level as it is directed towards the target. This results in a method of attack which places little stress on the muscles and joints of the arm. It was also found that there was no impediment to the action of an attack made from the low position regardless of whether the participant was wearing linen or metal plate armour, or no armour at all. The adoption of an oblique body posture also meant that the helmet did not inhibit the action regardless of whether it was open-faced or more enclosed.

Strikes delivered from the couched position similarly follow a natural motion of the arms. The right hand/arm is simply extended forward and the left arm straightens as the weapon extends. The action also places little stress on the arms, but due to the way that both arms are in different positions (the left forearm is vertical with the hand supporting the pike from underneath, while the right arm lies flat upon the shaft of the *sarissa* with the hand grasping the shaft from above), as the arms extend, the joints are forced to move in different ways which begins to place stress on the joints the further the arms are advanced. This does not occur in the low position as both arms are hanging down beside the bearer and both hands grasp the *sarissa* from above. Thus forward motion of the weapon has the same effect on both arms.

Additionally, the act of simply holding a 5kg pike aloft in a couched position does result in the arms becoming fatigued very quickly. It was also noted by many of the test participants that having a shield strapped to the right forearm made getting the pike into position difficult, it inhibited vision once it was in position, it added to the weight that the arms had to hold aloft, and made it harder to maintain the weapon in its extended position while data was recorded. The uncomfortable and inhibiting nature of this posture suggests that it is unlikely to have been used by the phalangite in combat for any considerable period of time. This was also found to be true of the elevated high position used by sixteenth century pikemen (see following).

The shield had other effects on the tests as well. When the couched and high postures were adopted, the way in which the left arm is raised meant that the *ochane* (the shoulder strap on the shield) was not bearing any of the weight of the pike and holding the weapon in these positions was simply a function of the muscles of the arms and shoulders. In the low position, on the other hand, the *ochane* helps support the weight of the weapon. Furthermore, the *ochane* inhibited just how far the weapon could be extended in every combative technique. When adopting the low 'ready position', for example, the *ochane* is pulled taut by the weight of the weapon it is supporting. Due to the way that the *ochane* was a set length of cord, this also meant that the left arm could only be extended so far before the *ochane* would prevent the arm from being extended any further. Importantly, the limited movement of the forearm dictated by the length of the *ochane* meant that even when an attack was made from the low position, the shield would not be totally removed from its defensive position across the front of the body. The angle of the shield (relative to any opponent to the front) is increased as the arm extends, but generally not to such a degree that it would leave the phalangite totally vulnerable. The use of cords of different lengths, on the different shields used by the test participants, also contributed to the variance in the measurements taken as, if the participant used a shield with an *ochane* that was too long for them, this allowed the pike to be advanced further and the shield moved out of its defensive position.

The limiting influence of the *ochane* also illustrates that the side-on posture could not have been used to adopt a close-order formation as some scholars suggest. As noted earlier, some scholars suggest that phalangites adopted an intermediate-order interval of 96cm per man to advance upon an enemy and then, when about to engage, moved into a close-order formation of 48cm per man (see pages 142-154). However, because the *ochane* was a cord meant to support the shield and pike when the phalangite was in an oblique posture, it had to be of a set length for each individual so that it would pass over the right shoulder but still be taut enough to support the shield and pike. Yet, if the phalangite moved into a side-on posture to conform to a different interval of the phalanx, because the right arm is extended further away from the body than it is when in the oblique 'ready position', the *ochane* would need to be longer in order to still support the shield and *sarissa* and to allow the shield to be moved into the appropriate position. Due to the way in which the *ochane* was of a set length, this prevents the left arm and shield from being moved into the correct position for the side-on posture. This also impacted on the test results as participants adopting the high posture naturally attempted to move their body into a more side-on position to carry the weapon but found that the length of the *ochane* on their shields prevented them from fully doing so. It seems unreasonable to assume that part of any suggested movement of the phalanx from intermediate order to close order (as is suggested by some scholars) required thousands of phalangites to put down their pikes, take off their shields, and adjust the length of their shoulder straps so that they were long enough to be of use in a side-on position before putting all of their equipment back on and commence fighting. It seems much more plausible that the *ochane* was set to the length for which it would be used the most – when the phalangite was in an oblique body posture and the phalanx was deployed in an intermediate order.

Similar to the observations made by the participants concerning the couched position, it was observed that the shield posed considerable problems when attempting to adopt the high posture with the *sarissa* held above shoulder height. It was found that this technique was very easy to accomplish if a shield was not in place (as per the sixteenth century pikeman), but the presence of the phalangite *peltē* made wielding the pike in this elevated position difficult. The weight of the *sarissa* makes the left hand want to roll back – allowing the shaft to be cupped in the hand and its weight taken by the bones of the forearm – but the rim of the shield prevents this from happening. Instead the weight of the *sarissa* forces the back of the hand onto the rim of the shield where it can be potentially injured. This again suggests that this posture was not used by phalangites in combat. It was also observed that depending upon how the pike was held in the high position, it is very difficult to advance the weapon forward following a natural movement of the arms. Instead the right wrist locks as the weapon moves forward which makes the technique both uncomfortable and inhibits how far the thrust can go. If the weapon is extended beyond this point, the construction of the wrist joint results in the weapon following a downward curving trajectory, rather than advancing directly ahead, and with an even more unnatural contortion of the joints (see following). This results in considerable stress placed upon the right arm, and particularly on the wrist. Some participants got around this by allowing the pike to lower as the strike was delivered. This however, moved the weapon into a position closer to the couched position than to an elevated one (see Plate 17). Other participants moved their weapons away from the body to deliver the strike. While this alleviated some of the stresses placed on the limbs, it also contributed to the variances in the data recorded.

The rigid shoulder sections of both the *linothorax* and the plate metal cuirass were found to inhibit the movement of the arms during strikes delivered from the high position. These restrictions were absent from attacks made from the low and couched positions. This was another contributing factor to the reduced *effective range* of the high thrust. Based upon the analysis of reach, it would seem that any theory that suggests that phalangites wielded their pikes above shoulder height is unfounded.

A further indication that the high position was not used to wield the *sarissa* can be seen in the results of the examination of the trajectory of the three combative techniques that were tested. For the purpose of this analysis, the trajectory of a pike thrust was defined as the change in the elevation of the tip of the *sarissa* between when it was in its 'ready position' and when the weapon was at full arm extension. The trajectory for strikes delivered using each technique shows that each method of using the *sarissa* results in the weapon travelling along a different path (Table 10).

Posture	Variation between the start and finish point of the tip for each posture per participant (centimetres)								
	A	B	C	D	E	F	G	H	Avg
Participant									
Low (waist height)	+10	-2	+5	+35	+5	+5	+15	-1	**+9**
Couched	-5	-30	0	+10	-40	+5	+15	-5	**-8**
High (shoulder height)	-9	-2	-13	+15	-10	-20	+5	+5	**-4**

Table 10: The trajectory of attacks made with the sarissa per posture and participant.

As can be seen there was a large variance in the path that the weapon followed for each of the tested positions across all of the test participants. As with the other tests, these differences were the result of a number of influencing factors:

1. If a participant leaned their body forward as the pike was advanced, this greatly influenced the path that the tip of the weapon followed compared to strikes delivered where the participant attempted to stay upright. This was particularly so for any of the elevated postures (see following).

2. The lengthy *sarissa* has a certain amount of flex in it (see Plates 1, 15-17). If a participant experienced trouble holding the pike in its extended position, the tip could move considerably which made the recording of an accurate measurement difficult.

3. Similarly, due to the rearward point of balance of the *sarissa*, if a participant was experiencing trouble holding the pike in its extended position and was adjusting their grip to try and compensate, only a small change in the angle of the wrist could alter the position of the tip by tens of centimetres at the moment of recording.

4. If a participant attempted to keep the pike level, or if they aimed at a particular point on the vertical scale, this affected the overall trajectory of their strike when compared to participants who allowed the weapon to travel in conjunction with the most natural motion of the arms.

Yet even with a considerable margin of error taken into account, it seems clear that strikes delivered from the low position will generally follow a slightly upward curving trajectory while those delivered from either the couched or high positions will follow a downward path. This is due to the mechanics of the arm in relation to the adopted posture.

From the low 'ready position', both arms are simply swung forward to effect the strike. As the arms are moved forward, the hands (and the pike they are holding) will naturally rise. Due to the limited amount of distance that the arms can be advanced because of the limitations imposed by the *ochane*, the amount of change in elevation is also restricted. If the *ochane* was not in place, the arms could potentially be swung all the way up to shoulder height, with the arms fully extended and the flex of the wrist allowing the weapon to be kept level (see Fig.15). This would result in a considerable change in the elevation of the tip of the pike from the 'ready position' to full extension of the arm. However, the limited distance that the arm can be extended prevents the arms and the pike from being raised too far. This resulted in an average increase in the elevation of the pike tip of only 9cm. The wrists could still be used to keep the weapon level if needed or, if a higher target presented itself in combat, the angle of the wrist could be adjusted to direct the tip towards that target (see Accuracy and Endurance when Fighting with the *Sarissa* from page 207). The small swinging motion of a low thrust did not cause any contortion of the joints or discomfort and the majority of participants commented on the ease of this kind of offensive action.

The downward movement of the arms during the delivery of strikes made from the more elevated positions, on the other hand, meant that the tip of the *sarissa* descends as the arms are extended forward. This motion continues until the joint of the wrist locks which, unless the body is leaned into the strike and/ or the weapon is consciously kept level, causes the weapon to follow a more accentuated downward trajectory until further forward motion is prevented by the length of the *ochane*. This means that the further a phalangite using one of these postures was from a potential target, the more difficult it would be for the weapon to be aligned with the target. Targets at the very limit of the weapon's *effective range* would be almost impossible to hit. Consequently, both the reach and the trajectory of strikes made from these elevated positions, and the negative influences of elements of the panoply on the delivery of such strikes, make such attacks very ineffective combative techniques and these are unlikely to have been used by the phalangite in combat.

Additionally, the natural and comfortable movement of the arms when employing the low technique makes it very easy to withdraw the weapon once an attack has been made. For the low technique, the arms simply have to be swung back to the 'ready position' – following the same path as the strike, but in reverse. This motion causes no muscular stress or contortion of the joints and is very easy to accomplish in preparation for the delivery of another strike. Attacks made from either of the elevated positions, on the other hand, are much harder to withdraw due to the contortion of the wrist and the downward path that the *sarissa* has followed which then has to be retraced to bring the weapon back to its 'ready position'. The ease or difficulty of retracting a committed weapon would have been a considerable concern for a phalangite whose weapon had become imbedded in an opponent during the course of a battle.

What the varied results of the examination of the reach and trajectory of attacks made with the *sarissa* demonstrate is that many of the more generalized statements made by modern scholars about the thrusting actions of the pike-phalanx are fundamentally correct.[15] With no one standing in front of him to inhibit his movements, a phalangite in the front rank of the phalanx could easily lean in to an attack (or 'lunge' as Pietrykowski refers to it) to gain better reach with his weapon. Alternatively, the front rank phalangite could also take a step forward with his attack, so long as sufficient room was available, to add body mass behind the thrust and drive the attack home, to attempt to push an opponent back, or try to move an opponent's shield out of the way (see The Penetration Power of Attacks made with the *Sarissa*). If, on the other hand, an opponent was standing not much further than the distance that the *sarissa* projects beyond the bearer when held in its 'ready position', or if that opponent was even pressed up against the tip of the *sarissa*, the phalangite could simply hold his position and jab at his target (as per Worthington) without further movement of the body until a more opportune target presented itself (see The 'Kill Shot' of Phalangite Combat from page 183).

Dickinson, citing the heroic duel between Dioxippus and Coragus/Horratus, suggests that the Macedonians considered the *sarissa* so manoeuvrable that it was considered adequate for single combat.[16] Regardless of whether this is true of the Macedonian mindset or not, the ease of withdrawing a strike delivered from the low position, and the ease of committing the actual attack itself, further suggests that this was how the *sarissa* was employed in combat. With no restrictions to movement, the use of this low position to wield the *sarissa* gave the front rank phalangite considerable offensive flexibility.

Through physical testing it can be concluded that the phalangite must have engaged an enemy by wielding his *sarissa* in a low position with the weapon held at waist level. Adopting an oblique body posture to wield the pike in this way allowed the shield to be kept in its defensive position across the front of the body, no undue stress was placed on the muscles or joints of the arm, and the ease of delivery and withdrawal of a committed attack made this the most efficient means of acting offensively within the massed confines of the pike-phalanx. Importantly, this massed nature of Hellenistic formations meant that it was only those in the front rank who were physically able to direct thrusts at an opponent on the battlefield. The test results also confirm the unlikelihood of the phalangite using either a couched or elevated posture to wield the *sarissa* as these methods cause the arms to become quickly fatigued and cause considerable contortion of the wrists and arms. The use of the low technique to wield the *sarissa* subsequently affected the basic mechanics of the phalangite in action – in particular which parts of an opponent's body the pike could be offensively directed towards.

Chapter Seven

The 'Kill Shot' of Phalangite Combat

How did a phalangite employ his *sarissa* during the actual fighting of an engagement in the Hellenistic Age? What target areas, or 'kill shots', did the phalangite aim at on his opponent? Such questions themselves raise a number of other issues about the nature of phalangite combat. For example, were the major battles fought by Alexander the Great such one-sided affairs as the reported casualty statistics suggest and, if so, what part did the weaponry play in making these engagements so decisive? Were the phalanx versus phalanx encounters of the Age of the Successors little more than two formations of men armed with lengthy pikes simply hammering away at each other, or was phalangite combat something more 'technical'? The identification of, and the examination of, the 'kill shot' of phalangite combat not only takes steps towards answering such questions but also provides a better understanding of the fundamental mechanics of the warfare of the Hellenistic Age. However, before any examination of the 'kill shot' of phalangite combat can begin, an understanding of two basic principles – how much of the phalangite was protected, or at least covered, by his panoply (this was especially important when the phalangite was engaged against another pikeman) and how much of an opponent a phalangite could actually see, and therefore target, when wearing his helmet – must be developed. The details of how well the phalangite was protected and how much he could see can then be compared to other sources of evidence to form the basis for the comprehension of the phalangite 'kill shot'.

THE PROTECTION PROVIDED BY THE PANOPLY

Unfortunately, the ancient sources provide no direct information as to how much of the body was protected by the armour of the time. This can only be determined using observations and calculations drawn from physical re-creation. A man 168cm tall and weighing 82kg presents a target area of approximately 6,241sq cm when not wearing any armour but with his lower legs and feet protected by the 'Iphicratid' style boots of the Hellenistic Age (Plate 18). This amount of exposed body area is then significantly reduced by the other elements of the phalangite's panoply and the stance adopted to wield the *sarissa*.

One of the main sources of protection for vital areas of the body was the helmet. The all-encompassing Corinthian style helmet of the Greek Classical Age, for example, provided a high level of protection – leaving only 181sq cm of the head exposed by the openings for the eyes and the split down the front which facilitated breathing, speech and, to a lesser extent, ventilation.[1] This represents a reduction in the vulnerable surface area of the head by 75 per cent. In the fourth century BC, design advances to the Corinthian helmet, such as the incorporation of openings for the ears, improved hearing but did not reduce the amount of protection given to the rest of the head against attacks coming directly from the front. The elongated cheek flanges of the Corinthian helmet alone, which did not dramatically alter in size or shape from the fifth century BC to the fourth, more than adequately cover the throat and neck from the front while the rest of the head is protected within the rest of the helmet.

Similarly, Hellenistic Era helmets which possessed large sweeping cheek pieces which covered the neck, throat and lower face provided excellent protection to the head of the wearer. The so-called Masked Phrygian style helmet, for example, with its embossed cheek guards/face-plate which covers the lower face, neck and throat, leaves only the 77sq cm around the eyes and nose, and 44sq cm of the lower neck, between the lower edge of the face-plate and the upper edge of the armour, exposed. The more open area around the eyes on a Phrygian or Thracian helmet, which lack the nasal guard of the Corinthian, allows more air to circulate within the helmet, thus improving ventilation, and gave the wearer a natural range of vision (see below).[2] The openings for the ears common to the Phrygian-style helmet, as per the later types of Corinthian helmet, also improved hearing without significantly compromising the overall level of protection given to the face. This then agrees with the generalisation made by Warry who states that helmets developed to improve ventilation, hearing and vision without a reduction in the amount of protection provided.

Helmets like the Attic or Chalcidian, the cheek pieces of which did not sweep around the front of the face to protect the neck and throat, while increasing the amount of ventilation experienced by the wearer, came at the cost of leaving approximately 209sq cm of the face and neck exposed. This equates to an increase in the amount of area of the head that is unprotected compared to the Corinthian, Phrygian and Thracian styles of helmet. Other than the improved ventilation of the Attic and Chalcidian helmets, there was no improvement to hearing or vision in comparison to the Thracian/Phrygian or the Corinthian styles used in the fourth century BC. Thus it seems that the choice to wear an Attic or Chalcidian-style helmet would have been a conscious decision to offset protection against style and comfort.

For phalangites who may have had a basic *pilos* style helmet issued to them by the state, such aesthetic considerations were not an issue. While basic in design, easy to make en masse, and certainly less costly than more elaborate styles of helmet, the *pilos* had the greatest disparity in terms of the amount of ventilation given compared to the protection provided. Unless augmented with the attachment of cheek guards, which then give it very similar protective qualities to the Attic/Chalcidian or even the Thracian and Phrygian-style helmet, the basic conical *pilos* left the neck, throat and face of the wearer totally exposed. This would have allowed for a great amount of air to circulate around the head, but left many vital areas vulnerable to attack (see following). If the phalangite was just wearing the *kausia* as some scholars suggest, then the head would have been given no protection at all unless it is assumed that the cloth from which this item of headwear was made was able to resist impacts from weapons like swords, spears, arrows and other missiles. Thus enclosed helmets like the Corinthian, Phrygian and Thracian gave the best protection but at the expense of reduced ventilation, while more open helmets like the Attic, Chalcidian and *pilos* provided greater ventilation at the expense of protection.

The phalangite was also protected by body armour. The bronze 'muscled' cuirass which seems to have been worn by some phalangites, and which covers the body from collar bone to waist, further reduces the exposed surface area of the chest and torso to the 420sq cm region around the lower abdomen and groin and the 1,103sq cm of the arms. The lower parts of the body would have been protected to some extent by the addition of a set of leather or linen *pteruges* to the panoply, suspended from the lower rim of the cuirass. Similarly, the composite *linothorax*, with its incorporated sets of *pteruges* and double-layer protection for the shoulders, would have also provided the wearer with coverage from collarbone to just above the knee

– leaving only 1,002sq cm of the arms and 1,020sq cm of the legs exposed between the boots and the *pteruges*.

When greaves, which the poet Alcaeus states provided protection against missiles, were worn to cover the shins, and in some cases the knees as well, the vulnerable areas of the legs were similarly lessened to the areas exposed above the greave and below the *pteruges*.[3] If the high 'Iphicratid' style boots common to phalangites were worn without any greaves, as often seems to have been the case based upon the way they are depicted in Hellenistic art, these would have provided coverage to the lower legs from just below the knee and a certain level of protection for the feet. While possibly able to stop, or at least limit, the amount of damage received from missile impacts, it is unlikely that the leather of the Iphicratid boot would have withstood the impact of a strong spear thrust. However, the fact that these boots seem to have been regularly worn without the additional coverage of a greave suggests that the boots in themselves could provide a certain level of protection against some impacts (or that the lower legs were infrequently targeted areas of the body). Thus a phalangite wearing a panoply incorporating an enclosed type of helmet, such as the Thracian or Phrygian, *linothorax*-style body armour and Iphicratid boots had only 2,143sq cm of their body unprotected by their armour. This is a reduction in the amount of exposed areas of the body of 4,098sq cm or 66 per cent (Plate 19).

The protection of the phalangite in combat was also augmented by the positioning of his shield in a defensive position across the front of his body. A phalangite bearing a *peltē* with a diameter of 64cm in an oblique body posture, with the shield supported on the left arm, and wielding a *sarissa* in both hands at waist level, greatly reduces the amount of exposed flesh further still. Carried in this manner, the shield completely covers the left arm, the lower throat, and adds an additional layer of protection to the chest and abdomen. This removes a further 501sq cm for the left arm, and 44sq cm for the lower throat, from the amount of area that is not protected by some form of defensive armour.

The adoption of an oblique body posture to wield the *sarissa* also places the right leg to the rear and positions it well back from any strike which would impact with the shield, effectively removing a further 510sq cm from any calculation of exposed area. The use of an oblique posture to wield the *sarissa* at waist level additionally places the majority of the right arm in a protected position behind the shield and to the rear beside the body, removing the surface profile of the right arm (approximately 501sq cm) from the amount of body area directly exposed to attacks coming from the front. It is only the forward areas of the hands, extending beyond the protective covering of the shield in order to help wield the hefty pike, that are the most vulnerable part of the arms. However, the exposed hands still offer only a small target profile of around 45sq cm in total. Thus a phalangite wearing the full panoply, with the shield supported on the left arm and wielding the *sarissa* in a low position, has only the 77sq cm of the face, 45sq cm of the hands, and approximately 510sq cm of the left leg exposed. This equates to a total exposed area of only 632sq cm, or just over 10 per cent of the entire body (Plate 20).

How much protection was awarded to the phalangite by the various elements of his panoply was also very much dependent upon the type of opponent he was facing. If the phalangite was facing an opponent armed with a short reach weapon such as a sword, axe or even the spear of the traditional hoplite, then his main piece of defensive armament was the *sarissa* itself. For example, a traditional hoplite, who had the best range with his primary weapon of any infantry warrior engaged in hand-to-hand combat in the ancient world except for the Hellenistic pikeman, had a total *effective range* of just over 2m with his spear (the *doru*).[4] Yet for the Hellenistic pikeman, even one armed with one of the shorter variants of the *sarissa* 12 cubits (576cm) in length, the pike projected ahead of him by around 4.8m even when the weapon was merely held in its ready position (See The Reach and Trajectory of Attacks made with the *Sarissa* from page 167). Thus the easiest way for a phalangite to defend himself against such an opponent was to press the tip of the pike into the shield of the man opposite him and use the length of the weapon to keep the opponent at a distance where he would not be able to bring his own weapons to bear. This is exactly what the Macedonians accomplished at the battle of Pydna in 168BC. Plutarch relates how the Macedonians used their pikes to hold the Romans at a distance where they were not able to engage with their short swords.[5] Thus when fighting against an enemy armed with a short reach weapon, the phalangite's primary offensive weapon also became his main defensive asset. With opponents unable to reach him under such conditions, the defensive attributes of the phalangite's shield, armour and helmet (regardless of whether it was of a more exposed open-face style of not) were somewhat secondary, and only really needed to protect him against enemy missile fire or in the event of the phalanx breaking up.

The reduction of the size of the shield carried by the phalangite, reduced from the 90cm diameter hoplite *aspis* to the 64cm diameter *peltē* as part of the reforms of Iphicrates in 374BC, therefore posed no risk to the phalangite if the phalanx was facing opponents with short-reach weapons so long as the formation remained intact and with its serried rows of pikes presented forward. The all encompassing Corinthian-style helmet common to the Classical hoplite which, along with the offset rim of the hoplite's *aspis* provided a double layer of protection to the neck, was also not a mandatory requirement for phalangite warfare due to the reduced chance of the phalangite being struck in the throat by an enemy melee weapon. It is because of these two basic facets of phalangite combat that more open-faced styles of helmet, such as the *pilos*, the Attic/Chalcidian and the Boeotian, were able to be used by phalangites so readily – especially in the early years of the Hellenistic Period when the enemies faced by the armies of Macedon were all armed with weapons that were easily outreached by the lengthy *sarissa*.

On the other hand, if the phalangite was facing another pikeman – such as occurred during the numerous battles of the wars of the Successors, the face and head was placed in considerably more danger. In such confrontations, a phalangite wielding a pike which extended ahead of him by 4.8m or more would easily find himself facing an opponent bearing their own *sarissa* with the exact same *combat range* – if not a greater range if the opponent carried a longer pike. It is in such instances that one would assume that protection to the head and lower face would have been paramount to the phalangite as the opponents he now faced were armed with weapons which could easily reach him. Interestingly, contrary to this initial assumption, helmets which provided a high level of protection to the face and neck, such as the masked Phrygian common to the age of Alexander, seem to have been less commonly used in comparison to more open styles such as the variants of the *pilos*. This in turn suggests that the regions of the neck and face were not regularly targeted areas in phalangite combat, particularly in the pike versus pike encounters of the Age of the Successors, otherwise a level of protection for these areas would have been retained. This in itself suggests that the main targeted areas on a phalangite, particularly in the later Hellenistic Period, were parts of the body other than the head.

THE AVAILABLE VISION OF THE PHALANGITE IN COMBAT

When bearing a panoply of equipment which included a shield, body armour and helmet, how much of an opponent could a phalangite actually see when engaged in combat? Did this in any way affect what part of an opponent a phalangite would aim his pike at? Most helmets available to the Hellenistic phalangite were relatively open-faced (such as the *pilos*, Attic or Chalcidian) or had rather large apertures for the eyes (the Thracian, Phrygian and Corinthian of the fourth century BC). Julius Africanus states that, when wearing the basic *pilos*, 'the face is uncovered and an unencumbered neck allows for an unhinded view of everything'.[6] Africanus similarly states that 'the vision of the combatants [within the pike-phalanx] was unobstructed through the use of the Laconian *pilos* in the Macedonian army'.[7] However, how much the Hellenistic man-at-arms could see when wearing an open-faced helmet, and how this in turn dictated his actions on the battlefield, has not yet been examined (until now).

To examine the range of vision available to a phalangite a simple test was conducted. A test participant, wearing a replica masked Phrygian style helmet, faced a horizontal scale attached to an erected screen while extending a 12 cubit (576cm) *sarissa* forward from the 'ready position' with the pike held at waist level. This placed the participant at a 'pike thrust' length of just over five metres from the screen. The gaze of the test participant was directed beyond the tip of the *sarissa*; to where an opponent would be located during an engagement. While the participant maintained this line of sight, markers were moved in from both the left and right until they could be seen at the margins of the participant's range of vision, as provided by the Phrygian helmet, without the participant averting their eyes from the point directly ahead. The angle from the markers to the centre of the face was then measured and combined to determine the overall range of vision.

The results of this test showed that, even with the enclosed protection of the Phrygian style helmet, a phalangite would still possess a natural range of vision of approximately 90° (45° to either side of the axis of the head) – the same as if not wearing a helmet at all. Furthermore, it was found that with the adoption of an oblique body posture to wield the *sarissa*, the body naturally rotates slightly to the right, but the head can be turned so that the axis of the head (and the head's line of sight) is directed towards the tip of the *sarissa*. With the body in such a position, the rotation of the head is stopped by the lower rim of the cheek guards connecting with the pectoral area of any armour being worn (see Plate 12). However, due to the rotation of the body and the angle of the lower rim of the cheek guard, this occurs when the head is looking directly towards the tip of the *sarissa*. As Africanus states, if a helmet with no cheek guards, such as the basic *pilos*, is worn, then there is no impediment to the rotation of the head at all other than the natural range of movement of the neck.

This further demonstrates that a side-on posture could not have been used to wield the pike. When standing in this way, and wearing a more enclosed helmet, the head is prevented from being fully rotated to the left by the features of the helmet and so the phalangite's vision is directed to his front-right, rather than towards the tip of his weapon (Fig.16).

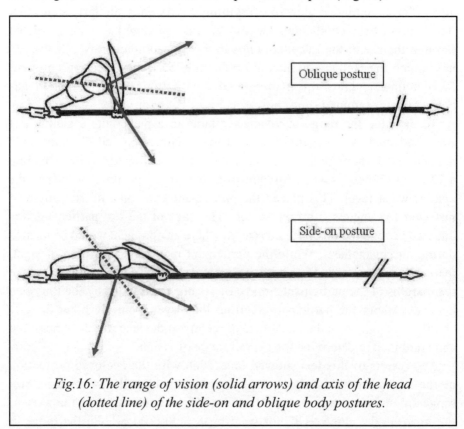

Fig.16: The range of vision (solid arrows) and axis of the head (dotted line) of the side-on and oblique body postures.

The range of vision provided by the adoption of an oblique body posture allows a phalangite to see, at a 'pike length' distance of around five metres, approximately 10m across the front rank of an opposing formation without either turning their head or averting their gaze from the point directly in front of them. This equates to the ability for the phalangite to see at least ten men across the front of an enemy formation if they were deployed, like the pike-phalanx, with an intermediate-order interval of 96cm per man or even more if the enemy formation was arranged in a closer order (Fig.17).

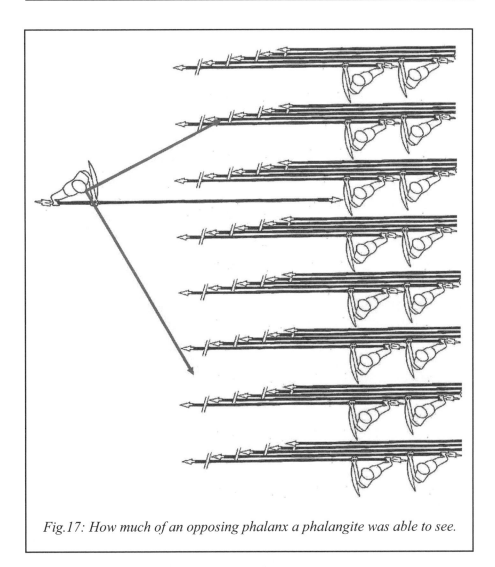

Fig.17: How much of an opposing phalanx a phalangite was able to see.

With the shield supported on the left forearm, the upper rim, which sits level with the left shoulder, in no way impedes this range of vision or the ability for the phalangite to look downwards. The visor of the Phrygian style helmet only marginally impedes the vertical range of vision unless the head is moved, but as an offset the visor shields the eyes from rain and sun, while still allowing for objects to be seen up to a height of 5m when they are a 'pike length' away.

It was found that the range of vision when wearing a replica fourth century Corinthian style helmet was the same as for its Phrygian counterpart. The nasal guard, a feature of the Corinthian helmet that is absent from many other styles, does not impede vision due to its close proximity to the face. The upper edge of the eye sockets of the Corinthian helmet only marginally impede the vertical range of vision, but also allows objects to be seen up to a height of 5m at a 'pike length' distance without moving the head while still providing a modicum of shielding against rain and sun. As the Corinthian is the most enclosed of the Greek helmets of the Hellenistic Period, it can be considered the 'worst case scenario' in terms of possible impediments to vision. Yet the wearer of such a helmet still possessed a natural range of vision. Tests were also conducted while wearing reconstructed versions of open-faced *pilos*, Boeotian and Chalcidian helmets. It was found that the range of vision available when wearing these helmets was the same as being bare-headed (due to the open-faced nature of the helmet) or when wearing either the Corinthian or Phrygian-style helmet. This means that a phalangite wearing any of the styles of helmet common to the Hellenistic Age would have had a natural range of vision and would have been able to see every part of an opponent, from the top of their head to their feet, without moving his head or redirecting his gaze when the two were standing a pike length apart.

A phalangite would have been able to register movement just beyond the margins of this range of vision, but would not have been able to recognize distinct features without turning his head or averting his gaze. Due to the slight rotation of the head to the right caused by the adoption of an oblique body posture, the phalangite would have seen slightly more to his right than to his left. This in no way endangered the phalangite as his left was covered, both physically and visually, by the man beside him in the formation unless he was at the very extreme left of the line. Even here, peripheral vision to the left would have only been required to recognize flanking manoeuvres by enemy troops and would not have been vital to the engagement of an opponent directly to the front.

This conclusion does not take into account atmospheric impediments to sight on the battlefield such as smoke, fog, dust, sun glare or precipitation.[8] However it can be concluded that, despite such hindrances, all styles of helmet available to the Hellenistic phalangite provided a sufficient range of vision for him to adequately gauge what was going on around them on the battlefield. More importantly, these same helmets allowed the phalangite to observe all potential target areas, armoured or otherwise, on their opponent at a 'pike length' distance of 5m or more.

ANCIENT REFERENCES TO THE 'KILL SHOT' OF PHALANGITE COMBAT

Attempting to correlate how well the phalangite was protected, and how much of an opponent he could see on the battlefield, with the concept of a primary target area in phalangite combat (or even if there was one) is problematic due to the very limited amount of available source material on this subject. Unlike the Greeks of the Archaic and Classical ages, both the Greeks and Macedonians of the Hellenistic Period have not recorded the same level of detail regarding armed conflicts as they had previously done. This dearth of information is consistent regardless of the medium employed.

In the art of the Archaic and Classical periods, combat and conflict, regardless of whether it was mythological, heroic or contemporary in its context, was a common illustrative motif in paintings, vase decorations and monumental sculpture. Within these images are found depictions of weapon impacts and wounds to various parts of the body ranging from wounds to the extremities (such as the vase illustration of Achilles bandaging the wounded left arm of Patroklos [Berlin F2278]) to injuries sustained to the chest, abdomen and upper thigh (the image on a *krater* showing the death of Sarpedon [NY 1972.11.10]) to hoplites piercing the armour of opponents (the image of a hoplite killing a Persian with a spear thrust to the body which penetrates the armour [NY 06.1021.117]). This rich corpus of artistic representations provides many insights into the dynamics of combat in the earlier ages of Greek history.[9]

However for the Hellenistic Period, this design element seems to almost completely disappear from the artistic record. There are very few representations of the phalangite within the art of the time, and those that do exist are primarily in the form of funerary *stele* and other grave monuments which generally show the individual being commemorated in a pose other than a combative posture (for example see Plate 8). Consequently, these representations are of little use in the examination of the broader mechanics of the fighting undertaken by the Hellenistic man-at-arms. The one clear depiction of the phalangite in action that does exist, the Pergamon Plaque, despite its combative context, also does not depict a moment of impact between the two opposing sides shown in the relief as the tip of the pikes wielded by the phalangites are obscured by other figures within the image. As such, this pictorial representation can only be used as a basis upon which to build an examination of the phalangite 'kill shot' by extrapolating where a weapon held in the manner shown on the plaque would point at an opponent standing directly opposite.

Additionally, there is little evidence available for the results of phalangite engagements in the archaeological record either. Details for the *polyandrion* at Chaeronea, the supposed burial site of the Theban Sacred Band, for example, was said to have contained many skeletons bearing signs of wounds when the site was excavated in the nineteenth century.[10] Such evidence would have provided a wealth of information about phalangite combat. Unfortunately, the forensic analysis of the remains does not seem to have been a priority for the excavators, photographs of the finds have never been published, nor has the initial excavation report by Stamatakis, and the site has not been re-excavated since.[11]

It is also not only remains of actual combatants from the Hellenistic Period that we have a lack of archaeological evidence for. During the time of the Peloponnesian War (431-404BC) the practice of dedicating arms and armour taken from a defeated enemy to religious sanctuaries seems to have gone into decline in Greece – possibly due to not wanting to glorify a conflict in which Greeks were fighting against fellow Greeks.[12] The depiction of combative scenes in Greek art also declines around this time, possibly for similar reasons.[13] Consequently, by the time of the Hellenistic Period, there are no examples of pieces of armour, which may bear the marks of impacts made with the *sarissa*, dedicated to cult centres like Olympia and Delphi. This also hampers the examination of the 'kill shot' of phalangite combat. Due to the lack of artistic and archaeological evidence for the nature of conflict in the Hellenistic world, any analysis must primarily base itself on other forms of evidence – in particular the literary record.

Yet even here the accounts for the Hellenistic Age are much more limited than they are for the earlier Archaic and Classical Periods. As far back as the works of Homer, Greek writers seem to have gone out of their way to emphasize the bloody nature of fighting involving hoplites. The poetry of Tyrtaeus, the narrative histories of Herodotus, Thucydides and Xenophon, and the plays of Aristophanes, Sophocles and Euripides, all contain graphic references to wounds and fatalities suffered at the hands of the hoplite's primary weapon – the *doru*. However, much like every other source of evidence available for the Hellenistic Period, the narrative histories recounting the events of the time contain a significantly reduced number of specific references to injuries inflicted by the *sarissa*. Instead, the ancient narratives mainly provide massed casualty statistics at the end of an account of a particular engagement – mainly because the focus of most narrative histories is on the key personalities of the time (who usually commanded cavalry rather than infantry), and/or because the sources used by the narrative historians, even if eye-witnesses to the encounters, were generally not part of the infantry engagement.

This is not to say that there is a complete lack of references to the actions of the phalangite in the ancient literature. In fact, while somewhat limited in number, the ancient literary record does contain a number of key passages which refer to the use of the *sarissa* against specifically targeted areas on an opponent on the battlefield depending upon the tactical circumstances and requirements of the engagement. Plutarch, for example, describes how, at the battle of Pydna in 168BC, the Macedonians pressed the tips of their *sarissae* into the shields of the Romans in order to keep them at a distance where they were unable to strike the phalangites with their shorter swords.[14] Later, in the same engagement, opposing troops who charged against the phalanx suffered great losses as the pikes are described of having the ability to 'pierce those who fell upon them, armour and all, as neither shield nor breastplate could resist the force of the *sarissa*'.[15] Livy similarly states that had the Romans made a frontal attack against the phalanx, they would have impaled themselves.[16] Earlier, in 327BC, Alexander's pikemen engaged a contingent of lightly armed mercenaries and 'thrust their *sarissae* through the *pletae* of the barbarians and pressed the iron point into their lungs'.[17]

Such references indicate a primary use for the *sarissa*, across the entire Hellenistic Period, where the weapon is held so that the tip points directly at the shield/chest of an opponent standing directly opposite. This correlates with the depiction of the phalangite in action shown on the Pergamon Plaque. Holding a pike at waist level, and using an oblique body posture to provide strong and stable footing, braces the weapon in position so that it can be more effectively used to hold an opponent back by pressing the tip into the opponent's shield (as per the battle of Pydna). A weapon supported in this manner is also wielded with sufficient strength so that a charging opponent will impale themselves upon the presented weapon. Additionally, a pike wielded at waist level can also be used to deliver a strike with sufficient power to penetrate the shield carried by an enemy and continue on into their body (as per the accounts of Alexander's men in 327BC). Such uses of the *sarissa* cannot be accomplished with a weapon held at shoulder height and/or carried in a side-on posture (see The Reach and Trajectory of Attacks made with the *Sarissa* [from page 167] and The Penetration Power of the *Sarissa* [from page 225]).

Furthermore, any weapon held at shoulder height will have the spike of its butt pointing directly at the face of the man in the rank behind. This would be extremely dangerous as any impact against the presented pike, such as an opponent charging upon it, runs the risk of forcing that weapon backwards with the result that the member of the second rank could actually be killed or severely injured by the weapon held by the man in front of him. This aspect alone indicates that it was unlikely that the *sarissa* was presented from shoulder height to engage in combat.

The only time in which a *sarissa* may have been held higher than waist level would have been if the phalangite was standing in an elevated position. Diodorus describes how Ptolemy, defending the Fort of Camels in 321BC, stood on the wall of the fortification and put out the eye of an attacking elephant with a *sarissa*.[18] Similarly, at Thermopylae in 191BC, pikemen deployed along the top of a wall used their weapons to strike downwards at those who were trying to scale it.[19] At Raphia in 217BC men fought with pikes from small towers mounted on the backs of elephants.[20] In such positions, it would be more likely that the *sarissa* was held higher than waist height as the more elevated deployment of the weapon would allow the phalangite to strike downwards at a rather acute angle, over crenellations and other protection, in order to engage targets that were below him. This would not be possible with *sarissae* held at waist level. However, it is important to note that a phalangite either on a wall or on the back of an elephant would not be part of a phalanx and, as such, his actions would not be dictated by the confines of that formation and he would be free to use the best means to attack his opponents. The weapons (in particular the butt) of phalangites in such positions would also pose little threat to those behind them (if indeed there was anyone positioned there). While the use of the *sarissa* in an elevated posture would have been the easiest form of offensive action for troops in elevated positions, it is unlikely that such a mode of deployment was used within the phalanx itself and weapons deployed as part of the standard battleline would have been wielded at waist height. Deployed at this level, the pike could then be used to either deliver an offensive strike to the chest, or used defensively to hold an enemy at bay by pressing the tip into the opponent's shield.

According to the ancient texts, at other times the *sarissa* was aimed at a more elevated target area of an opponent's body – the head. The references to surviving wounded visiting the Asclepion at Epidaurus, with wounds sustained from strikes delivered with the *longche*, which seems to be another term for the *sarissa*, suggests that the pike could be aimed at the face.[21] Curtius similarly describes how, at the battle of Issus, Alexander's phalangites, who were experiencing stiff opposition from the Persians as they tried to cross the river, 'were so densely packed that the wounded could not withdraw from the front, as on other occasions, and so they could only advance by killing the man in front of them'.[22] In order to quickly dispatch their opponents to expedite their advance, Alexander's phalangites directed their attacks at the faces of those arrayed against them.[23] Despite what this passage and the inscriptions from Epidaurus state, it is important to note that because Curtius goes out of his way to emphasize that Alexander's troops had to resort to striking at their opponent's face in order to advance suggests that strikes to the head were not the standard battlefield practice for phalangites.

Another indicator that the face and head were not generally targeted in phalangite combat is the number of references to either 'walking wounded' or to individuals suffering numerous wounds, in the ancient texts. Many of these passages are in relation to people who had faced a pike-phalanx but survived. At Chaeronea in 338BC, the Athenian dead had 'wounds all over' (*adversis vulneribus omnes loca*).[24] At Halicarnassus in 334BC, many of the Greek mercenaries facing Alexander's assault suffered 'frontal wounds' (τραύμασιν ἐναντίοις) while others were carried away from the battle unconscious.[25] Diodorus similarly says that the Spartan king, Agis III, received 'many frontal wounds' (πολλοῖς τραύμασιν ἐναντίοις) when facing the pike-phalanx of Antigonus at Megalopolis in 331BC, was carried from the field, but was still able to defend himself to the last.[26] Some 4,000 men of Antigonus' army, wounded at the battle of Paraetacene in 317BC, were sent to a nearby town.[27] Walking wounded from the battle of Sellasia in 222BC returned home.[28] At Pydna, the Roman Marcus, son of Cato, received 'many wounds' (τραύμασιν ὤσαντες).[29] Livy states that the number of Roman wounded at the battles of Magnesia (190BC) and Pydna (168BC) were somewhat greater than the number of fatalities he claims the Roman army suffered.[30]

What suggests that many of these reported injuries were not wounds sustained to the head or neck is that the region of the face, neck and throat contains many vital areas such as the jugular vein, the carotid artery, the thyroid cartilage and the jugular arch, as well as the eyes and many other nerves and ancillary blood vessels. When wearing any of the open-faced styles of helmet common to the Hellenistic period, all of these features are left exposed. Injuries sustained to any of these areas are incapacitating at best – if not immediately fatal. For example, a relatively deep penetrating wound to the eye, even delivered with nothing but a finger, can pierce the thin layer of bone at the back of the orbital socket, enter the brain, and kill the recipient instantly.[31] The severing of the carotid artery with a hit from a sharp implement like a pike-head will result in unconsciousness after approximately five seconds and the recipient of such an injury will be beyond medical help after about ten seconds – with death occurring shortly thereafter.[32] The crushing of the thyroid cartilage, through even a reasonably light blow which may not even break the skin, will cause the windpipe to swell and cause a suffocating death in a matter of moments.[33] A stronger blow which pierces the trachea spills blood into the windpipe and makes breathing impossible due to a reflex gagging action. A victim of this kind of injury will choke on their own blood in a matter of moments.[34] The sensitive and vital nature of the neck was an aspect of human anatomy well known to the ancient Greeks as far back as the Archaic Age. In the *Iliad*, the Trojan prince Hektor is killed by a wound to the throat, an area of the body described by Homer as a place 'where a wound is most quickly fatal' (ἵνα τε ψυχῆς ὤκιστος ὄλεθρος).[35]

Due to the rapid lethality of most wounds to the neck and throat, the references to walking wounded found in the ancient accounts of Hellenistic battles can only be descriptions of individuals who have suffered an injury to another part of the body. The unconscious combatants who were pulled from the front line at Halicarnassus in 334BC, for example, are most likely to have suffered other wounds – such as a blow to the head which was strong enough to knock out the individual, but not strong enough to pierce the helmet, or wounds to the chest which have caused the recipient to pass out due to shock and/or blood loss. It is unlikely that the unconscious condition of these wounded soldiers was the result of a wound to the neck or throat (despite the fact that many such injuries do cause a loss of consciousness) as the short amount of time between unconsciousness and death in such cases would make it more likely that such combatants would be classed as casualties rather than unconscious individuals as is stated by Diodorus. Those wounded that either returned home under their own power, or were transported to another place, are even less likely to have suffered an injury to the neck or throat.

Despite this, undoubtedly armies who faced the Hellenistic pike-phalanx suffered casualties – whether quick kills to the face or neck, or incapacitating (if not fatal) wounds to the chest.[36] Diodorus states how corpses piled up at the battle of Chaeronea in 338BC.[37] Many combatants fell on both sides of the pike-phalanx versus pike-phalanx clash (φαλαγγομαχεῖν) at Paraetacene in 317BC – a struggle which had gone on for 'a considerable time' (ἱκανὸν μὲν χρόνον).[38] A great mound of bodies and equipment is similarly said to have littered the field following the battle of Pydna in 168BC.[39] All of these passages show that the pike-armed phalanx could inflict considerable damage among those who faced it.[40] By correlating these descriptions to the other passages from the ancient texts which refer to a specifically targeted area in phalangite combat, aspects of the mechanics of these engagements can be better understood.

For example, the fact that the battle of Paraetacene is described by Diodorus as lasting 'a considerable time' suggests that the pikes employed by both sides were not primarily aimed at an area on an opponent such as the head which would result in a quick kill. If the *sarissa* was primarily used in such a manner, and almost every hit resulted in the death or incapacitation of a combatant on one side or another, it is hard to comprehend how a battle of pike-versus-pike could last so long as to warrant such a description. This, in turn, suggests that the pikes were being used in a manner that did not cause immediate injuries – most likely by pressing them into the shield of the man opposite to limit the advance of the enemy formation as occurred at the battle of Pydna. During such a struggle, pikes would have undoubtedly slid off some of the shields they were pushed into – either upwards into the opponent's face, or downwards into their legs. At other times, if the opportunity presented itself, a phalangite would have exploited an opening in an opponent's defensive posture or armour to direct a strike at a vital area of the body that was the least protected and which would result in a quick kill – such as the head as the Macedonians had done at Issus, or the chest as they had done at Pydna.

The *sarissa* was an effective penetrating weapon and could easily pierce the different types of armour common to the Hellenistic Period (see The Penetration Power of the *Sarissa* from page 225). Thus the chest, if the opportunity presented itself, was an easily targeted area on anyone facing a pike-phalanx. This is clearly illustrated by the account of the mercenaries fighting for the Persians who were faced by the Macedonians in 327BC, whom Diodorus states received wounds to the chest from weapons that passed through their shields. Diodorus also states that many of the Persian dead at the battle of Issus were not wearing any armour at all which suggests strikes delivered to the chest of lightly-armoured opponents.[41] Consequently phalangite combat, at least from the perspective of the 'kill shot', does seem to have been a rather technical exercise rather than simply having two opposing formations of heavily-armed men hammering away at each other with a frenzied storm of undirected blows. Rather, it seems that the *sarissa* was used, first and foremost, to hold an enemy formation in place. As an engagement progressed, opportune shots against available and specific target areas like the chest or head would have been exploited to wear an opposing side down depending upon the armament of the opponent, while other elements on the battlefield came into play to bring the encounter to its climax (see The Anvil in Action from page 375).

This then provides another reason for why there is a distinct lack of references to injuries and fatalities inflicted by the *sarissa* in the ancient texts – basically because the *sarissa* is primarily not a strike weapon. Rather, it is used more as a defensive implement to hold an opponent back until an opportunity to kill him is presented. Thus there is not a specific 'kill shot' in phalangite combat as such. Instead the *sarissa* is used in a defensive manner until an opportunity presents itself where it could then be used in an offensive manner. Based upon the length of the *sarissa* and the manner in which it was wielded, the most opportune target areas on an opponent would have been firstly the chest and secondly the head.

Again, it seems unlikely that the pike was wielded at shoulder height or in a couched position. Weapons carried in such ways point directly at the throat of an opponent standing directly opposite. If the two opposing sides were only a 'pike length' apart (i.e. the opponent was abutted up against the *sarissa* when it was held in its 'ready' position), any strike delivered from shoulder height would almost always impact the face or neck – most likely resulting in the slaying of any opponent who did not deflect the blow with his shield. This then goes against most references to impacts made by the *sarissa* against lower areas such as the chest and shield and against references to walking wounded. While it is possible to allow the tip of an extended pike held at shoulder level to dip, and so point at the chest and shield, the very nature of the stance and position used to deploy the pike in such a manner makes it almost impossible for it to account for the details of pikes penetrating shields, breastplates and flesh or of the weapon being used to hold an opponent at bay. Additionally, a weapon held at shoulder height but angled down towards the shield becomes something of a redundant strike weapon as, due to the length and weight of the pike, and the muscular fatigue caused by holding it in an elevated position for any length of time, it is very difficult to quickly alter the pitch of the *sarissa* and exploit opportune shots to the head or neck.

If the two opposing sides are separated by more than a 'pike length', and the combat is reliant upon individuals thrusting at their target(s), wielding the pike at shoulder height also presents problems. The accounts of walking wounded who had suffered impacts to the face would suggest that the pike was held level with this target area. However, whereas repositioning the tip of the pike from shoulder height to press into the shield of an opponent is unlikely, a *sarissa* held at waist level can easily be angled slightly upwards and still be an effective weapon against an opponent's head – especially if he was wearing a more open faced style of helmet (see following). Furthermore, the slightly upward swinging trajectory of a thrust executed from the low position automatically means that any attack made will rise towards the head of an opponent as the attack reaches the full *effective range* of the action (see The Reach and Trajectory of Attacks made with the *Sarissa* from page 167). On the other hand, attacks made from an elevated position will follow a downward curving trajectory and therefore move the tip of the pike away from pointing at an opponent's face to strike the shield or chest (Fig.18).

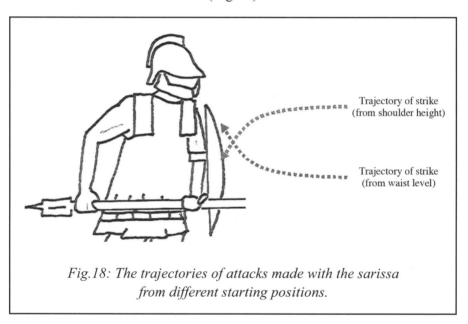

Trajectory of strike
(from shoulder height)

Trajectory of strike
(from waist level)

Fig.18: The trajectories of attacks made with the sarissa from different starting positions.

Thus weapons wielded at waist level can easily account for the two main target areas for which there are descriptions (the shield/chest and the head) whereas weapons held at shoulder height can only account for the head shots when both sides are abutted against each other. This then suggests that the *sarissa* was normally wielded for combat by carrying it at waist level.

WHY AIM AT THE CHEST AND THEN THE HEAD?
The simple fact is that chest/shield of an opponent was the easiest area to target because this is what the *sarissa* points at when it is held in its ready position. By keeping the pike lowered in this position, it could be used to press into an

enemy's shield to keep him at bay, be used to probe his defences with small jabs to try an move the shield out of the way and then, should an opportunity present itself, a stronger thrust could be delivered to the area of the chest. Importantly, at no stage in this process does the grip on the weapon change, nor is the weapon elevated or its pitch altered. This results in a method of combat that is the least taxing on the muscles and joints of the arm which would then allow the phalangite to engage for an undetermined period of time.

Another point to consider is that the chest of an opponent is simply a much larger target area than the head – it is the 'centre of the seen mass' to use modern military parlance. An alteration of the angle of a weapon pointed at the head by only a few degrees would be enough to make the strike miss the target. However, a weapon aimed at the chest and torso, with its pitch altered to the same degree, would still most likely strike the opponent – albeit not in the desired area and possibly not in a vital part of the body. The required alteration of the pitch of a weapon held at waist height to strike at the head – particularly a heavy weapon like the *sarissa* – also takes a brief moment of time. Weapons pointing naturally at the chest can simply be jabbed forward quickly as a target area presents itself. However, adjusting the pitch of the weapon, accounting for any flex that this movement has caused, and then thrusting upwards towards the face, all adds to the amount of time needed to deliver the strike. While this added time can be measured in fractions of a second, it may still be enough time for an opponent to realize what was happening and meet the strike with his shield. Thus strikes made towards the chest, especially if the opponent's shield has been forced out of the way and/or they are off balance, are the quickest and most efficient attacks to make with the *sarissa*.

At other times, again if the opportunity presented itself, the head would be targeted. Discounting 'accidental' strikes which were the result of pikes sliding up a shield into the opponent's face, there would have been several other reasons for why an 'opportune kill-shot' of phalangite combat would have been an attack to the head. One of the main reasons would have been that the area of the neck and head is the most vital area of the body that is close to where the tip of the *sarissa* is pointing when it is deployed in its 'ready' position or when it is being used to hold an enemy back. Should a phalangite need to dispatch his opponent quickly, as occurred at Issus, then an area that would result in a quick kill would be targeted. In this instance the phalangite would need to alter the pitch of his weapon by around 5° in order to make the strike. With the forward left hand acting as the fulcrum for the weapon, its deployed angle can be adjusted either by pushing downward with the rearward right hand (which then raises the tip), raising the left hand and keeping the right hand in place (which also results in raising the tip of the weapon), or a combination of both. While placing a small amount of muscular stress on the arms, this was simply the cost of dispatching an opponent quickly. However, the cumulative effects of doing this too often suggest that the head was not regularly targeted. Weapons held at shoulder height, on the other hand, require the application of significantly more muscular force to alter the pitch rapidly. At no time when altering the pitch of a weapon held at waist height is the phalangite's field of vision taken away from looking towards the tip of his weapon and the whole of the opponent remains visible while the pitch of the *sarissa* is adjusted and the strike delivered.

Additionally, very little force is needed for a bladed weapon, such as a pike-tip, to penetrate the flesh of the neck or the soft tissue of the eye to a fatal depth. Many of the vital blood vessels of this area are only millimetres below the skin. Some vital areas, such as the thyroid cartilage, can be fatally damaged with a blow so light that the skin is not even broken. It requires the application of only 2ft lb of pressure to cut flesh. Significantly more energy than this can be produced with a thrust made with a *sarissa* (see The Penetration Power of the *Sarissa* from page 225). This makes the regions of the neck and face quite susceptible to serious or fatal injuries. However, it is unlikely that this region was continuously targeted in combat as keeping the pike pitched upwards for a protracted period of time begins to make the arms fatigue much more quickly than simply holding the weapon level. Consequently, while a vital and vulnerable area of the body, the sheer weight of the *sarissa*, combined with how it was wielded at waist height, made it impractical for the region of the head to be the main target area in combat and attacks would only have been directed at the head sporadically. References to those who had suffered wounds to the jaw or cheek from the *longche* but survived can be considered accounts of those (lucky?) few who had received blows from this weapon to an area of the head that does not contain any vital blood vessels.

What further suggests that the head was only an 'opportunity shot', rather than the primary target of phalangite combat, is the nature of the protection given to this area. If the head and neck were regularly targeted in phalangite combat, then the use of open-faced styles of helmet like the Attic, Chalcidian and *pilos* would have been completely redundant – especially during the Age of the Successors when fighting was often pike-phalanx against pike-phalanx. In the warfare of Classical Greece, the neck was doubly protected behind the elongated cheek flanges of the common Corinthian helmet and the offset rim which ran around the circumference of the larger hoplite *aspis*. Yet in Hellenistic warfare, these two elements of protection to the head almost totally disappear. This suggests that the head was not the primary strike area for a phalangite in combat. The use of open-faced helmets and the reduction of other forms of protection for the neck and head also shows that the shoulder height position to wield the pike suggested by Smith could not have been used. In the confines of a phalanx, any *sarissa* held in this manner would have its spiked butt pointing directly at the face of the man in the rank behind. It is unlikely that such a comprehensive mode of warfare would develop where the weapons being used posed just as much danger (if not more so) to the rearward ranks of the formation as they did to the enemy.

However, it does seem that the Macedonians did recognize that the head could occasionally be struck as this would account for the continued use of certain styles of helmet which provided facial protection such as the so-called 'Thracian' helmet or the masked Phrygian helmet. The use of these helmets shows that, while the head/neck does not seem to have been the main targeted area on an opponent, it was clearly recognized by some that the area was both vital and vulnerable and therefore required some degree of protection.

Another point for consideration in relation to the reasoning behind a head 'kill shot' is that due to the small amount of penetration depth required to inflict a fatal injury to the area of the head or neck, the weapon used would be less likely to become lodged in the opponent if a quick kill was what was sought. If a bladed weapon is thrust into the chest, the elasticity of the skin causes the flesh to close around the penetrating weapon. Suction caused by the moist tissue of the body then causes drag on the blade and inhibits the extraction of that weapon. This is why modern soldiers are taught, once they have thrust a bayonet into an enemy, to twist the rifle (and therefore the bayonet as well) before attempting to extract the blade. By twisting the bayonet, the wound is 'opened', the suction is relieved, and the blade can be more easily withdrawn.[42] Such techniques were also known in the ancient world. The Roman writer Vegetius, for example, advises Roman legionaries to thrust the tips of their short swords only 2in into the body of an enemy.[43] Vegetius goes on to describe that, no doubt due to the width of the blade of the Roman *gladius*, a penetrating wound delivered to this depth was generally fatal. However, another benefit of such a shallow wound is that there would have been less surface area of the blade susceptible to the suction of the opponent's flesh, making the sword easy to recover.

Livy describes the horror experienced by Philip V and his army upon seeing the wounds inflicted by the *gladius* for the first time.[44] Livy states that Philip V's men were accustomed to the sight of wounds inflicted by spears and arrows due to their prior campaigns against the Greeks. What scared these men the most was the sight of dismembered and disemboweled bodies which had been the result of attacks made with the Roman short sword.[45] This then suggests that the Greeks, and probably the Macedonians as well, did not regularly target an area of the body, or strike it to a depth, which would cause such injuries. The head of the *sarissa* was also smaller than the width of the average *gladius*. Thus the majority of the fatalities that were the result of actions involving phalangites were likely caused by relatively shallow strikes to a vital area of the body, either the torso or the neck and head, which did not leave wounds of a significant size.[46]

Moreover, the Roman *gladius*, being a relatively short weapon, could also be twisted (much like the modern bayonet) to aid extraction. However, attempting to undertake a similar practice with a pike 5m in length or longer is somewhat more problematic. Due to the length and weight of the *sarissa*, it is difficult to apply enough rotational torque with the wrists to sufficiently rotate a blade embedded in a target. Additionally, as the *sarissa* seems to have come in two halves, applying any rotation to the rear half, which was gripped in the hands, would have little effect other than to make the rear end rotate within the coupling sleeve if the forward half was held rigidly in place. Consequently, it is impossible to twist a pike with such a configuration and so 'open' a wound on an enemy to help withdraw the weapon.[47] This again suggests that softer and/or thicker parts of the body such as the torso, and even fairly rigid pieces of armour like shields and breastplates, were not regularly struck deeply with the *sarissa*. This then corresponds with the concept of the *sarissa* being primarily a defensive weapon to help hold an opponent at bay unless they were very lightly armoured.

Furthermore, in the confines of a deployed phalanx there would be little or no room for a phalangite in the front rank whose weapon had become embedded in an opponent to step back and use his body weight to try and dislodge it. Due to the difficulties involved with attempting to recover a lodged weapon, it can only be assumed that enemies who impaled themselves on opposing pikes through the momentum of their charge, as occurred at the battle of Pydna, subsequently rendered those weapons completely useless as there would have been no means for a phalangite to extract a weapon which had sunk deeply into an opponent's chest or abdomen, and who had then fallen to the ground when they were slain. Under such circumstances a phalangite would have had little recourse except to resort to using his sword as a secondary offensive weapon if he was unable to fall back to the rear of the formation.

For opponents facing a pike-phalanx, one way to inhibit the offensive actions of the phalangites would be to allow them to press the tips of their pikes into your shield (assuming that it was sturdy enough to withstand the force of the weapon). By doing so, the tip would penetrate the wood of the shield to a shallow depth. This would then hold the entire pike in place and so prevent both accidental deflection into either your head or legs, and prevent the opponent from altering the pitch of their weapon to deliver an opportune strike. Thus an effective shield for engaging a pike-phalanx had to be able to allow very shallow penetration, but be sturdy enough to withstand the pressures placed upon it. To counter this, the phalangite would have pushed his pike forward, attempting to drive the opponent back so that the tip of the *sarissa* could be withdrawn from the surface of the shield, and a quick strike to the opponent's chest or head could be delivered while they were off balance.

A review of the evidence, combined with an understanding of the fundamentals of Hellenistic combat, allow for the questions over the concept of the main 'kill shot' of phalangite warfare to be addressed. It seems that the long *sarissa*, wielded at waist level, was primarily used to hold an opponent in place, preventing him from reaching the phalangite with a shorter reach weapon like a sword or spear. In engagements where a pike-phalanx fought against another pike-phalanx, these same initial principles applied with both sides using their weapons to keep their opponents at bay. During the course of the encounter the phalangite would have continued to use his weapon in the same manner – pushing forward with the weight of his body and the weapon to both hold the enemy at bay and to probe his defences by slightly adjusting the position of the pike to try and move the opponent's shield out of the way or force them back. When this occurred, certain areas of the opponent's body would become exposed for a brief moment. At this stage the phalangite could either direct an attack towards the opponent's chest which would require no alteration of the pitch of the weapon and would be slightly quicker to deliver, or take a short moment to alter the pitch of the weapon and strike at the opponent's head, to either kill or incapacitate him.

Chapter Eight

Accuracy and Endurance when Fighting with the *Sarissa*

If a phalangite managed to push an opponent's shield out of the way with his *sarissa* during combat, or if an opportune target presented itself for some other reason, could he have hit the desired target area? Was phalangite combat a frantic series of ill-aimed shots with most of the fatalities the results of luck, or was it more the result of the skill of the combatant with his primary weapon? Additionally, how long could the actions and physical exertions of pike combat have been maintained by the individual? The examination of these inter-related aspects of infantry warfare in the Hellenistic Age demonstrates that even an inexperienced conscript in the Macedonian army would have possessed enough skill to hit a small opportune target most of the time. This then suggests that with the extensive training that the soldiers of Hellenistic armies underwent from the time of Philip II onwards, seasoned phalangites would have been able to deliver attacks with even more accuracy, and for a considerable length of time. This, in turn, partially accounts for why the Hellenistic pike-phalanx was able to dominate the battlefields of the ancient world for so long.

All modes of combat, whether hand-to-hand, employing missile weapons, or even unarmed, are reliant upon two inter-related and inter-dependent criteria. The first is the ability to strike a desired target area on your opponent with the weapon being employed. The second is to engage in fighting using a method that does not cause undue fatigue so that both the offensive and defensive actions required to gain victory can be carried on for as long as possible (and hopefully longer than your opponent). Once a combatant begins to suffer from the effects of fatigue, the ability to use any weapon with a certain degree of efficiency is dramatically reduced. Thus what will determine the effectiveness of any combative action on the battlefield is a combination of the stamina of the combatant and his proficiency with his weapons.

Due to the dearth of detail in the literary and artistic record relating to the finer points of phalangite combat, modern research into the warfare of the Hellenistic Age has made no attempt to analyse how accurate a phalangite would have been with his primary weapon. This is in part due to techniques such as physical re-creation still being somewhat in their infancy in terms of research methods for the examination of ancient history. Subsequently, it is in areas such as this that physical re-creation and practical testing is able to fill in a gap in the knowledge base by investigating an aspect of ancient warfare that can be explored by no other means in the modern world.

Similarly, the ancient narratives and other texts are of only limited use when examining the levels of endurance that a phalangite was required to have to effectively engage in combat. Yet the idea that stamina played an important role in warfare was a subject well known by the Greeks. As far back as the Archaic Age, the poet Homer espoused that a man is unable to fight beyond the limits of his strength.[1] Pausanias states how a lack of sleep was a deciding factor in the fall of Eira during the Second Messenian War in the early seventh century BC, and Diodorus later recounted how fatigue was a contributing factor in the surrender of the Italian Greeks to Dionysius in 389BC.[2] Despite such statements, there are no references in the literary record of the Hellenistic Age (or in the Classical Age for that matter) which specifically details how long a warrior of the time could maintain the actions of combat. Rather, what the ancient texts do contain are often vague references to how long a phalangite battle lasted, and it is from these passages that any conclusion about the endurance of the Hellenistic man-at-arms must be drawn.

Diodorus, for example, states that the battle of Chaeronea in 338BC lasted for 'a long time' (ἐπὶ πολὺν χρόνον), as did Alexander's battle outside the walls of Thebes in 335BC.[3] Similarly, the battle of Megalopolis in 330BC is also said to have lasted 'a long time'.[4] At Gaugamela in 331BC, Alexander the Great rose late from his sleep, which suggests the battle may have begun in the early afternoon, and ended at dusk.[5] The clash of the phalanxes at Paraetacene in 317BC is said to have gone on 'for a considerable time' (ἱκανὸν μὲν χρόνον) with the fighting continuing at night under a full moon.[6] Plutarch states that Pyrrhus' confrontation with the Romans at Asculum in 279BC lasted 'a long time, or so it is said'.[7] This suggests that Plutarch was either unaware of accurate accounts of the battle, or was dubious of his sources in regards to the battle's duration. Plutarch's account of the battle of Pydna in 168BC states that this engagement lasted for only an hour.[8] Even if Plutarch's concerns are founded, what these passages demonstrate is that a Hellenistic commander could not be certain how long a battle he was about to fight would last.

What many of these passages also fail to consider is that, as well as the actual fighting, there was a considerable amount of pre-battle manoeuvring that took place before the first blow was even struck. Troops had to arm themselves, get in formation, march as a unit into position, relay orders and instructions, and then wait for the signal to lower pikes and advance. All of this would take place on a variety of terrains across the Hellenistic Period from alluvial mud flats to desert plains to hills and river valleys, and under all sorts of weather and temperature conditions. All of these factors would tax a soldier's energy and subsequently his effectiveness on the battlefield.[9]

Even once fighting had started, prolonged exposure to the environment and the traumas and toils of combat would sap a soldier's reserves of energy. Good commanders, leading well seasoned troops, could use this to their advantage. Frontinus states that at Chaeronea, Philip II intentionally prolonged the engagement as he knew that his men were accustomed to the trials of conflict due to their extensive campaigning while the Athenians, fielding a force of part-time citizen militia, were not. Frontinus goes on to state that when Philip noticed the Athenian forces becoming fatigued, he pressed his attack more vigorously and gained the victory.[10] Good commanders could also recognize when their troops had had enough. At Gaugamela, Alexander the Great put a stop to the fighting because many of their weapons had become blunted, their hands were sore, the troops were exhausted and night was setting in.[11]

What such passages demonstrate is that a phalangite, and more importantly, a commander of contingents of phalangites, had no way of knowing how long the battle he was about to fight was going to last, nor could he guarantee the conditions that it would be fought under or the terrain it was going to be fought upon. Consequently, for the phalangite way of war to be effective, it had to be possible for those engaged in the fighting to employ their primary weapons in an effective way that did not cause undue fatigue for an undetermined period of time and under all possible conditions.

As a person's energy decreases, their actions, both physical and mental, begin to slow. With the onset of fatigue, breathing becomes shallower and more rapid, depriving the muscles of much needed oxygen and causing lactic acid to build up in the muscles which, in turn, slows down the bio-mechanical processes of the body. This effect then begins to compound upon itself. Further action involving these muscle groups weakens them further, making the actions themselves less effective. Cognitive abilities are also impaired by oxygen deprivation and this combines to have a negative impact on the accuracy and efficiency of any form of action (martial or otherwise).[12] Extended periods of physically taxing labour, such as marching or fighting, a lack of sleep and/or minimal periods of rest similarly affect performance.[13] Any such debilitating effect on performance would mean that a phalangite who became fatigued on the battlefield may not have been able to direct his attacks at targets with the same level of accuracy as someone who was not fatigued. A fatigued phalangite may have also not have had the strength to penetrate the target to an injuring depth even if he did hit it, and if his cognitive abilities were low, may not have even recognized a potential opportune target when it presented itself. Thus the phalangite way of war had to incorporate techniques that were both effective on the battlefield and did not quickly fatigue the combatant. The potentially negative impacts of muscular stress and fatigue on the Hellenistic phalangite would have been partially offset by the extensive training that the infantrymen of the Macedonian army underwent.

Training develops skill, confidence, muscular strength and stamina. These were principles widely known to the professional military commanders of the fourth century BC. Iphicrates, for example, stated that 'the untrained lack endurance'.[14] For the Greeks and Macedonians, regular exercise and training in the *gymnasion* would have begun to develop strength and endurance in many who would go on to serve in the military from a young age. Other cultural pursuits, and the manual nature of most labour of the time, would have also resulted in high levels of general physical fitness.

There is little doubt that in the hand-to-hand combat of phalangite warfare both skill and stamina would have played important roles. But was one more important than the other? Aristotle states that mercenaries were better for both offensive and defensive operations due to their experience with their weapons and this resulted in them being like fully armed soldiers fighting against unarmoured opponents if that enemy were non-professionals with lower levels of experience.[15] This would suggest that Aristotle believed that technical skill was an important element in the successful conduct of war. Written around 340BC, not long after he had been hired as the tutor of the young Alexander, it is interesting to ponder just what sort of impact sentiments such as these from Aristotle's *Nicomachean Ethics* would have had on the future king and the Macedonian military in general. Justin states that Alexander chose to keep many of the experienced veterans who had served under Philip in his army as he wanted to campaign with the 'masters of war'.[16] Frontinus similarly states that Alexander conquered the world with only 40,000 men who were accustomed to the discipline that had been taught to them by Philip.[17] This would suggest that Alexander had taken Aristotle's sentiments very much to heart.

Centuries later, the military writer Vegetius would outline several military maxims which hold as true for the earlier Macedonians as they do for later Romans in which he states that skill and training produce courage and that those who fight with skill and courage are more likely to secure victory while an untrained force which lacks courage will either be slaughtered or turn in flight.[18] Vegetius further states (1.3) that those nurtured by a life of hard work under the sun, with a lack of acquaintance with luxury and bath houses, toughened by every kind of toil from wielding iron, digging trenches and carrying burdens, and those who are simple souled and content with little are the best in war.[19] Vegetius additionally adds that training and long-term service makes for better soldiers 'for men who stopped fighting a long time ago should be treated as recruits'.[20] These generalizations about what makes the ideal solider find many parallels with the extensive, and grueling, training undertaking by the Macedonian pike-phalanx.

When Philip II ascended the throne of Macedon he had to quickly replace the large portion of the army which had been defeated by the Illyrians in 359BC. Part of the large scale recruitment was to train the new Macedonian army with 'constant manoeuvres of troops under arms and competitive drills'.[21] Part of this training involved lengthy marches carrying weapons, armour, rations and equipment.[22] Philip's pike-phalanx also trained regardless of the season. Demosthenes states that the Macedonians were 'under arms all the time' in reference to this year-round regime of training and campaigning.[23] This level of training began to accustom the troops (especially the new recruits) to the trials of campaigning, to promote physical fitness, obedience, unit bonding, experience in weapon handling and drill movements, and to shape the army into a combat effective fighting force. This training seems to have been both intensive and rapid as Philip was able to successfully field an army of 10,000 men the following year.

Appian states that Macedonian pikemen became professional due to their long training and their experience in many campaigns.[24] This success was in part due to the initiatives put in place by Philip which became standard practice for most Macedonian armies of the Hellenistic Age. The troops of Alexander the Great, for example, carried their own equipment, and spoils, following the practices of Philip.[25] Rations were also carried by the troops of

Demetrius and Antigonus, and spoils seem to have been individually carried by the troops of Eumenes.[26] The elderly Silver Shields, who campaigned with commanders from Philip to Eumenes, are also said to have acquired great skill and courage from their constant campaigning.[27] Such passages clearly show the privations of campaigning that soldiers of the Hellenistic Age were subjected to as part of the intensive training to turn them into an effective and cohesive fighting force.

Such intense training had to be maintained if the army was to retain its edge and its combat effectiveness. Alexander is said to have kept his men busy with constant training in the use of their weapons and with tactical exercises.[28] If training was not maintained, the result could be disastrous for the army. Diodorus states that the quality of troops degrades if they are kept idle for even a few months.[29] Antiochus allowed his troops to remain idle for an entire winter in 196BC with the effect that, when the spring campaigning season arrived, he found them unfit for any kind of duty.[30] Such professionalism and constant drill is in stark contrast to the troops fielded by many of the Greek city-states of the time. The military institutions of states such as Athens were based upon part-time militias whose members only came out to fight when required and who possessed only limited training with their weapons and in formations.[31] Thucydides for example, in the words of the funeral oration ascribed to Pericles, states that the Athenians preferred to rely on the natural courage of the members of the *polis* than the blind obedience taught by military training.[32] Ducrey argues that this was because the Athenians saw compulsory military training as an infringement on the personal liberties of the citizen.[33] Despite such noble sentiments, it seems clear that a reliance on natural courage over skill and discipline imparted some harsh lessons on the Athenians and they reintroduced the institution of the *Ephebia* (compulsory, two-year, military service for youths) in 335BC following their defeat at the hands of the professional Macedonian army at Chaeronea in 338BC. It is little wonder that Aristotle stated that 'the sanguine are confident in the face of danger because they have previously won many victories over many foes'.[34]

Training brought with it many benefits other than just skill at arms. While the lifestyle of the Greeks, with manually based labour and time spent in the *gymnasion*, would have built up certain levels of fitness and stamina, the training undertaken by the professional Macedonian soldier would have similarly built up strength and a capacity for endurance. Where the difference lay is that the Macedonian phalangite would have built up these attributes in conjunction with their weapon handling skills – and so would have built up the muscular groups required for constant campaigning and efficiency in combat. The Greek hoplite, on the other hand, with the exception of the Spartans and the few professional units of soldiers found in the various city-states of Greece, would have possessed a certain level of fitness, but may not have been as proficient with their weapons.

Regular training and campaigning would have also made the Macedonian phalangite much more accustomed to the traumas of conflict – both physical and psychological.[35] This is the essence of many of the ancient passages relating to the professionalism of the Macedonian military. Acclimatisation to a dangerous environment like the ancient battlefield will affect how the body will physically respond to danger. The more accustomed an individual is to a particular situation, the way that their body reacts to the stress of that environment (i.e. the release of adrenaline and hormones, loss of bodily functions and cognitive abilities etc.) is significantly reduced.[36] Consequently, a seasoned veteran of constant campaigning would be able to better withstand the physical and psychological stresses of war. As Idzikowski and Baddeley observe, cohesive units of soldiers rarely panic.[37] This unit cohesion would have been built through the regular training of the Macedonian phalangite within his unit and the larger formation of the phalanx.

It seems clear that at any one time the Macedonian army would not have been made up totally of experienced soldiers. Right across the Hellenistic Age, from Philip II onwards, positions within the pike-phalanx would have been occupied by new recruits, raw conscripts and others who might not yet have received a 'baptism of fire' on the fields of battle. However, this would have made little difference to the offensive abilities of the pike-phalanx as a whole. It was only the members of the front rank of the formation who were capable of thrusting their primary weapon at an opponent, and the front rank was made up entirely of experienced officers. As such, those who constituted the 'cutting edge' of the pike-phalanx would have been those who were the most accustomed to the trials of campaigning and who were the most proficient with their weapons.

However, no commander could guarantee that the more inexperienced troops, who would have held positions in the more rearward ranks of the phalanx, would not be called on to engage. All manner of variables might come into play on the battlefield to cause the formation to break up, or to result in the loss of the front ranks, which would then call for the rawer members of the phalanx to use their *sarissa* in an offensive manner. Yet despite possibly having undergone extensive training in drill and weapon handling prior to his arrival on the battlefield, the question remains: how good would a phalangite have been at engaging a target with his *sarissa*?

TESTING ACCURARY AND ENDURANCE

How much difference experience and training had on the offensive abilities of the phalangite is an aspect of Hellenistic warfare not covered in any of the extant sources of evidence. An understanding of just how easily (or not) a phalangite could engage a target in a combat situation is essential for the broader understanding of the conflicts of the age. To gain such an understanding, a series of simple experiments were conducted. A group of 'non-professional' phalangites was recruited from among re-enactors and other interested individuals. All of the test participants were kitted out with full phalangite panoplies incorporating different Hellenistic accoutrements ranging from replica plate-metal cuirasses, linen corslets, both open-faced and enclosed helmets, and the *peltē*. Each participant was then tested individually with the same replica *sarissa* so that any influence imposed by the configuration of the weapon would be standardized across all of the tests.

Each participant stood before a target dummy of stacked hay bales 2m in height with a life-sized human silhouette attached to it representing a member of an opposing enemy formation. Two distinct target areas were marked on each silhouette – a 10 x 10cm square in the centre of the chest, and a 10 x 10cm square in the area of the throat. This ensured that all of the participants were aiming at the same targets areas in each test so that the data could be standardized across the collected results. All tests were done with the *sarissa* held in the low position to test the effectiveness of the most likely way the phalangite wielded his primary weapon. Each participant approached the target so that the tip of their pike when held in the ready position was roughly 20cm from the surface of the target. This simulated a phalangite having the tip of his pike pressed up against the shield of an opponent which was presented in a protective position across his front. Each participant then directed five consecutive strikes (hereafter called 'primary strikes') towards the target area on the chest of the silhouette (simulating an 'opportune shot' when the opponent had moved his shield out of the way). When these primary strikes had been delivered, the participant withdrew his weapon while the locations of the impacts in relation to the target area were recorded. The participant then reapproached the target and, from the low position, redirected five more primary strikes at the target area on the throat of the silhouette. The location of the impact of these strikes was also then recorded.

The participant then moved away and engaged in a period of 'simulated combat' (akin to shadow boxing) by 'engaging' an upright post with a pike for a period of five minutes or until they could no longer perform a thrusting action with the weapon due to fatigue. When the time had elapsed, or they had 'fought' for as long as they could, the participant returned to the target and made another five consecutive thrusts (hereafter referred to as 'secondary strikes'), first at the chest, and then at the throat. The amount of time that each participant was able to undertake the period of 'simulated combat', and the placement of their secondary strikes, was recorded for later comparison. The distance of the impact points of both the primary and secondary strikes from the target areas was then measured and analysed. Any comments that the participant made in relation to the test and/or the offensive actions performed, were also recorded. These tests allowed for several aspects of fighting with the *sarissa* to be observed: the accuracy of the primary strikes to two different targets on an opponent, the part fatigue played in the accuracy of subsequent attacks, the amount of time that a phalangite could have maintained offensive action.

It must be noted that there are several aspects of these tests which do not accurately reflect the conditions of the Hellenistic battlefield. The tests were conducted on a hot sunny day of 36°C. However, due to the year-round campaigning of the Macedonian military, Hellenistic battles could have been fought in any temperature. Due to the impact that high temperatures would have on people engaging in strenuous activities and while wearing armour and metal helmets, the day of the testing could be considered a possible 'worst case scenario' yet still one that was environmentally similar to some of the engagements of the ancient world.

As the participants were all 'non-professional phalangites' with little or no experience of wielding the *sarissa*, their level of performance would not be representative of the seasoned Hellenistic campaigner who had undergone extensive training and who may have previously fought many engagements. However, as noted, not everyone within the pike-phalanx would have possessed this level of skill and experience. Consequently, the test participants better reflect the abilities of a raw recruit in the Macedonian army, who was just beginning to learn how to use the *sarissa*, but who suddenly found himself in the front rank of the phalanx during a battle. This would also categorize the results of the tests using these participants under a 'worst case scenario' and the results of the tests can be considered as accurate a reflection of phalangite combat as is possible to replicate in the modern world.

The 'simulated combat' element of the tests also did not accurately re-create a combative environment. It is highly unlikely that a phalangite in the front rank of the pike-phalanx would have been continuously thrusting with the *sarissa* over the course of an engagement as the test participants did. Rather the phalangite would have been using his pike to keep an opponent at bay, to probe his defences and, should an opportune target present itself, then deliver a thrusting attack. However, any fatigue experienced by the test participants would have been partially offset by the lack of other combative aspects of the test, such as a lack of jostling within a massed formation, a lack of fatigue brought on by marching into position or engaging in defensive actions, and the lack of a dangerous environment, where the stresses and fear brought on by combat would sap a phalangite's reserves of energy. The tests were also devoid of other elements of the ancient battlefield such as noise, dust and the detritus of combat underfoot. A testing environment which simulated all of these combative aspects would be impossible to replicate without putting the test participants in real danger. It has been noted that it is impossible to gauge the effects of fear on test participants in controlled, ethical, experiments due to the sheer number of variables involved in such an exercise.[38]

Even the target that the test participants directed their attacks against did not accurately replicate the characteristics of an ancient battle. The human figure on the target was depicted in a front-on position whereas most opponents in a battle employing a shield and a hand-held weapon would have most likely adopted a more oblique body posture. Nor did the target have any area of the body which was depicted as being protected by a helmet, shield or armour. Due to the adoption of an oblique body posture and the use of a shield and armour, both the chest and throat target areas used in these tests may not have been visible on an enemy depending upon the way in which an opportune target presented itself during the varied nature of combat. The amount of time that such target areas may have been visible, as the opponent moved and defended himself, would have undoubtedly influenced how (if at all) a phalangite could have directed an attack against them during the course of a battle. Consequently, it must be conceded that these tests do not present a truthful re-creation of the accuracy and endurance of phalangite warfare but are an approximation of this style of combat based upon a set of standardized experimental criteria and tests conducted with modern best-practice safety standards. Regardless, the results of these tests do provide baseline data for the accuracy of attacks delivered from the low position with the *sarissa* and for how long such actions could be maintained.

THE RESULTS – ACCURACY

It was found that even a 'non-professional phalangite' with limited weapon handling experience would have been quite accurate. The results of the tests showed that primary strikes directed towards the chest of an opponent would have been more accurate that those directed at the throat (Table 11).

PRIMARY STRIKES Distance from target	At chest	At throat
Within target	54%	35%
1-3cm outside target	17%	14%
4-6cm outside target	20%	14%
7-9cm outside target	3%	3%
10+ cm outside target	6%	34%

Table 11: The proximity of 'primary strikes' to the target area.

The greater accuracy of attacks directed at the chest compared to those directed at the throat can be accounted for by a number of factors. When the pike is held in its low 'ready position', the weapon naturally points at the chest of an opponent standing directly opposite. In order to effectively engage the target, all that is needed is to swing the weapon forward while using the flex of the wrists to keep it level. Due to the small amount of distance that the pike needs to travel to connect with the target, the upward swinging trajectory of a low strike has only a marginal influence on the overall path that the tip of the weapon follows and this makes it quite easy for the shot to land on or very near to the target area. As the test results show, 91 per cent of all of the strikes delivered by the various test participants either hit the target or landed within 6cm of the target area. On the torso of an opponent, this means that almost every attack made with the *sarissa* from a 'pike length' of about 5m distance would have hit that opponent in the chest.

Attacks aimed at the throat of an opponent, on the other hand, were found to be a less accurate action. In order to engage a more elevated target such as the throat, a pike held in the low 'ready position' either has to first have its pitch altered so that it will point at the throat (which may potentially alert an enemy to what is coming and give him time to react), or have the pitch altered as the strike is committed. However, due to the length of the *sarissa* and the effect of the rearward point of balance of the weapon, even a slight variation of the angle of the wrists translates to a considerable movement of the forward tip. If the adjustment of the pitch is too great or too small, and if the upward swinging trajectory of the strike (which is more pronounced when aiming at an elevated target from the low position) is not taken into consideration, this makes it much harder to strike at any target area other than the chest.

Due to the smaller overall size of the throat and head compared to the torso, any attack which missed the designated throat target area by more than 6cm ran the risk of missing the target completely. There was found to be a much greater disparity between those whose strikes landed some 10cm outside of the target area for the chest than there were for the throat. In some cases the participant, not adjusting their weapon appropriately, missed the silhouette completely when their attacks were aimed at the throat and their strikes landed above the outline's head. This was the result of the alterations that were required to adjust the pitch of the weapon to point at the throat which then compounded through the actions of the strike with the result that the participant missed. Many of the strikes that landed 7-9cm from the throat target area, hit the silhouette in the ear or grazed the side of the head. While there is little doubt that such injuries would have caused considerable pain and damage in a real contest, they can hardly be regarded as a 'kill shot'. Other strikes which missed the throat target area still struck the head or the chest. While these would have resulted in serious injury, for the purposes of this test, they cannot be considered accurate.

Furthermore, the amount of flex inherent within the lengthy *sarissa* impacted on the overall accuracy of attacks which required the pitch of the weapon to be adjusted. When the weapon was moved as the strike was committed, the tip of the weapon would 'bounce' and this made it harder to land the strike on target. Attacks aimed at the chest, on the other hand, require no adjustment of the pitch of the weapon and so the amount of the flex in the weapon is limited to only that caused by the actions of the strike itself. As this is simply a forward extension of the weapon with minimal rise along the trajectory of the attack, there is little 'bounce' along the shaft and this contributed to making the chest attacks more accurate overall.

This amount of flex in the weapon would have had benefits for a formation of phalangites. As the members of the pike-phalanx moved into action with the pikes of the front ranks lowered and those of the rearward ranks angled over their heads, the simple act of advancing would cause the pikes to flex and bounce. Anyone witnessing this formation coming at them would see thousands of weapons all moving individually. This would add to the psychological impact of seeing the pike-phalanx in action as the entire formation, weapons included, would seem alive. It is no coincidence that some ancient writers ascribed the characteristics of highly animated objects to the phalanx – with Polyaenus referring to the formation as a 'beast', and Plutarch saying it advanced like the waters of a flood.[39] Additionally, as the two sides closed with each other, the flexing pikes levelled at an enemy would make it harder for the opponent to judge where an attack was coming from – thus making it harder to defend against. Thus the very bounce of the *sarissa* acted as a probing action which may have momentarily opened an enemy's defences.

Once engaged, the flex of the *sarissa* also provided benefits. The shaft of the weapon could bend with any pressure exerted against it if the tip was thrust into the shield of an opponent and the two sides continued to push against each other as occurred at the battle of Pydna in 168BC. This would have helped prevent the shaft of the weapon from breaking. If an opponent attempted to hack through the front of the weapon, the flex in the shaft and the rearward point of balance of the *sarissa* would mean that any impact against the forward end would simply force the front of the weapon downwards, negating much of the impact energy of the blow and reducing the likelihood of the attack severing the tip from the weapon.

Interestingly, the results for the secondary strikes showed that there was an overall increase in the level of accuracy following the period of simulated combat (Table 12).

SECONDARY STRIKES		
Distance from target	**At chest**	**At throat**
Within target	60%	42%
1-3cm outside target	22%	26%
4-6cm outside target	9%	9%
7-9cm outside target	3%	3%
10+ cm outside target	6%	20%

Table 12: The proximity of 'secondary strikes' to the target area.

For both target areas the number of strikes that landed either on target or within 3cm of the target increased – from 71 per cent to 82 per cent for attacks aimed at the chest, and from 49 per cent to 68 per cent for attacks aimed at the throat. The number of strikes which missed by more than 10cm for the throat also decreased following the period of simulated combat (from 34 per cent to 20 per cent), while the number of strikes greater than 10cm from the target for attacks aimed at the chest, while not improving, did not increase either. This suggests that the low posture used to wield the *sarissa* that was employed by the test participants in their sessions of simulated combat did not cause a significant level of fatigue and that during even the brief period of simulated combat the participants quickly learned how to compensate for the flex and angle of the *sarissa*, and develop sufficient muscle memory of the action, to make their secondary strikes more accurate.

THE RESULTS – ENDURANCE

The results of the sessions of simulated combat showed that attacks made with a *sarissa* wielded at waist height could be maintained quite easily. Most participants lasted the five minute period of the testing session with very little discomfort. This was due to the action of attacks delivered from a low position simply involving a forward swinging of the arms, with only a minor flex of the wrists if it was attempted to keep the pike level, to the impact point of the strike and back again. This resulted in very little muscular fatigue on the arms as most of the action and power of the strike comes from the rotation of the shoulders. Some participants observed that their hands had begun to ache by the end of the testing period due to gripping the weapon tightly, but this was not enough to prevent the participant from then executing their secondary strikes against the target, and it was further observed that when all participants returned to the targets following the period of simulated combat there was no reduction in the speed of their secondary strikes compared to their primary attacks (some even became more determined to try and thrust the *sarissa* right through the hay bales). Some participants may have grasped the weapon tightly due to their unfamiliarity with it. However, to effectively use the *sarissa* the shaft does not need to be gripped as tightly as possible, but only with enough force to keep it stable and to prevent it being knocked out of the hands if an enemy attempted to parry it. The increased level of accuracy demonstrated by the test participants even after only five minutes of simulated combat shows how the level of effectiveness with a weapon increases as the bearer becomes more used to wielding it and this kind of familiarity with the *sarissa* would have been another result of the extensive training undertaken by the Macedonian phalangite.

It was observed that some participants also made small steps forward and back with each of their attacks during their period of simulated combat or, if they did not step, rocked forward and back with the motion of the strike. This resulted in a natural rhythm being created which allowed the participants to work through any minor levels of fatigue that they may have experienced. Importantly, such motions could only have been undertaken by members of the front rank of a pike-phalanx whose movements were not inhibited by the presence of a man standing in front of them. This again suggests that if any pike fighting was undertaken during a phalangite battle, it was only done by the experienced officers in the front rank of the formation. If the engagment did not require a thrusting action, but simply involved the phalgite holding his pike in a lowered position in order to keep an opponent at bay, then the support for the weapon gained through the use of the *ochane* would mean that phalangites could maintain a defensive combative posture almost indefinately with very little stress placed on the muscles of the arms. Furthermore, a rocking or stepping action is not easily accomplished if the side-on posture to wield the pike that is suggested by some scholars is adopted which also suggests that this posture was not used to wield the *sarissa* (see Bearing the Phalangite Panoply from page 133).

The levels of discomfort experienced by the test participants when the reach of attacks made from either the couched or high positions was tested suggests that such postures, and particularly thrusting actions made from them, could not have been maintained for very long (see The Reach and Trajectory of Attacks made with the *Sarissa*). The amount of muscular stress placed on the arms by simply holding the weighty *sarissa* aloft quickly tires the arms and a phalangite wielding his pike in this manner would have found himself fatigued simply advancing on the enemy with his weapon deployed (let alone doing any fighting). This suggests that both of these postures were not used. Additionally, even the easier motions of attacks made from the low position were found to be taxing on the muscles of the arms due to the weight of the *sarissa* if the actions were performed continuously as they were in the sessions of simulated combat. While many of the test participants were able to perform through these levels of discomfort and fatigue, two of the test participants failed to complete the five minute test period. This would further suggest that the more taxing couched and high positions were not used to wield the *sarissa*, and that phalangite combat was not a series of continuous thrusting actions but was a method of fighting where the pike was simply held in place, holding an enemy at bay, with only occasional attacks directed at opportune target areas being made.

Physiologically speaking, modern humans are little different from those of the ancient world. The main difference between modern and ancient people is that many inhabitants of the modern world are not accustomed to the harsher and more manual lifestyle of the past. As a result it is unlikely that the levels of fitness and stamina in the test participants would be similar to that of a Hellenistic phalangite. Consequently, the results of these tests cannot be regarded as accurate representations of the levels of strength and endurance that would have been found in the soldiers of a Hellenistic army. However, by using the data gathered from these tests, and then viewing them on a comparative basis, it seems that even an inexperienced phalangite wielding the *sarissa* at waist height and aiming his weapon at the chest of an opponent would have been able to engage in combat for a considerable period of time. For any mode of combat to be effective, it has to be maintained for the duration of the battle while causing the least amount of fatigue on those fighting it. This would correlate with the low method of offensive action, one that could be maintained for an indeterminate length of time, with the varied durations of engagements outlined in the ancient literary sources.

It seems that the simulated combat, while not overly taxing on the arms when done from the low position, allowed participants to get used to the feel of the weapon, get their eye in, be able to judge distances to targets and to adjust for the flex of the weapon better when they made their secondary strikes. This would suggest that with further training and exercise as the professional Macedonian phalangite would have undergone, the level of accuracy would have correspondingly increased further. From these test results a number of conclusions can be made:

- If a new recruit with little or no experience at handling the *sarissa* could hit a small target with at least half of his attacks delivered from the low position, and many of those that missed would still impact an opponent's body, a more seasoned phalangite would have had a very high probability of hitting an opportune target if it presented itself on the battlefield and the phalangite had the time to exploit it. This would be particularly so if the targeted area was the chest of the opponent.
- Attacks aimed at areas other than the chest require the pitch of the

lengthy *sarissa* to be altered. While this does not discount the possibility that areas such as the throat and head were intentionally targeted in phalangite combat, the required movement of the *sarissa* to attack an area other than the chest of an opponent make this method of attack less accurate.

- Despite any problems with aiming at an area on an opponent other than the chest, it is interesting to note that the majority of strikes that missed the designated target areas in the tests still fell within an area smaller in size than the phalangite shield or the chest on an individual of average size. This, coupled with the ease of using the *sarissa* when held at waist height, suggests that the upper body of an opponent was the commonly targeted area in Hellenistic pike combat.
- It is unlikely that phalangite combat was a series of frenzied and random attacks made with the *sarissa*. Rather it seems to have been a mode of combat incorporating the best uses of the lengthy pike – to hold an enemy at bay and to occasionally engage him offensively by directing attacks at exploitable opportune targets if they presented themselves.

It is also interesting to consider just how often such thrusting attacks would have occurred over the course of a phalangite battle. If the objective of the pike-phalanx for a particular engagement was to simply present their pikes, press the tips of their weapons into the shields of an opponent, and hold them off until a flanking attack by light troops or cavalry could be executed, then attempting to exploit any opportune target with a thrusting attack would have been something of a secondary consideration. Under such circumstances, as were seen at the battle of Pydna in 168BC, the phalangites of the phalanx simply needed to hold their ground with their pikes lowered and due to the way that the *sarissa* was pressed into an enemy's shield, little thrusting with the weapon could have taken place. Undoubtedly, if a chance to slay the enemy facing him arose, a phalangite in the front rank of the phalanx would have tried to do so. However, if he was unable to kill his opponent this did not matter as, from the perspective of the phalanx, such combats were not encounters of attrition. Consequently, under such circumstances, even an inexperienced phalangite could have functioned effectively within the phalanx as all he was required to do was lower his *sarissa* and keep the enemy at a safe distance. This then raises the question as to how much of the training that is referred to in the ancient texts was the practice of drill and moving in formation, and exercises to build up stamina and endurance, and how much of it was actually learning how to fight with a lengthy pike. From the test results it seems that phalangite combat (and most likely the training for it as well) was a combative system that was very adaptable – not only to the tactical requirements of the situation, but to the skill level of the combatants as well.

Chapter Nine

The Penetration Power of the *Sarissa*

If a phalangite took advantage of an opportunity to deliver a killing blow to the chest of an enemy standing opposite him, could enough power be delivered with a thrust of the *sarissa* to overcome the armour of the time and injure or kill that opponent? The ancient literary sources seem to suggest that this was possible, but how easily this could be accomplished is not detailed. An examination of the penetration power of the *sarissa* demonstrates that this lengthy and somewhat unwieldy weapon possessed the ability to deliver tremendous killing power on the field of battle.

Many modern works which examine the warfare of the Hellenistic Age suggest, albeit sometimes generally, that the members of the pike-phalanx were able to easily dispatch an enemy through thrusting actions performed with the *sarissa*. Warry, for example, states that the pike 'must have given the formation greater thrusting power'.[1] Heckel suggests that the heads of the *sarissae* presented by the phalanx, which he incorrectly equates with the large 'head' found by Andronicos at Vergina, 'sliced through shields and armour like swords'.[2] Similarly, Gabriel states that the *sarissa* had sufficient power to penetrate an opponent's shield and armour.[3] Sekunda, following the premise that the *sarissa* had only a small head, suggests that it was specifically designed for piercing armour.[4] Worthington elaborates further on the penetrative power of the *sarissa* by stating that the head was designed 'not merely to damage an opponent's armour or wound him like a conventional spearhead, but to penetrate the armour and keep going into the enemy's body'.[5]

The ancient sources contain passages which reinforce these claims. At Halicarnassus in 334BC, the Greek mercenaries fighting against Alexander's troops received numerous frontal wounds.[6] This would suggest that the *sarissa* was capable of penetrating the bronze or linen armour commonly worn by the Greeks in the fourth century BC. In 327BC, Alexander's phalangites were able to thrust their weapons through the shields carried by those facing them and further into the opponent's lungs.[7] This suggests that the pike was capable of penetrating shields as well as armour with the one strike. Pausanias states that linen breast-plates were not useful for soldiers as they let the iron [head of a weapon] pass through if the blow is a violent one.[8] If this was true for both spears and pikes, this would correlate with the other statements which suggest the possession of a high penetrative ability by the *sarissa*.

Yet the ancient texts also contain numerous references to those who had faced a pike-phalanx, been wounded, and survived. Such references can be found across the Hellenistic Period from the Medizing Greek mercenaries who were carried unconscious from the battle of Halicarnassus in 334BC, to Agis III receiving 'many frontal wounds' at Megalopolis in 331BC, to Antigonus' 4,000 'walking wounded' following the battle of Paraetacene in 317BC, to the numerous Romans wounded at Magnesia and Pydna.[9] Such accounts suggest that the *sarissa* may not have been able to penetrate an opponent's shield and/ or armour as easily as other ancient passages would suggest. An examination of the penetrative power of an attack delivered with the *sarissa* helps put these passages into context.

In their experiments to determine the penetrative power of various pieces of ancient weaponry (other than the *sarissa*), Gabriel and Metz filmed a man 6ft (183cm) tall and weighing 180lb (82kg) using high-speed strobe photography of ten, thirty and sixty frames per second while using different replica weapons in front of a graduated scale.[10] Measuring the distance that a select point on the weapon (e.g. the tip of a spear or sword) travelled between each frame of the strobe photography allowed for the velocity of each action to be determined in feet per second. The velocity of the strike, and the weight of the weapon tested, was then inserted into the following formula, used by the US Army Ballistics Laboratory, to determine the amount of energy (in foot pounds [ft lbs]) that was delivered with each action:[11]

$$\text{Energy (ft lbs)} = \frac{(\text{weight of the weapon}) \times (\text{velocity})^2}{64 \,(\text{value of the gravitational constant})}$$

The 'killing power' of each weapon was then determined by examining the area of a wound that it would produce by measuring the circumference of the opening made and the size of its impacting edge or tip.[12]

Gabriel and Metz then calculated the amount of energy required for a weapon to penetrate bronze plate armour by conducting baseline tests using a bow and arrow, which is the easiest weapon to use which produces a standardized set of velocities over multiple impacts, against target plates of 2mm thick brass.[13] They concluded that an arrow needed to hit the target plate with around 75.7ft lbs (103j) of energy to penetrate to a 'killing depth' of 2in.[14] The results of this base-line test were then used to calculate the impact energy required for other weapons to penetrate the same target plate to a similar 'killing depth' by determining that, if the size of the impacting edge is the same for two weapons (e.g. the tip of an arrowhead and the tip of a spearhead), and it requires a certain amount of energy for one weapon (e.g. an arrow) to penetrate to a depth of 2in (A) resulting in an opening in the target of a certain size (B), then for a different weapon (e.g. a spear) to penetrate to the same depth but create an opening of a different size (C), that weapon must deliver an amount of energy equal to (A/B) x C.[15] Based upon the results of these tests and calculations, Gabriel and Metz determined that it requires 137.0ft lbs (186j) of energy for a spear to penetrate a 2mm thick metal plate to a 'killing depth', while it would require only 2.0ft lbs (3j) to penetrate flesh, 68.0ft lbs (92.0j) to break any bone of the body other than the skull (which requires an impact with 90.0ft lbs (122j) of energy to be broken), and an impact to the head of between 56.0-79.0ft lbs (76.0-107j) to produce unconsciousness.[16]

Using these same principles, the penetrative power of the *sarissa* was calculated. A test participant 5ft 5in (168cm) tall and weighing 180lb (82kg) was filmed performing attacking actions with a replica *sarissa* in front of a graduated scale at a rate of ten frames per second and while wearing a full phalangite panoply. Five different actions were filmed and the footage was then analyzed using computer software to determine the velocity of the movement of the tip of the pike in each thrust from its 'ready position' to its *combat range* (the most effective point of impact for the penetration of a target). The average recorded velocity for the delivered strikes was 20.9 feet (6.4m) per second. Using the US Army Ballistics Laboratory formula, and the weight of the *sarissa* used in the tests (11.9lb – 5.4kg), the amount of energy delivered with a thrust of the *sarissa* would be 81.2ft lbs (110j). This would suggest that a thrust from the *sarissa* would be incapable of penetrating the target plates used by Gabriel and Metz in their experiments. Yet the ancient literary sources suggest that the *sarissa* could easily pierce both shields and breastplates (and in some cases the body underneath). There are several factors which must be considered in relation to the penetration power of the *sarissa* and, when these are taken into consideration along with the parameters of Gabriel and Metz's tests, the results of these re-calculations demonstrate that the Hellenistic pike was a formidable piece of weaponry.

THE SIZE OF THE HEAD OF THE *SARISSA*

The size of the head on the spear used by Gabriel and Metz in their calculations was bigger than the head of the *sarissa*. The figure given by Gabriel and Metz for the circumference of an opening made by a spear penetrating to a 'killing depth' is 3.6in (9.1cm).[17] This would be the result of a weapon with a head that was approximately 4.5cm across at a point 5cm back from its impacting tip. However, the *sarissa* seems to have been configured with a head similar to that used on the spear wielded by the Classical Hoplite – the head of which was smaller. The average maximum width of the common 'J-style' spear head of the Classical Age is 3.1cm.[18] It also seems likely that the head of the traditional hoplite spear was reused in the creation of the pike as part of the reforms of Iphicrates in 374BC (see pages 89-90). Thus the *sarissa* would have possessed a head similar in size to that of the hoplite *doru*. The head on the replica *sarissa* used in this research had a similar width to that of the 'J style' hoplite spear head – 3.2cm at a point 5cm back from its impacting tip. Such a head would produce an opening of around 6.6cm (2.6in) in circumference when thrust into a target to a 'killing depth' of 2in. A recalculation of the energy required for the smaller head of a *sarissa* to create an opening in the target plates used by Gabriel and Metz with a circumference of 2.6in shows that an attack delivered with a reduced amount of energy equal to 98.9ft lbs (134j) would penetrate the target to a 'killing depth' of 2in. This is still greater than the estimated amount of energy that can be delivered with an attack made with a *sarissa* (81.2ft lbs or 110j) which still suggests that the Hellenistic pike would have been incapable of piercing Gabriel and Metz's target plates. However, there are factors other than just the size of the impacting head that affected how easily the pike could be used as an offensive weapon (see following). Importantly, such considerations demonstrate that the large 'head' found at Vergina, which has a width of 4cm at a point 5cm back from the tip, is unlikely to have come from a *sarissa* unless it is assumed that the pike possessed less penetrative power than the traditional hoplite spear.

Plate 1: Reconstructions of the equipment of the classical hoplite c.400BC (top), and the equipment of the reformed Iphicratean peltast, c.374BC (bottom).[1]

Plate 2: The butt of a Macedonian sarissa - Shefton Collection, Great North Museum, Newcastle upon Tyne #111.[2]

Plate 3: Photo (top) and reconstruction (bottom) of a depiction of a Hellenistic cavalryman from Kinch's Tomb - Lefkadia, Macedonia.[3]

Plate 4: Detail of the Alexander Mosaic (Naples Archaeological Museum #10020) showing Alexander wielding a cavalry lance.[4]

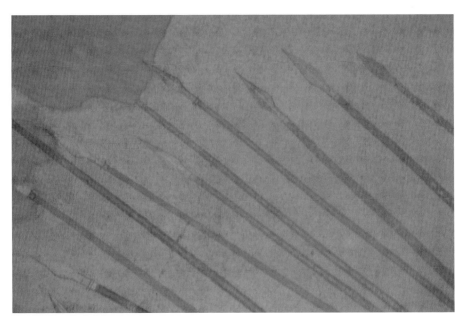

Plate 5: Detail of the Alexander Mosaic showing the heads of Macedonian sarissae.[5]

Plate 6: The terracotta mould for the facing of shields of Ptolemaic phalangites (third century BC) – Allard, Pierson Museum, Amsterdam #7879.[6]

Plate 7: Two phalangites (left) engage enemy cavalry and infantry – line drawing of a bronze plaque from Pergamon.[7]

Plate 8: Stele of Nikolaos, son of Hadymos,
showing a phalangite carrying the peltē.[8]

Plate 9: Different ways of using the ochane: as a wrist strap (top), as a hand grip (bottom).[9]

Plate 10: Replica helmets, headwear and swords of the Hellenistic Age.[10]
Back Row (L to R): Corinthian helmet, Attic helmet, Thracian helmet, Masked Phrygian helmet.
Middle Row (L to R): Cap helmet, rounded pilos helmet, Boeotian helmet, Conical Pilos helmet, Kausia.
Front Row (L to R): Xiphos sword, Macharia sword.

Plate 11: A 'shield coin' of Alexander the Great showing the facing of a Macedonian shield on the obverse side and a crested pilos helmet on the reverse side.[11]

METAL: AE DENOMINATION: Hemiobol
WEIGHT: 3.95g DIAMETER: 15mm REGION: Asia Minor
MINT: Miletus DATE: late 334BC or early 333BC
OBV: Macedonian shield with five sets of concentric crescents and five sets of quincunx dots in an alternating pattern around edge; encircled Gorgon's head in centre.
REV: Macedonian pilos helmet with transverse crest and cheek pieces (facing front); double axe in lower left; K (mint mark) in lower right; INSCRIPTION: B_A (ΒΑΣΙΛΕΩΣ_ΑΛΕΞΑΝΔΡΟΥ = 'of King Alexander'). DIE AXIS: CONDITION: VF SERIES IDENTIFICATION: Price 2064

NOTES: Plant (Greek Coins Types and their Identification (London, Seaby, 1979)) has no reference to a coin of this type in his catalogue. Price (The Coinage in the name of Alexander the Great and Philip Arrhidaeus Vol.I: Introduction and Catalogue (London, British Museum Press, 1991)) does not list a coin of this type with this die axis. Price (p.32) suggests that the Gorgon head can be identified with the head of Fear (Φόβος) which ties this design with the 'shield of Herakles' (see Hesiod, The Shield, 144) and the coinage to other Herakles based issues under Alexander. Coins of this type were found in the Caria Hoard (1986) which confirms the region of their issue. The coin may commemorate the actions of a unit of hypaspists at siege of Halicarnassus in late 334BC.

Oblique posture: can rotate the head to face the front and look directly ahead.

Side-on posture: cannot rotate the head to face the front and look directly ahead.

Plate 12: How body posture dictates the axis of the head when wearing a masked Phrygian helmet.[12]

Plate 13: Phalangite drill – moving a pike angled forwards at 45° (top) to lowered for combat (bottom).[13]

Plate 14: Marching with a sarissa separated into two parts and sloped over the shoulder.[14]

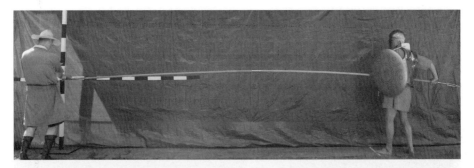

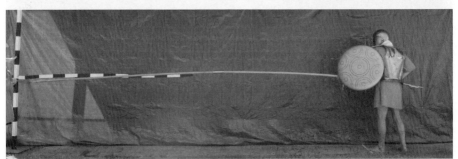

Plate 15: Recording the effective range and trajectory of an attack delivered with a sarissa held at waist height (top: 'ready position'; bottom: at full extension).[15]

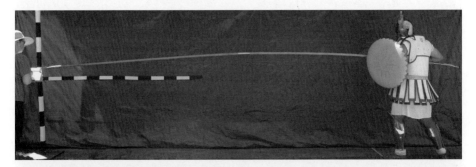

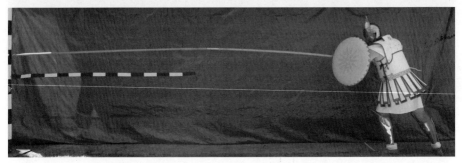

Plate 16: Recording the effective range and trajectory of an attack delivered with a sarissa couched into the armpit (top: 'ready position'; bottom: at full extension).[16]

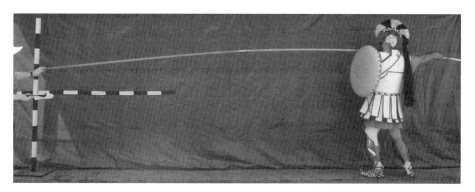

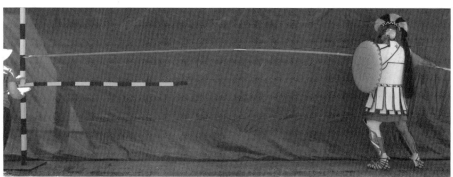

Plate 17: Recording the effective range and trajectory of an attack delivered with a sarissa held above shoulder height (top: 'ready position'; bottom: at full extension).[17]

Head & Neck: 723cm^2

Chest & Torso: 2,477cm^2

Arms & Hands: 1,103cm^2

Upper Legs: 1,938cm^2

Plate 18: The exposed areas on a phalangite when wearing no armour.[18]

Head & Neck: 121cm^2

Chest & Torso: 0cm^2

Arms & Hands: 1,002cm^2

Upper Legs: 1,020cm^2

Plate 19: The exposed areas on a phalangite when wearing a helmet and armour.[19]

Head & Neck: 77cm^2

Chest & Torso: 0cm^2

Arms & Hands: 45cm^2

Upper Legs: 510cm^2

Plate 20: The exposed areas on a phalangite adopting an oblique posture with shield and pike.[20]

THE THICKNESS OF THE TARGET

The bronze plate corslets that may have been worn by the Greeks who faced Philip II's pike-phalanx at Chaeronea in 338BC or Alexander's troops at Halicarnassus in 334BC were not 2mm thick like the target plates used by Gabriel and Metz in their calculations. The thickness of both helmets and armour from the Classical Age averages only 1mm.[19] The linen armour used by the Greeks and Macedonians in the fifth and fourth centuries BC, which may have been up to 10mm thick, possessed at least the same protective abilities as 1mm bronze plate.[20] The thickness of the armour worn by a combatant facing a pike-phalanx will influence how easy it would have been for a *sarissa* to penetrate it. In its most basic terms, the thicker that the armour is, the greater the amount of energy is required to pierce it. The thickness of a specific type of armour can be incorporated into the calculation of the killing ability of the *sarissa* through the inclusion of an 'energy multiplier'. Williams, in his examination of the effectiveness of medieval armour, provides a scale of 'energy multipliers' that can be used to account for the varying thicknesses of different pieces of armour (Table 13):

Thickness of Armour	1mm	1.5mm	2mm	2.5mm	3mm	3.5mm
Energy Multiplier	1.0x	1.9x	2.9x	4.1x	5.5x	7.0x

Table 13: The 'energy multiplier' required to calculate the penetration of armour of different thicknesses. [21]

Based upon the figures provided by Williams, it takes almost three times the amount of energy for a weapon to penetrate a target with a thickness of 2mm than it does to penetrate a target only 1mm thick to the same depth. Thus the amount of energy needed for the *sarissa* to penetrate a 2mm thick target to a killing depth of two inches (98.9ft lbs or 134j) can be recalculated to only 34.1ft lbs (46.2j) needed to pierce the armour worn by some Greek hoplites, who were possibly the best armoured of any of the opponents that the Hellenistic pike-phalanx would face other than the Romans. This is far less than the amount of energy generated by an attack made with the *sarissa* (81.2ft lbs or 110j) which in turn suggests that the *sarissa* was an effective killing instrument.

THE ANGLE OF IMPACT AND THE CURVATURE OF THE ARMOUR

Additionally, both the angle at which the pike impacted with the target, and the curvature of the armour itself (which also determines the overall angle of impact) will have bearings on the ease (or difficulty) that an attack delivered with a *sarissa* could penetrate a target. The arrow impact tests undertaken by Gabriel and Metz to gather their base-line data involved projectiles striking flat target plates at an approximate angle of 45°. This was done to simulate the downward trajectories of lobbed volleys of missile fire and the downward trajectories of swung hand-held weapons like the mace or axe.[22] The surfaces of most armour worn in the Hellenistic Period were not flat like the target plates of Gabriel and Metz's tests but the curvature of the armour itself may have still resulted in an equivalent angle of impact up to 45° (or possibly greater if the curvature was combined with a strike delivered at an angle to the target). The helmet of the Hellenistic Period, for example, could have numerous curved sections depending upon the style being worn (see Plate 10). The bronze 'muscled' cuirass had stylised musculature depicted in high relief on the front plate of the corslet which meant that there were very few flat areas on the armour. Even the less elaborate *linothorax* possessed a curved barrel-shaped torso and curving shoulder guards. Williams states that due to the curvature of the medieval chest plate, which is similar in many regards to the curvature of the *linothorax*, even a flat thrust from a weapon such as a spear will impact with the armour at an angle equivalent to 0-45° from the perpendicular depending upon where on the armour the strike actually lands.[23]

Thus in a 'best case scenario', where an attack made with the *sarissa* from the low position followed a fairly flat trajectory and hit an enemy almost squarely in the chest, the weapon would impact at an angle not far from perpendicular to the surface of any armour that opponent might be wearing or shield he might be carrying. It would only be the curvature of the armour itself which would alter the angle of this impact. Additionally, as the location of the impact moves around the body, the curvature of the armour (relative to the angel of impact) increases and this would then make it harder for the weapon to penetrate.

Similar to the way an 'energy multiplier' can be used in any calculation of penetration power, a different set of 'energy multipliers' can be used to account for the angle at which any attack will strike a particular target (Table 14):

Angle of impact away from the perpendicular	0°	20°	30°	40°	45°	50°	60°
Energy Multiplier	1.0x	1.1x	1.2x	1.3x	1.4x	1.6x	2.0x

Table 14: The 'energy multiplier' required to calculate the penetration of armour at different angles. [24]

Based upon these figures, it requires forty per cent more energy for a weapon to penetrate a target at a 45° angle of impact (either actual or equivalent based upon the curvature of the target) to a certain depth than it does to penetrate a target struck perpendicular to its surface to the same depth. When the 'energy multipliers' for both the thickness of the armour and the angle of impact are considered, the adjusted figure of 34.1ft lbs (46.2j) of energy required for the *sarissa* to penetrate a 1mm thick target at a 45° angle of impact can be recalculated (Table 15).

Angle of impact away from the perpendicular	0°	20°	30°	40°	45°	50°	60°
Thickness of the target	Energy required for penetration – ft lbs *(joules)*						
1mm	24*(33)*	27*(36)*	29*(39)*	32*(43)*	34*(46)*	39*(53)*	49*(66)*
1.5mm	46*(62)*	51*(69)*	56*(75)*	60*(81)*	65*(87)*	74*(100)*	93*(125)*
2mm	71*(95)*	78*(105)*	85*(114)*	92*(124)*	99*(133)*	113*(152)*	141*(191)*
2.5mm	100*(135)*	110*(148)*	120*(162)*	130*(175)*	140*(189)*	160*(216)*	200*(269)*
3mm	134*(181)*	147*(199)*	161*(217)*	174*(235)*	188*(253)*	214*(289)*	268*(361)*
3.5mm	171*(230)*	188*(253)*	205*(276)*	222*(299)*	239*(322)*	273*(368)*	341*(460)*

Table 15: The energy required for the sarissa to penetrate armour of different thicknesses and at different angles.

Thus an attack made with a *sarissa*, delivering energy of 81.2ft lbs (110j) would be capable of easily penetrating the 1mm thick plate armour worn by the Greek hoplite even at angles greater than 60° from the perpendicular. If the calculations of Aldrete, Bartell and Aldrete are correct, and 11mm thick linen armour was the equivalent of 1.8mm thick metal plate, the *sarissa* would still be able to inflict a killing blow on a victim so long as the attack was delivered at an angle less than 30° from the perpendicular. Due to the ability to keep the pike level through a flexing of the wrists when an attack is made from the low position, all attacks will strike their target roughly perpendicular – with the only variance in the angle being caused by the curvature of the actual armour the target was wearing. But even here, angles of impact up to 50° would not prevent a *sarissa* from delivering a killing blow so long as the armour worn by an opponent was less than 1.5mm thick (or equivalent). Even with a given amount of flex that would absorb some of the impact energy, it would seem likely that the *sarissa* would have been capable of penetrating linen armour as well as bronze plate and some shields.

In another test, a replica *sarissa* was used against a sheet of 1mm thick corrugated iron (which was held perpendicular to the ground at chest height by two brave volunteers) to test the weapon's penetrative abilities. Participants wearing full phalangite panoplies engaged this target sheet with attacks delivered from the low position. In almost every case, the tip of the *sarissa* easily penetrated the corrugated sheet to a killing depth of 2in or more with very little effort. In one case, the entire head of the weapon and part of the shaft passed through the sheet – a depth of penetration of at least 16cm which would have caused significant injury, if not immediate death, to any victim. This level of penetration occurred regardless of where upon the corrugated surface of the metal sheet the weapon struck. At no time did the curvature of the sheet cause the pike to glance off or fail to penetrate. This basic practical experiment correlates with the calculations provided above that an attack made with the *sarissa*, following a relatively flat trajectory, was capable of easily penetrating plate metal armour at almost any angle of impact.[25]

Such conclusions then shed light on some of the ancient passages which recount the combative use of the *sarissa*. At Issus for example, Alexander's pikemen, who were in the riverbed as they attempted to cross, found it difficult to slay the Persians holding the higher bank above them and so were forced to thrust their *sarissae* at the faces their opponents.[26] This would suggest that attempting to thrust upwards at an angle with the *sarissa* (as would happen with a phalangite in a river bed thrusting at an enemy on an elevated bank) could either not generate enough force to penetrate armour and shields due to the awkwardness of the action, caused too great an angle of impact for the penetration of armour and shields, or both. In this case, Alexander's troops were then forced to direct their attacks at a more vulnerable area of their opponent's body, the face. The ease with which a more level attack made with the *sarissa* could penetrate armour would also confirm the accuracy of the passage of Diodorus which recalls how Alexander's phalangites were able to thrust their pikes through enemy shields and into their bodies.[27]

THE EFFECT OF MOMENTUM

All of the data outlined above was gathered from filming a stationary participant wielding the *sarissa* who did not lean into, or step forward with, their attacks as their movements were recorded. The ability of a member of the front rank of the phalanx to either step or lean into any attack they may commit would add substantial body mass to that attack and so increase the amount of energy delivered through the impacting tip of the weapon.

The transference of the body's potential energy to the impact of a weapon occurs when the link between the two is sufficiently rigid.[28] If the pike is simply held in its ready position, braced in place by the two hands wielding it, and a phalangite leant forward to try and force an opponent back or move his shield out of the way, a considerable amount of his body weight is applied to the attack. Such short leaning attacks would have occurred when an enemy was pressed up against the tip of the phalangite's weapon and when there would have been little room for the swinging motions of a thrusting attack. If a phalangite weighing 80kg and wearing 20kg of equipment, could get even a fraction of his body weight behind this sort of drive, it would add considerable weight to the calculation of how much energy could be delivered through the tip of the weapon – not enough to force the tip through a piece of defensive armour such as a thick shield or breastplate, but possibly enough to either knock the opponent over or force them backwards.[29]

Considerably more energy could be 'delivered' if the pike was held in place and an enemy then charged upon it, impaling themselves. Plutarch states that at the battle of Pydna in 168BC the Pelignians and Marricinians charged upon the presented weapons of the phalanx 'with animal fury' and that the *sarissae*, which were held firmly in place with both hands by the Macedonians, pierced 'those that fell upon them, armour and all, as neither shield nor breastplate could resist the force of the Macedonian pike'.[30] If the Roman auxiliaries who charged onto the phalanx weighed a total of 100kg (220lb), and hurled themselves onto the tips of the *sarissae* even at a brisk walk of 5km/h (4.4ft/second), then the amount of energy of the collision between a charging Pelignian and the tip of a presented pike would be around 665ft lbs (902j). This is nearly double the amount of energy required to penetrate armour 3.5mm thick at an angle of impact of 60°. Had the charge been quicker (as the 'animal fury' described by Plutarch would suggest) then the amount of impact energy would be even higher. This would account for Plutarch's statement for how, at least in this instance, the *sarissa* was able to easily penetrate shields and armour as the very momentum of the charging assailant was the cause of their own demise. This would also account for the passage of Lucian which describes how a pike, rigidly braced in position, was used to kill both a horse and rider which charged upon it.[31]

THE EFFECT OF FATIGUE

As a battle wore on and a phalangite's level of energy decreased, this would have a detrimental effect on the effectiveness of any of his offensive actions to penetrate the armour of an enemy he was facing. The first committed attack made (i.e. not a short jab designed to probe an enemy's defences, but a solid thrust aimed at delivering a killing blow) would undoubtedly be the strongest. As time wore on, and the muscles of the arms began to tire, the actions of a thrust would be slower and weaker, thereby imparting less impact energy on the target.

However, a phalangite could not guarantee that his first committed attack would be successful – it could miss, be parried, taken on an opponent's shield or evaded in some other way. Additionally, even if the phalangite was able to slay his opponent with his first attack, there would undoubtedly be more enemies to face him. As such, phalangite warfare had to involve a method of fighting that could be maintained for as long as possible and, most importantly, be as effectively as possible. The energy that could be delivered with a *sarissa* from the low position – the posture that causes the least muscular stress on the arms – can accommodate a 70 per cent reduction in the amount of energy delivered due to the onset of fatigue and yet still be capable of penetrating 1mm thick armour to a killing depth so long as the target is struck perpendicular to its surface. This, combined with the ability to simply hold the *sarissa* in place and keep an enemy at a safe distance, with committed attacks against opportune targets only being made when (and if) they presented themselves, would have allowed phalangites to operate effectively on the battlefield for considerable periods of time.

The *sarissa*, for all of its weight and bulk, was an extremely effective battlefield weapon. The amount of energy that could be delivered with a committed attack could easily pierce any of the armour used in the Hellenistic Period. The ability to simply hold the weapon in position allowed the *sarissa* to be used in a number of ways which allowed contingents of phalangites to engage in combat for protracted periods of time, and may have even resulted in the slaying of enemies brash enough to charge upon them. It can subsequently be concluded that the numerous references to walking wounded who had survived confrontations with a pike-phalanx were either the result of weakened attacks made by the tiring phalangites (attacks that may not have penetrated an opponent's defensive armour to a serious depth), the result of 'accidental' attacks where a pike that was pressed into the shield of a facing enemy slid off its surface to impact with the face or legs, or were the result of being pushed back by the lengthy pike which may have resulted in some form of minor injury. Regardless, the *sarissa* appears not to have been the cumbersome weapon that it may seem to have been. Rather, it was an impressive piece of weaponry, one which allowed the Macedonians to dominate the battlefields of the ancient world, and which allowed its length and weight to be used to their full advantage so long as it was employed within the safe confines of the massed formation of the phalanx.

Chapter Ten

The Use of the Butt-Spike in Phalangite Combat

Much like the *sauroter* affixed to the rear of the hoplite *doru*, the large, heavy, metallic butt-spike on the rear end of the *sarissa* served a number of purposes. Its primary function was to provide the weapon with the correct point of balance. Additionally, the butt-spike protected the end of the shaft from the elements and its pointed end allowed the weapon to be thrust securely upright into the ground when not in use. This is quite easy to accomplish with the hoplite *doru* as its relatively short length and low weight allow it to stand unsupported in a vertical position simply by thrusting the *sauroter* into the ground.

However, such an action is much harder to accomplish with a *sarissa*. The weight of the *sarissa* (5kg) was considerably more than that of the *doru* used by the hoplite (1.5kg). Furthermore, the *sarissa* was considerably longer (some 5.7m compared to 2.5m). As such, it is much harder to get this weapon to stand up on its own. Only a minor deviation away from the vertical will cause the weapon to tip over – especially if the ground is soft and/or the spike on the butt is not thrust into the ground sufficiently. This is where many of the design elements of the *sarissa* butt, which differ from the hoplite *sauroter*, came into play and may partially explain the variances in design between the example of the winged *sarissa* butt found at Vergina and the traditional hoplite *sauroter* which came in the form of a conical or pyramidal spike. The wings on the *sarissa* butt recovered from Vergina would help stabilise any weapon thrust upright into the ground regardless of the consistency of the soil. If the earth was relatively hard packed, then the weapon could be thrust into the ground at least as far as the elongated spike on the end of the butt - a depth of 14.5cm. From here the small wings, acting almost like four feet splayed out on the corners of the butt, would then help keep the weapon vertical. If the soil was quite soft, sand for example, then a *sarissa* with a winged butt could be thrust even deeper. The splayed wings then provide a greater surface area for the soil to purchase onto – providing greater resistance against toppling and helping to maintain the weapon in a vertical position.

The Newcastle butt, while shaped in a conical spike similar to many examples of the hoplite *sauroter*, is considerably longer than the spear butts used in the Classical Period on the lighter *doru*. This means that more than half the Newcastle butt's 38cm length is nothing but a spike that could be thrust deep into soft soil or as far as it could go into more hard packed surfaces. In either case, the sheer length of the spike would aid in supporting an upright weapon. Thus is seems that the design of the *sarissa* butt, and not just its weight, had an important role to play in the overall functionally of the weapon. However, were balance, protection of the end of the shaft, and the ability for the *sarissa* to be thrust upright into the ground when not in use the only functions of the butt?

THE USE OF THE BUTT IN A DEFENSIVE CAPACITY

Lucian outlines at least one defensive use of the *sarissa* butt in a combative situation. He explains that one of the tactical uses of the butt was to plant it into the ground so that the pike could be braced into position and then used defensively against an enemy horseman.[1] In his account of this action, Lucian describes how the pike transfixed both the horse and rider at the same time. Lucian also provides details of the position that the pikeman (in this case a Thracian) adopted and exactly where the pike pierced both the rider and his mount:

> [A] Median (Arsaces) was charging with his 20 cubit (εἰκοσάπηχύν) lance in front of him; the Thracian knocked it aside with his *peltē*; the point glanced by; then he knelt, received the charge on his pike, pierced the horse's chest – the spirited beast impaling itself by its own impetus – and finally ran Arsaces through groin and buttock.[2]

From such details we can gain a further understanding of how such an action was accomplished. Depending upon the exact breed and size of the horse being ridden, the region of Arsaces' groin and buttocks would be approximately 1.7m off the ground as he sat astride his mount. If the pike being carried by the Thracian was, for example, the same length as the *sarissae* being carried by the phalangites of Alexander the Great (12 cubits or 576cm long), this then allows, using basic trigonometry, for the angle of the pike to be calculated so that it will pierce both the horse's chest and the rider's groin as is stated by Lucian (Fig.19).

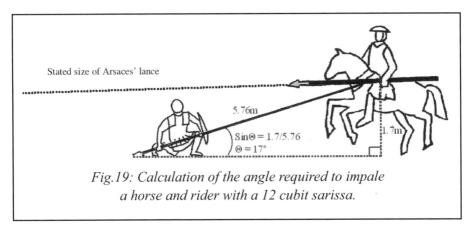

Fig.19: Calculation of the angle required to impale a horse and rider with a 12 cubit sarissa.

The relatively small angle at which a pike of this length has to be deployed to be directed at the targets detailed by Lucian (17°) explains why the Thracian was required to kneel. Such an angle could not have been accomplished if the Thracian had remained standing. Furthermore, it becomes clear that, for a similar action to be taken against infantry, the angle would have to be decreased even further in order for the tip of the pike to descend below head height. This, in turn, suggests that such a position was not adopted to receive an attack of enemy infantry, and it is more likely that in such circumstances the *sarissa* would have simply been wielded at waist level so that the enemy could be engaged directly – as is shown on the Pergamon plaque (see Plate 7). Interestingly, the Pergamon Plaque shows phalangites engaging both infantry and cavalry. Consequently, it seems that it was not a necessity of phalangite combat for the combatants to kneel in order to engage mounted troops and either posture (standing or kneeling) would have been used depending upon the tactical situation of the engagement.

Additionally, in assuming such a defensive position as a kneeling posture, the hands do not have to be shifted from their correct placement. All the phalangite has to do to adopt such a posture is to step forward slightly with their left foot as they bend down and ram the spike of the butt into the ground. The kneeling posture means that, with the hands kept in their original position, the pike will automatically be repositioned at an angle of around 17°. The hands can then simply be used to maintain the angle of the pike and allow the momentum of the incoming opponent to do the rest. As the position of the hands does not have to be altered, and therefore no more of the pike extends behind the bearer than would have normally been the case, such a defensive posture could be adopted by every member of every rank of a phalanx deployed in an intermediate order of 96cm per man.

THE USE OF THE BUTT IN AN OFFENSIVE CAPACITY

Some scholars have suggested that the butt of the *sarissa* could be used as an ancillary weapon if the head and/or shaft of the pike broke during the course of a battle.[3] The use of the hoplite *sauroter* in such a capacity is also a part of many theories on the mechanics of hoplite combat, but this concept seems to be in error.[4]

Markle cites a passage of Polybius (6.25.9) as evidence for the use of the *sarissa* butt as a secondary weapon. Polybius states that the problem with the Roman cavalry lance is that it does not possess a spike on the rear end of the shaft and that, if the head, breaks off, the weapon becomes useless whereas the Greek lance, which did have a butt, could still be used. Polybius states that Philopoemen used the butt of his cavalry lance to kill the Spartan Machanidas at the battle of Mantinea in 207BC.[5] However, Polybius' description of the advantages of the Greek cavalry lance can in no way be interpreted as saying that, because the *sarissa* carried by infantry (rather than the cavalry lance as is specifically mentioned by Polybius) did have a spike on the end of its shaft, that it could therefore be used as a secondary weapon in the case of breakage. It must be remembered that the cavalry lance had a butt very different in configuration to the infantry pike – more like a large spearhead than an elongated spike (see Plates 3-4).

Furthermore, even though the Greek and Macedonian cavalry lance did have a large butt, it does not seem to have been regularly used as a secondary weapon anyway – even when the primary head broke off. When the lances carried by Alexander's cavalry broke at the battle of the Granicus in 334BC, Arrian and Plutarch state that these broken weapons were abandoned in favour of swords or replacement lances; the butt-spike appears not to have been used as a secondary weapon.[6] Diodorus, in his account of the battle, states that when the tip broke off the lance that Alexander himself was carrying, he continued to thrust the jagged end of the shaft at the faces of his opponents rather than turn the lance around and engage with the butt.[7] This suggests that, despite what

Polybius states, even when equipped with a weapon which would allow them to do so, Macedonian cavalry rarely resorted to the butt of the lance as a secondary weapon.[8] Importantly, such a passage can in no way be extrapolated as being a reference to the use of the *sarissa* butt as a secondary weapon by phalangites in the massed confines of the phalanx as some scholars assert.

BREAKING THE *SARISSA*

Any theory which proposes the use of the butt of the *sarissa* as a secondary weapon is reliant upon the assumption that the head of the weapon could break off during combat and force the bearer into a position requiring recourse to an alternative means of offence. The references in Arrian, Plutarch and Diodorus to the head of the Macedonian cavalry lance breaking clearly demonstrates that such a thing could occur to mounted troops. Cavalry, by its very nature, regularly charges into combat. Consequently, the momentum of the charge, and the collision with the enemy, would have been a contributing factor to the breakage of many cavalry lances. However, under what conditions, and how regularly, would this have occurred among the heavy infantry?

The ancient sources provide a few accounts of events where the front of the *sarissa* was broken off. Frontinus, for example, states that, at the battle of Pydna in 168BC, the Roman commander Aemilius Paulus ordered his cavalry to ride across the front of the opposing phalanx while covering themselves with their shields so that the heads of the Macedonian *sarissae* would be broken off.[9] In this instance it can be assumed that the cavalry were riding from left to right, with their shields covering their left-hand side, and using the edge and face of the shield, in concert with the momentum of the moving horse, to hit the extended pikes of the phalanx and sever the heads.

In another example, Diodorus relates how, in the heroic duel between the Athenian Dioxippus and the Macedonian Coragus who advanced into battle with a leveled *sarissa* only to have Dioxippus, who was armed with a club, hit the end of the pike and shatter it.[10] Similarly, Plutarch states how, again at the battle of Pydna, some of the Romans tried to grab the tips of the *sarissae* in their hands.[11] Plutarch says that the Romans had tried something similar against the pikemen of Pyrrhus' phalanx at Asculum in 280BC.[12] During the battle of Plataea in 479BC, the Persians had similarly attempted to grab the tips of the hoplite spears arrayed against them, most likely in their left hand, and then use a one-handed weapon wielded in the right hand to try and hack off the end.[13] Plutarch's accounts of Pydna and Asculum may be a reference to something similar whereby some of the Romans may have been attempting to grab the *sarissa* in one hand and break off the end with a weapon held in another as he states that the Romans were also trying to 'knock aside the *sarissae* with their swords'.[14] Sheppard suggests that the large hacking *macharia* style of sword used by the Greeks and Macedonians was a post-Alexander addition to the panoply of the phalangite to be used to try and cut the tips off opposing pikes.[15] Interestingly, in the Age of the Successors, much of the combat was pike-phalanx against pike-phalanx where a weapon capable of severing the tip of an enemy's pike would have been of great benefit.

It is unlikely that Plutarch's reference to the Romans trying to grab the *sarissa* in their hands at Pydna was an attempt to simply try and snap the shaft. Tests have shown that it is physically impossible for an individual to break a spear shaft 25mm in diameter, even when employing both hands in the attempt.[16] It would be even more impossible to snap something thicker such as the 35mm diameter shaft of the *sarissa*. Thus any attempt to grab the *sarissa* and render it ineffective would need to make use of another means of severing the head such as striking it with a weapon.

In order to accomplish such an action, certain conditions have to be assumed. Firstly, the weapon used to try and sever the *sarissa* or spear would have been held in the right hand. Consequently, in order to grab the shaft of the weapon, it must be assumed that no shield was being held in the left hand. This would then allow someone to grab the shaft of the pike/spear with the left hand, pivot inwards and to the left, and bring down a weapon to hack up to the first metre off the end of the weapon. Grabbing hold of the shaft, while potentially dangerous to the person doing so, holds the weapon in a relatively rigid manner which then allows the end to be broken off through the force of the blow from the weapon in the right hand. If the shaft is not grabbed, any blow against the shaft is more likely to simply deflect it rather than break it – this is particularly so for the *sarissa* with its great length and inherent flex. However, the Romans at Pydna are unlikely to have abandoned their shields in order to undertake such a move. As such, Plutarch's reference to the Romans trying to 'knock aside the *sarissae* with their swords' must be a description of attempts to parry the incoming weapons rather than an attempt to sever them.

Any attempt to sever the tip from an opposing *sarissa* with a hand-held weapon like a sword would have required speed – especially in a massed combat situation. It is unlikely that the Persians at Plataea, for example, grabbed the Greek weapons while they were held in their 'ready' position and then tried to cut through the shaft as this would have brought any grappling Persian dangerously within range of the spears of the second rank of the Greek phalanx. The Persians could only have attempted to grab the Greek spears at the moment when a hoplite of the front rank had extended their spear into the attack and before they withdrew it again (which would have pulled any Persian holding it onto the spears of the second rank). Consequently, any severing of the Greek spears would need to have been accomplished rapidly; again suggesting the use of a weapon to break the end off. For individual combatants like the duelists Coragus and Dioxippus, this was less of an issue. So long as Dioxippus could safely move outside the danger zone directly ahead of the *sarissa* (i.e. by either moving 'inside' its reach and hacking at the pike from the side, or stepping back as it was thrust forward and striking at it while it was at its full *effective range*) then he would have been both safe from potential injury and in a position where he could have used his club to snap the head off Coragus' pike. The fact that Dioxippus is only armed with a club, and bereft of a shield, would have meant that, had he moved 'inside' the reach of the pike, he would have been easily able to grab the weapon and then bring his club down upon it. If, on the other hand, Dioxippus simply swung his club at Coragus' *sarissa*, for it to shatter the end suggests that the wood had been weakened in some regard or that the head itself was simply knocked off the weapon.

Another possible way of interpreting Plutarch's account of Pydna and Asculum is that some of the Romans may have been grabbing the Macedonian pikes in an attempt to detach the front half of the *sarissa* from the rear half. However, this could only be accomplished if a certain set of battlefield conditions were met. For example, in the press of a battle where, as Plutarch describes Pydna, the Macedonians had the tips of their *sarissa* pressed hard up against the shields of their opponents, it can be assumed that both sides were pushing forward to try and maintain their line and dislodge their opponent, and that the Romans were being held at bay by the length of the Macedonian pikes at the same time. What this would also mean would be that the men of the subsequent ranks on both sides would be pressed almost into the backs of the men in front as the intervals between the lines compressed and the formation as a whole strove to advance. If the intervals of the Roman formation were not compressed in such a way, then any Roman facing the advancing pikes of the phalanx would be simply forced backwards, as the pike was driven into his shield, onto the man behind him by the strength of the Macedonian advance. This would also cause the intervals between the Roman front ranks to compress. Alternatively, if the Romans were advancing and the Macedonian phalanx was attempting to hold its position, then the Roman ranks would still compress as the front rank was stopped in its tracks by the extended *sarissae* of the Macedonian formation and the rear ranks of the Roman formation continued to press forward. In any scenario, due to the compression of the Roman lines, any legionary in the front rank who managed to take hold of the forward end of a *sarissa* he was facing would not be able to step backwards in order to provide himself with sufficient room to try and pull the front half of the pike off. Additionally, the idea that an opponent would even attempt to pull the front half of the *sarissa* off a weapon he was facing must automatically assume that they had abandoned any weapon that they had been carrying in their right hand (and possibly their shield in their left as well). The unlikelihood of this happening, especially if that opponent was another *sarissa*-armed pikeman, makes such a prospect entirely improbable.

The only way in which such an undertaking would work would be if both formations were relatively static, and with the Romans deployed in something akin to the Macedonian intermediate order of 96cm per man or greater (see The Anvil in Action from page 374). This would then provide the legionary in the front ranks sufficient room to grab onto an opposing weapon which may be pressed against their shield, step back slightly and use the slight increase in room to separate the two halves of the *sarissa* thus rendering it unserviceable. However, such conditions were rare in phalanx warfare. In almost every engagement from the time of Alexander onwards, at least one (if not both sides) was in motion when the lines met regardless of whether the opponents were Persians, Macedonians, Greeks or Romans, and regardless of whether they were infantry, cavalry, chariots or elephants. Thus the configuration of the *sarissa* into two halves, while providing numerous benefits to logistics, could not be exploited as the conditions required for an opponent to be able to separate the two halves rarely materialized on the battlefield.

Another way in which the *sarissa* could lose its head would be if the head itself broke during the course of a battle. This was a fairly common occurrence with the head of the hoplite *doru*. Euripides states that spears could break as a result of a powerful thrust (ἀπὸ δ' ἔθραυσ' ἄκρον δόρυ).[17] The Greeks at Thermopylae (480BC) resorted to using their swords only once their spears had been broken during the combat of the preceding two days.[18] Xenophon's account of the aftermath of the battle of Coronea (394BC) lists broken spears among the detritus of the confrontation.[19] Diodorus states that the tip of the spear which pierced Epaminondas' chest at Mantinea (362BC) broke off after it had entered his body.[20] All of these passages suggest that it was common for the hoplite spear to break during combat – most likely from impacts with a hard surface such as an opponent's shield or armour, and that the force of these impacts would be enough to break the head off the weapon if resistance to penetration was high enough. Something similar is likely to have happened to the phalangite *sarissa* as well.

Unfortunately, such passages as those of Herodotus, Xenophon, Euripides and Diodorus do not detail what caused the spear head to break off. A spear, for example, may have splintered at a weak point that had developed in its wooden shaft. Alternatively, a rivet or adhesive that was holding the head in place may have weakened or broken through the actions of combat and this may have resulted in the head simply falling off. It has also been estimated that if the head of a spear was held in place (such as having it penetrate a shield or body), the application of a lateral force of 119ft lbs (160.7j) would be enough to wrench the shaft from the socket.[21] Due to the length of the *sarissa*, and the leverage that this length and the rearward placement of the grip on the weapon creates, it would be very easy to generate this much force. Another possibility was that the spearhead itself may have broken at a weak point where the blade of the head meets the socket.[22] Finds from Olympia contain several spearheads from the Classical Period broken in just such a manner suggesting that this was a common weak point in the design of many spearheads and may be what both Herodotus and Euripides are referring to.[23]

If all of these conditions of hoplite warfare, spears breaking due to severing by an enemy or through the rigours of combat, are also taken as elements of phalangite warfare (and there is no reason to assume that they are not), then it seems that, under certain battlefield conditions, the phalangite's *sarissa* could actually break, or be broken, in his hands. However, this does not automatically mean that, if this did occur, that the phalangite would resort to the use of the butt on the broken *sarissa* as a secondary weapon and an examination of how such a broken weapon could be used (or not) indicates the unlikelihood of it being used offensively.

THE OFFENSIVE CAPABILITIES OF A BROKEN *SARISSA*

Could a broken *sarissa* be of any use in the massed combat of the phalanx? Could the butt of a broken weapon actually be brought to bear against a target? And could it have reached that target? Due to the varied nature of how a *sarissa* could break or be broken, the remaining section of weapon could be of any length. Weapons that merely lost their head, either through it breaking or falling off, would lose between 20 and 30cm of their overall length. If the *sarissa* fractured at a weak point in the wood, the amount of remaining weapon would be dictated by the location of that weak point anywhere along the shaft. In the case of the spears hacked through by an opponent, the length of the weapon may have been reduced by as much as a metre.

For the pikes from the time of Alexander the Great with an overall length of 12 cubits (576cm), if they were hacked through by an opponent, the remaining weapon would have been approximately 476cm in length, 45cm of which would be the length of the butt itself. The removal of the head and around 70cm of shaft dramatically alters the balance of the remaining section of the weapon which would weigh in the vicinity of 4.5kg. If this broken weapon was turned around so that the butt acts as the impacting tip, the sheer length and mass of the shaft (acting to offset the weight of the butt) relocate the point of balance to 204cm behind the new forward tip of the weapon, a shift of 108cm away from the butt and just behind the mid-point of the weapon (Fig.20).[24]

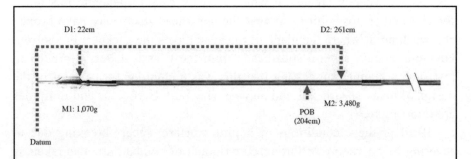

Point of Balance (POB) = (M1 x D1) + (M2 x D2) / (M1 + M2)
Where:
M1 = the mass of the butt-spike
M2 = the mass of the remaining shaft
D1 = the distance from the datum (the end of the weapon) to the point of balance of the butt-spike
D2 = the distance from the datum (the end of the weapon) to the point of balance of the remaining shaft
Thus:
POB   = (1,070 x 22) + (3,480 x 261) / (1,070 + 3,480)
          = 23,540 + 908,280 / 4,550
          = 931,820 / 4,550
          = 204 (cm from the datum)

Fig.20: Calculation of the point of balance of a broken sarissa.

A weapon such as this, if it were wielded correctly with the left hand placed at the point of balance and the right hand positioned about 96cm further back, leaves over 1.7m of shaft projecting behind the bearer and only 2m to their front. Due to the large weight of the butt which would now be acting as the forward tip, it is almost impossible to wield a broken *sarissa* by gripping it further rearward than its new point of balance as the muscular stresses caused by the pressure required to keep the weapon level for a protracted period of time become too great.[25] Thus even a *sarissa* which had only lost its head, leaving an extra metre of shaft attached to the broken weapon, could not be held much further back than 2.5m from the tip of the butt.

It is almost impossible to reposition a weapon balanced in this way so that the butt can be used offensively when deployed in a phalanx. Regardless of where and how the *sarissa* breaks, the butt will be pointing towards the rear when the weapon fractures. The leading left hand, holding the *sarissa* at its correct point of balance, will be approximately 96cm forward of the butt. In order for the butt to be brought to bear against a target as a secondary weapon, the broken *sarissa* has to be spun around 180° and the grip repositioned by more than a metre. It would require a considerable amount of room for a phalangite to rotate a broken weapon 476cm in length or longer, without it becoming entangled with his own equipment or that of the other members of the phalanx. This is much more space than is available in even the most open order described by the ancient military writers.

Nor could the broken weapon be raised vertically (so that the broken end of the shaft pointed up and the butt was towards the ground) and then reverse the grip on the weapon and feed the butt forward into a combative position. To do so would require a large amount of the shaft to be angled backwards as the weapon is fed forward – again risking entanglement with the weapons of those in the rearward ranks, especially those which were angled forward over the heads of those in front. In the confines of a phalanx, regardless of whether they were hard pressed up against an enemy or not, there is simply no way in which the phalangite would be able to bring the butt of a broken weapon to bear against a target.

Furthermore, even if a phalangite was able to somehow reposition his shattered weapon so that he could strike forward with the butt, a broken weapon 476cm in length and gripped at its new point of balance 204cm behind the tip, affords a combat range of only 240cm and an overall effective range of only 252cm when used in an intermediate-order formation. This is less than half the reach of an intact *sarissa* used in the same order (*effective reach* 540cm). Thus if the phalanx was facing an opponent with a short reach weapon (e.g. Romans or Persians), the opposing formation would most likely be held back at the distance that the fully intact pikes of the phalanx extend forward of the line (on a 12 cubit (576cm) weapon this equates to a distance of 10 cubits or 480cm). Consequently, even if a phalangite within this formation could thrust a broken weapon forward, if both he and his opponent maintained their positions in their respective formations, the reduced reach of the broken weapon would mean that any strike made by the phalangite would fall short of its target by more than a metre. If the phalangite with the broken weapon was facing another phalangite, who may have the tip of his weapon pressed into his shield, thus preventing any movement of the left arm, the phalangite with the broken *sarissa* would not be able to thrust at all and would only have a reach equivalent to the amount of broken weapon that projected (204cm) ahead of him. Thus, if those on either side of him were still using intact pikes, and engaging an enemy at a distance of nearly five metres, a phalangite would not be able to reach an opponent with a broken weapon unless he stepped forward of the ranks and advanced into the 'no man's land' between the two formations in order to engage. It seems unlikely that a phalangite would have advanced forward of his position in the phalanx to attack, as the very basis of phalanx-based combat was the maintenance of the formation, nor could he advance further forward in order to engage if an opponent's pike was pressed into his shield.

Additionally, due to the 1.7m of broken shaft projecting behind him, a phalangite wielding a broken *sarissa* at its correct point of balance could not maintain an intermediate order spacing of 96cm between himself and the man behind him. The jagged end of the shaft also makes wielding a broken weapon in any formation both dangerous to surrounding phalangites and liable to entanglement with the equipment of other members of the formation (Fig.21).

The rearward projection of so much of the broken shaft, and the physical inability to wield a shafted weapon with a heavy tip any further back, clearly indicates that a *sarissa* broken in such a manner could not have been used as an improvised weapon in the massed confines of the pike-phalanx. Consequently, any suggestion that the butt of a broken *sarissa* was used as a secondary weapon in the press of battle can only be regarded as completely untenable.

At such a distance from an opponent, a broken *sarissa* could only have been used in a defensive capacity to fend off attacks. The *sarissa* could still be held

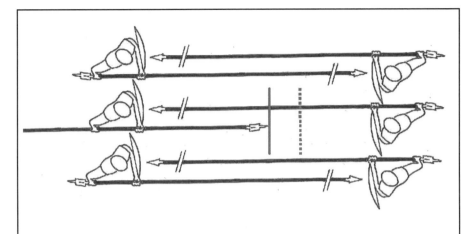

Fig.21: The rearward projection of the shattered shaft when a broken sarissa is held at its altered point of balance and the reduced effective range of a broken sarissa (dotted line = if the weapon can be thrust forward; solid line = if the left arm is pinned in place).

in place with the jagged end of the shaft projecting towards an opponent simply to keep him at bay. In doing so the *sarissa* does not have to be rotated and only a minor adjustment to where it is held made to compensate for the altered point of balance. If the wooden shaft of the spear had fractured too close to the butt to make the weapon of any use in this capacity, it would have simply been abandoned on the field – much like Alexander's broken lance was. In this case a sword may have been drawn if the phalangite possessed one, although he would be unlikely to reach his opponent with this weapon either, or be able to even swing it without becoming entangled in the weapons of the other ranks unless he simply slashed downward across his shield or just thrusted directly ahead. It is more likely that a phalangite in this situation, especially if facing another phalangite, simply stood his ground and absorbed or parried any opponent's attacks with his shield.

In a confused melee, such as when a formation began to break up and/or rout, either of which may not have conformed to the intervals of the phalanx, any phalangite who did not possess a sword would have clearly used any weapon available to defend himself. Under these conditions it is possible that the butt of a broken *sarissa* could have been repositioned and used in an offensive manner. Similarly, the jagged end of the shaft could be jabbed, thrust or swung at an opponent, although it is unlikely that it would penetrate a shield or armour and could only have been used against unprotected areas of an opponent's body in the same way that Diodorus describes Alexander's actions at the Granicus. In such a frenzied environment anything is possible. However, these would have been acts of desperation rather than a standard battlefield use of the butt as a secondary weapon.

Another thing to consider is whether or not the spike of the butt of a *sarissa* would be offensively useful, even if the broken weapon could be repositioned so that it could be used in such a manner. During the testing of the penetrative abilities of the *sarissa*, participants used the butt on the replica *sarissa* against a sheet of 1mm thick corrugated iron. The weapon used was not broken and was held in a location slightly forward of centre towards the front end of the weapon. This was not the exact location of the weapon's point of balance, but accurately simulated a phalangite whose weapon had broken attempting to use as much of the length and reach of the remaining *sarissa* as possible.

Due to the amount of energy that can be delivered with a thrust made with the *sarissa* from the low position (see The Penetration Power of the *Sarissa* from page 225), it was found that even the spike of the butt, which has a much larger impacting point than the head does, could penetrate the 1mm thick plate with only minimal effort. The wing-like flanges of the butt prevented it from penetrating too far, however the butt still penetrated up to almost the entire length of the spike – an average depth of penetration of approximately 10cm. This would have easily resulted in severe injury to anyone who received such an impact regardless of whether the area of the body hit was protected by armour or not. This penetration occurred regardless of where on the corrugations of the plate the butt impacted, thus showing that even the curvature of any armour that a recipient of such an attack might be wearing was insufficient to resist this kind of impact. It is further likely that such impacts would cause considerable damage to shields and would penetrate the *linothorax*. This suggests that, if it was ever employed as a secondary weapon, the butt of the *sarissa* would have been rather effective as an offensive weapon. However, the difficulty of repositioning a broken weapon within the confines of the pike-phalanx so that the butt could be used in such a manner suggests that it was never designed to be used in such a way.

All of these considerations show that there is no basis of support for theories which propose a use for the butt of the *sarissa* as an alternative weapon. If the phalangite's pike broke during combat there would be little reason to retain the broken section except for the possible eventuality that the combat may develop into a situation where a secondary weapon was necessary. If that did not occur, the butt on the end of a broken *sarissa* could not have been brought to bear against a target in the confines of the phalanx, and could not have reached a target even if it was done so. In these instances, any weapon which broke would most likely have been kept in place and used to fend off an attacker while the battle was continued by those whose pikes were still intact. It is the combination of these factors which make it unlikely that the butt was ever used as, or even designed to be, a secondary weapon.

THE USE OF THE BUTT TO DISPATCH A FALLEN OPPONENT

Another proposed offensive use of the butt of the *sarissa* is for the dispatching of a fallen opponent who lay prone at a phalangite's feet.[26] Similar to the use of the butt as a secondary weapon, the use of the butt to attack downwards at an opponent on the ground seems to have been extrapolated from a suggested use of the *sauroter* affixed to the hoplite *doru* - a theory that has no supporting evidence.[27] However, while the *sauroter* of the hoplite spear could not be used in this manner, was it possible for the butt of the *sarissa*?

In order to effect such a strike, the *sarissa* has to be moved into a vertical position from where the butt can be thrust downwards at the target. However, would conditions which allowed this to happen easily manifest themselves on the field of battle? There are no literary descriptions or artistic representations of the use of the butt of the *sarissa* in this manner and whether it could have been used in this way or not can only be determined by examining who, if anyone, was able to engage a prone opponent in this manner. If the phalanx was deployed sixteen ranks deep, and with the pikes of the first five ranks leveled for combat while the remainder were angled at a 45° angle over the heads of the men in front as per the descriptions of the phalanx found in Arrian, Aelian, Asclepiodotus and Polybius, then hardly anyone within the formation would be in a position where they could easily use the butt of their weapons against a prone opponent.

The phalangite in the front rank, for example, may have had the tip of this *sarissa* pressed against the shield of his opponent – as the Macedonians are described as doing at Pydna in 168BC.[28] For an opponent to appear prone at the phalangite's feet the phalanx must be advancing while the enemy is also moving under one of two possible scenarios: either the enemy formation is withdrawing – marching slowly backward, but still facing the enemy, in order to maintain the integrity of their line, or the enemy is simply in panicked flight – in which case the phalangite may not have the tip of his *sarissa* pressed against an opponent.

If the phalangite of the front rank was still engaged, he would be unlikely to raise his weapon to engage a prone enemy with the butt of the *sarissa* as this would provide the opponent he was facing with a momentary reprieve and an opportunity to move in on the phalangite. Due to the way that the *sarissae* of the first five ranks are serried and stepped back from each other by a distance of 2 cubits (96cm) per rank due to the size of the interval that each man occupies, an advancing opponent would most likely not be able to move forward any more than about a metre before the tip of the weapon held by the man in the second rank was thrust against his shield. However, once this had happened, even if the man in the first rank had used his butt to engage a prone opponent, he would now no longer be able to redeploy his weapon as the distance between him and the opponent was now less than the length of the weapon he was carrying.

Furthermore, it would be difficult for any member of the first rank to raise their *sarissa* vertically without the weapon becoming entangled in those held by the rear ranks that were angled over his head. The whole purpose of having the rear ranks deploy their pikes at this angle was to help shield the forward ranks from missile fire.[29] Thus it can be assumed that these weapons were positioned above the men of the front five ranks in order to provide such protection. However, this protection would then greatly limit the ability for the members of these forward ranks to use their own pikes in any manner other than having them pointed directly at an enemy or similarly angled upwards at 45°.

Similar problems would be encountered by those in ranks two to five, who would also have their pikes leveled for combat. Indeed, those in ranks four or five would experience even more difficulties in raising their pikes as those that were held at an angle over their heads by the more rearward ranks would, due to the angle, be even closer to them – in the case of those in rank five the angled pikes would literally be just above their heads and would greatly hinder any attempt to raise a pike vertically. Furthermore, as each man in the formation occupies a space of 2 cubits (96cm), and those of the rear ranks have their shields pressed hard against the butt of the *sarissa* held by the man in front, this greatly limits the amount of room available to members of the second to fifth ranks to raise their pike vertically and then bring it down at an opponent laying prone at their feet. It would be questionable, due to the press of men around them on all sides, whether members of these ranks would even see such a target until they were literally right on top of it. Finally, it must also be considered whether an advancing formation would simply step over a prone, but still living, opponent so that members of a more rearward rank could dispatch them with the butt of their weapon. This would run tremendous risk that that opponent would be able to use their position, and a short reach weapon like a sword or dagger, to attack the advancing phalangites from below (stabbing at the legs or groin, for example) as they advanced over them and while their pikes, extending both above and beyond them, were unable to be brought to bear.

This also makes it unlikely that the members of ranks six to fifteen, who were holding their pikes angled over the heads of those in the front ranks, were the ones who could have engaged a prone opponent with the butt of their *sarissae*. While pikes held in this angled position are easier to raise vertically, they would still run the risk of becoming entangled with all of the other angled pikes of the formation, and it is further unlikely that an advancing formation would allow the men of at least the first five ranks to pass over a prone opponent only to allow the man in rank six (or even further back) to engage.

The only member of the phalanx who was in any position to engage a prone opponent with the butt of his weapon was the file-closing *ouragos* in rank sixteen. Due to the lack of further pikes being positioned at an angle behind him, the man in the rear rank would encounter no impediment to raising his weapon vertically and then using the butt to try and dispatch an enemy lying at his feet. However, while such a move for the *ouragos* is physically possible the question still remains: would fifteen ranks of men simply walk over a danger such as a live opponent so that the man in the rear rank could then dispatch him? This seems highly unlikely.

Thrusts brought directly downwards can generate a great deal of force. Clearly this style of attack could do considerable damage against an unprotected area of the human body, enough to easily kill or at least severely injure, and could penetrate armour and damage shields. However, the inability of the members of most ranks of the phalanx to even move their pikes into a position where a downward attack with the butt could be made, indicates that any prone opponent had to be dispatched by another means.

In the standard deployment of the pike-phalanx, it can be assumed with some certainty that any enemy combatant who was knocked down during the course of a battle would have been done so by an attack delivered by the men in the front rank of the phalanx. If the opponent was knocked down but not killed, this would then make them one of the prone targets who would then be later dispatched with the butt of the *sarissa* in some modern theories. Yet, there would be no need for the target to be killed in such a way. If the phalanx was slowly advancing, the man in the front rank would not continue to engage an opponent who had fallen as the opponent would move inside the *effective range* and/or *combat range* of the front rank phalangite as the formation advanced. However, as the members of the more rearward ranks also had their pikes leveled, and staggered by 96cm intervals, the prone target could easily be dispatched by one of the members of ranks two to five simply by having them dip the tip of their *sarissa* downward to engage a target who was on the ground but still ahead of the formation and allow the slow momentum of the phalanx's advance help impale him or drive him back.

Due to the serried nature of the weapons held by the first five ranks, if an opponent was knocked down and remained more or less where he fell, he could easily be dispatched by members of ranks two or three as the line rolled forward. If the opponent was knocked down, but crawled forward (i.e. towards the advancing phalanx in an attempt to get under their pikes) then the crawling enemy could be easily taken out by members of ranks four or five. This was the whole essence of having a serried line of pikes; those from the more rearward ranks could still engage an opponent who got 'inside' the weapons held by the forward ranks – regardless of whether that opponent was prone or standing. Thus the different ranks served different purposes in a combat situation: rank one to directly engage the enemy, or keep him at bay or push him back with the pike; ranks two to five to cover the gap between the files and to engage prone or standing opponents who got within the reach of the front rank; and ranks six to sixteen to provide protection for the formation against enemy missiles. Importantly, at no point is any member of the phalanx required to engage a prone opponent with the butt of their *sarissa*. It is possible that, if the file-closing *ouragos* did encounter a prone opponent as the phalanx moved forward who, although having already been engaged by members of at least two different ranks, was still alive, he could deliver a *coup de grace* with the butt of his pike to simply finish him off. However, there is no reference to this ever occurring and the chances of someone surviving being stabbed several times with pikes and then being trampled over by a whole file of phalangites would have been slim.

Modern theories suggesting the offensive use of the *sarissa* butt are not supportable. The mechanics behind the use of the butt as a weapon conflict with all models and accounts of phalangite combat. A broken weapon could not be reoriented to use the butt as the point of impact in most cases. A broken *sarissa*, with a heavy butt still attached, may have been used to parry blows under the most dire of close combat conditions, but is unlikely to have been consciously regarded as an alternative weapon. Similarly, there is little evidence to support claims that the spike could be used to finish off a fallen adversary. It can therefore be concluded that the butt was only designed to function in its other capacities: to protect the end of the shaft from the elements; to balance the *sarissa* correctly; to allow the pike to be thrust upright into the ground and to add weight to the mass of the weapon. All of these characteristics would play a part in how the *sarissa* was used in the massed formation of the pike-phalanx.

Chapter Eleven

The Phalanx

Like the hoplite armies of the Greek city-states, the Macedonian pikemen of the Hellenistic Age fought in the massed formation of the phalanx. The careful arrangement of the pike-phalanx was an essential part of overall battlefield strategy as how it was deployed subsequently dictated how any ensuing combat would play out. It is for this very reason that a number of ancient writers devoted considerable time to research and compose examinations and instructional manuals on the operational techniques for marshalling a Hellenistic army. Yet despite this detailed corpus of literature recounting the organizational workings of the pike-phalanx, how this formation was arranged and maintained across the Hellenistic Period has been a contentious issue among scholars almost since it first took to the field. By examining the literary accounts of the pike-phalanx, coupled with an analysis of how the panoply of the phalangite dictated, and in some ways limited, how the pike-phalanx could be deployed, it becomes clear that the Hellenistic pike-phalanx was a very adaptable formation – one that could be configured and reconfigured to meet the varied environment of the ancient battlefield.

Many modern works on Hellenistic warfare contain only a very basic analysis of the formations used in this style of fighting, how these formations were constructed or the reasons behind their varying deployments. For many scholars, the basis of Hellenistic warfare is a simple and rigid 'block' formation – a carefully arranged mass of ranks and files of heavily armed pikemen we know as the phalanx.[1] Similarly, when an ancient text uses a generalisation to describe how an army arranged itself 'in line of battle' (κατὰ τὴν τάξιν), it is clear that some form of commonly understood formation, one that did not require further elaboration for the ancient reader, is being referred to.[2] This is most likely the basic 'block' phalanx unless something more specific is mentioned. Yet for the ancient military writers, the structure of the pike-phalanx was much more complex than a simple mass of ranks and files. Rather, it was a sophisticated military instrument based upon the careful arrangement of units and sub-units, and with a structure following precise mathematical models.

THE SIZE OF THE FILE
(aka THE DEPTH OF THE PHALANX)

The three main ancient works on Hellenistic formations that have survived to the present day are those of Asclepiodotus, Aelian and Arrian. Additionally, Polybius devotes a section of book eighteen of his *Histories* to a discussion of the workings of the pike-phalanx. It is from these four sources that the bulk of our knowledge about the structure of the pike-phalanx is derived. While Polybius was writing at the end of the Hellenistic Era, and the manuals were all written much later (the first and second centuries AD) than the time-period they are discussing, all of these writers used earlier military treatises as their source material. In many cases this source material included works that were contemporaneous with the Hellenistic Age – some of which were written by some of the major military personalities of the time. Aelian for example, who states in the dedication of his work that the contents outline 'Alexander of Macedon's manner of marshalling his army', provides an exhaustive list of earlier works he has consulted as part of his research into Hellenistic warfare.[3] Aelian cites reading texts written by Stratocles, Frontinus, Aeneas, Cyneas the Thessalian, Pyrrhus of Epirus and his son Alexander, Clearchus, Pausanias, Evangeleus, Polybius, Eupolemus, Iphicrates, Poseidonius and Brion. Sadly, many of these works have not survived the passing of the centuries.[4]

In the manuals of Asclepiodotus, Aelian and Arrian, and in the examinations of Polybius, the 'block' phalanx is used in all of the initial discussions of the structure of Hellenistic pike formations and other configurations of the phalanx are only considered following the lengthy assessment of the structure of the 'block'. As such, any modern examination of the mechanics of the Hellenistic pike-phalanx must begin with an understanding of how the ancient writers viewed the arrangement of the simple 'block' formation and a comparison of these views to current scholarship.

All of the manuals agree that the smallest unit within the phalanx was the file – variously known as the *lochos* (λόχος), the *dekad* (δεκάδος), the *enomotia* (ἐνωμοτία) and several other names.[5] The manuals further state that the files of the phalanx (regardless of what name they went by) could come in a variety of forms:

Asclepiodotus, *Tactics*, 2.1-2:
    ...some have formed the file [*lochos*] with eight men, others with ten, others with twelve, and yet others with sixteen men, so that the phalanx will be symmetrical both for doubling the depth of its units...so that it may comprise of thirty-two men [in each file], and for reducing it by half to eight men....The file was previously called a row (*stichos*), an *synomoty*, and a *decury*, and the best man and leader of the row was called a *lochagos*, while the last man was called an *ouragos*. But later on when the row was reorganized, its parts received different names; for the half-file is now called a *hemilochion* or the *dimoiria* – the former being the term used for a file of sixteen men, and the latter for one of twelve, and the leader [of a half-file] is now called a *hemilochites* and a *dimoirites* [respectively]...the quarter-file is called an *enomotia* and its leader an *enomotarch*.

Aelian, *Tactics*, 4, 5:
    To organize men into 'files' [λόχοι] is to arrange men one behind the other...The numbers that make up a file vary – for some make it with eight men, some with twelve, and others with sixteen... because it allows the depth of the phalanx to be easily increased to thirty-two men or lessened, and so decrease the depth of the phalanx, to eight men... The best man of every file is positioned in the front and is known as the 'file leader' [*lochargos* – λοχαργός], the 'commander' [*hegemon* – ἡγεμών] or the 'fore-stander' [*protostates* – πρωτοστάτης]. The man at the rear of the file is known as the 'rear commander' [*ouragos* – οὐραγός] or the 'bringer up'. The file as a whole is called a *dekad* [δεκάδος] or an *enomotia* [ἐνωμοτία]. There are those that call only one fourth of the file an *enomotia*, and the commander of such an *enomotarch* [ἐνωμοτάρχης], while two *enomotiae* are called a *dimoiria* [διμοιρία] under the command of a *dimorites* [διμοιρίτης] – in other words, a half file is sometimes called a *dimoiria* led by a *dimorites* and this unit covers the rear of the line.

Arrian, *Tactics*, 5, 6:

> The number of men...is called a *lochos*. Some make the number in the *lochos* ten [men], others add two to the ten, and some even [make it] sixteen...If it is necessary to double the depth to thirty-two men, the formation will be proportionate [if based on files of sixteen]. If the depth is set to eight at the front, the phalanx will not entirely lack depth...[The commander] is called both 'fore-stander' and *hegemon*. Some call the *lochos* a *stichos*, some a *decury* (since for them the *lochos* happens to be of ten men). Concerning the *enomotia*, [this issue] is ambiguous as some say that this is another name for the *lochos*. Others call the *enomotia* a quarter of a *lochos* and call the leader an *enomotarch*. Two *enomotiae* are [called] a half-file and its leader a *dimorites*.

Across the three military manuals we are given a variety of terms and configurations for the most basic unit of the phalanx, all use files of sixteen in their analyses of the formation's structure. All three writers also use the term *lochos* as a generic term for the file but outline other names, presumably referred to in their source material, under which the *lochos* had gone by – *stichos*, *synomoty*, *dekad/decury*, *enomotia*. All of the writers additionally outline a standard arrangement for the file of sixteen deep, which could be redeployed either to a 'double depth' of thirty-two or a 'half depth' of eight. A file leader (of varying name) and a file closing *ouragos* are also mentioned regardless of the file's configuration. A sixteen man file could also be divided into half files of eight – known as a *dimoiria* (διμοιρία), and commanded by a *dimorites* (διμοιρίτης), according to Aelian and Arrian, while Asclepiodotus states that *dimoiria* was the name for the half file only when the whole file consisted of twelve men, while the name of a half file of eight (from a total file of sixteen) was a *hemilochion* (ἡμιλόχιον) led by a *hemilochites* (ἡμιλοχίτης). The file could also be arranged in quarter files of four men each which, according to Aelian and Asclepiodotus, were known by some as an *enomotia*, each commanded by an *enomotarch*, rather than that being the name for the file as a whole and its leader. Interestingly, Aelian states that a deployment by quarter-files was not a standard practice – suggesting that it was more of an organisational aspect rather than a functional one. Arrian goes as far as to state that a deployment only four deep gives the resultant formation no depth, suggesting that it was either not a common practice or, if it was, it was of very limited tactical value.[6] Regardless of how commonly used some of these other configurations were, it is clear from the sources that even the basic sixteen-man file of the phalanx could come in a variety of forms (Fig.22).

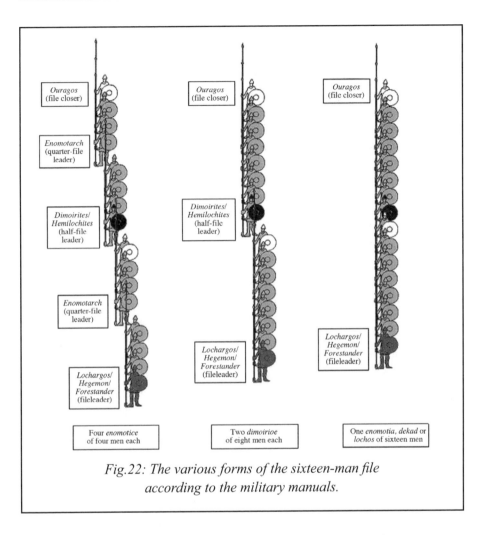

Four *enomotice*
of four men each

Two *dimoirioe*
of eight men each

One *enomotia, dekad* or
lochos of sixteen men

*Fig.22: The various forms of the sixteen-man file
according to the military manuals.*

These varying descriptions of the file show that its size and arrangement was not constant across the Hellenistic Period. This has caused considerable contention among scholars attempting to understand the dynamics of the conflicts of the Hellenistic Age at particular points in time. Scholars have forwarded numerous theories about the developmental processes which took place to increase the size of the file, who may have initiated these changes, and when they may have occurred. Some suggest that the Macedonian phalanx began as an eight deep formation and that its depth was later increased to an eventual depth of sixteen.[7] This would seem to follow the sequence set down by Asclepiodotus and Aelian (Arrian only notes a deployment to a half-depth of eight rather than a standard one). While many of these theories fail to provide a specific time frame for when they believe a standard eight deep formation may have been in use, such conclusions may be based upon Callisthenes' account of Alexander the Great's engagement at Issus in 333BC. Polybius, in his critique of Callisthenes' account, describes how, as Alexander's infantry advanced onto the narrow coastal strip upon which the battle was fought, they were initially deployed thirty-two deep. Then, as the army advanced further and the ground continue to open out, allowing for a longer battle line to be deployed, they redeployed to sixteen deep and then finally to eight deep.[8] Curtius also refers to the initial deployment of thirty-two deep at Issus.[9] He then states that, as the terrain of the battlefield widened, the troops were able to deploy in their 'usual order'. This then correlates (at least in part) the account of Curtius with that of Callisthenes and may be the basis for the conclusion for Alexander's army using a standard file of eight men.

Unfortunately, Curtius does not state the depth of the 'usual order' he refers to – it can only be assumed that it is either the sixteen or eight-deep deployment mentioned by Callisthenes. Regardless, the accuracy of such statements seems certain when the source material used by these writers is considered. Callisthenes (the writer whose work Polybius was commenting on) was Alexander's official chronicler for the expedition against the Persians and was present at the engagement. This would then place very little doubt over the details that Callisthenes provides. Curtius' main source, on the other hand, was Cleitarchus – a Greek writing in Alexandria sometime after 310BC – and so is not contemporaneous with the events he describes. However, Cleitarchus drew heavily from eyewitness accounts to complement the existing Alexander narratives and many of the details that we receive third hand through Curtius can also be considered correct (although Curtius does tend to exaggerate other things such as Alexander's vices). Despite the validity of the source material used by Polybius and Curtius, does Callisthenes' reference to an eight-deep deployment at Issus mean that this was the standard depth of the file in the time of Alexander the Great?

Anaximenes, through his use of the term *dekad* to describe the file, indicates that it was comprised of ten men in the time of Alexander II.[10] There is also clear evidence to indicate the use of sixteen-man files by Alexander the Great by at least 334BC. Alexander also formed improvized units on the phalanx based upon files of sixteen in 324BC (see following). Yet if the regular file of the pike-phalanx was ten deep no later than the time of Alexander II's death in 368BC, was sixteen deep in 334BC, was then reduced to eight deep at Issus in 333BC as some scholars interpret the passage of Callisthenes, only to later be enlarged to a standard depth of sixteen by 324BC, then it must be assumed that the pike-phalanx had followed an evolutionary path which saw its size fluctuate dramatically. Head calls this proposed model 'unduly complex' and suggests that the eight deep formation referred to at Issus by Callisthenes is a description of a phalanx deployed by half files – which would then make the standard depth of the file sixteen.[11] Such a 'half-depth' deployment is outlined in all three of the military manuals, which base their examinations of the phalanx on base files of sixteen men, but which state that the formation could be arranged eight deep if required. This 'required' eight-deep deployment seems to be what has occurred at Issus.

Arrian states that Alexander's entire deployment at Issus was made to occupy the complete breadth of the battlefield, stretching from the sea on his left to a range of hills on his right, so that the Persians would not be able to outflank the Macedonians with their superior numbers.[12] Consequently, the necessities of the impending engagement dictated the order of battle. If keeping the pike units of his phalanx deployed in a sixteen-deep formation would not provide enough frontage to adequately cover the battlefield and avoid encirclement, then it would make sense for Alexander to bring the rear half files of his units forward. This is a process known as 'doubling' (διπλασιάζω) in the military manuals as, while moving the rearward half files forward essentially halves the depth of the formation, it simultaneously doubles the number of men across the front of the line.[13] Interestingly, Polybius disputes Callisthenes' description of the use of an eight deep formation at Issus due to his calculation that a formation of this configuration would not actually fit onto a battlefield of the size stated by Callisthenes.[14] Thus Polybius must assume that Alexander's phalanx was arrayed in larger files – most likely sixteen deep – the depth he also uses for his own examination of the structure of the pike-phalanx.

McDonnell-Staff suggests that the Macedonians regularly formed up sixteen deep but then, to actually engage the enemy, their standard practice was to reduce the depth of their line by moving the rear half files forward.[15] Hammond also claims that the depth of the pike-phalanx in the time of Philip and Alexander was between eight and ten men.[16] While this may have been what has happened at Issus (and may be what McDonnell-Staff and Hammond are basing their conclusions on), it cannot be taken as a standard practice under Alexander, who does not seem to deploy in this manner again, or for any other Hellenistic commander except, possibly as at Issus, when a particular characteristic of the impending battle may have required a more extended line.[17] On the other hand, if Polybius' critique is accepted, such a deployment seems to have never occurred at Issus at all. The extension of the line is one of the specifically stated purposes of the process of 'doubling' in the military manuals, but they do not state that this was done at every encounter in order to engage. Indeed, their use of a sixteen-deep file for their examinations of the phalanx would suggest that this was the standard arrangement of the file, while eight-deep deployments were more improvised dispositions out of tactical necessity. The important thing to note is that, regardless of whether the 'doubling' of the file was a standard practice or a manoeuvre undertaken only out of tactical necessity, if the result of this 'doubling' manoeuvre was a deployment eight deep as Callisthenes states happened at Issus, then the standard arrangement for the files of Alexander's phalanx was still sixteen deep. Thus it seems clear that the army of Alexander the Great was based upon files of sixteen men by at least 333BC.

Arrian's narrative also shows that Alexander used sixteen deep files throughout his campaign. In 324BC Alexander began integrating Persian troops into the pike-phalanx. These troops were organized into files of sixteen – although the original name of the ten-man file, the *dekad*, seems to have been retained:

> ...he enlisted them [i.e. the Persians] into the Macedonian ranks, with a Macedonian *dekadarchos* leading each *dekad* and, following him, a Macedonian *dimoirites* and a 'ten-stater man' [*dekastateros*] – so named after his pay – which was less than that of the *dimoirites*, but greater than that of the soldiers with no supplement.[18] Added to this were twelve Persians and, last in the *dekad*, a Macedonian who was also a 'ten-stater man'; so within the *dekad* there were four Macedonians, three of whom were on increased pay, the commander of the *dekad*, and twelve Persians.[19]

This passage may be the basis for models which suggest that the file was increased in size to sixteen later in the time of Alexander the Great – these models basically accept Callisthenes' reference to an eight-deep deployment at Issus as the standard size of the file at that time and, therefore, Alexander's use of a sixteen-man file in 324BC is some kind of organizational reform. Heisserer, for example, argues that Arrian's description of Alexander's Perso-Macedonian files cannot be applied to the Macedonian army as a whole, but can only be attributed to Alexander's units from 324BC onwards.[20] However, if Callisthenes' reference to an eight-deep deployment is seen as a description of the phalanx deploying by half files at Issus as the evidence would suggest, then the concept of a reform which doubled the size of the file to sixteen deep in 324BC is unlikely.

What Arrian's later passage does provide is a detailed account of the internal structure of the file (and the wages of the men within it). First it demonstrates a retention of the nomenclature associated with the file as it was under Alexander II – the *dekad*.[21] As Arrian states that this terminology was also applied to a file, and its commanding officer, when based upon sixteen men rather than ten, the use of this terminology should not be seen as indicative of the number of soldiers within the file in the time of Alexander. Rather the use of the term *dekad* in this context should be seen as a generic use of the term to refer to the file as a whole regardless of its size. Arrian's passage additionally outlines how the regular soldier in Alexander's army received some form of wage (but no supplement), and provides confirmation of the names and number of officers found within each file of the phalanx. Furthermore, Arrian provides details of each officer's rank in relation to each other based upon their respective pay grades.

However, despite Arrian's passage being a singular reference to many of the pay grades within the Macedonian army in 324BC, it cannot be automatically assumed that these levels did not exist before this time and were only created as part of the 'mixed phalanx' that incorporated the Persians into the Macedonian army as Heisserer suggests. Arrian, for example, refers to officers such as Abreas, whose rank is given as a *dimoirites*, who had come to Alexander's aid in India prior to the creation of the Perso-Macedonian units in 324BC.[22] Therefore, such a position as the rank assigned to Abreas cannot have been created as part of any supposed later reform to enlarge the size of the file.

From the other sources we know that the positions of file-leader and file-closer had existed well before 324BC as well. Thus it can only be assumed that the internal structure of the file, as outlined in Arrian's passage for 324BC, was followed across the period of Alexander's reign and that it was only the nationality of those who took up positions within the file which were altered. Furthermore, the authors of Hellenistic military manuals, some of whom claim to be writing about the army of Alexander, all refer to the same officers within the files of Hellenistic armies as Arrian outlines in his narrative. This would then correlate with Alexander's pike-phalanx being based upon files of sixteen, with a corresponding number of officers, as early as the battle of Issus in 333BC, if not earlier. This, in turn, suggests that the positioning of these officers within each file of the phalanx was a standard organizational arrangement.

Within the file, Arrian outlines the presence of a file commander (called a *dekadarchos*), three subordinate officers (1 x *dimoirites* and 2 x *dekastateroi*) and 12 regular infantrymen (making the base strength of the file 16 men).[23] Aelian, in his examination of the Hellenistic phalanx, similarly refers to files of sixteen heavy infantry arranged in half files with the four corresponding officers – a structure outlined in the works of the other tactical writers.[24] According to Arrian, the *dekadarchos* was located at the head of each file with the *dimoirites* the leader of a half file.[25] Thus the *dimoirites* must have been located in the ninth position (of a sixteen-deep file) in command of the last half of the unit. This would also correlate with the possible deployment of Alexander's pike-phalanx in half files of eight men at the battle of Issus in 333BC as is suggested by Callisthenes.

Arrian, Asclepiodotus and Aelian also refer to a 'file closer' (*ouragos*) who can be associated with one of the *dekastateroi* that is mentioned in relation to Alexander's Perso-Macedonian files of 324BC.[26] As both half files would require a file closer, each of the two *dekastateroi* described by Arrian would be located in the rearward position of each half file (i.e. in the eighth and sixteenth positions when deployed sixteen deep). Between the file/half-file leaders and the file/half-file closers, would have been positioned an equal number of regular infantry (six to each half-file) thus giving the half-file a strength of eight men, and a full file a strength of sixteen (and both half files would have an officer at their front and rear). As such, the configuration of a sixteen deep file of Alexander's integrated Perso-Macedonian phalanx in 324BC would have been (Table 16):

POSITION IN FILE	RANK	ROLE
1	*Dekadarchos*	File leader
2-7	6 x Persians	Regular infantry
8	*Dekastateros*	Half-file closer
9	Dimoirites	Half-file leader
10-15	6 x Persians	Regular infantry
16	*Dekastateros*	File closer

Table 16: The distribution of officers and troops within a file of the phalanx.

It is also likely that this was the configuration for the Macedonian phalanx prior to 324BC as well – but with standard Macedonian infantry taking the place of the Persian conscripts in positions 2-7 and 10-15. Some have interpreted Arrian's description of the new units of 324BC as meaning that there was a *dekadarchos* at the head of the file (position 1) then a *dimoirites* immediately behind him (position 2), then a *dekastateros* (position 3), then 12 Persian infantry men (positions 4-15) and then another *dekastateros* closing the file (position 16).[27] Such a configuration seems unlikely. Through the process of 'doubling', a sixteen-man file would be reconfigured into two half files of eight men each by simply having the rear half of the file move forward to either the left or right to take up a position beside the front half file (Fig.23).

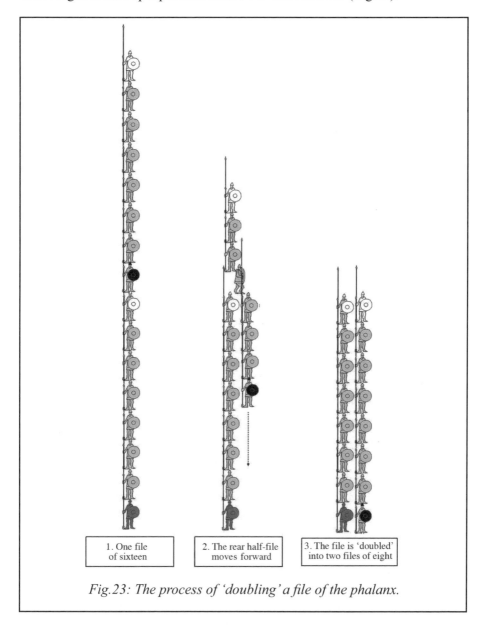

| 1. One file of sixteen | 2. The rear half-file moves forward | 3. The file is 'doubled' into two files of eight |

Fig.23: The process of 'doubling' a file of the phalanx.

Importantly, for understanding the process of 'doubling', it must be remembered that many texts state that officers were positioned at the head and rear of the file.[28] This must have been the case regardless of whether the file was in its original configuration or when it was 'doubled' so that experienced men would always be located across the front and back of the formation.[29] As such, there are only certain ways in which the officers of the file could be positioned and the process of 'doubling' undertaken.

The first thing to note is that Arrian specifically states that the *dimoirites* was positioned at the head of the half file.[30] This statement seems to be ignored by scholars who interpret his description of the configuration of Alexander's files in 324BC as meaning that the *dimoirites* was second in the line (and with one of the *dekastateroi* in position three in some models). This alone would suggest that models which place officers in the first three positions in the file are incorrect. Furthermore, if officers occupied the front three positions of the file as is suggested, and the file was then 'doubled' by having the rearward half file move forward, officers would not be present in both of the forward positions, nor in both of the rearward positions, of the two redeployed half files (Fig.24).

Thus, if three of the four officers of the file are positioned in the first three positions of the file, a redeployment of an entire formation through the process of 'doubling' its files by moving the rearward half file forward results in every second file of the reconfigured formation being led by a standard infantryman rather than an officer. Furthermore, every alternate half file would not have an officer in the position of a file closer either. This contradicts what is stated in the ancient texts about the officers being located across the front and rear of the formation and further indicates that models which suggest that officers took up the first three positions in the file are incorrect.

In relation to Alexander's combined Perso-Macedonian files of 324BC, if such a configuration was followed for the structure of the phalanx, when the files were 'doubled' each successive half file in the formation would be led by a Persian conscript – who had occupied position nine in the sixteen man file prior to the doubling being undertaken. This seems unlikely and suggests that position nine was occupied by an experienced Macedonian officer acting in the capacity of half-file leader – this would have been the *dimoirites* just as Arrian states.

Connolly suggests that the *dimoirites* (whom he places in position two of the file) was the half-file leader and that a *dekastateros* (whom he places in position three) was a half-file closer.[31] It is uncertain how officers in these posts could take up their appropriate positions once the file was 'doubled' or how a

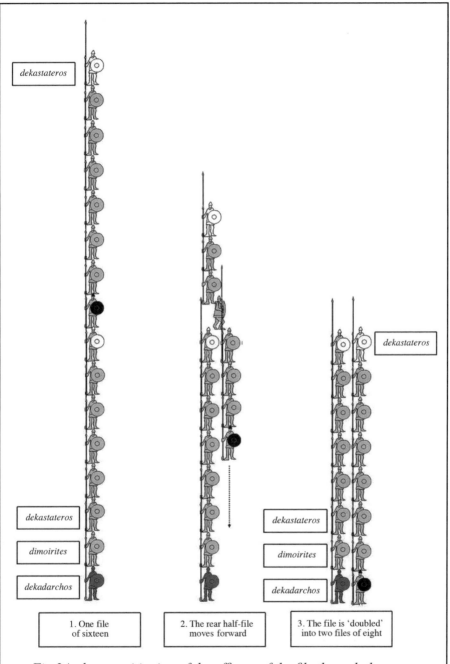

dekastateros

dekastateros

dimoirites

dekadarchos

1. One file
of sixteen

2. The rear half-file
moves forward

dekastateros

dekastateros

dimoirites

dekadarchos

3. The file is 'doubled'
into two files of eight

Fig.24: the repositioning of the officers of the file through the process of 'doubling' by moving the rear half file forward if the file is configured as per some interpretations of Arrian (Anab. 7.23.3-4).

dekastateros in position three can even be considered a half-file closer when most of the remaining men in the file are positioned behind him.

If the file was configured with the officers in the first three positions, but the process of 'doubling' was accomplished through each second man in the file side-stepping to either the left or right, and then both half files shifting forward to close up the intervals between each man, this would place the *dimoirites* at the head of the new half file as Arrian states, but the supposedly file-closing *dekastateros* in position 3 would maintain his place in the original file – in effect now taking up the second position in the half file – which leaves no officer at the rear. Consequently, a 'doubling' manoeuvre undertaken in such a manner and with a file configured with officers occupying the first three spots in the file would still result in a lack of officers positioned across the front and rear of both of the rearranged files – again, contrary to Arrian (Fig.25).

The only way in which the file of the phalanx can be 'doubled', yet still maintain officers in both the forward and rearward positions, is if the officers are positioned at the head and rear of each half-file. In this case, when the 'doubling' manoeuvre was undertaken officers would still be located in the forward and rearward positions (Fig.26).

Thus it seems clear, from an understanding of the required positions for the officers within the file, that not only were the four officers within the file that are mentioned by Arrian positioned in a way that would ensure that experienced men would form the front and rear of the file when it was 'doubled', but also that the process of doubling had to involve the rearward half-file moving forward into a new position as the depth of the formation was halved.

The military manuals also contain another reference to the positioning of men within the structure of the file which is somewhat confusing. Asclepiodotus, for example, states that: 'the leading man [of the file] has been given the title of 'fore-stander' (*protostates*), while the one who follows him is called a 'follower' (*epistates*), so that in the whole file there come first a *protostates*, then an *epistates*, then successively a *protostates* and then an *epistates*, and so on, one after another, until you reach the file closer (*ouragos*).'[32] Both Aelian and Arrian make similar statements in their works.[33]

Such passages should not be seen as a reference to the positioning of the officers and/or regular soldiers within the structure of the file. The first thing to note is that while the term *protostates* is one of the names given to the file leader in the ancient military manuals, if every second man in a sixteen-man file was given this designation, there would then be eight men with the title of *protostates* in each file. Yet the manuals outline the presence of only four

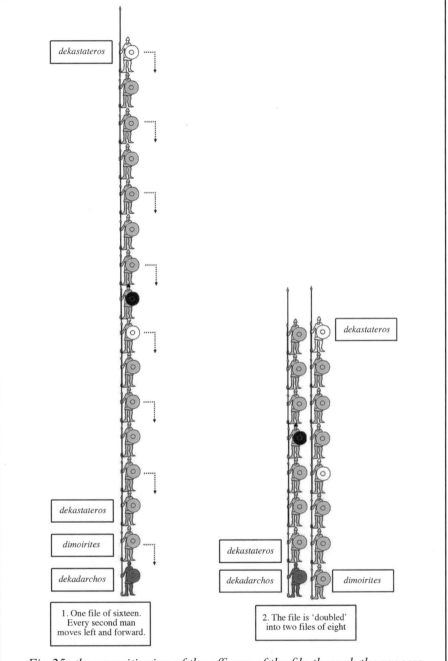

Fig.25: the repositioning of the officers of the file through the process of 'doubling' by moving every second man forward if the file is configured as per some interpretations of Arrian (Anab. 7.23.3-4).

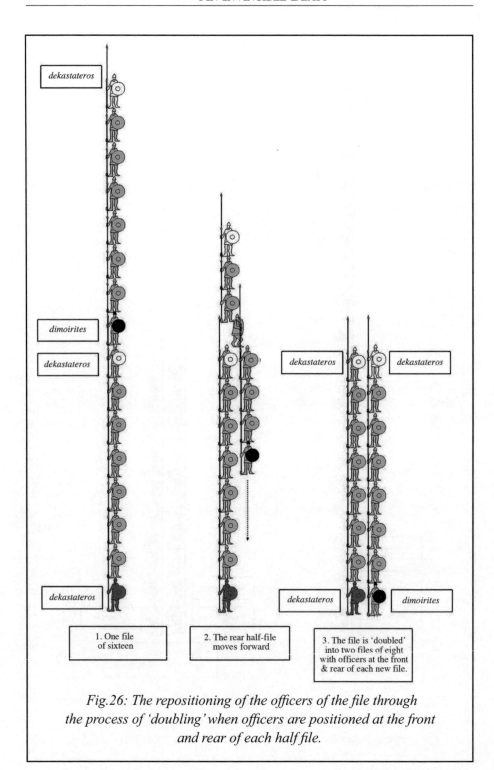

Fig.26: The repositioning of the officers of the file through the process of 'doubling' when officers are positioned at the front and rear of each half file.

officers in the file – and only one of these is actually called the *protostates* in reference to his rank and position.

Nor can the designation of men as either a *protostates* or an *epistates* be seen as a reference to how the file was 'doubled'. If, for example, it is assumed that the process of 'doubling' required each *epistates* to move forward to their left or right to create a new half-file, the same problems are encountered in regard to the positioning of the offers in each of the new files, when doubling is undertaken by having each second man in the file move into a new position (as per Fig.25).

Additionally, Asclepiodotus states how, when the entire phalanx was deployed, the whole first rank would comprise of men with the title of *protostates*, the second rank would be made up of men with the title *epistates*, the third would be another rank of *protostates*, and so on all the way back to the file closer.[34] If the file was 'doubled' by having each *epistates* move forward, the resultant half files of the phalanx would be alternatively a half file of men only with the title of *protostates*, then a half file of men only with the tile of *epistates*, then another half file of men with the title of *protostates*, then *epistates*, and so on. This would then mean that each rank across a 'doubled' phalanx would contain men of each designation in alternating order – which is not where Asclepiodotus says they should be. This again demonstrates that the files of the phalanx were not 'doubled' by having each second man move forward to a new position.

However, if the officers are positioned at the head and rear of each half file, and if the 'doubling' procedure is then undertaken by having the rearward half-file move forward, both leaders of the newly deployed half files would hold the designation of *protostates*, both file closers would have the title of *epistates*, and each of the men in each new half file would hold the same designation as the man to either their left or right. Such correspondence of titles across the ranks of the formation holds true for the phalanx as a whole regardless of whether the files are 'doubled' or not, and even when the phalanx is deployed to a double depth of thirty-two. This must therefore be what Asclepiodotus is describing. Importantly, this retention of ranks of men with the same titles following the process of 'doubling', combined with the correct positioning of the officers at the front and rear of each half file, demonstrate that the officers could not have taken the first three places in the file and that 'doubling' had to have been conducted by moving the rearward half files forward (Fig.27).

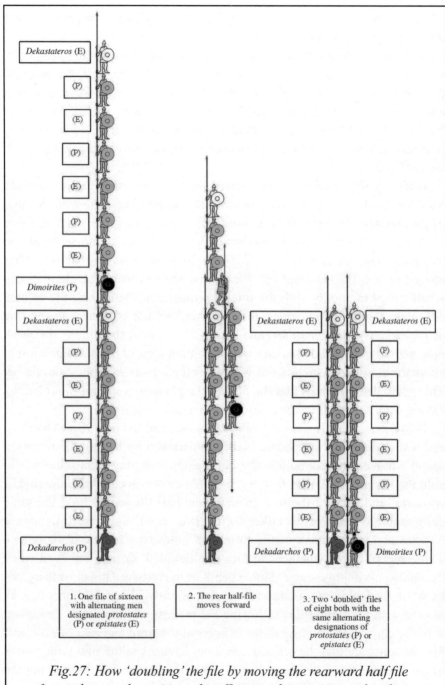

Fig.27: How 'doubling' the file by moving the rearward half file forward correctly positions the officers and maintains ranks of men titled protostates and epistates.

The retention of men positioned beside each other with the same title, even when the file is 'doubled' will occur no matter what depth the phalanx is deployed at so long as the numbers of each file are even. This explains why the various depths of the file that are outlined in the military manuals are all divisible by two as this ability to create half files of the same size, with men in positions corresponding to their title, and with officers at the front and rear, is all part of the symmetrical nature of the phalanx on which the manual writers place so much emphasis.

Thus it seems that, rather than being a reference to the location of officers, the designation of men as either a *protostates* or an *epistates* within the structure of the file seems to merely be a way for each member of the file to know his place (and for commanders to know how the files/phalanx could be reconfigured if necessary) in a similar way to which members of units in modern armies count off their number 'one-two-one-two....' regardless of how many people there are in the unit. If there was another, possibly tactical, purpose for designating the men of the files as either a *protostates* or an *epistates*, the military manuals unfortunately do not detail what that was. Cascarino suggests that the designation of the men of the file as either a *protostates* or an *epistates* was an organizational aspect of the phalanx common only to the Successor Period.[35] However, there is nothing that indicates that Alexander's troops were not organized along similar lines other than the assumption that most of what is provided in the military manuals is only reflective of a later part of the Hellenistic Age – a conclusion for which there is no firm basis. Additionally, such conclusions run contrary to the time period which Aelian specifically states he is examining in his work.

Importantly, the arrangement of the files of Alexander's phalanx in 324BC, with files of sixteen divided into half files of eight, each with their own leading and closing officers, correlates not only with the structure of the file as provided in the later manuals (as per Aelian's claim), but also agrees with a deployment of Alexander's phalanx to a half depth of eight at the battle of Issus in 333BC. This then demonstrates a continuity of a sixteen deep structure for the phalanx as the standard deployment across the period of Alexander's campaign in Asia. If Alexander's files were a standard sixteen deep in 333BC, this would further suggest that the pike-phalanx had been initially ten deep under Alexander II as recounted by Anaximenes, but that the depth had been increased to sixteen sometime prior to Alexander the Great's departure for Asia in 334BC. This model of a standard sixteen deep formation, which seems to more closely fit with the descriptions of the pike-phalanx in both the military manuals and the narrative histories, is followed by many scholars for battles across the Hellenistic Period for the time of Alexander the Great onwards.[36]

Sheppard offers that the file was still a unit of ten under Philip II and then expanded to sixteen under Alexander.[37] This also seems unlikely. Anaximenes' use of the term *dekad* to describe the files of the phalanx of Alexander II show that this formation began with files of ten, and Frontinus indicates that Philip's army was still based on units of ten for at least part of his reign when he describes how each *dekad* was given one servant to carry their heavy equipment.[38] When, then, did the change to sixteen-deep files occur? Alexander had fought only a few major battles prior to his engagement at Issus in 333BC which is the first clear reference to a standard sixteen-man file within the phalanx (regardless of whether this is then 'doubled' to eight deep or not). Alexander's previous encounter at Granicus in 334BC had involved only about half of his army and the depth of his deployment is not mentioned in any of the ancient narratives.[39] This battle, and his Thracian campaign prior to his invasion of Asia, seem unlikely to have been the catalyst for a change to the organizational structure of the pike-phalanx. Indeed, the accounts of Alexander's Thracian campaign suggest that other elements of his army were structured in a manner similar to that outlined by the military manuals as early as 335BC (see following). It seems more likely that an increase in the size of the file from ten to sixteen occurred after the reign of Alexander II yet prior to the reign of Alexander the Great – with the change most likely occurring when Philip II 'organized' (as Diodorus put it) the Macedonian phalanx early in 358BC.

If the pike-phalanx was ever a standard eight deep as some scholars suggest, it would have been when it was first created by Iphicrates in 374BC.[40] While Iphicrates altered the equipment carried by the traditional Greek hoplite to create a pike-wielding *peltast*, both Diodorus and Nepos make no mention of an alteration to the deployment of troops as part of this series of reforms. As such, it seems likely that they still deployed in the standard eight deep formation for hoplites at that time.[41] An eight-deep pike formation may then have been what was initially introduced to Macedon under the reigns of Amyntas and Alexander II between 374BC and 368BC – a formation which was then enlarged to ten deep under Alexander II some time before his death. This would then correlate with Anaximenes' account of Alexander II (re)organizing the phalanx into files of ten and 'other commands' – the reorganization was part of a restructure of the army following, and possibly altering, the reforms of Iphicrates which increased the depth of the file.

Another point that requires consideration is that all three military manuals state that the pike-phalanx could be deployed with a depth of twelve – with Asclepiodotus going so far as to assign different names to both the file and its officers when deployed twelve deep as opposed to sixteen. Asclepiodotus infers that a standard deployment to a depth of twelve was a development that occurred after the common use of a sixteen-man file when he states that '...but later on, when the row was reorganized, its parts received different names; for the half-file is now called a *hemilochion* or the *dimoiria* – the former being the term used for a file of sixteen men, and the latter for one of twelve...'.[42] This would suggest that sometime after the reign of Alexander the Great, during which time sixteen-deep files were the standard, the use of twelve-deep files became a common practice. Strangely, the narrative histories for the time of the Successors do not detail the use of a twelve-man file at any of the major engagements of the time. This suggests the continued use of sixteen-man files with the phalanxes in these engagements.

Even among the armies of the Classical Greek city-states, there is only one reference to a hoplite formation possibly being deployed twelve deep – that of the Spartans at Leuctra in 371BC – yet even here the depth of the formation is only given in the vague term of 'not more than twelve deep'.[43] However, the Spartans lost the battle of Leuctra, an outcome which sent shockwaves across the ancient Greek world and shattered the myth of Spartan invincibility, and so their deployment is unlikely to have been the inspiration for the development of the pike-phalanx over the following years. The writers of the military manuals all clearly believed that such deployments did exist otherwise they would not have written about them.

Even if there is doubt about whether the phalanx was originally eight deep following the reforms of Iphicrates or not, it seems clear that the file was at least ten deep by the reign of Alexander II and was sixteen deep by Alexander III. This depth seems to have then been used as a standard means of deployment up until the end of the Hellenistic Period. As such, if the file was ever twelve deep in the later parts of the Hellenistic Age as Asclepiodotus suggests, it can be regarded as either a short term experiment which was not continued, or as some kind of improvized formation, possibly dependent upon the number of men present in certain armies. In either case it cannot be regarded as a standard structure for the pike-phalanx. Importantly, what becomes clear from a critical examination of the available evidence is that from the very beginning of Alexander the Great's campaign in Asia, the file of the phalanx was arranged in a manner which closely correlates with the structure of the file as is given in the later military manuals. A review of the histories of Alexander's campaign shows that many other features of his pike-phalanx also mirrored the details of the larger formations found in these later works.

THE LARGER UNITS OF THE PHALANX

According to the manuals, it was upon the file that all other larger units within the phalanx were based. The manuals describe how, through a simple process of combining two units together to create a new and larger one, a basic file could be expanded further and further to eventually create a formation known as a quadruple-phalanx (*tetraphalangarchia* – τετραφαλαγγαρχία) of more than 16,000 men (Table 17).

Name of Unit	Unit Commander	Number of Files	Number of Men
lochos, dekad or *enomotia*	*lochargos, hegemon* or a 'fore-stander'	1	16
dilochia (διλοχία)	*dilochites* (διλοχίτης)	2	32
tetrarchia (τετραρχία)	*tetrarch* (τετράρχης)	4	64
taxis (τάξις)	*taxiarchos* (ταξίαρχος)	8	128
syntagma (σύνταγμα)/ xenagia (ξεναγία)	*syntagmatarch* (συνταγματάρχης)/ *xenagos* (ξεναγός)	16	256
pentacosiarchia (πεντακοσιαρχία)	*pentacosiarch* (πεντακοσιάρχης)	32	512
chiliarchia (χιλιαρχία)	*chiliarch* (χιλιάρχης)	64	1,024
merarchia (μεραρχία)/ *telos* (τέλος)	*merarch* (μεράρχης)/ *teleiarch* (τελειάρχης)	128	2,048
*phalangarch*ia (φαλαγγαρχία)/ *strategia* (στρατηγία)	phalangarch (φαλαγγάρχης)/ strategos (στρατηγός)	256	4,096
di-phalangarchia (διφαλαγγαρχία) / *meros* (μέρος)		512	8,192
tetraphalangarchia (τετραφαλαγγαρχία)		1,024	16,384

Table 17: The organizational structure of the pike-phalanx based on files of sixteen.

Cascarino suggests that such a configuration is only applicable to the phalanxes of the later Successor Period rather than to the time of Alexander.[44] Connolly, on the other hand, suggests that such a formation as is described by Asclepiodotus, Aelian and Arrian never existed in reality at all and that what the manuals are outlining are merely exercises in mathematical philosophy.[45] Sekunda similarly suggests that the army that Asclepiodotus describes is that of the Seleucid army of the second century BC that his source, Poseidonius, may have served in, but that many of the units he details are 'fanciful'.[46] Indeed, despite the standardized series of ever increasing units provided in the military manuals, problems are encountered when attempting to tie this stated structure of the pike-phalanx to examples of the size of armies given in the narrative histories for the campaign of Alexander the Great and/or the time of the Successors as the numbers do not readily correlate. For example, Alexander may have had 12,000 pikemen in his army early in his campaign against Persia (see following). Antigonus had 8,000 troops armed in the 'Macedonian fashion' for the battle of Paraetacene in 317BC, while Eumenes had 6,000 men similarly armed in the 'Macedonian fashion' for the battle of Gabiene the following year.[47] Ptolemy IV had approximately 56,000 phalangites for the battle of Raphia in 217BC, and Perseus had 21,000 pikemen for the battle of Pydna in 168BC.[48] While there is little doubt that these are rounded figures rather than accurate troop numbers, not all of them are readily divisible, even closely, by one of the larger units outlined by Asclepiodotus, Aelian and Arrian unless it is assumed that these armies were arranged upon multiples of units no bigger than the *chiliarchia* of roughly 1,000 men. Some figures, such as Antigonus' 8,000 men for the battle of Paraetacene or Ptolemy's 56,000 men at Raphia, may be representative of armies arranged on the larger units of the *phalangarchia*, or even the *di-phalangarchia*, but this can clearly not be the case for all of these listed forces.[49]

It is also interesting to note that some of these given figures would correlate to an arrangement of the units of the phalanx, based upon those given in the manuals, but only if the files that they were based on were comprised of twelve men rather than the standard sixteen. Eumenes' 6,000 men at Gabiene, for example, is close to the total figure of 6,144 for the members of a *di-phalangarchia* made up of 512 files of twelve men each. While many of the accounts of these forces do not go into such detail as the depth of the deployment, the figures may be too close to be pure coincidence. If this is the case, then not only would such an arrangement correlate with what Asclepiodotus states about the use of the twelve-man file occurring after the regular use of the sixteen-man file, but Antigonus' deployment of an 8,000-man phalanx (which would equate to a *di-phalangarchia* based upon files of sixteen) would suggest that the two different modes of deployment were in operation at the same time – possibly due to the number of men available to a commander. It should also be noted that Alexander's 12,000 men is very close to the figure for a *tetraphalangarchia* of 1,024 files of twelve men each (a total of 12,288 men). However, various passages in the ancient texts demonstrate that Alexander's army was based upon files of sixteen from the beginning.

Due to the way that the phalanx was divided into units and sub-units, there are numerous ways in which the structure of an army could be accounted for with the figures given for some of these encounters and yet still correlate with the structure given in the manuals based upon a standard file of sixteen. One possibility is that the army in question contained units that were either under or over the strength given in the manuals. For example, was Eumenes' 6,000 phalangites at Gabiene an over-strength *phalangarchia* or an under-strength *di-phalangarchia* with correspondingly over- or under-strength sub-units within? Were some of the sub-units missing? This seems to have been the case with the organization of Alexander's army early in his campaign (see following).

Another possibility was that the army was arranged in asymmetrical combinations – did Eumenes' 6,000 man formation comprise one 4,000 strong *phalangarchia* and one 2,000-man *merarchia*? Yet another possibility is that Eumenes' army was organized into units no bigger than six *chiliarchiae* of 1,000 men each. Given the strong emphasis placed upon symmetry within the structure of the phalanx in the military manuals, it seems unlikely that armies such as those of Eumenes were based upon asymmetrical combinations. This leaves the only possibilities being either under-strength units or missing ones. In either case, it can only be concluded that such forces were arranged on the largest possible unit or units which would allow for an even distribution of troops across the line. Unfortunately, as the narrative histories do not go into such detail (but only provide a total for the number of phalangites present at the engagement), nor are the ranks of the various commanding officers provided, it is almost impossible to determine the organizational structure of any of these armies to a great deal of certainty. Such examinations can only deal in probabilities rather than absolutes – at least in regard to comparing their structure to the military manuals.

And yet the ancient texts also contain references which suggest that the larger organizational structure of the phalanx that is outlined in the manuals may have been followed by some. Philip V, for example, had a phalanx 16,000 strong for the battle of Cynoscephalae in 197BC.[50] Antiochus the Great similarly

had a phalanx 16,000 strong, which Appian states was armed in the fashion
of Alexander the Great and Philip II, at the battle of Magnesia in 190BC.[51]
While again undoubtedly rounded figures, the stated size of these forces bears
close similarities to the number of men within the largest unit outlined in the
manuals – the *tetraphalangarchia* or quadruple phalanx. While it cannot be
ruled out that Philip's and/or Antiochus' armies may have been based upon
units no bigger than sixteen *chiliarchiae*, the size of both these forces and the
size of the full quadruple-phalanx as given in the manuals may be too close
to dismiss as pure coincidence. Similarly, much later (AD217), the Roman
emperor Caracalla, thinking himself to be the new Alexander, 'organized
a phalanx, composed entirely of Macedonians, 16,000 strong, and called it
'Alexander's Phalanx' and armed it with the weapons that warriors had used
in Alexander's day'.[52] Again, the size of this force closely corresponds with
the size of the *tetraphalangarchia*. Caracalla's phalanx may have been based
upon the information written down in the military manuals of Asclepiodotus,
Aelian and Arrian a century earlier which (in Aelian's case at least) purport to
be describing the structure of the pike-phalanx in the time of Alexander – the
very person Caracalla was emulating.

Thus while it seems that some of the principles set down in the later military
manuals were put into practice by some commanders, it is also clear that these
principles were not set in stone. Rather, it seems that the organizational structure
of the larger pike-phalanx as a whole was adaptable to the number of men at
hand and an army could be based upon the largest tactical unit or units possible
which would still allow for a relatively symmetrical deployment. This would
then mean that many, if not all, of the sub-units of the phalanx as detailed in
the manuals would have been in use across the Hellenistic Period in one form
or another. The development of this adaptability of the phalanx through the
use of symmetrical units and sub-units was no doubt part of the evolutionary
processes that the formation underwent during the course of its period of usage.
These developmental changes affected the structure of the pike-phalanx at
almost every level – right down to the most basic units of the formation.

Anaximenes' reference to '*lochoi* and *dekads* and other commands'
indicates that, back when the pike-phalanx was first adopted by Alexander II,
it comprised more than just files of men.[53] The use of the term *lochos* is most
likely the use of a general term for the file while the reference to the *dekad*
is the use of a name for the file that is indicative of the number of men it
contained. This is similar to the way in which the later manuals provide varying
names for the file; some generic and others specific. Anaximenes' reference to
'other commands', however, clearly indicates that the early pike-phalanx was

structured upon larger units as well – possibly along similar lines, at least in part, to the more elaborate configuration of the later phalanx as is detailed in the manuals. However, it must also be noted that, if the phalanx under Alexander II was based upon files of ten rather than sixteen, even if the structure of the phalanx was exactly the same as the later manuals in terms of the different units and the number of files contained therein, the numbers within each unit as a whole would differ from the manuals due to the smaller size of the file.

Daniel claims that the *lochos* in the time of Alexander the Great, rather than being a file, was a unit of 120 men – similar to the *taxis* outlined in the later manuals.[54] Daniel bases this claim on Arrian's account of Alexander's withdrawal from Pelium in 335BC in which his infantry were drawn up 120 men deep.[55] Daniel states that 'this can only have meant that Alexander's [infantry] were first deployed by *lochoi* (companies) in single file'.[56] Sheppard and Cascarino, on the other hand, suggest that the *lochos* in the time of Alexander the Great was a unit of sixteen files of sixteen men each, or 256 men in total – a unit which was later renamed the *syntagma*.[57] English and Sekunda claim that the *lochos* was a unit of thirty-two combined files (based upon sixteen men per file) creating a unit with an overall strength of 512 men.[58] This would then make the early *lochos* the same as the *pentacosiarchia* found in the works of Asclepiodotus, Aelian and Arrian. Cascarino argues against such a conclusion and suggests that a unit of this size was only ever called a *pentacosiarchia*.[59] How are such claims to be reconciled? Clearly they cannot all be correct for the time of Alexander unless it is assumed that there was considerable structural reorganizing taking place across the breadth of his entire campaign.

The evidence indicates that all of these theories are incorrect. The term *lochos* is only ever used by ancient writers recounting Alexander's campaign in relation to the file.[60] Consequently, it is uncertain upon what the claims for a larger size for the *lochos* are based.[61] As the *lochos* and the file are synonymous, the size of the *lochos* only varied with the increasing size of the file from a unit of ten men to one of sixteen, and the only qualifier for its size is the time period at which the file is being examined. There is no need to assume that the *lochos* was a unit independent of the file that was similar to one of the other stated units in the manuals, although the name of the *lochos* may have been changed as is suggested by some. Nevertheless these other interpretations of the size of the *lochos* have had significant implications for subsequent examinations of the pike-phalanx depending upon what size is attributed to this unit.

Sheppard, for example, further suggests, based upon her interpretation of the early *lochos* as a unit of 256 men, that a formation known as a *taxis* under Alexander was a unit of six combined *lochoi* or 1,536 men.[62] A similar conclusion is reached or used by other scholars who state that the army of Alexander was organized into units of 1,500 men which they also call the *taxis*.[63] Cascarino suggests that each of these *taxeis* was made up of three combined units of around 500 men – the *pentacosarchia* outlined in the later military manuals.[64] This *taxis* of 1,500 men is different to the unit of the same name given in the later manuals which contains only 128 men (as noted earlier, Daniel suggests that this was the size of the *lochos*).

The conclusion that Alexander's pike-infantry were divided into six *taxeis* of 1,500 men, is due to the interpretation of the troop numbers given by Diodorus for the size of the army when Alexander crossed into Asia in 334BC. Diodorus, who is the only source to provide a detailed breakdown of the troops of Alexander's army at this time, states that there was a total of 12,000 Macedonian infantry in the expeditionary force.[65] It is assumed by some that 3,000 of this number were the elite *hypaspists* – despite Diodorus not specifically mentioning this unit in his breakdown – and that Alexander therefore had 9,000 regular pikemen in his army.[66] Other scholars, in a variant of this conclusion, state that the 3,000 *hypaspists* were also armed as phalangites and that, therefore, Alexander did have 12,000 pikeman but that the *hypaspists* were organized differently to the rest of the pike-phalanx – thus still leaving a total of 9,000 men for the pike-phalanx.[67]

Arrian provides the details of six specifically named senior infantry commanders in Alexander's army for the battle of the Granicus River in 334BC.[68] Thus the 9,000 men of the regular pike-phalanx would have been divided among these officers into units of 1,500 men each. However, this unit does not seem to have been specifically called a *taxis* as some scholar suggests. Rather it appears that Arrian's use of the term *taxis* should be seen, in this instance, in its generic form to simply mean 'unit'.[69] Throughout his narrative, Arrian uses the term *taxis* in this generalised manner to refer to a contingent (or contingents) of troops. Cascarino suggests that this generic use of the term *taxis* only applied to units larger than the file.[70] This conclusion seems to be supported by Arrian himself who at times calls the leader of a *taxis* a *taxiarch* or a *strategos* – which although the specific name for the commander of a unit of 4,000 men in the manuals, most likely is also used as a generic term to simply mean 'leader' or 'commander'.[71] Further evidence of the use of the word *taxis* in its generic form can be seen in Arrian's use of the term to refer to contingents of cavalry, mounted javelineers, skirmishers and archers.[72] Diodorus also uses the term *taxis* to occasionally just mean 'unit'.[73] In an expanded use of the generic form of the term, Plutarch uses the word *taxis* to refer to the whole Macedonian formation at the battle of Pydna in 168BC.[74]

If these units under the command of Alexander's six senior infantry commanders are not a *taxis* of 1,500 men, what then are they? The key to identifying these units comes from later in Arrian's narrative. At Gordium in 334BC, Arrian states that Alexander received 3,000 'Macedonian' reinforcements.[75] These troops are most likely to have been phalangites due to the way that they are described as 'Macedonian'. Yet for the battle of Issus the following year, there is still only six senior infantry commanders – Craterus, Meleager, Ptolemaeus, Amyntas, Perdiccas and Coenus.[76] Due to this consistent number of senior commanders, it seems clear that these additional 3,000 troops received at Gordium were not organized into two new '*taxeis*' of 1,500 men each under new officers but were distributed among the contingents already in existence. The addition of 3,000 men to the 9,000 phalangites who had fought at Granicus in 334BC would bring Alexander's total pike-infantry to a strength of 12,000.[77] This would then mean that each of the six officers named in Arrian's account of the battle of Issus were in command of a contingent of 2,000 men. This is the same size as the *merarchia* outlined in the later manuals. Whether the terms *merarchia* and/or *merarch* are used or not is irrelevant. What is of importance is to understand that Alexander's pike-phalanx was, by 333BC, based upon one of the larger units detailed in the manuals and this was not a unit of 1,500 men.

Bennett and Roberts suggest that 3,000 phalangites were sent to Asia as part of the advance force in 336BC.[78] Unfortunately, none of the ancient sources detail the break down of the troops that were sent and Polyaenus, the only writer who does provide a figure, gives only a total of 10,000 for the number of men in the expedition.[79] However, if 3,000 phalangites were dispatched, it seems clear that this contingent were made up of one *pentacosiarchia* of 500 men drawn from each of the six existing infantry *merarchiae* of the army that Alexander was later to take with him. This would then account for why the army that crossed into Asia with Alexander in 334BC contained under-strength *merarchiae* of 1,500 men each and why, when 3,000 reinforcements were received at Gordium, these troops were folded back into the existing *merarchiae* to bring them back up to full strength.

Other evidence also suggests that the *merarchiae* may have already existed prior to Alexander's departure for Asia. Arrian states how, in 335BC, Alexander was forced to deploy his phalanx 'with a depth of 120 files', and with 200 cavalry on either wing, in order to conduct a drill display to try and frighten the opposing Taulantians – thus allowing him to withdraw from Pelium.[80] As noted, Daniel suggests that what Arrian is referring to is a *lochos* (which he claims was a unit of around 120 men) standing in single file.[81] Similarly, Hammond suggests that Arrian is referring to all 12,000 men of Alexander's army being arranged in 100 files of 120 men each.[82] However, these conclusions seem unlikely as deploying in such a manner would require considerable space – space that Arrian records Alexander did not have as the terrain was 'narrow and forested, bordered on one side by the river and on the other by a lofty mountain'.[83] Furthermore, a deployment 120 men deep does not conform to any of the stated structures of the phalanx and it is unlikely that Alexander would have his men deviate from a standard formation. The validity of both the description of the terrain and the deployment in files of around 120 men each seems certain: Arrian's sources for this part of his history were Ptolemy and most likely Aristobulos as well.[84] Consequently, Alexander's deployment has to conform both with the stated nature of the terrain and with one of the known arrangements of the pike-phalanx. As such, Arrian must be referring to *merarchiae* of the pike-phalanx deployed in column rather than in an extended line. In other words, each *merarchiae* has been rotated 90° to fit into the available space and the 128 files of the formation are now its ranks – giving each *merarchia* a frontage of sixteen men and a depth of 128 – or '120 files deep' as Arrian puts it in rounded figures.[85] Such rotated deployments were not unknown to Hellenistic commanders. Aelian describes the process of *paragogē* (παραγωγή) where 'the phalanx [or unit] marches with the flank leading'.[86] Alexander's *merarchiae*, formed up in just such a manner to impress the Taulantians, seems to be what Arrian is referring to.

If it is assumed that this is not a reference to a *merarchia* deployed in column, then another way in which a formation 120 ranks deep could be achieved is by deploying four *pentacosiarchiae* of thirty-two files of sixteen men each, a unit that Curtius states was in use in Alexander's army early in his campaign, similarly rotated and in column one behind the other. Whether this was the organizational structure of the phalanx at this time or not, grouping four *pentacosiarchiae* together still creates a unit with the same size and configuration as a *merarchia*. Whichever way the 120-deep deployment is viewed, it suggests that prior to his invasion of Asia in 334BC, Alexander's army was based upon at least one of the larger units outlined in the military manuals.

At the battle of Gaugamela in 331BC, the structure of Alexander's army does not seem to have changed from being based upon six *merarchiae* as six senior infantry officers are again named by Arrian – Craterus, Simmias, Polyperchon, Meleager, Perdiccas and Coenus.[87] Curtius states that Alexander celebrated his victory at Gaugamela with a series of games and initiated a structural reform aimed at altering the internal organization of the *merarchia*:

> [Alexander] appointed judges and established prizes of a unique kind for a contest based upon military courage. Those judged to possess the greatest valour would win command of individual units of 1,000 men and be called *chiliarchs*. This is the first time the Macedonian troops had been divided numerically, for previously there had been units of 500, and command of them had not been granted as a prize for valour.[88]

Clearly, not all of Curtius' statement can be accepted at face value. The creation of the *chiliarchiae* of 1,000 men each cannot have been the first time that the sub-units of the Macedonian pike-phalanx had been arranged numerically when Curtius himself states that there had previously been units based upon groupings of 500. Furthermore, the file seems to have been based upon standard units of sixteen as early as 333BC, if not earlier, and Anaximenes refers to files based upon groups of ten men and 'other commands' during the reign of Alexander II.[89]

One of the most important elements of Curtius' passage is that prior to the battle of Gaugamela, the pike-phalanx had been arranged into units (Curtius uses the Latin term *cohortes*) of around 500 men each. Thus each of the 2,000 man units commanded by the six named senior officers at both Issus and Gaugamela would have been made up of four 500-man units combined. These 500 man units are similar to the *pentacosiarchiae* detailed in the manuals (again, regardless of whether they are actually called this or not).

This then places the 1,500 strong *taxis* that is suggested by some scholars, and the number of Alexander's reinforcements, into context. In 334BC, each of the six named commanders in Arrian is, in effect, commanding a three-quarter strength *merarchia* made up of only three of the requisite four *pentacosiarchia* of 500 men each. Then prior to the battle of Issus, when the 3,000 reinforcements, presumably organized into six *pentacosiarchiae* of 500 each, arrive, these troops are integrated into each of the under-strength *merarchiae* to bring them to their full number. This shows that the use of the term *taxis* by ancient writers like Arrian in regards to Alexander's larger units and their officers can only be seen as a use of the term in its generic context, in this case to refer to a unit the same as the 2,000-man *merarchia*, rather than a description of a specific unit of 1,500 men as some scholars suggest.

By the battle of the Hydaspes in 326BC Alexander's pike-infantry were still based upon 2,000-man *merarchiae*. The number of named senior commanders had increased from six to seven with many changes of command having taken place since the battle of Gaugamela. Craterus, for example, was elevated to the overall command of the Macedonian camp at the Hydaspes, and a contingent of cavalry, but also oversaw infantry contingents (Arrian again uses the generic term *taxis* throughout his narrative here) commanded by Alcetas and Polyperchon.[90] According to Arrian, on the bank of the river were positioned Meleager, Attalus and Gorgias with 'the mercenary cavalry and infantry'. Due to the ambiguity of the passage, it is uncertain exactly what sort of infantry these three officers commanded. Fuller suggests that they were all pike units who later crossed the river behind Alexander and took part in the major confrontation with Porus.[91] Both Perdiccas and Coenus had also been elevated to cavalry commands, and Alexander initially crossed the river with two infantry units under the command of Cleitus the White and Antigenes.[92] Thus the total number of named senior infantry commanders had increased to seven. By this time, however, Alexander's pike-infantry numbered around 14,000.[93] Thus the seven named infantry commanders at the Hydaspes (Alcetas, Polyperchon, Meleager, Attalus, Gorgias, Cleitus and Antigenes) are still *merarchs* in command of units of 2,000 men. This shows a continuance of the use of this larger unit as the basis for Alexander's army right across his campaign.

Strangely, following Alexander's celebratory games at Gaugamela, Curtius goes on to name all eight of the recipients of the command of a *chiliarchie*:

> The first prize went to Atarrhias...Antigenes was judged second, Philotas the Augaean was judged third, and fourth place went to Amyntas. After these came Antigonus, then Amyntas Lyncestes, with Theodotus gaining seventh and Hellanicus last place.[94]

Each of these new *chiliarchs* would be in command of two of the four 500-man *pentacosiarchia*e which had made up part of each *merarchia*, but which were now combined into *chiliarchiae* of 1,000 men each for the first time (in effect, these new officers would therefore be commanding half of a *merarchia*). The *pentacosiarchia*e still existed within the organizational structure of the phalanx as a sub-unit of the *chiliarchia*, but every two of these units now fell under the command of a single officer of a rank higher than that of the *pentacosiarch*.

The interesting thing to note is that Curtius details the commanders of eight new *chiliarchiae*. Yet if the army was arranged into six larger *merarchiae* (as the naming of the more senior officers and the number of troops under their command indicates), then the halving of these units into smaller *chiliarchiae* would necessitate the awarding of a total of twelve commands – two *chiliarchs* per *merarchia*. The reason behind the appointment of only eight *chiliarchs* is due to the way in which the officers of the phalanx were distributed across the line. Following the restructure of the phalanx in 331BC, each of Alexander's six *merarchiae* was divided into two *chiliarchiae*, but each of the twelve *chiliarchiae* was not led by an officer of the rank of *chiliarch*. Rather, one of the *chiliarchiae* in each of the *merarchiae* would be led by an officer with the rank of *merarch* – in overall command of the two combined *chiliarchiae* as a whole unit. The other *chiliarchia* would be led by the more junior *chiliarch* who would just be responsible for the command of this unit within the larger formation of the *merarchia* (see 'The Officers of the Phalanx' following at page 296). Thus out of the twelve *chiliarchiae* of Alexander's phalanx, six would be led by the named *merarchs*. Of the eight named recipients of the title of *chiliarch*, six of these would have then been given command of the other halves of each *merarchia*.[95]

This then leaves two of the new *chiliarchs* unaccounted for. These would have been assigned to the *hypaspists* who must have undergone the same organizational restructure in 331BC as the pike-phalanx. The 3,000 men of the *hypaspists* would have been, like the pike-infantry, based upon standard units of 500 men prior to 331BC. This would constitute a three-quarter strength *phalangarchia* of *hypaspists* – similar to the way the *mararchiae* of the pike-infantry had been initially at three-quarter strength. This *phalangarchia* would have fallen under the command of a single senior officer, Nicanor.[96] With the restructuring of the infantry units into *chilarchiae* in 331BC, Nicanor would have still retained overall command of the *hypaspists* as a whole, and specific command of one of its constituent 1,000-man *chiliarchiae*. The other two *chiliarchiae* would have fallen under the command of the remaining two recipients of the new rank of *chiliarch*.

This structure for the *hypaspists* contingents is confirmed through Alexander's integration of Persian troops into his infantry units in 324BC. According to Diodorus, 1,000 Persians formed a new unit of *hypaspists*.[97] Arrian, who uses the term 'Silver Shields' to refer to the *hypaspists*, also refers to the incorporation of a new unit (he again uses the generic term *taxis*) of Persian troops into the *hypaspists*.[98] This addition of 1,000 Persian troops would have increased the number of the *hypaspists* from a three-quarter strength *phalangarchia* of 3,000 men, to its full strength of 4,000. Following the structural reforms that had occurred earlier in 331BC, this complete *phalangarchia* of *hypaspists* would not have been based upon units of 500, but would have been made up of four *chiliarchies*. As a result, the total complement of *hypaspists* would fall under the command of an officer holding the rank of *phalangarch* who would also be in direct command of one of the four *chiliarchiae*, one *chiliarchia* would be commanded by a *merarch* in overall command of one half of the *phalangarchia* but still subordinate to the *phalangarch*, and the remaining two *chiliarchiae* would be commanded by *chiliarchs* – the same number that had been appointed to the *hypaspists* back in 331BC.

Heckel suggests that all eight recipients of the title of *chiliarch* were assigned to commands of units of *hypaspists*.[99] This seems unlikely for a number of reasons. Firstly, the presence of eight *chiliarchs* would suggest that the *hypaspists* numbered 8,000. However in 331BC the *hypaspists* numbered only 3,000 and their number was not increased to 4,000 until 324BC. Heckel attempts to avoid this issue by suggesting that four of the named recipients were actually given the position of *pentacosiarchs* rather than *chiliarchs*.[100] This goes against Curtius who specifically states that all eight men were give the title of *chiliarch*. Furthermore, there is little evidence that all of these men were assigned to the *hypaspists*. Heckel states that Antigenes' command at the Hydaspes, for example, was 'clearly not [of] pezhetairoi' and therefore he had to have been in command of a unit of *hypaspists* (citing as proof Arr. *Anab.* 5.16.3 and Curt. 8.14.15). This interpretation is clearly incorrect. Arrian states that Antigenes, among others, was placed in charge of part of the 'phalanx' at the Hydaspes, while Curtius goes on to have Alexander address his commanders prior to the engagement, and in a reference to their potential against the Indian elephants opposing them, state: 'Our spears, which are very long and strong, will never serve us better than against these beasts and their drivers.'[101] This can only be a reference to Antigenes commanding a unit armed with the *sarissa* – that is, a contingent of the pike-phalanx as Arrian states.

Of the eight named recipients of the title of *chiliarch* in 331BC, little else is known about Amyntas, Antigonus, Amyntas Lyncestes and Theodotus. As noted, Antigenes is listed as a senior infantry commander at the battle of the Hydaspes in 326BC and it seems clear that he was given command of one of the *chiliarchiae* of the pike-phalanx in 331BC only to be promoted at a later stage.[102] Philotas the Augaean also seems to have been the commander of a unit of pike-infantry in 329BC.[103] Hellanicus is mentioned by Arrian in his account of the fighting at Halicarnassus in 334BC where he is positioned alongside another officer by the name of Philotas (whose identity is difficult to determine).[104] Yardley and Heckel suggest that Hellanicus was a member of the *hypaspists*, but the limited number of references to this individual make the determination of his command exceptionally circumstantial. Atarrhias is also mentioned in relation to the siege of Halicarnassus where he earned special distinction for his conspicuous actions.[105] In particular he is noted for encouraging his men which shows that he was in command of troops of some kind.

As both the *hypaspists* and the pike-infantry were initially arranged in units of 500, it is difficult to determine who was in command of what type of troop following the restructure of 331BC. However, the postings for some of the recipients of the rank of *chiliarch* can be determined. For example, as Antigenes was a pike-infantry *merarch* at the Hydaspes in 326BC, it seems unlikely that he would have been assigned to the *hypaspists* in 331BC and so must have been awarded the position of *chiliarch* within the pike-infantry. The praise for Atarrias comes from Cleitus the White, another pike-infantry *merarch* at the Hydaspes, which would suggest that Atarrias was also made a *chiliarch* of the pike-infantry – possibly subordinate to Cleitus. Additionally Atarrias is described as 'old' (*senex*) by Cleitus. This would suggest that Atarrias was a veteran of the Macedonian army and may have been a man who had worked his way up through the ranks of the pike formations under Philip II. If Philotas the Augaean was also a pike-infantry commander in 329BC, this would also suggest that he was made a *chiliarch* of the pike-infantry back in 331BC. If Yardley and Heckel are correct in their conclusion that Hellanicus was a member of the *hypaspists*, it would make sense that he had been one of the two new *chiliarchs* appointed to the command of one of these elite units. This then leaves the positions of one *chiliarch* of the *hypaspists* and three *chiliarchs* for the pike-infantry unaccounted for. Unfortunately, due to the limited amount of details we have for the other winners of Alexander's contest in 331BC, it is almost impossible to assign them with any degree of certainty.

It is interesting to note that Arrian also details the death of Adaeus, whose rank is given as a *chiliarch*, at the siege of Halicarnassus in 334BC – three years before Curtius says the title even existed.[106] Adaeus is only briefly mentioned in the ancient sources. During a sally by the garrison of Halicarnassus, Arrian states that 'Ptolemaeus, the Royal Bodyguard, met them, bringing up the units of Adaeus and Timander.'[107] Later, when accounting the losses for the engagement, Arrian gives the Macedonian casualties as 'about forty of Alexander's forces, including Ptolemaeus the Bodyguard...[and] Adaeus, a *chiliarch*...'.[108] The identity of Adaeus is somewhat problematic due to the ambiguity of the passages in which he is mentioned, and the interpretation of his position and stated title have significant implications for the understanding of the structure of Alexander's infantry in 334BC. Tarn, for example, claims that due to the way in which the Royal Bodyguard Ptolemaeus leads the contingents of both Adaeus and Timander into battle, these units had to have been contingents of *hypaspists*.[109] Berve, on the other hand, also suggests that the men being led are *hypaspists*, but that Adaeus is actually in command of two '*taxeis*' of men (each assumed to be 500 strong) rather than one *chiliarchia* as, according to Curtius, the office of *chiliarch* would not be invented until 331BC.[110] Alternatively, Milns suggests that because there is no specific reference in Arrian as to what type of troops Adaeus' unit is comprised of, he may be commanding anything – even mercenaries.[111]

Milns' interpretation seems unlikely.[112] Rarely are the structural details of the mercenary contingents in Alexander's army given other than in overall numbers. While this in itself cannot dismiss the possibility, it would be more likely that Adaeus was the leader of a unit of troops for which references to their structure are more regularly given in the ancient narratives. This would make the troops under Adaeus' command either *hypaspists* or regular phalangites. This then leaves a number of problems in relation to the structure of Alexander's army regardless of which type of troop Adaeus is assumed to be leading. The 3,000 men of the *hypaspists* are generally considered to have been arranged in three units of 1,000 men each. This is similar to the *chiliarchia* of the later manuals and correlates nicely with the attribution of Adaeus as a commander of a 1,000-strong contingent of *hypaspists*. While there is no specific reference to the *hypaspists* being organized into *chiliarchiae* early in Alexander's reign, Arrian does mention them being formed into these units of 1,000 later in the campaign.[113] Importantly, all of the references which detail the *hypaspists* specifically being organized in *chiliarchiae* all occur in Arrian's narrative after 331BC – the time when Curtius says the office of *chiliarch* was created.

If the office of *chiliarch* did not actually exist unit 331BC as per Curtius, it is uncertain how either the *hypaspists* or the regular phalangites could have been arranged in such units in 334BC. It is possible that they changed, along with the pike-phalanx, following Alexander's reforms of 331BC. Units of 500 existed for the pike-phalanx in 334BC and it seems most likely that the *hypaspists* were arranged in a similar manner at this time as well. If Adaeus is assumed to be in charge of a 1,000-man unit of *hypaspists*, rather than phalangites, in 334BC, it is interesting to note how many scholars see the *hypaspists* as being arranged in this manner while they assume that the rest of the pike-phalanx at this time is not (with many assuming that the pike-phalanx was arranged in '*taxeis*' of 1,500). This is even more curious when it is considered that some scholars also suggest that the *hypaspists* may have been armed as phalangites and were a part of the pike-phalanx. Regardless of their armament, it would make more organizational and operational sense to have all of the Macedonian infantry, both *hypaspists* and phalangites, organized in the same manner – based upon units of 500.

This leaves two possibilities: either Arrian's use of the term *chiliarch* is anachronistic, or it is a generic reference to an officer commanding a contingent of 1,000 men.[114] In this case the *chiliarchia* can only be seen as something of an improvised unit made up of two of the standard formations of 500. Regardless of whether the term is seen as generic or not, or anachronistic or not, such a conclusion has strong implications. Again, whether a specific term, in this case *chiliarch*, is used or not is irrelevant. If Adaeus was in command of a contingent of 1,000 men at the siege of Halicarnassus, even if this was just two combined *pentacosiarchia*e as Berve suggests, and even if such a unit was not actually called a *chiliarchie* until 331BC, units of 1,000 still had to exist, even if only in an improvised fashion, within the structure of Alexander's army as early as 334BC.

Ultimately, whether the office of *chiliarch* officially existed in 334BC or not is a matter of semantics. What is important for understanding the structure of Alexander's army early in his campaign is that two standard units of around 500 men each may have been organizationally combined and under the leadership of a single officer. Whether these were still viewed as separate units of 500, or a combined unit of 1,000, is again irrelevant. Regardless of what it was called, a unit in every way similar to the *chilarchia* seems to have been in use by Alexander from the beginning of his campaign. Thus another aspect of Alexander's army was based upon a division of troops that is outlined in the later military manuals (albeit possibly only in an impromptu manner).[115] This also correlates with the interpretation of the six senior officers named by Arrian at the start of Alexander's campaign being in command of a *merarchia* of around 2,000 men each, regardless of whether it was made up of two units of 1,000 or four units of 500, rather than the otherwise unattested *taxis* of 1,500.

However, the narratives for Alexander's campaign also include the specific use of the term *taxis* (or *taxiarch*) to refer to a unit of around 120 men or its leader – as it is described in the later manuals. At the battle of Issus in 333BC for example, Arrian states that Ptolemy, son of Seleucus, whose rank is given as a *taxiarch*, was killed along with 'about 120 Macedonians of distinction' (καὶ ἄλλοι ἐς εἴκοσι μάλιστα καὶ ἑκατὸν τῶν οὐκ ἠμελημένων Μακεδόνων).[116] During the battle, Ptolemy's unit, which had been positioned in the centre of the line, had been unable to keep up with the rapid advance of Alexander's right wing and found itself in dire straits while it tried to cross the river as Greek mercenaries fighting on the Persian side 'attacked where they saw that the phalanx had particularly broken up'.[117] The fact that Ptolemy died with a unit of around 120 men, which is identified as a *taxis* by Arrian through his description of Ptolemy's rank, confirms the existence of units of this size within Alexander's army as early as 333BC.

Arrian also states that at the battle of Gaugamela in 331BC the file commanders (*lochargoi*) were instructed to encourage the men of their respective files (*lochoi*), while the *taxiarch*s were instructed to encourage the units that they were in command of.[118] This suggests that the file was part of the *taxis*. Both the *Suda* and the military manuals state that the larger *syntagma* had a trumpeter and standard bearer attached to it to relay commands and other information across the din of battle (see following).[119] However, Arrian's reference to words of encouragement being passed to the members of the *taxis* would suggest that this unit was small enough not to warrant the use of trumpets and standards to deliver information and instructions. It must also be noted that both trumpets and banners cannot be used as a means of relaying words of encouragement or detailed speeches. This then suggests that the *taxis* was smaller than the *syntagma*. Polybius used the term *taxis* to refer to the Roman century.[120] This suggests a correlation of the term to generically refer to a unit of around 100 men. Both the *Suda* and the manuals state that the *syntagma* was made up of two *taxeis* of 128 men each.[121] This smaller size of the *taxis* correlates with the ability to pass verbal words of encouragement to the men of the unit, and the numbers involved further correlate with the size of Ptolemy's *taxis* at the battle of Issus. Thus the specific references to the *taxis* in the narratives for Alexander's campaign match with the other sources to show that yet another organizational aspect of the Macedonian army was the same as those outlined in the later manuals almost from the very beginning.

The literary sources also indicate the presence of units the size of the *syntagma*, quite distinct from the smaller *taxis* or the larger *pentacosiarchia*, in Alexander's army as well. Curtius states that at the battle of Issus there were standards positioned across the front of Alexander's line and that he had to restrain his men from advancing too far beyond them.[122] The positioning of a line of standards across the front of the formation correlates with the reported structure of the *syntagma*. Curtius also states that, prior to the battle of Gaugamela in 331BC, Alexander's phalangites had been organized into units (Curtius uses the Latin term *cohortes*) of around 500.[123] This figure is close to the 512-man size of the *pentacosiarchia* given in the later manuals which was made up of two combined *syntagma*e. Even if the term *syntagma* itself is not used in connection with these deployments, it is clear that the men of Alexander's phalanx could be organized into units which matched the structure of this unit as it is described in the manuals. According to the manuals, the *syntagma* was a square unit of 256 men arranged in sixteen files of sixteen.[124] The manuals additionally detail how each *syntagma* had five supernumeraries attached to it:

> In each *syntagma* of 256 men there are five super-numeries [*ektaktoi* – ἔκτακτοι]: a standard bearer [*semeiphoros* – σημειφόρος], a rear commander [*ouragos* – οὐραγός], a trumpeter [*salpigktēs* – σαλπιγκτής], an aide-de-camp [*huperetēs* – ὑπηρέτης], and a herald [*stratokērux* – στρατοκῆρυξ].[125]

The *Suda* states how, of these five supernumeraries, all but the rear commander were positioned at the front of each *syntagma*:

> In former times the unit had these, just as their name shows, because they were supernumeraries of the company. They are five [in number]: herald, trumpeter, standard-bearer, aide-de-camp, rear-commander. But nowadays there are officers with this name both in the *syntagma* and in the other [units]. These units, and especially the *syntagma*, must have them: the first one to communicate the orders with his voice, the second one [to do so] by the standard, if the voice is not heard because of noise [of battle]; the trumpeter, whenever they cannot see even the standard because of dust; and the aide-de-camp, so as to carry over some of the things that are needed. Moreover, the special-duty rear-commander [is needed] to lead back any stragglers to the company; [it is he], rather than the four above-mentioned, who is stationed behind the front of the formation.[126]

Thus the standard-bearer would be positioned at the front of the formation (most likely at its front-right or front-left next to the commanding officer of the unit – see following). This would then agree with Curtius' reference to the standards being arrayed across the front of Alexander's line at Issus. Connolly and Wrightson place all five of the supernumeraries behind their associated unit.[127] Not only does this disagree with the ancient texts, but locating a standard bearer behind their formation would place them in a position where any signals given would not be seen by the members of the unit. This makes such a deployment of the supernumeraries, and the standard bearer in particular, to the rear unlikely. A positioning of the standard bearer to the front-right or front-left of the formation, on the other hand, next to its commanding officer who can then issue instructions, and visible to most of the unit, would also indicate that the two phalangites depicted on the Pergamon Plaque are the commanding officer, identifiable by his plumed helmet, at the front-right of a formation next to the standard, and the leader of the file next to him – a man of lesser rank as indicated by his bare helmet (see Plate 7). A similar standard, in the shape of a square banner of cloth suspended from a staff, can be seen in the Alexander Mosaic – said to be a representation of the battle of Issus.

Nylander has argued against what he calls the 'universally accepted' idea that it is a Macedonian standard shown in the Alexander mosaic – offering instead that it is a representation of the Persian royal banner.[128] However, this conclusion seems to be incorrect for a number of reasons. Firstly, the staff upon which the banner is suspended is situated behind the shaft of one of the Macedonian pikes. This would place the banner within the Macedonian formation in the background of the image rather than as part of the Persian contingent in the foreground. As the Macedonian troops in the image on the mosaic are marching from left to right (as one looks at the mosaic) the positioning of the banner would place it at the front-right of the advancing formation – exactly where it is supposed to be. Assuming that the pike overlaying the standard is accurate and not an artistic error, the additional weapon, which would be held by a man standing on the right-hand side of the standard bearer, must belong to one of the other supernumeraries of the unit – possibly the *aide-de-camp* as neither the trumpeter nor the herald are likely to have carried one, and the rear-commander would be at the back of the formation.

Furthermore, the banner seems to be held by a figure wearing a metal helmet. This identifies the bearer as a Macedonian due to the way that the Persians in the mosaic are all clearly distinguished by their yellow cloth tiaras. This would then mark the standard as a Macedonian banner rather than a Persian one.[129] Lastly, Nylander argues against the banner in the mosaic being Macedonian on the grounds that there appears to be some kind of emblem on the cloth (commonly reproduced as a bird of some kind) and that, while there are numerous sources describing the insignia on the Persian royal standard, there is no literary reference to similar insignia being placed upon the standards of the Macedonians. Such an argument from literary silence is, at best, highly problematic – especially when other forms of evidence, such as the Pergamon Plaque, clearly show a Hellenistic standard of the same shape and with a design placed upon it (a star or sunburst in the case of the Pergamon Plaque). It thus seems more likely that what is being shown on the Alexander Mosaic is a representation of one of the standards attached to one *syntagma* of Alexander's pike-infantry.

Serrati observes that the *syntagma* is the smallest unit to have officers that operate outside of its ranks (e.g. the standard bearer and the *aide-de-camp*).[130] According to the manuals, the *syntagma* is the smallest tactical unit of the pike-phalanx and is a semi-independent formation unto itself. Importantly, if a standard bearer was attached to each *syntagma* of the pike-phalanx as the manuals state, then Curtius' reference to standards being deployed across Alexander's battleline at Issus would suggest that units of this size were being used in Alexander's army as early as 333BC. Each of these *syntagmae* would constitute half of each *pentacosiarchia* which, in turn, were each a quarter of the six *merarchiae* which made up Alexander's army early in the campaign – led by the six named officers detailed by Arrian.

The *syntagma* seems to have been a standard operational division of troops across the Hellenistic Period – although the actual term *syntagma* is rarely found outside the later military manuals. Connolly observes that Polybius regularly uses the term *speira* (σπεῖρα) to describe elements of the phalanx.[131] Feyel suggests that the *speira* was another term for the *syntagma*.[132] Polybius uses the term *speira* to refer to the Roman maniple which was the smallest tactical unit of the legion – similar to the way that the *syntagma* was the smallest tactical unit of the pike-phalanx.[133] Plutarch also uses the term *speira* to refer to units within the phalanx.[134] The term *speira*, and the unit's commanding *speirarch,* are also mentioned in the Amphipolis inscription from the time of Philip V which outlines fines for the loss of phalangite equipment.[135] Interestingly, the term *syntagma* is never used in connection with a pike-army within Europe, while the term *speira* is never used in connection with an army in the east. According to Connolly, *speira* was the Greek/European name for the eastern *syntagma*.[136] Connolly further suggests that the use of the *syntagma*/*speira* in all of the armies of the Successors shows that it must have had a common predecessor.[137] This could have only been a formation used in the army of Alexander the Great if not earlier under Philip II, Perdiccas III or even as one of the 'other commands' of Alexander II. If this is the case, then the reference to the *speira* in later sources like Polybius and the Amphipolis inscription, and the representation of the commanding corner of a *syntagma* on the Pergamon Plaque, demonstrate the use of this unit, under two different names, as a tactical unit of the pike-phalanx across the entire Hellenistic Period.

Interestingly, Appian states that Antiochus the Great's 16,000 strong phalanx at Magnesia in 190BC was deployed in ten units of 1,600 men each – fifty-two men across and thirty-two ranks deep – with thirty-two elephants positioned between each of these contingents.[138] This passage contains several important elements for understanding the structure of the later pike-phalanx. Firstly, it describes a deployment of sixteen-man files to a double depth – indicating the use of files of this size late in the Hellenistic Period (contrary to what Asclepiodotus states about the use of twelve-man files in this period). Secondly, contingents of 1,600 men do not correlate with any of the standard divisions of the phalanx that are outlined in the military manuals. While the size of these dispositions is close to the supposed size of the *taxis* in the time of Alexander that is offered by some scholars, other elements of Antiochus' forces, such as the double depth deployment thirty-two deep, suggests that the organisation of his phalanx more closely followed the arrangements set down in the military manuals. As such, Antiochus' units of around 1,600 men each would be the deployment of three *pentacosiarchiae* together, with each containing around 512 men, to give a total around 1,536 men per contingent (plus an additional thirty supernumeraries). Sekunda suggests that each of these 1,600-man units was led by a *merarch*.[139] This must therefore assume the use of under-strength *merarchiae* – a standard organizational arrangement outlined in the manuals. It is additionally important to note that Antiochus' deployment at Magnesia was not a standard operational arrangement, but an improvised manner of arranging his men in the available space of the battlefield while allowing enough of an interval between each contingent of infantry to position thirty-two elephants. Thus Antiochus' deployment at Magnesia should not be read as a divergent structure of the pike-phalanx in the late Hellenistic Period, but as an example of how flexible the arrangement of the phalanx could be due to its use of standard units and sub-units.

The military manuals also outline larger units within the phalanx based upon the *merarchia* such as the 4,096-man *phalangarchia*. This formation could then be combined to create the 'double-phalanx' and then again to create the 'quadruple-phalanx' – the largest unit outlined within the ancient manuals. None of these larger contingents are specifically referred to in the narratives for Alexander the Great, although it seems clear that Alexander's pike-infantry was arranged into three *phalangarchia* as early as 334BC (see following). Connolly, despite suggesting that such formations never existed, places them in his discussion of later Hellenistic armies – a time for which the ancient sources indicate that many of the organizational elements of the phalanx were also in use.[140] Yet the possibility of groupings in these larger formations remains for any part of the Hellenistic Age so long as the numbers within the armies being discussed permit. Whether these units were labelled using the terminology of the later manuals is somewhat redundant. If Alexander's army contained units of around 120, 250, 500, 1,000 and 2,000 men from the beginning of his campaign in Asia, it really makes no difference whether they are called a *taxis*, *syntagma*, *pentacosiarchia*, *chiliarchie*, and *merarchia* respectively or not.

Similarly, whether the 128-man *taxis* was, as per the manuals, divided into quadruple-files, double files and then single files of sixteen men is equally redundant.[141] The term *tetrarchia* is not used in connection with the organization of Alexander's infantry and is only used once to refer to a contingent of cavalry.[142] However this is not the issue. The important thing is that the number of men within the units and sub-units of the Macedonian army seems to have been a standard organizational aspect of the pike-phalanx right across the Hellenistic Period from Alexander onwards and that armies were based, not upon the largest formation outlined in the manuals, but on the largest unit or units that would allow for the army to deploy with a certain level of tactical adaptability, organizational symmetry and mutual support. Regardless of whether an army in the time of Alexander the Great or the Successors was based upon a unit of 500, 1,000, 4,000 or even 16,000, and regardless of what these contingents were actually called at the time, both the narrative histories and the later manuals show that each of the units and sub-units of the pike-phalanx would have been led by their own officers.

THE OFFICERS OF THE PHALANX

Similar to the way in which the units and sub-units of the phalanx were arranged upon an ever increasing numerical system, the officers of the phalanx were also positioned according to precise models. Within the perfect construct of the formations outlined by the manual writers, the officers that commanded specific units across the broader phalanx as a whole were positioned based upon precise mathematical criteria. Both Asclepiodotus and Aelian discuss how the placement of the officers commanding their respective units was set to achieve a numerical symmetry and a balance of command abilities across the formation of any deployed force.

There is little doubt that the leaders of the *lochos*, the smallest sub-unit within the phalanx, would have been positioned at the head of the file of men that they commanded.[143] However, as sub-units were combined to make larger and larger tactical and administrative bodies, the location of those who were in command of these bigger contingents varied to provide a certain level of adaptability to the formation and to spread the abilities and experience of the officers evenly across the phalanx's front. Aelian, for example, begins his discussion of the positioning of the officers of the phalanx with the *phalangarchs* – those in command of two combined *merarchiae*, a total of 4,096 men, or one quarter of the quadruple-phalanx. Thus in the perfect quadruple-phalanx as outlined in the manuals, there would be a total of four *phalangarchs*. These officers were assigned a numerical value based upon their experience and command abilities:

the senior and most experienced *phalangarch* was given a value of one; the next most experienced a value of two; the third in order of experience a value of three and the least experienced *phalangarch* a value of four.[144]

When each of the four *phalangarchiae* were deployed side-by-side in order of battle, the first ranked *phalangarch* was positioned at the head of the right-hand file on the very right of the line (i.e. leading the most right hand file of the most right-hand *phalangarchia*). The second ranked *phalangarch* was similarly positioned, but at the other extreme end of the line commanding the most left-hand file. The third ranked *phalangarch* was in command of the *phalangarchia* second from the left, but was positioned at the head of its most right-hand file – towards the centre (known as the 'navel' (*omphalos* – ὄμφαλος), the 'mouth' (*stoma* –στόμα) or the 'disection' (*araros* –ἀραρός)) of the whole formation. Similarly, the fourth ranked *phalangarch* commanded the *phalangarchia* second from the right, but was positioned at the head of its most left-hand file – also towards the centre of the formation (Fig.28).

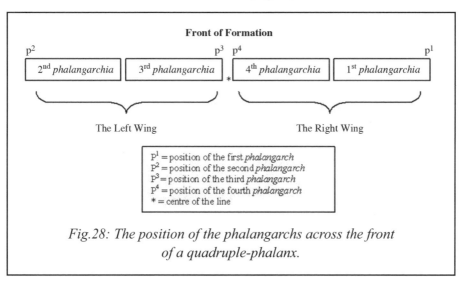

Fig.28: *The position of the phalangarchs across the front of a quadruple-phalanx.*

The distribution of the officers in this manner achieved a balance of command abilities across each wing of the quadruple-phalanx. The two *phalangarchs* of the right-hand wing, for example, had a combined command value of five (1+4). The *phalangarchs* of the left-hand wing also had a combined command value of five (2+3). In other words, there are two 'average' officers in command of the two *phalangarchiae* on the left wing, while the best commander and the most inexperienced commander hold the two units on the right wing.[145] Thus any deficiency in the command skills of the officer leading the fourth *phalangarchia* is offset by the superior abilities of the lead *phalangarch* commanding the unit next to him. This, in effect, means that both wings have a relatively equal, albeit average, level of command with the two most experienced officers holding positions on the vulnerable wings of the formation while the centre is held by the two most inexperienced. In this arrangement, the fourth ranked *phalangarch* is also positioned immediately next to an officer of slightly higher quality and would be able to gain support from that officer while both of them held the centre of the line. Thus the whole of the line (both wings and the centre) was well commanded.

The positioning of the *phalangarchs* at each end of the two wings of the quadruple-phalanx also meant that, if one wing advanced forward independent of the other, the files on each end of each wing would be led by a high ranking officer in the same way that each end file was when the two wings were combined. This would have made each wing of the quadruple-phalanx a self-supporting army unto itself which could, to all intents and purposes, operate on its own on the battlefield.

Each of the *phalangarchiae* were also divided into two *merarchiae* of around 2,000 men each. The *merarchs* in command of these units were also distributed across the front of the line in accordance with the same mathematical model. Even though the perfect quadruple-phalanx was made up of eight *merarchiae*, there would have only been a total of four *merarchs* in this formation rather than eight. This was because four of the *merarchiae* were actually led by officers with the rank of *phalangarch*, in command of their respective larger units (see following). As with the *phalangarchs*, the *merarchs* were given a numerical value, based upon their experience and command abilities, from one to four. Aelian explains their positioning within the formation as:

> The *merarchs* are positioned in a similar way. He who is ranked highest is posted in command of the *merarchia* on the left-hand side of the first *phalangarchia* on the right wing and is positioned at the head of the most left-hand file of that unit. The second ranked *merarch* is posted in command of the *merarchia* on the right-hand side of the second *phalangarchia* on the left wing and is positioned at the head of the most right-hand file of that unit. The officer ranked third is posted in command of the *merarchia* on the left-hand side of the third *phalangarchia* and is positioned at the head of the most left-hand file of that unit. The fourth ranked *merarch* is posted in command of the *merarchia* on the right-hand side of the fourth *phalangarchia* and is positioned at the head of the most right-hand file of that unit[146] (Fig.29).

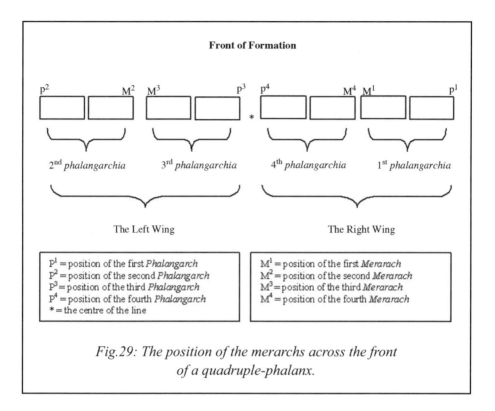

Fig.29: The position of the merarchs across the front
of a quadruple-phalanx.

Again the positioning of the *merarchs* based upon assigned command values creates a balance of experienced officers across the front of the line on each wing. The command values of the leading officers on each wing add up to ten (on the left 2+2+3+3, and on the right 4+4+1+1). Additionally, not only are the *merarchs* in positions where they can support each other, but they are also at opposite ends of each *phalangarchia* from their respective *phalangarchs*. This continues the practice of assigning symmetrical levels of command ability across the front of the line and means that, should one *phalangarchia* advance independently of the others, it would still have its most senior officers positioned at either end.

Aelian goes on to state that this same principle for the balanced distribution of officers carries on to even the smaller combined sub-units of the phalanx: the two combined *tetrachiae* of the *syntagma*, the two combined *dilochiae* of the *tetrachia*, and the two combined files of the *dilochia*:

The leaders of each of the four files in a *tetrarchia* [i.e. the *lochargoi*] are similarly arranged. The leader of the first file [on the right] is he who has the most experience, and the fourth ranked officer holds a position next to him [i.e. to his left]. The second ranked file leader commands the left hand file, while the third ranked file leader holds the position next to him. By these means, the *dilochiae* within the *tetrarchia* have an equal share of strength – for the right-hand *dilochia* has leaders of the first and fourth rank while the left-hand *dilochia* has commanders of the second and third rank... As there are four *tetrarchiae* in every *syntagma*, each *tetrarchiae* is drawn up in a similar way so that, in every *syntagma*, the commander of the first *tetrarchia* is placed on the right, and ranks first. The fourth *tetrarchia* is positioned to his left, then comes the third *tetrarchia* and then the second. The same proportions also exist in the higher levels of command.[147] (Fig.30).

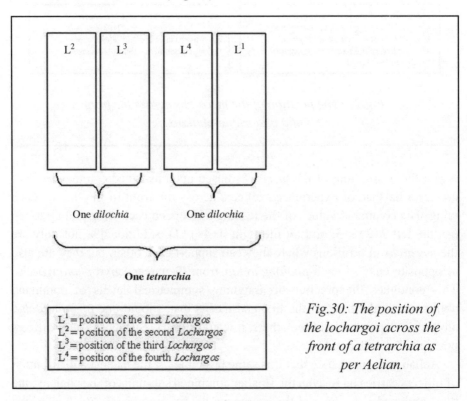

One *dilochia*   One *dilochia*

One *tetrarchia*

L¹ = position of the first *Lochargos*
L² = position of the second *Lochargos*
L³ = position of the third *Lochargos*
L⁴ = position of the fourth *Lochargos*

Fig.30: The position of the lochargoi across the front of a tetrarchia as per Aelian.

An inscription from Greia, dated to 181BC, refers to both the *tetrarch* and 'those of the first *lochos*'. Wells suggests that each *lochos* within a *tetrarchia* was numbered one through four from right to left.[148] Thus, according to Wells, the 'first *lochos*' would be on the right hand side of the unit, commanded by the tetrarch. While this position of the officers would be true of a *tetrarchia* on the right-hand end of the line, such a conclusion would not hold true for one positioned on the far left – which would have its commanding officer, and presumably its 'first *lochos*' as well, on the left side of the unit. Well's conclusion also does not consider that any *tetrarchia* that was part of a larger formation may not have been commanded by a *tetrarch*, but would have been led by a more senior officer (see following). Additionally, according to Aelian, the files of the *tetrarchia* were not numbered one through four from right to left (or left to right), but were arranged with a symmetrical distribution of commands. Thus it can only be concluded that the Greia inscription, which is a petition to Philip V for land by a group of soldiers, is referring to the men simply in terms of their position within the *tetrarchia*, and their status of belonging to the 'first *lochos*', and not as part of a larger military formation.

By using the principle of an even distribution of officers across the formation in order to create symmetry, the positioning of the officers in each of the increasing sub-units of the larger phalanx can be identified. It is important to note that, contrary to what is outlined in the manuals, each file of the phalanx may not actually be led by an officer with the rank of *lochagos*. Similarly, each of the larger units of the phalanx may not be led by an officer directly associated with that unit either (e.g. a *merarch* with a *merarchia*). Rather the rank of the leader of a file or unit would be dictated by where each file or unit was within the larger formation that it was a part of.

For a single file, identifying the position of its leader is relatively simple – he is located at the very front of the file. However, if two files are joined together to create a *dilochia*, each file within this slightly larger formation is not lead by a *lochargos*. Instead, the very right-hand file is led by a *dilochites*, commanding the combined unit as a whole, while the left-hand file is led by a *lochargos*. From this point, as larger units are combined to make even larger ones, the positioning of the officers follows a fairly simple premise. For example, when two *dilochiae* are joined to create a larger *tetrarchia*, the *dilochia* which takes up the left-hand position within the new formation becomes a mirror image of its former self – with its commanding *dilochites* leading the left-hand file. The *dilochia* on the right of the new formation, on the other hand, retains its original configuration with the unit leader at the head of the most right-hand file. However, this officer is not a *dilochites*, but is a more senior *tetrarch* commanding the combined unit as a whole (Fig.31).

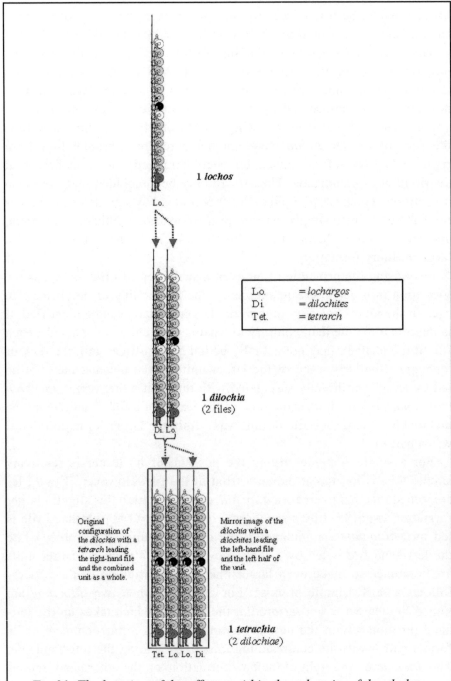

Fig.31: The location of the officers within the sub-units of the phalanx
(lochos to tetrarchia).

This same principle of reversing the original configuration of the unit which makes up the left side of a new and expanded unit, yet retaining the rank of its former commander, while retaining the original configuration of the unit on the right, but under an officer of the higher rank appropriate for the new combined unit, carries on to every subsequent evolution of the phalanx (Figs.32-33).[149]

From this standard and symmetrical system of positioning officers within the various units of the phalanx, the number of officers, and their respective ranks, per *phalangarchia* can be determined (Table 18):

Officers (in front rank)	Number per *Phalangarchia*
Phalangarch	1
Merarch	1
Chiliarch	2
Pentacosiarch	4
Syntagmatarch	8
Taxiarch	16
Tetrarch	32
Dilochites	64
Lochargos	128
TOTAL (front rank)	**256**
Officers (in mid/rear ranks	**Number per *Phalangarchia***
Dimoirites (half-file leader)	256
Dekastateros (half-file closer)	256
Ouragos/Dekastateros (file closer)	256
TOTAL (mid/rear ranks)	**768**
TOTAL (all officers)	**1,024**

Table 18: The number of officers per phalangarchia.

This total figure of 256 officers stationed across the front of the *phalangarchia* agrees with the number of files that the military manuals state made up this formation. When it is considered that each file of the *phalangarchia* would be closed by the rear-rank *ouragos*, this then correlates with the statements made by the ancient writers about how the entire front and rear ranks of the phalanx were made up entirely of officers.[150]

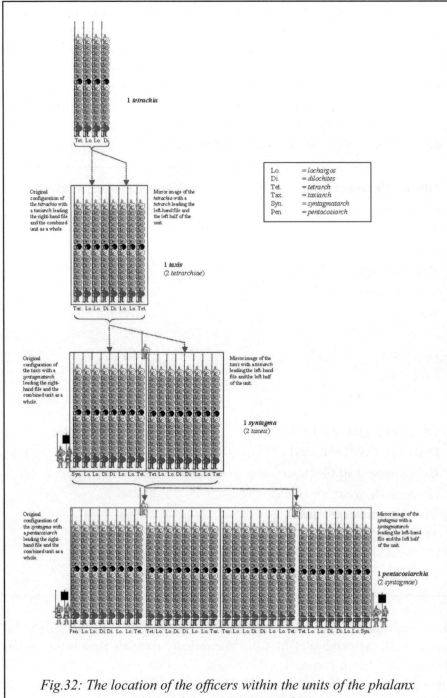

*Fig.32: The location of the officers within the units of the phalanx
(tetrarchia to pentacosiarchia).*

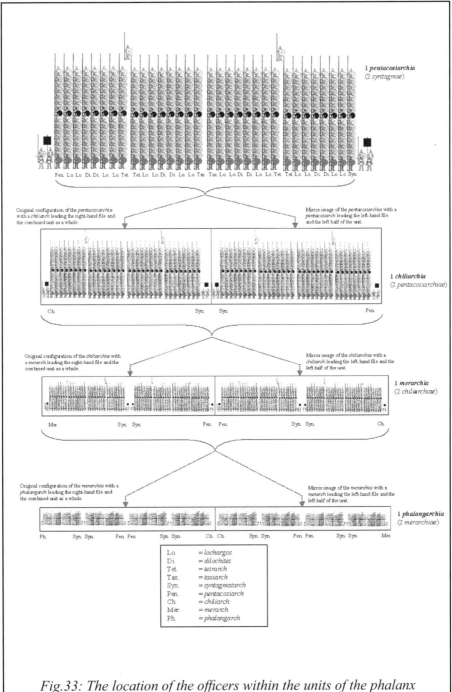

Fig.33: The location of the officers within the units of the phalanx
(pentacosiarchia to phalangarchia).

If the numbers in the above table are multiplied by four, this then provides the total number of officers in the perfect quadruple-phalanx outlined in the military manuals. Importantly, this multiplication would result in there being only four *merarchs*. This correlates with Aelian who only details the positioning of this number of *merarchs* despite the fact that a quadruple-phalanx would actually contain eight *merarchiae*. This shows that, as outlined above, when units were combined into larger ones within the construct of the phalanx, at least half of them were led by officers of a superior rank. This also shows why, when the office of *chiliarch* was invented in 331BC, only six of the recipients of this title could have been assigned to command of a pike-formation while the other two *chiliarchs* would have been assigned to units of the *hypaspists*.

The only way in which a *tetraphalangarchia* would contain eight *merarchs* is if the formation was not actually merged into the quadruple-phalanx but was left as eight independent *merarchiae*. In this case, each unit would be commanded by the highest grade of officer – a *merarch*. This would occur no matter what scale the phalanx was organized into. For example, if a formation was organized into units no bigger than the *chiliarchia*, then every senior command position across the line would be occupied by a *chiliarch*. Thus the exact number of officers within the army as a whole was dependent upon the biggest operational and organizational unit that the army was formed into.[151] How closely this mathematical principle was put in to practice by commanders of the Hellenistic Age, or whether such arrangements were, as Connolly calls them, exercises in mathematical philosophy, is uncertain as the exact location of named officers is rarely given in the ancient texts other than stating that a particular officer was in command of a particular unit, and that that unit was deployed at a specific point within the larger battleline. However, the correlations between the models given in the manuals and the descriptions of officers making up the forward and rear ranks would suggest that such principles were at least partially followed.

Even if an army was not based upon the evenly numbered evolutions for increasing the size of the units that is set out in the manuals (i.e. doubling the size of the sub-unit to create a new one so that each half of a larger formation will always be comprised of an even number of the new units), a symmetry of command could still be achieved. Alexander the Great's army, for example, contained six *merarchiae*. This constitutes three *phalangarchiae* in Alexander's battleline where the manuals state that the normal configuration was two *phalangarchiae* (or one 'wing'). If these units were arranged side-by-side, and the same process for the allocation of commanding officers is followed, then the six *merarchs* would be distributed in the following way (Fig.34):

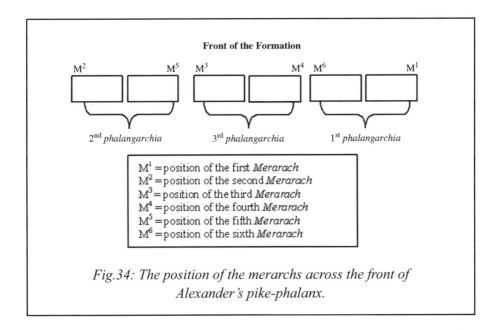

Front of the Formation

M²   M⁵   M³   M⁴   M⁶   M¹

2nd phalangarchia   3rd phalangarchia   1st phalangarchia

M¹ = position of the first Merarach
M² = position of the second Merarach
M³ = position of the third Merarach
M⁴ = position of the fourth Merarach
M⁵ = position of the fifth Merarach
M⁶ = position of the sixth Merarach

Fig.34: The position of the merarchs across the front of Alexander's pike-phalanx.

If the formation was simply divided into two wings by drawing a line down the centre as is done with the standard formation with even numbers of units outlined in the manuals, the command values of each wing would be unequal: a value of ten on the left (2+5+3) and eleven on the right (4+6+1). While this is not a great disparity, it does go against the concept of command symmetry which seems to have been a high priority for the arrangement of Hellenistic armies. However, the command values do equalise if the formation is divided into thirds – that is, into the largest units that are able to be made, following the concepts contained within the manuals, with the number of men at hand – in Alexander's case, three *phalangarchiae*.[152] If the formation is divided in this manner for command purposes, the left-hand *phalangarchia* has a command value of seven (2+5), as does the central *phalangarchia* (3+4), and so too does the right-hand *phalangarchia* (6+1). This demonstrates that, so long as the configuration of the army is based upon the largest possible formation that is outlined in the manuals without deviating from its set structure, all of the concepts of arrangement of both men and officers will hold true.

Arrian lists the six named senior infantry officers of Alexander's army at the battle of the Granicus in 334BC as, from left to right, Craterus, Meleager, Philip, Amyntas, Coenus and Perdiccas.[153] If these officers, each commanding a *merarchia* of troops, were positioned according to the numerical models, their ranks relative to each other, and their exact position within the formation, can be determined (Fig.35):

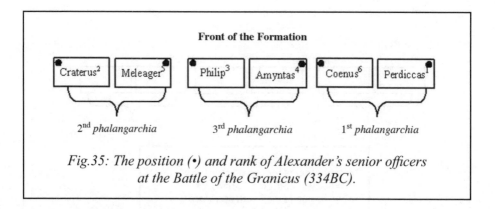

Front of the Formation

| Craterus² | Meleager⁵ | Philip³ | Amyntas⁴ | Coenus⁶ | Perdiccas¹ |

2ⁿᵈ *phalangarchia*   3ʳᵈ *phalangarchia*   1ˢᵗ *phalangarchia*

*Fig.35: The position (•) and rank of Alexander's senior officers
at the Battle of the Granicus (334BC).*

Arrian also provides the names of six senior infantry commanders for the battle of Issus the following year as, from left to right, Craterus, Meleager, Ptolemaeus, Amyntas, Perdiccas and Coenus (Fig.36).[154]

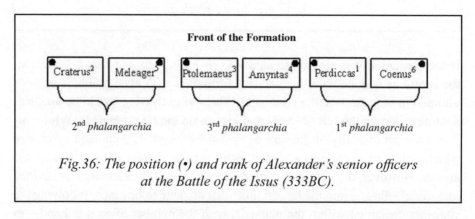

Front of the Formation

| Craterus² | Meleager⁵ | Ptolemaeus³ | Amyntas⁴ | Perdiccas¹ | Coenus⁶ |

2ⁿᵈ *phalangarchia*   3ʳᵈ *phalangarchia*   1ˢᵗ *phalangarchia*

*Fig.36: The position (•) and rank of Alexander's senior officers
at the Battle of the Issus (333BC).*

This is not overly different to the deployment at Granicus: Craterus and Meleager are in their respective positions on the left, Ptolemaeus (who had replaced Philip following his death) and Amyntas retain the central commands, and Coenus and Perdiccas have changed places with respect to each other, but still command the two most right-hand *merarchiae*. Arrian states that at Granicus in 334BC the phalanx had deployed with each unit 'under the commanders in the order of precedence for the day'.[155] This suggests that the positions of the commanders were not set in stone and that they were deployed on some form of rotating roster. English offers that this was designed to provide more junior officers the opportunity to gain experience.[156] However, it is strange to note that for the accounts of both Granicus and Issus, it is only Coenus and Perdiccas whose positions seem to change. Many of the *merarchs* also hold the same positions at the battle of Gaugamela a few years later (see following). Assuming that this is not just pure coincidence with the timing of these three battles, and additionally assuming that Arrian's passage is referring to the *merarchs* and not the leaders of the sub-units within each *merarchia*, it seems that it was only the commanders of the two right-wing units who may have traded places. This would explain why their positions for the battle of the Granicus seem to have been reversed for the battle of Issus.

It is also interesting to note that for the battle of Issus, Arrian states that the entire right wing of the line, which included pike-units, cavalry, light-troops and *hypaspists*, ended with the unit of Perdiccas on the left. Alexander himself commanded the right side of the right wing. Thus by swapping the positions of Perdiccas and Coenus for the battle at Issus, Alexander placed his most senior pike-infantry commander at one end of the wing, while he himself commanded the other. This, in effect, made the right-wing a semi-independent formation with strong commanders at either end. The exact location for the division between the two wings is somewhat harder to identify for the later battle at Gaugamela, but it is interesting to note that both Coenus and Perdiccas hold the same positions as they do at Issus.

At Gaugamela in 331BC, the infantry commanders are listed as, from left to right, Craterus, Simmias (replacing the absent Amyntas), Polyperchon, Meleager, Perdiccas and Coenus.[157] Diodorus and Curtius do not mention the troops commanded by Simmias but do refer to a unit under the command of Philip in this position.[158] This is most likely another transcription error in the text of Curtius. According to Arrian, Simmias was the son of Amyntas who, in turn, was the son of Philip (although this is also an error as Amyntas was actually the son of Andromenes). Arrian describes this unit as being that of 'Amyntas son of Philip; this was led by Simmias since Amyntas had been sent to Macedonia'. It seems that Diodorus' and Curtius' sources (or the writers themselves) may have confused who was actually in command during the battle. Curtius also omits any reference to the unit of Meleager. Regardless, most of the sources still list six senior infantry officers commanding at Gaugamela (Fig.37).

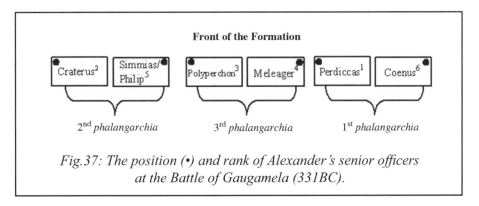

Fig.37: The position (•) and rank of Alexander's senior officers
at the Battle of Gaugamela (331BC).

In all three engagements Craterus retains his position on the far left. Simmias, fairly new to command at this level, is allocated command of the unit next in the line to the right of Craterus' unit. This would normally be commanded by an officer of the fifth rank (as opposed to Craterus' second rank) and Simmias' inexperience explains why he is put in this position. Craterus is additionally described as not only commanding his own unit, but also the whole left wing of the infantry line.[159] This would comply with the descriptions in the manuals of how the senior officer on the left of the line also commanded that entire wing (which also shows that the interpretations which place him in the centre of the line at Issus are incorrect). Thus Craterus was one of the more senior and experienced of the six named officers and this accounts for his positioning on the left of Alexander's formation (also as per the manuals). At this level, Alexander's army deviates from the structure outlined in the military manuals in that an infantry 'wing' would normally be two *phalangarchiae*. However, if Alexander's pike-formation is divided evenly into two halves, each wing would actually contain one-and-a-half *phalangarchiae* due to the total number of *mararchiae* within his army. Whether Craterus and Coenus actually held the rank of *phalangarch* or not is unknown. The army as a whole (that is pike-infantry, *hypaspists*, cavalry, light troops and mercenaries) was divided in halves for operational control with Alexander himself in command of the right wing and with Parmenion commanding the left. This dual level of command would then follow an arrangement for the leadership of a *tetraphalangarchia* of infantry as outlined in the manuals despite Alexander's 'wings' containing an assortment of troops. It is also likely that, as had occurred at the battle of Issus, the left-hand side of Alexander's right wing at Gaugamela was the infantry unit commanded by Perdiccas. Regardless of these larger fields of command, the arrangement of Alexander's pike-infantry seems to be based upon the *merarchia* and the *phalangarchia* as the two largest operational units.

Due to Simmias' move into the fifth rank command position on the left at Gaugamela, the other commanders of the pike *merarchiae* are also moved within the formation.[160] Meleager, for example, who had held the fifth rank position on the left at Issus, is 'promoted' to the fourth rank position in the centre for the battle at Gaugamela – no doubt making way for the inexperienced Simmias to be positioned beside the more senior Craterus. Ptolemaeus' unit on the left of the centre is taken over by Polyperchon, and the units of Perdiccas and Coenus are positioned in the same way as they were for Issus. The positioning of these officers across Alexander's infantry line, in three separate engagements, suggests that the distribution of his officers followed the mathematical principles set down in the later manuals, albeit with variances in positioning due to the operational divisions of the whole formation into two wings, and a continuance of this policy from the time of Alexander across the entire Hellenistic Period. It was this mathematical principle for the balance of command and organization which influenced most aspects of the pike-phalanx on campaign.

DEPLOYING THE PHALANX

The symmetrical distribution of officers across the line also had implications for how the units of an army marched in column and deployed on the battlefield. The ancient texts outline two different methods for the deployment of a pike-phalanx on the field or battle. Arrian, for example, states of Alexander the Great's deployment at Issus that:

> ...as long as the defile on every side remained narrow, he led the army in column, but when it grew broader, he deployed his column continuously into a phalanx, bringing up unit after unit of hoplites (ἄλλην καὶ ἄλλην τῶν ὁπλιτῶν τάξιν παράγων).[161]

This description suggests that, as the terrain of the battlefield opened, a unit of troops (Arrian uses the generic term *taxis*) positioned behind the leading unit of the column was brought up beside it and then, as the terrain opened further, subsequent units were also brought up, until the entire pike-phalanx was deployed into its battleline. What Arrian does not elaborate on is the exact units that were brought up in each successive manoeuvre – so we do not know if Alexander's army deployed by *merarchia*, by *chiliarchia* or even by *syntagma*. Bosworth suggests that individual files moved forward as the terrain widened, however Arrian's use of the generic term *taxis* would suggest units larger than this.[162] Curtius states that Alexander's forces marched into the defile in units thirty-two ranks deep.[163] While this shows that the formations were arranged at double their normal depth, it is still not certain as to the size of the units that were arranged in this manner. However, due to the orderly arrangement of the phalanx into units and sub-units, any model which might be applied to large contingents like the *merarchiae* would also hold true for smaller units such as the *syntagma* – the only differentiating factor for either model would be how rapidly the terrain opened out which would, in turn, dictate the size of the units that could be brought forward.

How each successive unit was brought forward, and in what order, is also uncertain. Burn, for example, states that 'since Greek armies invariably formed line, from column, to the left, the shielded side, this meant that Alexander [at Issus] automatically had with him on the right wing his favourite light troops and his guards'.[164] This suggests that Alexander's column was led by his light troops and guards, with the units of the pike-phalanx following behind, and, as the defile widened, pike units moved forward to take up positions to the left of the light troops and guards to form an extended battle line. Arrian states that Alexander's cavalry were positioned at the rear of the column and were only brought forward once the ground had sufficiently widened.[165] Polybius elaborates further by stating that the baggage train was positioned behind the cavalry.[166] This was not the first time that Alexander had undertaken such a manoeuvre either. In 335BC, Alexander's army was crossing a river to engage the Taulantians and his troops were ordered to 'extend [their line] to the left [as each unit moved up] so that the phalanx might appear solid the moment they crossed'.[167] These references show that such extensions of the line to the left were a normal part of phalangite drill and deployment for battle.

As each infantry unit of Alexander's column moved forward and left to begin forming a line of battle as the defile at Issus widened, the remainder of the column would have moved with them – in effect making the leading unit of the column the most left-hand unit in the developing line. This would result in each successive unit having a lesser distance to travel to take up its new

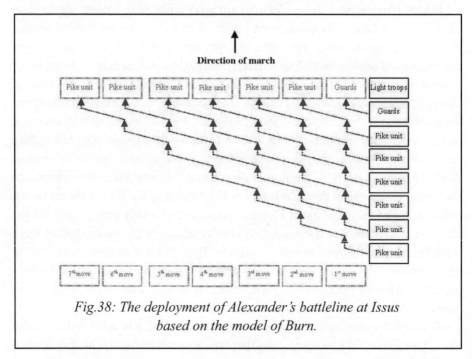

Fig.38: The deployment of Alexander's battleline at Issus based on the model of Burn.

position in the line as the ground opened further, as it would have if the column had remained formed behind the most right-hand unit (Fig.38).

Deploying a column into line by having units reposition themselves to the left of those preceding it was a manoeuvre used by other Hellenistic armies as well. At Cynoscephalae in 197BC, for example, Livy states that once the line had been formed, the whole phalanx then closed up any gaps that had been left by this manoeuvre by having entire units shuffle to the right.[168] This indicates that the units in the column had redeployed to the left of the leading unit (or the 'right marker' to use modern military terminology) and that, once the extended line had been formed, any gaps resultant from the manoeuvre were closed by the right-hand unit maintaining its position while those to the left shuffled across.

While this procedure initially looks quite straightforward, there was substantially more to arranging the column, which then allowed for the correct battleline to be created. Using the six *merarchiae* of pike-infantry of Alexander's army at Issus as an example, if these units were arranged in column in their numerical order according to the 'rank' of their commander (i.e. with *merarchia* #1 under Perdiccas leading, followed by #2 under Craterus, #3 under Ptolemaeus and so on) and, as the terrain widened at Issus, and each succeeding unit was brought up beside the one ahead of it, then once the battleline was formed, the units would not be in their correct order (Fig.39).

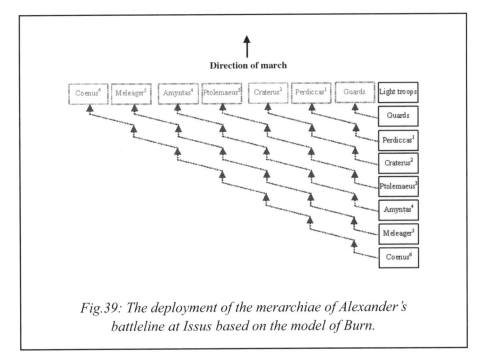

Fig.39: The deployment of the merarchiae of Alexander's battleline at Issus based on the model of Burn.

Such a method of deployment would result in a line where the units were arranged, from left to right: Coenus, Meleager, Amyntas, Ptolemaeus, Craterus and Perdiccas. This goes against the deployment outlined by Arrian who lists the arrangement as, from left to right: Craterus, Meleager, Ptolemaeus, Amyntas, Perdiccas and Coenus. This arrangement would also go against the symmetrical distribution of officers and command abilities across the line as detailed in the manuals.

It is unlikely that units arranged in a column in numerical order redeployed in line one-by-one as the terrain widened and then, as it widened further, side-stepped certain distances to the left or right so that succeeding units could take up their correct positions in what would be the eventual battle order. For example, it is unlikely that the unit commanded by Perdiccas, which would be at the head of the column according to Burn's model, but needed to be positioned second from the right according to the deployment outlined by Arrian, would have initially moved forward and left as the defile began to widen and then, as the terrain widened further, have Craterus' unit deploy on the far left and then, as the defile opened up further still, have Craterus' unit side-step to the left to create a gap which would keep his unit on the far left of the developing line and allow other units to take up positions in the centre.

Cascarino, on the other hand, based upon the four units of the 'perfect formation' found in the manuals, suggests that the second ranked unit of the formation moved forward and far to the left to leave enough space for later units to move up.[169] This cannot be what occurred at Issus as the narrow terrain did not allow for such gaps to be left in the line. Such deployments would also leave certain units particularly exposed until other units were brought up in the centre. Even on more open terrain, such a manoeuvre seems overly dangerous and complex. It is more likely that, again based upon the four units of the 'perfect phalanx', that the fourth unit, at the rear of the column, would have moved ahead first to take up a position to the left of the first unit. The third unit would move up next, and then the second unit would move to the far left of the forming line. This would place the units in their correct position for a symmetrical deployment.

Yet at Issus, with six *merarchiae* of pike-infantry on the march rather than four, such movements would be both considerably complicated and time consuming. It seems more likely that a simpler method of deployment, one that did not require complicated spatial calculations, or slow down Alexander's rate of advance, would have been undertaken. This suggests that the pike units of Alexander's column had been arranged in the order that they were to later deploy in within the battleline and that, as the defile widened, each unit simply took up its correct position (Fig.40).

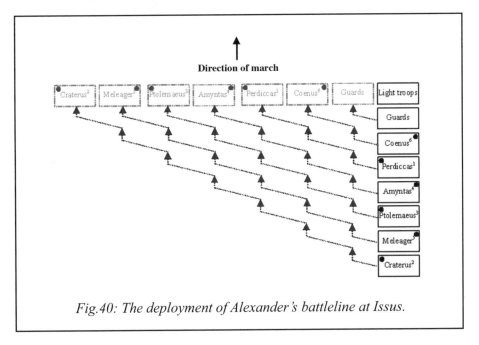

Fig.40: The deployment of Alexander's battleline at Issus.

Such an arrangement of the units in the column, so that they could deploy in their allotted position for the ensuing engagement, must have been based upon a preconceived order of battle that had been formulated before Alexander had even begun his march towards Issus. Curtius suggests that this is the case when he describes how Alexander had ordered his men to 'refresh themselves and to be ready and under arms by the third watch'.[170] Later, when Alexander reached the defile that his army was to march through to reach the battlefield, and was told that the Persians were still some distance away, Curtius tells us that Alexander 'ordered the line to halt and...arranged the battle order'.[171] This suggests that there was some predetermined order of march that had been arranged or considered by Alexander and that, upon reaching the narrowing terrain, Alexander then put that plan into effect. This demonstrates a high level of forward tactical planning on the part of Alexander and the professional nature of his troops who could operate within their specific units and sub-units as part of a broader stratagem.

Another point to consider is that Alexander's column cannot have marched straight down the central axis of the defile. Doing so would have meant that, while the terrain would have opened up to both the left and right, only the available space on the left side of the column would have been utilized for redeployment. This would have left the right-hand side of the column vulnerable and would not have been an efficient use of the widening terrain. It is more likely that Alexander's column marched along the right-hand edge of the defile, obliquely to the Persian deployment ahead of them, which would have not only provided some protection to the right-hand, un-shielded, side of the column, but would have also used all of the available space as the terrain opened up which would have allowed for the battleline to be formed more quickly.

Fuller suggests that, for each unit to be brought forward into line, the entire army would halt, the next unit in the column would then move forward and left into position with the remainder falling in behind and then, once this partial line had been formed, the army would advance again until the terrain sufficiently widened to allow for the army to halt again and the next unit be brought forward.[172] Fuller offers that, due to this series of advances, halts and redeployments, the amount of time that it would have taken Alexander to cover the 19km distance from his camp to the battlefield at Issus could have been as much as ten hours.[173] It seems more likely that, as part of this movement, as the terrain widened, any units that had already formed into line on the right would have slowed their advance, possibly to a half-pace, rather than stop completely, while the remainder of the column shifted to the left at their regular pace to both catch up with the units ahead of them and to form the extending line in one fluid movement. Once in position, the army could then resume its march at its regular pace until the terrain opened up and the process was performed again. This would have cut several hours off the timeframe of the army's march that is suggested by Fuller and would have allowed Alexander to get his troops to the battlefield in the quickest and most secure manner while simultaneously moving each unit of his army from column into line.

Furthermore, the symmetrical distribution of officers across the line to provide a balanced level of command abilities means that the commanding officers of the units marching within the column had to be arranged a certain way. It is most likely that each unit was arranged in the manner that it was to deploy and fight in – that is, each successive unit within Alexander's column had its officers arranged as a mirror image of the unit ahead of it in the column. In this way, when all the units were finally positioned in line, the commanding officers would be in their correct positions (Fig.41).

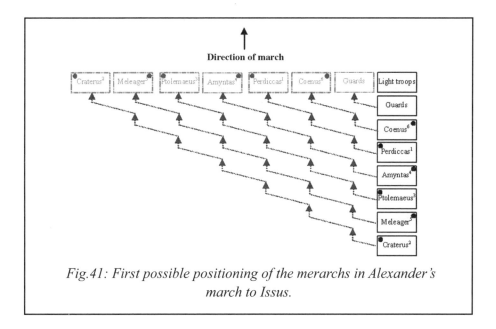

Fig.41: First possible positioning of the merarchs in Alexander's march to Issus.

Another possibility is that all of the *merarchiae* were arranged so that all of the commanding *merarchs* were on the right-hand side while they were marching in column. However, once deployed in line, this would mean that every second unit would not be in its correct configuration for the even and symmetrical distribution of officers and command abilities across the line (Fig.42).

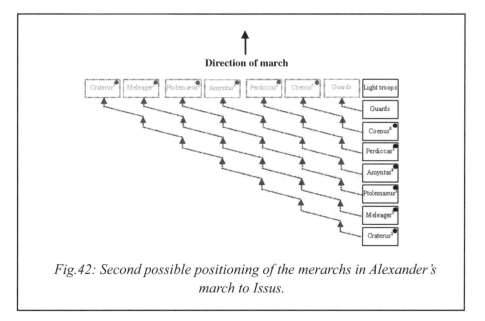

Fig.42: Second possible positioning of the merarchs in Alexander's march to Issus.

If this method was used, once deployed into line, the units of Craterus, Ptolemaeus and Perdiccas would be required to reverse their internal structure, moving the commanding *merarch* and his file to the left-hand side, and ensuring that the remainder of the unit followed suit, so that all of the units of the line were properly configured. This could be accomplished through the manoeuvre of 'counter-marching'. According to the manuals, the process of counter-marching was employed to change a unit's facing by 180°. There were three different methods of counter-marching: the 'Macedonian', the 'Lacedaemonian' and the 'Persian/Cretan/Choral' countermarch.[174] Each method had the same result, but was undertaken in a different manner and required different free space to perform.

The Macedonian counter-march, for example, was a process whereby a stationary formation could alter its facing by occupying the ground in front of it. To accomplish such a manoeuvre, each file leader would wheel about 180° and look back down between the files. The remainder of each file, which would still be facing forward, would then advance past the stationary file leader. As each man reached a position equal to his normal interval behind the front of the line, that man would wheel about, join the newly forming file, and face back in the direction from which he had come. This process would continue until the *ouragoi* took up their regular position at the rear of the formation. Thus the file leaders would resume their position at the head of the files, and the *ouragoi* would once again be in the rear, but the formation would be facing the other way (Fig.43).[175]

The Lacedaemonian counter-march differed from the Macedonian version in that the space utilized to execute the manoeuvre was behind the formation rather than in front of it. In this procedure, all the men in each file initially about face. The file leaders, now at the back of the formation, would then move down between the files and, as they passed the first man, that man would fall in behind the file-leader and continue down the interval. This process would continue as the file-leaders, and those following them, passed each man in turn; with the train following the file leader growing as each man is passed. When the file leaders reached the *ouragoi*, they would continue to advance forward to a distance equal to the depth of the file in its respective order (most likely determined by knowing how many paces it took to cover the depth of a file in various orders). When this distance was reached, the file leaders and those following them would halt and the *ouragoi* would simply fall in at the back of the file. The unit would have then changed its facing by 180°, but also moved into the open ground that had previously been to the rear of the formation (Fig.44).[176]

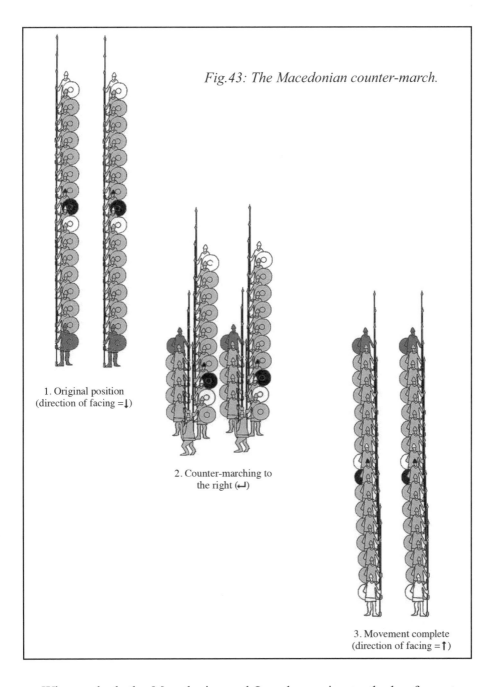

Fig.43: The Macedonian counter-march.

1. Original position
(direction of facing =↓)

2. Counter-marching to
the right (↵)

3. Movement complete
(direction of facing =↑)

Whereas both the Macedonian and Lacedaemonian methods of counter-marching required a certain amount of free space to be undertaken, the Persian/Cretan/Choral method used the space that the formation already occupied. All the file leaders would wheel to either the left or right and then march down

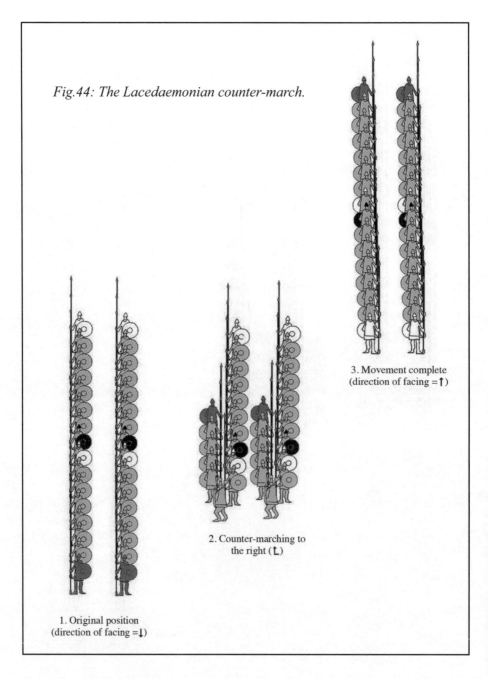

Fig.44: The Lacedaemonian counter-march.

3. Movement complete
(direction of facing = ↑)

2. Counter-marching to
the right (L)

1. Original position
(direction of facing = ↓)

between the files. However, rather than remain stationary and wait for those changing direction to pass them as in the other models, the file would advance forwards. As each man reached the point where their file leader had originally stood prior to the counter-march taking place, they too would then wheel about, using the file leader's original position as a pivot point, and follow along down between the files in the opposite direction. When the leader of each file reached the point beside the position that the *ouragos* of their respective file had initially occupied (most likely determined by knowing how many paces had to be made to cover the distance from the front to the rear of the line) the file leaders would halt. By this stage (if each man had maintained his interval and the pace count had been correct) the *ouragoi* should have advanced to the position originally held by the file leaders. The *ouragoi* would then simply wheel about to join the back of each file and create a formation that was now facing the other way but had required no more room for the transition than it had originally occupied (Fig.45).

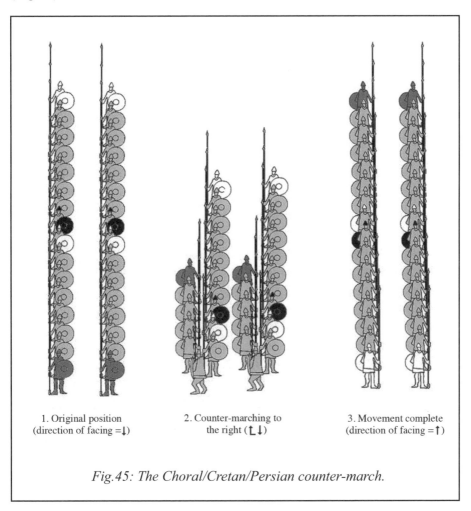

1. Original position
(direction of facing =↓)

2. Counter-marching to
the right (⌐↓)

3. Movement complete
(direction of facing =↑)

Fig.45: The Choral/Cretan/Persian counter-march.

This last method, the Persian/Cretan/Choral counter-march, could be employed within a deployed battleline to arrange (or rearrange) the units within it in their proper order so that the officers were distributed evenly and symmetrically across the line. Using the *merarchiae* of Perdiccas and Coenus at Issus as an example, if the column had initially been arranged with all of the *merarchs* and their files on the right-hand side as the column advanced, once it was deployed in line, Perdiccas' *merarchia*, to the left of Coenus', would need to alter its internal structure so that Perdiccas and his file were on the left. To accomplish this, once the formation had halted, every member of the formation would turn to face to the right (in effect making its ranks now its files).[177] The unit would then conduct a Persian/Cretan/Choral counter-march which would alter the unit's facing by 180° (meaning everyone within it would now be facing to the left) and Perdiccas and his file would now be on the left-hand side of the formation. Once the counter-march had been completed, all members of the formation would then turn to face the front, resulting in a formation correctly deployed as a mirror image of Coenus' one beside it. Such a process could be undertaken by any unit, of any size, which needed to alter its configuration, independent of all other units within the line, without requiring any additional room or compromising the integrity of the formation as a whole.

The ancient narratives also outline another way that the pike-phalanx could be moved from column into line. Polybius tells us that, at Mantinea in 207BC, the Spartan commander Machanidas advanced his troops in column and then, at the appropriate time, wheeled the entire formation to the right to deploy them in line.[178] The wheeling of a formation by 90° was a movement known as an *epistrophē* (ἐπιστροφή) in the military manuals.[179] Again, while such a process seems rather simple upon initial reading, it required specific placement of the officers within each unit, and specific configurations of each unit as well, for it to be undertaken easily, quickly, successfully, and with limited disruption.

For example, when the units of Machanidas' army were marching in column, they could not have been in their normal 'block' configuration with the long sides (i.e. their ranks) running left to right across the formation and their files running back to front. Instead each unit, regardless of what size it was, must have been rotated by 90° – marching with their short sides leading – a process known as *paragogē* (παραγωγή) in the ancient manuals.[180] In this way, when the entire column wheeled into position, the basic structure of the army, with the long sides of each formation facing towards the enemy, would be correct. However, the troops within each formation would, after initially completing the

wheeling manoeuvre, all be facing to the right. As such, to prepare to engage
the enemy, the members of each unit would need to turn in position 90° to their
left so that they were facing forward (Fig.46).

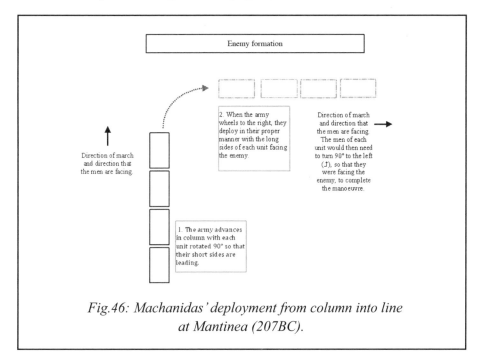

*Fig.46: Machanidas' deployment from column into line
at Mantinea (207BC).*

Even this seemingly simple transition from column into line requires careful
pre-planning for the positioning of the officers and their respective files within
each of the larger units of the column. If, on the one hand, the commanding
officers were positioned at the front-right of each unit as they were marching in
column (in effect, having their file stretch across the front of each unit to their
left) then, when the entire column wheeled into line, the officers would be at the
rear of their respective files/units (Fig.47).

This positioning of the officers could be corrected by having each unit
turn to face the front (towards the enemy) and then conduct a Macedonian
or Choral-style counter-march to reverse the files and bring the officers to
the front of the line. The symmetrical distribution of officers across the
line could then be accomplished by having the appropriate units conduct
further changes of facing and counter-marches to move the commanders
and their respective files to the left side of their units. Alternatively, some
of the commanding officers and their files may have been positioned at the
rear of their respective units when initially marching in column. In this way,

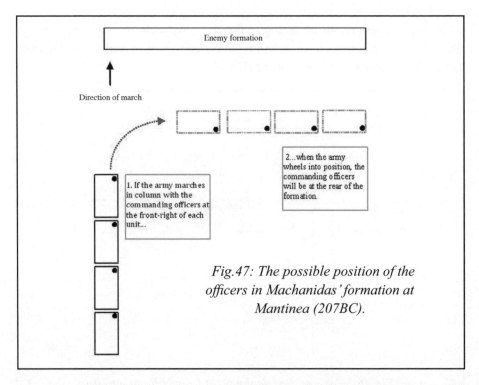

Fig.47: The possible position of the officers in Machanidas' formation at Mantinea (207BC).

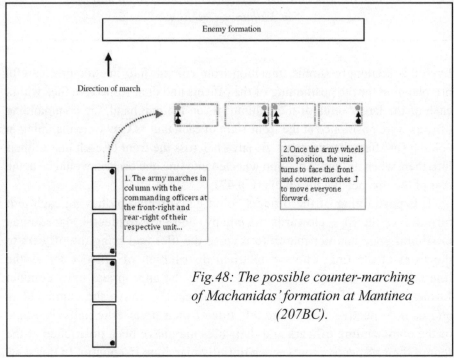

Fig.48: The possible counter-marching of Machanidas' formation at Mantinea (207BC).

once the formation had wheeled in line and the necessary counter-march conducted, all of the officers and men within each unit would be in their correct positions (Fig.48).

However, this process involving several changes of facing and counter-marches seems overly complex and not something that an army would undertake when in close proximity to an enemy as Machanidas' phalanx was at Mantinea. This then suggests, on the other hand, that the officers were arranged in a way so that, once the wheeling of the column into line was complete, the entire phalanx was immediately ready to engage or be engaged. This would mean that the commanding officers of each unit, rather than being on the right-hand side when marching in column, would have been on the left and in their alternating positions with some at the front-left and others at the rear-left. Once the formation had moved into line, all that would need to done would be for each unit to turn and face to the left and the entire line would then end up in its correct order and configuration and facing the enemy (Fig.49).

This more simple process is more likely to have been the manner in which such a manoeuvre was conducted as it involved less steps in the process and, importantly, meant that the formations, or parts thereof, were

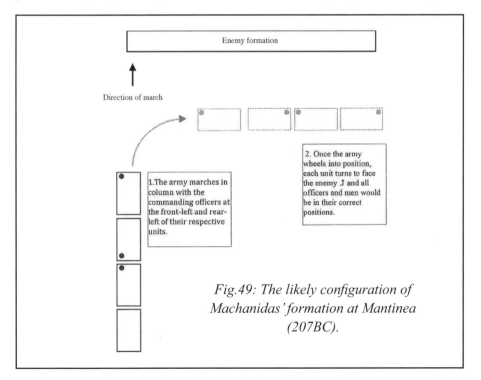

Fig.49: The likely configuration of Machanidas' formation at Mantinea (207BC).

not conducting drill movements such as counter-marches while close to an enemy who may exploit any opportunity to launch an attack while the line was not fully formed.

This arrangement also demonstrates a certain level of pre-planning on the part of the overall commander to organize his men in a way that will result in the formation being immediately ready for battle once any preconceived manoeuvres had been completed. Indeed, such an arrangement is quite easy to undertake. For example, prior to marching out in column, Machanidas' army would have been formed up in an extended line, with all of its officers and files in their proper locations. However, in respect to the position of the enemy, the initial formation would be at right angles to it – i.e. facing to the left rather than towards the enemy as per the diagrams above. This deployment would most likely have been done at some distance from the enemy when the flanks would not be exposed or vulnerable. The entire formation would then simply turn its facing towards the direction they were going to march – changing the formation from a phalanx in line to a phalanx in column but without altering the position of anyone within the formation. The column could then march forward to the battlefield and then, at the appropriate moment, wheel to the right to deploy in line, turn to face the enemy and be ready to fight. Importantly, at no time does this more simplistic model require counter-marches which disrupt the integrity of the line nor require any more changes of facing other than that to turn towards the enemy. If the plan called for the column to wheel to the left instead of the right, due to the nature of the terrain for example, this could also be accomplished by having the phalanx initially form up facing to the right (rather than to the left) before turning into column. Regardless of whether this more simplistic method of deployment was employed, or a more complex variant, or even the method of bringing up individual units one at a time as Alexander did at Issus, it is clear that the deployment of a pike-phalanx from column into line of battle required careful pre-planning. The ability to undertake such movements was in part due to the way in which the pike-phalanx was divided into units and sub-units which meant that such operations could be made on any scale depending upon the terrain, the plans of the commander and the tactical requirements of the impending confrontation.

Chapter Twelve

The Pike-Phalanx in Battle

THE 'RIGID' PHALANX

The merging of units and sub-units to form what the writers of the military manuals would describe as the 'perfect phalanx' highlights an element of the pike-formation's performance on the field of battle that has seen its fair share of scholarly contention: how rigid was a whole line of pikemen once the army had deployed in line? Some scholars suggest that the Hellenistic pike-phalanx lacked the ability to adapt to varying terrain and changes in the tactical situation. Adcock, for example, states that the pike-phalanx was 'muscle-bound and without flexibility'.[1] Pietrykowski similarly calls the pike-phalanx 'ponderous' and states that the formation was 'fatally flawed at its core' due to its rigidity 'which, in turn, inhibited tactical flexibility'.[2] Somewhat contradictorily, Anson says that the phalanx was organized into 'interdependent blocks of infantry' requiring 'less emphasis on the ability of the individual soldier and encouraging unit solidarity' while also stating that the formation 'lacked flexibility'.[3]

Other scholars suggest the exact opposite, claiming that there was an inherent level of adaptability within the structure of the Hellenistic pike-phalanx. Burn plainly states that 'Alexander's formations were, naturally, flexible'.[4] Featherstone states that it was Alexander who broke the pike-phalanx into smaller units, turning it into a 'spear-hedged', highly mobile formation 'rather than a juggernaut of moving spear points'.[5] Warry calls the pike-phalanx 'a highly flexible unit, capable of assuming various formations'.[6]

Other scholars offer that the early phalanx did possess tactical flexibility but that this ability was later lost during the time of the Successors. Bosworth suggests that flexibility was what mattered most to the phalanx.[7] Gabriel states that it is often mistakenly believed that the pike-phalanx lacked flexibility and that it was, in fact, less prone to breaking up than the Classical Age hoplite phalanx until the armies of the later Successor Period expanded the size and depth of the formation to a point where it could hardly manoeuvre at all.[8] Tarn states that, by the second century BC, the flexibility of Alexander's phalanx had been lost and rough or hilly ground always disrupted these later, rigid, formations.[9] Pietrykowski states that 'relentless training, brilliant leadership and a forgivingly flexible construction rendered the phalanx of Philip [II] and Alexander [the Great] far better suited to an irregular battlefield than the Hellenistic armies of later decades; suggesting that it was the longer pikes of the Successor period which made later phalanxes 'less flexible and less manoeuvrable'.[10]

A review of the ancient descriptions of the pike-phalanx in action tends to agree with those scholars who advocate a certain level of flexibility within the phalanx. As far back as the early reign of Philip II, and well into the Successor Period, the sources detail pike formations which are both adaptable in deployment and flexible in operation. At Chaeronea in 338BC, for example, the Macedonian army deployed and 'individual units were positioned where the situation required'.[11] At Thebes in 335BC, Alexander divided his army into three separate groups, some of which must have contained units of pike infantry, one to attack the palisade which had been erected in front of the city, another (which included a contingent of Macedonians) to face an advance by the Thebans, and another which was posted in reserve behind the lines and who advanced to support the units facing the Thebans when they became hard pressed.[12] This shows that not only could sections of the pike phalanx be deployed separately from each other, but that reserve units could be brought up and into the fight when required.[13] This can only be considered a use of pikemen in a line that was not rigid. Units of pikemen, rather than the whole phalanx, were also regularly used in many of Alexander's sieges and assaults as independent contingents.[14]

While the use of individual contingents, rather than the whole battleline, would seem an obvious use of manpower in siege warfare, the literary sources also detail the use of individual units of the phalanx in open field engagements as well. At Gaugamela in 331BC, for example, Alexander advanced his right wing, which contained some pike units, in an oblique line towards the Persian left.[15] Curtius states that the entire line, rather than just the right wing, advanced at an angle.[16] It is almost impossible for a pike-phalanx, either in part or as a whole, to advance at an angle if the formation is a solid, rigid line. However, it is much easier to advance individual units obliquely. This suggests that Alexander's phalanx was not an inflexible battle line. Rather it seems that the pike-phalanx was comprised of a number of independent 'combat groups' – made up of the various units and sub-units of the phalanx – who acted in concert with each other within the parameters of any predetermined plan or spontaneous command while attempting to limit any fragmentation of the line as a whole which may occur as different units advanced and/or engaged. Later, when a gap had opened in the Persian line opposite the Macedonian right, Alexander directed his attack into this opening using his cavalry and 'the section of the phalanx stationed there'.[17] This shows that the pike units on the right of the line could operate independently from the rest. This independence is confirmed by Arrian who describes how the *merarchia* commanded by Simmias, on the left of the line between the *merarchiae* of Craterus and Polyperchon at Gaugamela, 'halted their unit [Arrian uses the term 'phalanx' in its generic sense] and were fighting where they stood as the Macedonian left was reported to be in trouble'.[18] Thus it seems clear that each of the *merarchiae* of Alexander's line were operating somewhat independently on the field. Again, this indicates that the phalanx was not one rigid battleline.

This flexibility of the phalanx was also used to arrange the deployment of an army to meet certain conditions on the battlefield and to cater to an overall plan of battle. At Asculum in 279BC, Pyrrhus posted units of native levies in between the units of his phalanx.[19] At Sellasia in 222BC, Antigonus similarly posted units of allied infantry between the units of his pike-phalanx to provide the line with enough flexibility to assault up a hill while simultaneously maintaining unit cohesion.[20] At Magnesia in 190BC, Antiochus had his pike-phalanx divided into ten divisions with thirty-two elephants positioned between each contingent.[21] Thus, right from the onset, all of these deployments of the phalanx would not have been a rigid line of pikemen as some scholars suggest.

Even while marching, the phalanx operated as interdependent units. Any redeployment from column into line conducted in the manner of Alexander's advance towards Issus (see pages 311-317) can only have been undertaken on a unit by unit basis. Similarly, the wheeling of Machanidas' phalanx at Mantinea in 207BC could only have been conducted in units rather than as a whole line.[22] At the same engagement, Polybius describes how Philopoemen, commanding the opposing Achaean forces, deployed his phalanx in contingents with spaces left between.[23] As the battle began, the first section of the Achaean troops wheeled to the left onto clear ground and then advanced into battle without breaking ranks.[24] All of these descriptions can only be taken as references to a pike-phalanx that is both flexible in its arrangement and adaptable to varying tactical situations due to it operating as semi-independent combat groups. Similarly, during the parade at Daphne in 168BC, Antiochus IV is said to have ordered 'those to advance, those to halt, and assigned others to their positions, as occasion required'.[25] This again suggests a pike-phalanx which operated as independent units rather than as a solid, rigid, line.

Aelian also details a large number of other ways that the phalanx could be configured other than just in a horizontal or oblique line. These include such formations as the concave or convex crescent (*meneoides* [μηνοειδής]), with a variant also known as the *kyrtē* [κυρτὴ] formation), the open-ended square (*epicampios emprosthia* [ἐπικάμπιος ἐμπροσθια]), the wedge (*embolon* [ἔμβολον]), and the hollow square (*plaision* [πλαίσιον]).[26] In Illyria in 335BC, Alexander the Great formed part of his troops into a wedge, and formations such as defensive squares were adopted by Eumenes' troops at Gabiene in 316BC and by Antiochus III's phalanx at Magnesia in 190BC.[27] These formations are difficult, if not impossible, to form with a single, rigid, line, but can be accomplished by simply arranging the units and sub-units of the phalanx into the desired configuration. At Issus, the pike units on the right wing, finding the Persians who were opposing them in flight, wheeled inwards to attack troops who were holding the centre of the phalanx at bay at the river's edge.[28] Again, this shows that the units of the phalanx could operate independently of each other and, if the need arose, come to each other's aid. Dickinson suggests that the phalanx could change formation easily.[29] This can only be based upon an acknowledgement that the phalanx was composed of several individual units which could be rearranged with ease. Such references also illustrate a continued use of varied arrangements of the pike-phalanx across almost the whole of the Hellenistic Period.

Regardless of whether Aelian's descriptions of alternative formations are taken as indicative of the variable deployments of the early phalanx under Philip and Alexander (which Aelian suggests he is writing about), or of the later formations of the Successor Period, or both, it is clear that the phalanx was highly adaptable and not inflexible. Importantly, if these descriptions are seen as references to the formations of the later Successor Period – and the references to the adoption of defensive squares at Gabiene and Magnesia suggest that, if not wholly a Successor Period concept, then the use of such formations may simply be a continuance of earlier phalanx configurations – it is clear that the use of longer pikes in this time period did not inhibit the adaptability of the phalanx as some scholars suggest. This flexibility in the deployment, manoeuvring and operation of the phalanx also impacted on how it moved into the attack regardless of its initial configuration.

THE ADVANCE OF THE PHALANX

It is often stated by modern scholars that the Hellenistic pike-phalanx could only operate on certain terrain. Anson, for example, states that the phalanx required 'level and clear ground with no obstacles'.[30] Snodgrass similarly says that the phalanx could be clumsy on rough terrain.[31] Markle states that 'the *sarissa* armed phalanx could only be effectively used on a level plain unbroken by streams, ditches or rivers' and that uneven or broken ground was likely to cause gaps to form in the line.[32] English suggests that Alexander's troops flattening a grain field with their pikes in Illyria in 335BC (Arr. *Anab.* 1.4.1) was to create a level battlefield which suited the phalanx best.[33] Griffith, in a fairly generalized statement, claims that 'acting in big formations on fairly level and unbroken ground, the [phalanx] was very formidable'.[34]

Such claims, either referenced or otherwise, are based upon a passage by Polybius. In his examination of the pike-phalanx, Polybius states:

> The phalanx...requires ground that is level and bare which has no obstacles such as ditches, clefts, ravines running together, ridges, [or] flowing rivers; for all of the aforementioned are sufficient to impede and fragment such a formation.[35]

The sentiment of this passage can also be found in Livy (who used Polybius as a source) who declares that 'the slightest unevenness in the terrain renders the phalanx ineffective'.[36] Plutarch's account of the battle of Pydna, the event being described by both Polybius and Livy, also includes a declaration that the phalanx 'required firm footing and smooth ground'.[37]

However, there are several issues with the use of Polybius' passage to account for the operation of the pike-phalanx. Firstly, as Morgan points out, the word used by Polybius to describe the most preferable ground for the phalanx to operate on (*epipedos*), while regularly translated as 'flat' (as in the above quote) can also mean 'sloped'. Indeed, Polybius himself uses the word *epipedos* on a number of occasions to refer to ground with anything from a gentle undulation to a fairly steep incline such as a description of the peninsula of Sinope (4.56.6), the lofty ridges at Sellasia (5.24.3), and the valley of Leontini (7.6.3).[38] Consequently, it is possible to interpret Polybius' description as meaning that the phalanx required terrain that was 'sloped/undulating and bare...'. Furthermore, even if the passage is taken as meaning 'flat', just because this was the most preferable ground for the phalanx to operate on does not necessarily mean that it was the only ground that it could operate on. Hammond suggests that while 'ideally' the phalanx fought on flat ground, it could also fight on difficult ground.[39] In fact, some of the major pike-phalanx engagements of the Hellenistic Age were fought on ground which Polybius would have considered unsuitable.

As far back as the dawn of the Hellenistic Age, the phalanx was fighting on terrain that was anything but flat. In 350BC, Phocion held the high ground against Philip II and waited for the Macedonian attack.[40] This shows that the early Hellenistic pike-phalanx was capable of attacking up an incline. High ground was also employed at the battle of Chaeronea in 338BC, but this time by Philip himself. Polyaenus tells us how the units of the Macedonian phalanx on the right of the line withdrew to high ground to their rear, enticing the Athenian left flank to advance. Once in possession of this high ground, Philip's troops re-engaged causing the Athenian line to break.[41] Thus it seems clear that at least part of Philip's pike-phalanx could effectively engage an opponent on sloping ground and, with the advantage of fighting downhill, fight quite effectively.[42] Hammond goes as far as to suggest that the counterattack from the high ground was the key to securing Macedonian victory at Chaeronea.[43] If this was the case, then uneven ground could actually work in the phalanx's favour.

High ground was also a dominant feature of the battle of Sellasia in 222BC. Antigonus Doson had invaded southern Greece to aid the Achaean League against Cleomenes III of Sparta. Cleomenes secured the heights of two hills, called Olympus and Euas, where the road to Sparta and the river Oenous ran between the hills.[44] Both positions were occupied with contingents of the Spartan pike-phalanx and were additionally reinforced with a ditch and palisade.[45] On the right of their line, the attacking Macedonians formed up their phalanx with units of allied infantry in between each contingent.[46] The slopes of Euas, which range from a 15° incline near the base of the hill to around 20° nearer to the summit, were much steeper than that of neighbouring Olympus.[47] It has been suggested that Antigonus' alternating deployment of the phalanx units on his right wing may have been to provide the line with greater flexibility to cope with this steep ground.[48] On the Macedonian left, on the other hand, where the incline of Olympus was less, Polybius specifically states that the phalanx was deployed without any gaps or intervals, but does state that it was arranged in a double depth due to the narrowness of the front.[49]

The Macedonian right assaulted up Euas and, according to Polybius, their line was unbroken when it reached the Spartan position on the summit.[50] If correct, this shows that, with the insertion of more mobile troops to act as hinges between the units of the pike-phalanx, this formation could effectively advance up fairly steep terrain and possibly even fight over a defensive ditch and palisade. In another indication of how steep terrain could favour the pike-phalanx, Polybius outlines how he believed that one of the major mistakes made by the Spartans at Sellasia was to not have their phalanx on Euas attack down the slope and engage the advancing Macedonians using the advantages of the high ground which, if need be, would have provided them with a means of retreat.[51] Such claims by Polybius seem odd when it is considered that in his own account of the battle of Pydna, he supposedly claims that the pike-phalanx can only operate on 'flat' ground. This in itself, suggests that Polybius' passage should be better translated as 'sloped' or 'inclined' ground at best.

On the Macedonian left at Sellasia, the pike-phalanx also assaulted up the slopes of Olympus, but here Cleomenes ordered his phalanx to tear down part of their protective palisade and attack downhill (as Polybius says they should have also done on Euas).[52] Polybius states that both sides lowered their pikes and engaged each other on the slopes of the hill – showing that the pike-phalanx could engage on such ground while moving in either direction.[53] However, the impetus of the downhill Spartan attack forced the Macedonian line back onto the plain below.[54] Plutarch states that the Macedonian right, after defeating those holding Euas, moved back down the hill and attacked Cleomenes and his Spartans – most likely from the rear – and secured the victory.[55]

In this engagement, pike-phalanxes from both sides functioned effectively on anything but level ground. Both wings of the Macedonian phalanx attacked up fairly steep hills with the right wing apparently attacking over the ditch surrounding the Spartan position on the summit of Euas. Furthermore, the Macedonian left wing seems to have been able to withdraw down the slopes of Olympus without breaking the line or devolving into a rout, the Spartan right wing was able to very effectively attack downhill, in an action which must have also involved them crossing the ditch they had dug in front of their position, and (at least according to Plutarch) the Macedonian right was able to turn about and advance back down Euas to engage the Spartans from behind. Importantly, if Plutarch's statement is accurate, the Macedonian right-wing phalanx would have also needed to have crossed the terrain below the hills in order to engage. Plutarch describes the valley as 'full of ravines, water courses and generally irregular'.[56] None of these features seem to have impeded the functioning of the phalanx which indicates that it could operate on steep and/or broken ground.

The battle of Cynoscephalae in 197BC was also fought up and over high ground which Plutarch describes as 'the sharp tops of hills lying close beside each other'.[57] Polybius calls the ridge 'rough, precipitous and of considerable height'.[58] Pietrykowski calls the ridge upon which the battle was fought 'a true liability' to the 'ponderous phalanx'.[59] Morgan, on the other hand, points out that the terrain does not seem to have been a deciding factor in the outcome of the battle.[60] Both of these claims seem to be only partially correct as, depending upon the exact nature of the ground on certain parts of the ridge, the phalanx was either hindered (which ultimately led to its defeat) or not.

The Macedonian army of Philip V, and the Roman army of Titus Flaminius, were both encamped on either side of the ridge at Cynoscephalae. Philip is said to have considered the ground unsuitable and unfavourable for a major engagement but, following initial contact and skirmishing between advance units from both sides, and the receipt of favourable reports from the ridge above which stated that the Romans were in retreat, Philip began to commit more troops to the action including elements of his pike-phalanx.[61] Units of Philip's right-wing phalanx surmounted the ridge at a run: a manoeuvre which must have necessitated their pikes being held vertically.[62] Livy says that, once in position and arranged in double depth, the phalangites were ordered to drop their pikes and fight with swords because the length of the weapons was a hindrance.[63] Both Polybius and Plutarch, on the other hand, state that the phalanx engaged with its pikes lowered.[64] Indeed, there are several reasons why Livy's account should be considered incorrect in this matter. Firstly, Livy later states that the phalanx was unable to turn about to face an attack from the rear.[65] While this is true of a phalanx with its pikes lowered, it can be easily accomplished by one just fighting with swords. This suggests that the Macedonians were using the *sarissa*. Secondly, Livy also states that, at the end of the battle, parts of the phalanx signalled their surrender by raising their pikes.[66] It is unlikely that the members of the phalanx had put away their swords, picked up their pikes – which would have been somewhere uphill behind them as the sources all state that the Macedonian right wing pushed the Romans down the slope – and then used them to signal their surrender. It is more likely that the phalanx had been using their pikes all along.

The phalanx units on the Macedonian right wing effectively engaged the Romans using the advantages of the high ground to their fullest. Plutarch states that the Romans facing these units could not withstand their attack.[67] It was a different story on the Macedonian left, however, and it was in this quarter that the nature of the terrain may have hampered (and eventually defeated) the pike-phalanx. Livy says that additional pike units were brought up in column – a formation he says is better suited to a march than a battle – rather than in extended line. The ground here may have been more broken than on the right and this caused large gaps to open in the phalanx as it deployed: gaps which the more mobile Roman maniples were able to exploit to defeat the Macedonian left and then swing around to attack the remaining units on the Macedonian right.[68] Polybius states that this fracture of the phalanx on the left was due to some units already being engaged, others only just making the top of the ridge, while others were in position but were not advancing down the hill. Interestingly, none of these factors have much to do with the nature of the terrain itself and, as such, the extent to which the ground caused the fragmentation of the Macedonian line at Cynoscephalae cannot be conclusively determined. However, it seems clear that it is not the incline of the battlefield which is a hindrance to the operation of the pike-phalanx, but whether or not the line can be maintained on the terrain that the battle is fought upon. This again goes against Polybius' claim that the phalanx could only operate on 'flat' ground.

Another feature which Polybius states could disrupt the function of the phalanx is the presence of a ditch, ravine or river on the battlefield. Yet Alexander the Great's two battles at Granicus in 334BC and Issus in 333BC were both fought with the pike-phalanx advancing over watercourses and both resulted in Macedonian victories. Arrian's account of the battle at Granicus contains few details of the infantry action but he does describe the banks of the river as 'very high, in some places like cliffs' and how the Persians waited on the far bank so that they could fall upon the Macedonians as they emerged from their crossing.[69] Arrian and Plutarch also note how the current of the river was strong enough to pull Alexander and his cavalry downstream when they started to cross.[70] In another comment on the banks of the river, Diodorus similarly states that the Persians held the far side as they thought that they could easily defeat the Macedonians when their phalanx was disrupted as they crossed.[71] Plutarch also notes the uneven slopes of the river banks, but also states that the phalanx crossed the river and quickly put the Persians to flight.[72] Polyaenus simply states that the phalanx fell upon the enemy and defeated them.[73] There are few recorded casualties for the infantry in this engagement.[74] This would suggest that, regardless of how the banks are described or how fast the river was flowing, the river posed little obstacle to the advance of the phalanx.

At Issus, Arrian again describes the banks as, 'in many places, precipitous'.[75] Unlike at Granicus, however, at Issus the banks of the river do seem to have caused some problems for the phalanx. Arrian states that gaps began to form in the line as sections of the phalanx negotiated the crossing and while some units were held up by a stronger resistance than in other areas.[76] This finds many similarities with the fragmentation of the phalanx at Cynoscephalae. Other units of the phalanx at Issus were able to cross easily – aided by the fact that the

Persians opposing them had fled – reform and hit the remaining Persians in the flank.[77] Markle suggests that both Ptolemy and Callisthenes, the main sources for Arrian, exaggerated the difficulty of the crossing at Issus to make the victory more impressive.[78] Yet the fact cannot be dismissed that Alexander suffered more casualties among his pike-phalanx at Issus than he did at Granicus. The one thing that could have been a deciding factor in this outcome was the nature of the river that the phalanx had to cross and how it caused part of his line to lose cohesion. Furthermore, Markle claims that the nature of a watercourse at Issus would have made using the *sarissa* untenable and, as such, suggests that this indicates that Alexander's troops were not using the *sarissa* at all.[79] For the members of the pike-phalanx, it would be far easier to use the longer reach afforded by the *sarissa* to engage opponents on an elevated riverbank, as they crossed a river, than it would have been if they were only using more traditional length spears. Curtius' description of Alexander's infantry at Issus aiming their weapons at the faces of those arrayed on the bank above them seems to confirm the use of a lengthy weapon from the riverbed – most likely the *sarissa*. This further suggests that Alexander's troops were using lengthy pikes at both Granicus and Issus.

In his critique of Callisthenes' account of the battle of Issus, Polybius ponders:

> how did a unit of phalangites mount the river's bank which was both steep and covered with thorny bushes? For this would seem contrary to reason. One must not attribute such an absurdity to Alexander who acquired experience and training in warfare from childhood. Rather, one should attribute it to the historian who, on account of his own inexperience, is unable to distinguish the possible from the impossible in such matters. So much for...Callisthenes.[80]

Polybius also declares 'what can be less prepared than a phalanx advancing in line but broken and disunited'.[81]

While such accusations seem initially valid, Polybius is missing some important considerations in his claims. Firstly, Alexander had successfully attacked with the pike-phalanx across a river at Granicus the previous year so there would be no reason to assume that he thought he could not do it again at Issus. Secondly, the river does seem to have posed problems to the phalanx as it crossed (at least in Arrian's account) which could be attributable to the slopes and bushes cited by Polybius. Lastly, despite Polybius claiming that such factual errors belonged to those with no experience, unlike Polybius,

Callisthenes was actually present at the engagement and may have been an eye witness to the crossing. Consequently, there is little reason to place doubt on the description of the operations of the pike-phalanx at Issus that have come down to us second-hand though Arrian. Strangely, even though his account of the battle is much briefer than that of Arrian, Plutarch states that Fortune had presented Alexander with the ideal terrain for the battle.[82] Despite the much harder struggle at Issus, it is clear that Alexander's phalanx could advance and engage across a watercourse with only limited disruption to the line.

Ditches and rivers were also features of other battles during the later Hellenistic Age. At Mantinea in 207BC, for example, the Spartan pike-phalanx of Machanidas attempted to fight its way across a defensive ditch and was defeated.[83] However, the important thing to note is that the force holding the other side of the ditch was another pike-phalanx and not some other form of infantry. The ancient texts time and again outline the benefit of using pike-armed phalangites with the benefit of high ground and, while Polybius does state that the Spartan line at Mantinea fragmented as it crossed the ditch, part of the reason for their defeat would have undoubtedly been trying to overcome the lengthy weapons of the opposing formation arrayed against them from elevated ground. Other phalanxes, for example those of Alexander at Granicus and Issus, seem to have had only minor difficulties in crossing a river against opponents who were not armed with long pikes. A similar outcome occurred when Eumenes faced off against Antigonus at the river Pasitigris in 317BC. In this engagement, Antigonus attempted a river crossing but Eumenes and his troops 'withstood him, joined battle with him, killed many of his men and filled the stream with dead bodies'.[84] Yet again, it is the fact that the higher ground (or in this case bank) was held by opposing forces which seems to have been a more decisive factor in the outcome rather than just the influence of the terrain.

When Pyrrhus attacked the city of Sparta in 272BC, the city's defences were reinforced with a ditch 6 cubits (2.88m) wide and 4 cubits (1.92m) deep.[85] Plutarch says that the freshly turned earth of this defensive work made it hard for the attackers to gain a firm footing.[86] It is interesting to note that a lengthy *sarissa*, held in its ready position, projects ahead of the bearer by up to nearly 12 cubits (5.76m) if it is of the 14 cubit variety common to the late Hellenistic Period and is held by the last 2 cubits of its length. Plutarch recounts how Hieronymus considered the Spartan entrenchment as being rather small.[87] The validity of this statement can be seen in the fact that a phalangite standing on one side of the ditch would easily be able to reach an opponent standing on the other due to the length of his pike. Thus the only danger to the advancing pike-phalanx would have been when they actually chose to cross the ditch.

Had the phalangites simply remained on their side, the Spartans would have undoubtedly been forced to pull back as they would not have been able to engage their opponents. The key element of the perils of this defensive work as it is described by Plutarch is that it was the freshly turned earth, rather than the nature of the ditch itself, that was an impediment to the advance of the phalanx.

Channels and watercourses also played a part in the battle of Pydna in 168BC: although not in the way, or to the extent, that Polybius would suggest in his critique of the phalanx. Plutarch observes that the plain of Pydna was cut by the rivers Aeson and Leucus but that, at the time of the battle (the end of summer) the water was not deep.[88] This comment about the apparent lack of depth of the rivers suggests that both rivers were substantially deeper at other times of the year and, as such, the banks of the streams at the time of the encounter could have been rather steep. Plutarch states that the Macedonian commander, Perseus, thought that these would disrupt the Roman formations.[89] This suggests that the initial Macedonian strategy was to hold a defensive position and attack the Romans when they advanced. Hammond, in his analysis of the battle concludes that the two opposing battlelines ran from the southwest to the northeast and, as such, the two rivers on the battlefield, which ran roughly west to east, dissected both formations.[90] As Plutarch notes, Paulus waited for the sun to pass its zenith so that it would not be shining into the faces of his troops when he attacked.[91] This would only make sense if both armies were deployed north-south (with the Romans facing east). While Hammond's orientation of the lines in this manner conforms with the statement made by Plutarch, it must also be considered that in doing so parts of both armies had to have advanced down, or at least across, sections of a riverbed in order to engage. However these obstacles do not seem to have impeded either army to any great extent.

As if to confirm this tactic of a Macedonian defensive action, Plutarch describes Paulus as being reluctant to go into action against a phalanx that was 'already drawn up and fully formed'.[92] What happened to this initial strategy is not stated but, not long afterwards, the Macedonians are said to have quickly gone on the offensive and advanced almost up to the Roman camp.[93] This advance would have included parts of the pike-phalanx crossing both rivers which do not seem to have posed much of an impediment to the formation as it moved forward. However, as the battle continued, dangerous gaps began to form in the Macedonian line which allowed the Romans to attack individual phalangites from the flank and rear. Plutarch partially attributes this fragmentation of the phalanx to the terrain, which he says was uneven, but also to the very length of the phalanx and to the fact that some units were hard pressed while others were continuing to advance.[94] This finds similarities with both Alexander's battle at Issus and Philip V's engagement at Cynoscephalae. Yet it must also be noted that, prior to this occurrence, and while the phalanx maintained its cohesion, Plutarch states that the phalanx was 'everywhere unassailable', that Paulus was frightened by the sight of the organized phalanx, and that, due to the losses his troops were suffering, he rent his clothes in despair.[95] Further evidence that some of the fighting took place in and across the water channels is found in Plutarch's statement that the following day the waters of the Leucas were stained with blood.[96] Consequently, it is yet again clear that it is in the maintenance of the line (or not) where the deciding factor in the outcome of the battle involving a pike-phalanx lies rather than the terrain upon which it was fought.[97] Livy attributes the Roman victory to this factor alone in his account of the battle.[98]

Indeed, the maintenance of the line, or not, was the major contributing factor to the success or failure of the pike-phalanx in action. The use of units and sub-units, while providing the pike-phalanx with a certain level of tactical flexibility, also created the potential for the greatest danger to the formation: the opening up of gaps in the line. There are numerous ancient texts which contain generalised comments to the effect that, if the cohesion of the line is retained, a pike formation is almost unbeatable.[99] Such claims would also have to assume that the vulnerable flanks of the phalanx were protected as well.

How these gaps formed was due to a number of combined factors. Polybius states that whether the pike-phalanx advances and puts an opponent to flight, or is itself turned, the line will fragment as, in one scenario, units of the phalanx will pursue the routed enemy while, in the other, fleeing phalangites will not maintain their formation.[100] Plutarch states that the uneven terrain at Pydna resulted in the phalanx being unable to maintain 'as close a formation as possible' (using the term *synaspismos* in its generic sense) and that this also caused gaps to form in the line.[101] Plutarch also states that gaps will form in the phalanx when parts of it are hard pressed while other sections continue to advance.[102] This is exactly what happened at both Issus and Cynoscephalae which shows that this was a vulnerability that existed within the pike-phalanx across the entire Hellenistic Period.

The only way to avoid such perils would be for the entire line to halt its advance if part of it became hard-pressed. Arrian states that the integrity and safety of the whole line depended upon each man holding his position.[103] This would not only relate to each phalangite keeping his position within his file, but to each file and sub-unit keeping their respective positions across the line as a whole. However, due to the semi-independent nature of the sub-units of the phalanx, and the sheer size of some of the battlelines of the Hellenistic Age, there was not enough command and control to keep the entire line intact and, as a result, gaps would inevitably form.[104] In some instances unit commanders were able to gauge what was happening around them and halt their formation to prevent further fragmentation of the line. We see this in the actions of the units within the *merarchia* of Simmias at the battle of Gaugamela who 'halted their unit and were fighting where they stood as the Macedonian left was reported to be in trouble'.[105]

When gaps formed in the line, this was when the phalangite was most vulnerable. By exploiting these gaps, more mobile opponents could move inside the sets of serried pikes, negating the advantage that such a lengthy weapon provided, and attack at close-quarters with short reach weapons, or attack the phalangites from the side. In either case the panoply of the phalangite placed him at a considerable disadvantage. The lengthy *sarissa*, for example, was practically useless as an individual combative weapon due to its size and weight and the formation, not matter how fragmented, made it difficult for the phalangite to individually turn and face a direct threat from the side or rear. Markle offers that the only place that a *sarissa* did function effectively was within the confines of the massed phalanx.[106] Plutarch also comments on the small size of the phalangite sword and shield and that these made them no match for the Romans in hand-to-hand combat once a gap had been exploited.[107]

The creation of such gaps posed little threat when facing another pike-phalanx, the members of which could not exploit it and, even if an inevitability, the success of the pike-phalanx against more mobile opponents suggests that any deficiencies in the formation were countered by many of its other advantages in most cases. So long as enough of the line remained intact to pin the enemy in place for the amount of time required for a flanking cavalry attack to work, then the phalanx was reasonably secure – as per Alexander's battle at Issus. If, on the other hand, the protection for the flanks of the phalanx was removed, or any flanking attack against the enemy line failed to achieve its purpose, then the gaps that had formed in the line as a result of the phalanx's advance could result in its undoing. Yet the separation of the line did not mean immediate defeat. On the contrary, the ancient literature shows that the pike-phalanx could even operate reasonably well on terrain which could do nothing but cause the line to fragment.

In Polybius' critique of the phalanx he states that the formation could be interrupted by obstacles. However, even generally rough and/or wooded terrain did not always negatively impact the function of the pike-phalanx. At Asculum in 279BC, the Romans engaged Pyrrhus of Epirus on terrain which Plutarch describes as 'rough ground where his [i.e. Pyrrhus'] cavalry could not operate,

and along the wooded banks of a river...'[108] Pyrrhus engaged on this ground
with his whole army, including his pike-phalanx, and secured a long and hard
fought victory.[109] Polybius states that Pyrrhus deployed with alternating units of
phalangites and allied troops across his line.[110] This arrangement may have been
to ensure that his line was more flexible, and therefore less likely to have gaps
form, as it advanced across the difficult terrain. Polybius makes the comment
in his examination of the pike-phalanx that it is almost impossible to find clear
and obstacle free land for the pike-phalanx to fight on.[111] While generalised in
nature, such comments fail to consider that if obstacle free ground was so rare,
yet crucial to the proper functioning of the pike-phalanx, why anyone would
have adopted such a style of fighting in the first place or have been able to use
it to such great effect for almost two centuries up to the time that Polybius was
writing. Clearly, obstacles of any shape or form, whether they be rivers, ditches,
ridges or trees did not have a significant impact of the operation of the pike-
phalanx so long as adequate precautions were taken to ensure the flexibility and
maintenance of the line.

Even a narrow fronted battlefield seems to have been little impediment
to the pike-phalanx. In 191BC, Antiochus III occupied the narrow pass of
Thermopylae in central Greece against the army of the Roman consul Marcus
Acilius Glabrio.[112] Part of this army included a contingent of *sarissaphori*
who were positioned as a second line of infantry, against a series of defensive
works that had been constructed.[113] In the narrow pass, Antiochus' phalangites
engaged the Romans 'with their *sarissae* presented in a massed order, the
formation with which the Macedonians from the time of Alexander [the Great]
and Philip [II] used to strike terror into enemies who did not dare to engage the
thick array of pikes presented to them'.[114] The phalangites seem to have held
their own in the initial stages of the battle – Livy says they 'easily withstood the
Romans'.[115] When they became hard pressed, the phalangites withdrew inside
the fortifications and 'made what amounted to another palisade with their pikes
thrust out in front of them' and the Macedonians used this position to easily fend
off the attacking Romans.[116] This resistance continued until a force of Romans
managed to get around behind the Macedonians, using the same path that the
Persians had used to outflank Leonidas and his Spartans back in 480BC, which
caused the formation to rout and flee for their camp.[117]

Frontinus states that the phalanx did not like fighting in cramped quarters.[118]
Hammond offers that phalangites were not well suited to fighting in narrow areas
like streets or in siege warfare.[119] However, this is precisely the kind of terrain
where a unit of pikemen would have been at their best. In narrow fronted areas
like streets or passes like Thermopylae, where the flanks of the formation were

protected by buildings or natural features, a contingent of pikemen could easily hold their own against any opponent with shorter reach weapons so long as the high ground was not occupied by the enemy who could then use this position to rain missiles down upon the phalangites.[120] If the files could be maintained, the pikes strongly presented, and the formation not crowded by opposing troops, pikemen could simply either advance forward and drive or kill everything before them or hold their position and let the attacking enemy impale themselves on their pikes. However, if the formation was not maintained on such narrow terrain, then any warrior was in trouble regardless of the type of weapon he was using. Plutarch describes in detail the confusion of urban warfare in the Hellenistic Age in his account of Pyrrhus' assault on Argos in 272BC. In this attack, Pyrrhus' troops were driven out of the town square, where Plutarch says there was plenty of room to fight and give ground, only to have his formations spoiled by reinforcements coming the other way. In the confusion, orders could not be heard, several elephants went wild and trampled all underfoot, and 'once a man had drawn his sword or aimed his spear it was impossible for him to sheathe [it] or put it up again, but it would pierce whoever stood in its way'.[121] It is this type of cramped and confused environment where the use of weapons was limited, rather than any reference to the nature of the terrain itself, that Frontinus must be referring to. Had Pyrrhus' troops been able to withdraw without such confusion, much of the carnage mentioned by Plutarch could possibly have been avoided.

Despite such accounts, phalangites did take part in many assaults on urban areas – particularly ones that were taken through breaches in the walls rather than through the use of assault ladders. At both Tyre and Gaza, for example, some of the troops who stormed the cities were from units of Alexander's pike-phalanx.[122] This suggests that either these troops were employing their pikes in an urban environment, or may have been using other, smaller weapons, such as the front halves of their *sarissae* or just their swords. In any case, these units would have been fighting in a narrow, enclosed space, but with their flanks somewhat protected. Indeed, if contingents of Classical hoplites such as the 300 Spartans, arranged in close order, were able to hold the Thermopylae pass for two and a half days in 480BC, and inflict substantial casualties among their enemies, then even a single *syntagma* of phalangites could have held the same position with their longer pikes almost indefinitely. This shows that the narrowness of the terrain was also not an impediment to the functioning of the pike-phalanx and, depending upon the circumstances, could actually make the position even stronger.

Markle suggests that *sarissa*-armed infantry were never led across rivers or ditches except by incompetent commanders.[123] However, a review of the literary accounts of some of the major confrontations of the Hellenistic Period shows that in many cases experienced commanders from Philip II to Perseus successfully engaged opponents with pike-phalanxes on this very type of terrain. In other cases, pike formations fought effectively on hilly or narrow ground – contrary to what Polybius would have us believe in his examination of the phalanx. Anderson suggests that Polybius' description of the unsuitability of the phalanx on rough terrain may simply be an exaggeration to illustrate his point concerning the loss at Pydna.[124] Based upon the different terrains that the phalanx successfully fought upon, if Polybius' passage is not translated in another way, this seems more than likely. Thus it seems clear that the use of units and sub-units within the structure of the pike-phalanx made it a very flexible formation and one that could be configured and used on almost any battlefield of the ancient world.

COMMAND AND CONTROL OF THE PHALANX

A deployed pike-phalanx could be a lengthy formation. A single *syntagma* of 256 men arranged in sixteen files (for example, with each file occupying an intermediate-order interval of 96cm per man) would possess a frontage of around 16m. Each *chiliarchia* of the phalanx, if all four of its constituent *syntagma*e were arranged side-by-side with no interval between, would possess a frontage of 64m. Each *merarchia*, with its four constituent *chiliarchiae* similarly arranged side-by-side with no interval between, would possess a frontage of 256m. Thus a pike-phalanx such as that of Alexander the Great, comprised of six separate *merarchiae*, when at full strength and deployed in a continuous line in its standard depth, would stretch for more than 1.5km. The effective control of such a lengthy formation in battle came down to a number of factors across all levels of command. The effective control of the phalanx was not just due to the positioning of officers across its front rank and the symmetrical distribution of their command abilities to support each other, but through careful planning, clearly understood commands with an easy means of delivering them, and the adaptability of the pike-phalanx in general.

In many instances how the phalanx was to form up and fight was decided well before the commencement of hostilities at a pre-battle council of senior officers. Alexander the Great held just such a council meeting at Granicus in 334BC where, upon receiving reports from his scouts on the Persian positions, he immediately gave all necessary orders in preparation for a battle.[125] This was not a one sided discussion and Alexander's senior general, Parmenion, was able to offer his own advice – advice Alexander readily dismissed.[126] If Arrian is correct in stating that the deployment of the *merarchiae* at Granicus followed a rotating roster of command to some extent, it would seem likely that these dispositions were confirmed in such a pre-battle council as well.[127] Alexander's advance from column into line at Issus the following year must have also been determined well beforehand as is suggested by Curtius (see pages 311-317).[128] At Tyre in 333BC, officers of varying ranks were present at strategic conferences.[129] Machanidas' deployment from column into line at Mantinea in 207BC must have similarly followed a preconceived plan and operational deployment (see pages 322-326).[130]

A major pre-battle council was also held by Alexander prior to the battle of Gaugamela in 331BC which involved commanding officers of all of the different contingents within the army – all of whom seem to have been able to contribute to the discussion. Arrian states that:

> ...he stopped his phalanx [i.e. army]...and again summoned the Companions, generals, squadron commanders and the leaders of the allied and foreign mercenaries, and posed the question whether he should advance his phalanx at once from this point, as most of them urged, or, as Parmenion thought best, encamp there for the time being, reconnoitre the whole of the terrain...and make a thorough survey of the enemy's positions. Parmenion's advice prevailed and they camped there, in the order in which they were to go into battle.[131]

This passage provides a great deal of information about what took place at these pre-battle councils. Firstly, it is carried out at a time which allows for the options for the timing of the operation to be debated: in this case either to encamp and fight on another day or engage immediately. Secondly, Arrian's passage shows that the order of battle was pre-arranged at these meetings. Finally, these councils involved representatives of all the units within the army. This would allow the commander of each contingent to know its exact place in the battleline for the ensuing battle and what it was expected to do on the day. As Hammond notes, the discussions and conclusions reached at the pre-battle councils had to envisage how the ensuing battle would unfold as all of the senior commanders present at them would be in the thick of the fighting in their own individual sectors of the battlefield once hostilities had commenced.[132] Importantly, even if it was decided to go into battle immediately, it can only be assumed that enough time would have been left for these orders and dispositions to be relayed down through the chain of command to each respective sub-unit of the formation. If it was decided to encamp, then presumably orders for the night watch would have been issued, rosters for eating, resting and other duties would have been finalized, and these too would have to be filtered down to each and every member of the army.[133]

Following this council at Gaugamela, Arrian states that, following an inspection of the army, Alexander summoned the officers again and issued further instructions for how the army was to conduct itself on the day of the battle.[134] Parmenion is said to have gone to Alexander again following this second council to offer other options such as an immediate night attack on the Persian position – advice which, again, Alexander rejected.[135] This shows that, once plans had been set, they could be refined further, and possibly even altered, with discussions at the highest command level.

At the Hydaspes, orders were given to certain units regarding what to do if the battle progressed in a certain way. Craterus, for example, commanding units holding the riverbank to keep the Indians of king Porus in check, was instructed by Alexander:

> not to attempt a crossing until Porus and his army had left his camp to attack Alexander's forces [who were attacking from a different direction], or until he had learnt that Porus was in flight and [Alexander] victorious; 'but should take a part of his army and lead it against me', Alexander continued, 'and leave another part behind in his camp with elephants, still stay where you are. If, however, Porus takes all of his elephants with him against me, but leaves some part of his army behind in his camp, cross with all speed...[136]

It is unlikely that most battle plans were as elaborate, or as reliant upon certain conditions, as those given to Craterus at the Hydaspes. It is more likely that, as with Alexander's council at Gaugamela, the main concern with the plan would have been where each unit was to deploy in the line. From here, the majority of operational tactics for the pike-phalanx seem to have simply been for an order to be given for the phalanx to 'lower pikes' (καταβαλοῦσαι τὰς σαρίσας) and advance into action (see following).[137] This simplicity of direction would ensure that instructions were not confused or misinterpreted as they were passed down the line. It also partially explains why Alexander used the exact same tactic – a frontal advance with the phalanx while the right wing cavalry attacked from the flank – at each of his four major battles with only slight variations on a theme.

Once battle had been joined, a complex system for relaying commands was used to capitalise on the flexibility of the phalanx and to exploit any tactical opportunity that presented itself. Other commands put into effect orders that had been pre-planned. Arrian states that at Gaugamela the units of Alexander's phalanx had been ordered to open gaps to counter an attack by Persian scythe-bearing chariots. Arrian uses the past tense 'they had been ordered' (ὥσπερ παρήγγελτο αὐτοῖς) here which suggests that this had been part of the plan developed from Alexander's reconnaissance of the enemy position and the considered method of their attack. This relay of commands would have been carried out by four of the supernumeraries attached to each *syntagma* of the pike formations: the herald (*stratokērux* – στρατοκῆρυξ), the aide-de-camp (*huperetēs* – ὑπηρέτης), the standard bearer (*semeiphoros* – σημειφόρος), and the trumpeter (*salpigktēs* – σαλπιγκτής).

According to the *Suda*, it was the role of the herald to relay orders by voice.[138] During a battle, each officer above the rank of *syntagmatarch*, in command of their respective unit or sub-unit within the broader formation of the pike-phalanx, would have the aid of a herald. Thus, if the preconceived battle plan required the unit to conduct a particular movement such as an advance, a feigned withdrawal or a counter-march when either a certain point in time was reached or a certain criteria was met, the commander, once that moment had arrived, could instruct his herald of the new order who would then call it out so that it could be heard by the men in the unit. At Gaugamela, part of Alexander's standing instructions was for his troops to obey orders quickly.[139] This shows that, even though a pre-conceived battle plan may have been formulated, parts of it were either not put into effect immediately once the battle had begun, and/or that, due to the many variables of combat, plans could change and new orders issued at a moment's notice.[140]

If any such an operation was conducted by a relatively large unit, such as a *merarchia*, it is highly unlikely that the voice of the herald standing next to the *merarch* would carry far enough to be clearly heard by every member of the larger unit – even without the din of battle being considered. As such, an instruction announced by the commanding officer's herald must have been relayed across larger units by the other heralds within it. Due to heralds being attached to each *syntagma*, this would mean that each instruction would only have to be passed a distance of 16m from one herald to the next. Even if a phalangite on the far side of his *syntagma* did not hear the order called out by his respective herald, he still would be in a position to hear the instructions as it was relayed by the herald attached to the adjacent unit. This, in part explains why the supernumeraries were attached to the *syntagma* rather than a larger unit and why this unit was the smallest tactical formation of the phalanx.

Under such circumstances silence was paramount. Aelian and Arrian devote entire sections of their examinations of the phalanx to the importance of silence within the ranks so that such orders could be heard.[141] At Gaugamela, Alexander ordered his troops to advance in total silence and to only raise their warcry in unison when ordered to do so.[142] This would have ensured that any instructions given up to that point would have been easily heard and promptly followed as the sheer size of the formations involved caused considerable problems for the relaying of information. Appian states that one of the problems of large armies is that they are often so big that people cannot see what is going on and if one section of the line gets into difficulties, the news of this trouble is intensified as it is transmitted down the line and this can lead to a general panic forming.[143] The avoidance of such potential calamities explains why Hellenistic armies employed a whole range of methods to pass information and instructions between commanders and their units.

Verbal instructions could also be delivered by a runner. This was most likely one of the roles of the aide-de-camp. The *Suda* states that the role of this supernumerary was to 'carry over some of the things that are needed'.[144] The most likely form of these 'needed things' would have been the delivery of information. At the Hydaspes, the units of Meleager, Attalus and Gorgias were instructed to cross the river as separate units.[145] It is unlikely that Alexander personally delivered such instructions, although it cannot be ruled out that they were part of the already disclosed pre-conceived battle plan. Rather Alexander would have relayed these orders to the individual unit commanders via a runner and these orders would then be passed down through the respective formations. Arrian describes this in his account of the battle itself. Once across the river, Alexander is said to have 'ordered the infantry to follow in good order and at a marching pace...He directed Tauron, the commander of the archers, to lead them on with the cavalry, also at full speed.'[146] Once the fighting had commenced 'Coenus was sent to the right...and ordered to close with the barbarians from behind...[and] Seleucus and Antigenes and Tauron were put in command of the infantry phalanx, with orders not to take part in any action...'.[147] Again, it is unlikely that Alexander, who was himself engaged on the right of the line, would have delivered all of these instructions personally and would have used a series of messengers to deliver the orders.

Many of the pike units themselves seem to have had a considerable amount of command autonomy on the battlefield, especially at the higher levels. At Granicus, the units of Alexander's right wing wheeled to the left to strike the Persians in the flank. Such a manoeuvre could have only been undertaken if the commanders of each of the two *merarchiae* involved had co-ordinated with each other, most likely through the use of orders delivered by runners, the exact moment that this change in direction was going to take place. Had both units not co-ordinated their movements, they ran the risk of entangling with one another, possibly as one unit continued to advance while the other wheeled, or of one unit being left dangerously exposed while the other changed its path. At Gaugamela, Parmenion, in command of the left wing, also sent messengers to Alexander to inform him that the position was becoming hard pressed.[148] This not only demonstrates the use of messengers and runners to relay information and requests for assistance across the battlefield, but that some of these messengers were highly mobile, possibly mounted, and could cover considerable distances to locate specific individuals amidst a battle containing thousands of men. All of this combined to make the pike-phalanx, and pike-phalanx tactics, rather fluid and adaptable.

Yet, across the din of battle, general instructions such as the order to advance or retreat may not be heard for any number of reasons. This was where the other supernumeraries attached to each *syntagma*, the standard bearer and the trumpeter came into play. The positioning of standards across the front of the pike-phalanx could not only be used to help the formation retain a relatively level frontage, as Curtius states they were used for at Gaugamela, but could also be used to relay orders to the troops of their respective units.[149] The *Suda* states that one of the functions of the standard bearer was to relay orders 'by the standard, if the voice is not heard because of noise [of battle]'.[150] Thus the role of the standard bearer was to facilitate the issue of commands by delivering them visually. It can only be assumed that, when a unit received such instructions, they were relayed both verbally by the herald and simultaneously visually by the standard bearer so that each member of the unit would have a chance of receiving the orders in at least one form.

This, in turn, suggests that there was a standard set of drill movements and actions that the standard bearer could perform and that the meaning of these instructions was easily recognizable by every member of the unit. Unfortunately, none of the extant literature outlines what these drill movements or actions were. However, it must be concluded that there was an action (or series of actions) which translated into every basic command that a unit might be given – raise pikes; lower pikes; advance; halt, wheel, counter-march and so on – otherwise the standard bearer would have limited value to the command structure of the unit.

As with the herald, because there was one standard bearer attached to each *syntagma* of the phalanx, the signal that was given needed only to have been visible from a distance equal to the diagonally opposite side of the unit – a distance of some 20m. Even if those on the far side of the unit were unable to see their respective standard bearer due to dust, smoke, rain or any other visual impediment, they might have still been able to see the standard belonging to the adjacent unit, or simply relied on the verbal outbursts of the herald for their instructions.

Furthermore, there had to be both a verbal command and a visual signal which signified that the command that was to follow was for a specific unit to avoid confusion. For example if the men on the far side of one *syntagma* could not see their standard nor hear their herald, but the adjacent unit was given the order to advance while theirs was not, those men could not simply follow the orders relayed by the supernumeraries of the adjoining unit lest they incorrectly obey an order that was not meant for them. This suggests that each order that was relayed was preceded by a call and/or signal which alerted the men of the required unit that the information that was to follow was for them, and for them alone. Regardless of whether this was for a sub-unit such as an individual *syntagma*, or was an order that governed an entire *merarchia* but was being relayed to each sub-unit within it, as each successive unit was forwarded the command, the men inside it would have had to have been alerted that an order was coming.

Standards and banners, in a variety of forms, relayed information at all levels in battles across the breadth of the Hellenistic Age. Pyrrhus' army at Asculum in 279BC, for example, employed a series of 'hoisted signals' to relay instructions.[151] Standards were also used in the army of Philip V at Athens in 200BC and by the army of Antiochus III at Thermopylae in 191BC.[152] At Sellasia in 222BC, Antigonus himself began the attack by raising linen and scarlet flags to order different wings of his line to advance.[153] Polybius says that, once these signals had been given, the officers commanding the appropriate units then relayed these instructions down to their men.[154] This not only illustrates the use of other means for the officers of each unit to relay the order to advance, but also that a pre-conceived plan had been put into effect, most likely determined at a pre-battle council, where the officers were informed of what to do once the flags had been raised.

Yet under certain conditions neither the cries of the herald, nor the signals of the standard bearer, could be heard or seen. This was where the last of the supernumeraries – the trumpeter – was used.[155] As with both the herald and the standard bearer, commands issued for each unit would have been delivered though a variety of different trumpet blasts to relay specific orders. Thus, like the standard bearer, there must have been a certain set of blasts (or series of blasts) each of which translated into a specific command that was readily understood by each member of the phalanx. Additionally, and also like the standard bearer, there must have also been a set of specific notes that could be played to alert individual units to the incoming orders. Polybius states that at Sellasia a contingent of light troops was recalled by the sounding of a trumpet.[156] This indicates that there was a preliminary blast to alert just the light troops to the incoming order as, had the trumpet call simply meant 'withdraw!' any unit may have followed the order.

Other, more general orders, seem to have been able to be delivered by trumpet as well. Alexander's army, for example, was ordered to both attack and withdraw at Halicarnassus in 334BC, Issus in 333BC, Gaugamela in 331BC, and at the Hydaspes in 326BC, by the sound of a trumpet.[157] The entire army of Eumenes was recalled 'with the sound of the trumpet' at the Hellespont in 321BC, as was Polyperchon's army at Megalopolis in 318BC.[158] Demetrius' army was similarly recalled 'by a trumpet call' at the Nabataean Rock in 312BC.[159] At the battle of Beth-Zachariah in 162BC, the signal for the whole Seleucid army to advance into battle was given by the blast of a trumpet.[160] Such passages show the use of trumpeters to relay commands across the entire Hellenistic Period. Undoubtedly these instructions would have been passed among the units by their various trumpeters and more than one 'sound of the trumpet' over the entire execution of the command must be assumed.

This leads to the question of the advance itself; once the signal to advance had been given, how fast did the phalanx actually move and how was this pace maintained so that the cohesion of the line was not lost? Sheppard suggests that the advance of the phalanx was sometimes conducted at the run.[161] Adcock, on the other hand, offers that the phalanx relied more on a steady advance than it did on a rapid charge.[162] Pietrykowski suggests the phalanx advanced at the 'quick step'.[163] In modern military terminology, this equates to a rate of around 120 paces per minute – with each pace being about 75cm in length (from heel to toe).[164] If this is the rate of march that Pietrykowski is referring to, it would seem a bit quick for a formation of men encumbered with heavy armour and carrying a 5kg *sarissa* who were trying to maintain a compact formation.[165] When the maximum possible rate of march was tested using small groups of re-enactors equipped as phalangites, it was found that it is almost impossible to maintain any semblance of order within the ranks when the formation is moving at anything other than a brisk walk. This would suggest that a slightly slower pace would have been used by much larger pike-phalanxes.

Indeed, there are only few accounts in the ancient literature where the phalanx seems to have moved at anything like a running pace. One instance of a rapid advance occurred at Cynoscephalae in 197BC.[166] Yet even here this advance was to move troops into position quickly, and to seize high ground, rather than an actual advance into combat. The pike units on Alexander's right wing at Gaugamela are also said to have advanced against the gap in the Persian lines at the run (δρόμῳ).[167] However, it is unlikely that a rapid rate of march could have been conducted in a manner which would allow the pike formation to be maintained. Consequently, it is more likely that the Macedonians advanced at a brisk trot at best rather than a flat out run. In a clear instance of a rapid advance into contact, at Massaga in 327BC, Alexander's phalanx was said to have counterattacked 'at the run'. However, even here Arrian states that 'the mounted javelineers, the light armed Agrianians and the archers first raced forward and joined battle while Alexander himself kept the phalanx in formation'.[168]

This suggests that, not only is the term 'phalanx' being used here in its generic sense to mean 'army', but also that the pike units were brought into action at a slower pace than the lightly armed troops to maintain their cohesion. This is also likely what happened at Gaugamela where the gap in the Persian line was assaulted by cavalry at the charge (hence the use of a term meaning to 'run' or 'advance quickly') with pike units moving up in support but more likely at a slower pace to maintain their lines. Additionally, it would have taken a short amount of time for Alexander's phalanx at Massaga to conduct another counter-march and advance upon the enemy in their proper order – more time than would be required for the cavalry and light troops who could simply turn about. This again suggests that the pike units advanced more slowly.

Other passages suggest that a rather rapid advance of the phalanx was possible. At Chaeronea, for example, Philip's phalangites delivered a 'committed attack'.[169] At Gaugamela, Alexander's phalanx 'rolled forward like a flood'.[170] At Pydna, the Macedonians are said to have attacked so swiftly that the first Roman was killed not far from his camp.[171] While all of these accounts clearly describe an advance of the pike-phalanx that was seemingly quite rapid, none of them specifically state that they were conducted at a running pace and a simple fast walk cannot be discounted. It can also not be discounted that these descriptions are of the ferocity of the attacks made at Chaeronea and Gaugamela, rather than their speed, and that the cause for the seemingly rapid attack at Pydna was in part due to the unpreparedness of the Romans. Furthermore, it is possible that the descriptions of the running move of the phalanx up the ridge at Cynoscephalae are merely a literary motif, utilizing a description of something that the pike-phalanx rarely did, to emphasize the seriousness and urgency of the situation. This in itself would suggest that the pike-phalanx advanced into action at a much slower pace.

The few ancient passages that provide more details of the movement of the phalanx all suggest that the formation was moved at a slow pace that would allow the line to be maintained. Arrian, for example, states that, at the battle of Issus '[Alexander] was leading [his men] still in line... step by step so that no part of this phalanx should vary and break apart [as it would] at a quicker pace'.[172] Even formations of more mobile hoplites in the Classical Age rarely charged at the run and, as Arrian states, it would be almost impossible to maintain a formation of thousands, or tens of thousands, of men armed with lengthy pikes at such a speed.[173] Julius Africanus, in a reference to the combat of the Classical Age, states that 'running in hoplite equipment was infrequent and not prolonged; it is instead quick and of the sort that might be used when one is in a hurry to get inside the trajectory of an arrow'.[174] It can therefore only be assumed that variations of a 'quick-step' pace, and possibly one of a slower 'half-pace', were used to keep the phalanx (both Classical and Hellenistic) together depending upon things like terrain and the tactical necessities of each situation.

Philip II's feigned retreat at Chaeronea, for example, if accomplished by having his pike units march backwards, rather than having them conduct a counter-march, could have only been done at a slow pace so that the formation could be maintained, and the pikes kept presented, while the unit moved back and uphill away from an advancing enemy.[175] Alexander is said to have conducted a similar feigned retreat at Massaga in 327BC but, according to Arrian, this manoeuvre was clearly undertaken using a counter-march rather than by marching his phalanx backwards.[176] At Magnesia in 190BC the phalanx formed a defensive hollow square, with light troops and elephants positioned in the centre, and attempted to withdraw 'step-by-step'.[177] Appian specifically states that the defensive square was arranged in an intermediate-order and that thickly set *sarissae* projected from all four sides of the formation.[178] Parts of this square would have been definitely marching backwards and a slow pace would have certainly been used to keep the formation together.

The ability for the phalanx to advance or retreat with relative ease and still maintain the integrity of their formation also demonstrates that the side-on posture suggested by some scholars for how the *sarissa* was wielded, and the close-order interval of 48cm per man, could not have been used within the pike-phalanx. When standing side-on it is impossible to move at anything faster than a shuffling side step. While this could be used for a slow advance, it could not be used for anything more rapid and the inability of the weapons of the phalanx to be deployed while in a close order would simply be compounded if the formation was attempting to move forwards or backwards at the same time. Thus it seems clear that the pike-phalanx operated on the battlefield with its members wielding the pike using an oblique body posture, within an intermediate-order interval of 96cm per man, and moving at a brisk walk at the very best.

The pace of the march could have been maintained through the use of a cadence of some kind. Hoplite armies of the Classical Age regularly used a chanted 'marching song' (ἐμβατήριος παιάν), or *paean*, to keep in step and hold the formation together, just as many modern armies do.[179] While there are no specific references to the army of Alexander using the *paean*, and Aelian emphasizes the importance of silence within the phalanx, Alexander's order to advance in silence at Gaugamela suggests that on occasion the phalanx could be quite boisterous. This suggests the use of a sung cadence at times, and the references to units or phalanxes moving 'step-by-step' additionally suggest the use of a cadence of some form.

If a called or sung cadence was not used, another possibility is that a musical one was. Again, hoplite armies of the Classical Age were regularly accompanied by musicians, whose beats played on anything from drums to trumpets to pipes and lyres, helped keep the formation in step and maintain the phalanx's cohesion.[180] The trumpeter attached to each *syntagma* of the Hellenistic pike-phalanx could have similarly been used to blast a series of short notes at regular intervals to help keep the formation moving at the desired measured pace. If a new order was received, which then needed to be passed on to the men of the unit, the trumpeter could simply halt the playing of the cadence and deliver the instructions. The brief cessation of the blown cadence would have the added benefit of alerting the men within earshot that another order was about to be relayed.

Dickinson claims that, apart from wheeling, there is no evidence for intricate drill movements being made by Alexander's phalanx.[181] However, the literary evidence shows that the phalanx was capable of varied and detailed movements, both in the time of Alexander and the later Successors. Furthermore, it is clear that the use of the herald, the standard bearer, the *aide-de-camp* and the trumpeter as means of delivering instructions and to maintain the formation, allowed the commanders of pike-phalanxes to carry out the details of any battle plan, adjust them when necessary, and adapt the flexible units of the pike-phalanx to the changing nature of the battlefield, both easily and with a certain level of sophistication.

THE 'SERRIED WALL OF PIKES'

Once the phalanx had been deployed in its proper order and configuration, the weapons held by the individual phalangites had to be positioned in order to engage an opponent. Polybius details how phalangites were given orders to 'lower pikes' (καταβαλοῦσαι τὰς σαρίσας).[182] The order to lower pikes is also contained in a list of general drill commands found in the military manuals.[183] Both Polybius and the manuals also outline how the weapons of the forward ranks of the phalanx projected ahead of the line once they had been lowered for combat.[184] Polybius calls this arrangement a 'closely packed [hedge] of pikes' (συμφράξαντες τὰς σαρίσας).[185] Livy similarly describes the phalanx as 'closely packed and bristling with extended pikes'.[186] Plutarch says that the pikes of the phalanx were 'set at one level' (ταῖς σαρίσαις ἀφ᾽ ἑνὸς συνθήματος κλιθείσαις).[187] But how were the pikes of the phalanx actually arranged once they had been lowered and set in position? It all depended upon how the file itself was configured.

The vast majority of modern works which examine the warfare of the Hellenistic Age contain passages and/or diagrams which show the members of each file of the phalanx with their pikes lowered and parallel with each other. This seems to be a commonly accepted convention amongst scholars and no analysis of other possible configurations, or of the further implications of such arrangements, seems to have been made other than presenting a generalized statement on the matter. Fuller, for example, simply states that 'the phalangite wielded his *sarissa* with both hands, keeping it carefully aligned with the weapons of his comrades' (Fig.50).[188]

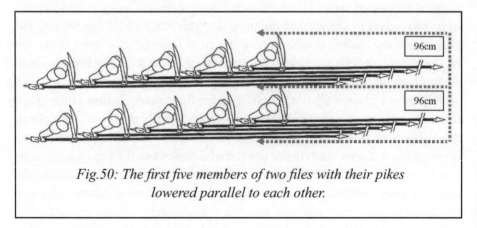

Fig.50: The first five members of two files with their pikes lowered parallel to each other.

Such an arrangement of the file conforms with the use of the 96cm intermediate-order interval outlined in the military manuals for how the phalanx was configured. Plutarch's description of all of the lowered weapons of the phalanx being 'set at one level' indicates that those weapons that were presented for combat were side by side and not positioned one above the other and the arrangement of the men in the file with their pikes held parallel to each other conforms with this description. This, in turn, means that the phalangites of each adjacent file could not have moved closer to each other and, as such, the intermediate-order was the smallest interval that the phalanx could adopt while still being capable of movement with relative ease and of presenting an offensive posture (see: Bearing the Phalangite's Panoply from page 133). Such an arrangement also has other implications for understanding the internal structure of the pike-phalanx.

In order for the weapons held by the members of the first five ranks to be lowered and positioned beside each other in a parallel manner, the men wielding these weapons have to be slightly offset to the right in relation to the man standing in front of them. This allows a small, yet sufficient, amount of space to be utilized which permits each subsequent man in the first five ranks to lower his weapon for combat. Importantly, it is this partial offset of each man in the first five ranks which actually makes that section of the file conform to the 96cm intermediate-order (see Fig.50).[189]

The gap between the sets of weapons held by two adjacent files equates to approximately 60cm – the lateral width of the 64cm *peltē* when it is held at a slight angle across the front of the body due to the angle of the left arm when it is used to carry the *sarissa*. Thus any enemy who was carrying a shield with a larger diameter – the Greeks at Chaeronea bearing their 90cm diameter *aspis*, for example – would not be able to force their way into this gap between the pikes without removing their shield from its protective position across the front of their body and so exposing themselves to attack by the phalangites in the phalanx. However, someone with a smaller shield, or a light-armed skirmisher, might be able to quickly move 'inside' most of the projected pikes and then use close-quarters weapons to attack the otherwise defenceless phalangites. Due to the way that all of the lowered pikes are parallel to each other, and the most right-hand weapon is abutted against the shield of the leading man in the adjacent file, should an enemy try to force his way into the gap, the weapons held by more rearward ranks are prevented from being swung across to either cover the gap or to engage that opponent.

The only way in which this gap could be covered is if the men in ranks three to five raised their pikes slightly. This would then create a small gap in the presented weapons and remove any impediment for the man in rank two to swing his pike to the right and so attempt to engage the oncoming opponent.

If the opponent had managed to force his way further between the lines, the same technique could be employed by the men in ranks four and five to allow the phalangite in rank three to engage. Importantly, if members of rearward ranks did raise their pikes to allow a more forward man to engage, those men would not be able to lower their pikes again until the opponent had been dispatched and the more forward man had moved his weapon back into a position parallel with that of the front rank or all of the weapons would become entangled. Thus the gap in a formation where the pikes were held parallel to each other was not always adequately protected, required particular movements to be undertaken in order to cover it or engage an enemy advancing into it, and had the potential to prevent further engagement by the rearward ranks. This left the files, and the formation as a whole, with certain vulnerabilities when engaged against more mobile opponents.

On the other hand, due to the slight angle of the shield, any lengthy weapon that was thrust into the gap between the files – the pike of an opposing phalanx during one of the battles of the Successor Period, for example – would be more easily deflected into the space occupied by the weapons of the adjacent file. This means that, when engaged against another pike-phalanx, the members of which could not individually exploit the gap between the files, the phalangite was relatively secure.

The other possibility is that the pikes lowered by the first five ranks of the phalanx were splayed rather than parallel to the formation's line of advance (Fig.51).[190]

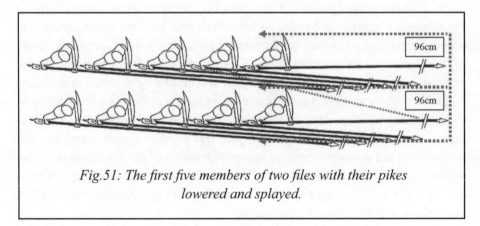

Fig.51: The first five members of two files with their pikes lowered and splayed.

Phalangites in this configuration are lined up one behind the other rather than slightly offset to the right as is required for the pikes of the file to be deployed parallel. Due to this arrangement, the presence of each man immediately ahead of those behind prevents the pikes held by those in the rearward ranks from being lowered into a parallel position and thus dictates that they have to be angled to the right in a splayed manner. Yet the pikes can still be set level with each other as per Plutarch's description of the phalanx, the points of the first five ranks still project ahead of the formation as per Polybius' account, and the weapons are still 'closely packed' together as per Livy. Interestingly, due to the way that the straight file of men occupies less lateral space, the greater lateral space required for the deployment of splayed pikes is offset and a file arranged in this manner still complies with the 96cm intermediate-order outlined in the manuals.

If the weapon carried by the man in the front rank is pointed directly ahead, but those held by the men behind him are angled to the right (as per Fig.51), this reduces the gap between the sets of pikes for each file to around 50cm. This would make it even more difficult for an oncoming enemy to try and force his way between the weapons. Even if faced with a more mobile opponent, if the men in ranks three to five raised their pikes slightly, this would allow the man in the second rank to swing his weapon further to the right up to the point where it touched the weapon held by the leading man in the adjacent file (the dotted grey line in Fig.51). Thus the gap in a formation with their weapons arrayed in this configuration can be completely covered (or even closed) by the men of the second rank – truly presenting a 'serried wall of pikes' towards an advancing enemy. This would allow a pike-phalanx arranged in intermediate order to effectively engage an opponent arranged in a close-order of only 48cm per man.

In order to position the rearward pikes in such an angled manner, each man would have to rotate his body and stance slightly more than 45°. However, the increase in rotation is not great, no more than an extra 10°, and this in no way impedes the forward movement of the phalangite or the advance of the formation as a whole. The increased rotation of the body does mean that the shield will be more perpendicular to any attack coming from the front, but it is still sufficiently positioned to either deflect incoming blows into the gap occupied by the weapons between each file, or to be rigidly held in place to attempt to absorb the blow.

Due the variable nature of hand-to-hand combat and the nature of the terrain that most battles of the ancient world were fought on, even if a phalanx had initially been arranged with their pikes parallel, there is little doubt a certain amount of splaying of the pikes would take place during combat as members of the file were jostled, the formation advanced over semi-broken ground, the natural flex of the weapon, and individual phalangites sought out opportune targets once they were engaged (if they were not arranged in a splayed manner to begin with). Due to the way that both arrangements occupy the 96cm intermediate-order interval, files of the phalanx could splay, or not, with the vagaries of combat without compromising the integrity of the line. Additionally, in both methods of arrangement, the projecting pikes held by the rearward ranks of each file would be on the right-hand, un-shielded, side of any enemy who tried to force their way into the gap. This would be a very risky movement which would leave the attacker quite vulnerable. Both of these configurations for the file work, and both comply with the descriptions of the phalanx provided in the ancient literature. Which one was used as an initial form of deployment is uncertain due to the limited nature of the detail contain within these descriptions. Both arrangements have their advantages and disadvantages and it can only be concluded that, in the chaotic environment of massed formation fighting, both configurations for the file would have been in effect – even if unintentionally so.

The file, and indeed the pike-phalanx as a whole, was a varied creature across the period of its common usage. Initially arranged on files of ten men, and then later on files of sixteen (which could, in turn, be 'doubled' to a half depth of eight), the file came in a variety of configurations across the Hellenistic Period. It was these files which were joined together to form the larger sub-units of the pike-phalanx. A critical review of the ancient references to the sub-structure of the phalanx shows that many of the units that are detailed in the later military manuals were in use by the early stages of Alexander the Great's campaign against Asia in 334BC – even if some of them may have only been ad hoc or temporary in nature and/or went under a different name. Other units found in the manuals were clearly in use by the end of Alexander's campaign. This supports the claims made by writers such as Aelian that in their works they were discussing the nature of the Macedonian army in the time of Alexander.

Another aspect which demonstrates a close correlation between Alexander's army and that described in the manuals is the positioning of the officers in set positions, based upon a mathematical model, to provide a symmetrical distribution of commanding officers across the line to achieve a balance of command abilities. The use of this principle subsequently allows for the identification, and quantification, of the officers and their respective positions within Alexander's infantry formations. This requirement for the officers of varying ranks to occupy certain positions also dictated the way in which the pike-phalanx could deploy on the battlefield – with only certain methods being possible if the arrangement of the men and files was to be maintained. This set method for the arrangement of the pike-phalanx, with its officers positioned at key points in the line, coupled with the use of units and sub-units within the broader formation, made the pike-phalanx very adaptable to terrain and tactical plans and, as a result, exceptionally effective on the field of battle.

DEEP AND SHALLOW FORMATIONS

Another facet of the phalanx in combat that needs to be considered is why, in some cases, the phalanx was deployed in its standard depth of sixteen men while at other times it was deployed to a half depth of eight or a double depth of thirty-two. It seems clear that the commanders of armies which contained pike-armed units possessed a substantial degree of latitude when it came to the deployment of their forces. Sixteen deep seems to have been the 'standard' deployment, but attestations to the use of other depths indicate that phalangite warfare was not limited to just one type of formation. When, where, why and how formations of different depths were used is a contentious issue amongst ancient and modern writers alike and various reasons and advantages of different configurations have been offered.

One of the main things that could affect how the phalanx was deployed for battle was the terrain of the battlefield itself. This influence could be due to the ground that the army had to cross in order to get into position, to the manner in which the army advanced, to the ground being either too narrow or too broad to adequately accommodate the army in its standard arrangement. Alexander the Great's advance to the battlefield at Issus in 333BC, for example, can be viewed as the narrow terrain of the defile through which the army had to pass dictating that the units be arranged in a deep column. While there is no specific reference to how the constituent units and sub-units of the phalanx were arranged for this march, from the perspective of the army as a whole, the formation was exceedingly deep.

Likewise, Machanidas' advance in column onto the plain at Mantinea in 207BC can be considered a deep deployment for the army. Again, there is no reference to how the individual units of the army were arranged. However, if Machanidas' forces were organized so that, once they wheeled into position they would be in their standard deployment ready for battle with each *merarchia* side-by-side, and with each sub-unit within these larger formations also side-by-side, then it can be assumed that, while in column, each *merarchia* possessed a frontage of sixteen men and a depth of 256 men (see pages 322-325). The interesting thing to note is that, unlike Alexander's move on Issus, Machanidas' advance in column at Mantinea was not dictated by the terrain. Rather, the arrangement must have been a conscious part of the overall plan to advance in column and then wheel the entire army into position.

Once on the field of battle, an army could have been deployed deeply due to the presence of only a narrow frontage upon which they were meant to engage. The deployment of the Macedonian left wing thirty-two deep at Sellasia in 222BC, for example, was required due to the narrowness of the slopes of Olympus up which this formation was expected to advance. Similarly, at Megalopolis at in 330BC, the terrain of the battlefield was so narrow that 'it would not allow a full-scale engagement of the two sides, so there were more spectators than there were combatants, and those beyond missile range shouted encouragement to their respective side.'[191] Conversely, if the battlefield was too broad, a commander might deploy his phalanx to a half-depth of eight so that his whole battleline stretched far enough to avoid encirclement. Arrian states that the phalanx should be arranged shallower 'whenever it needs to be deployed thinner, [and] if the ground makes that more useful'.[192] At Issus, Arrian states that Alexander deployed his forces shallow so as to occupy the entire plain and to make his flanks secure – protected on one side by the sea and on the other by a range of hills.[193] Thus at Issus there was a tactical necessity for Alexander to deploy his phalanx in a shallow manner so that his men formed an unbroken line stretching from one side of the battlefield to the other. It is this stated purpose with which Polybius, in his critique of Callisthenes' account of the engagement, finds issue. Polybius states that the deployment of Alexander's phalanx eight deep would have required a plain forty *stades* (approximately 7.25km) in width when arranged in the intermediate-order interval.[194]

It is interesting to note that Polybius states that Alexander's infantry were in intermediate order when arranged to a half depth. If it is assumed that the move to a half depth was undertaken by 'doubling' the files – that is, bringing the rear half-files forward – and if Polybius/Callisthenes is correct in that, following this redeployment the men were in intermediate order, then prior to the 'doubling' movement, the army had to have been in the open order of 192cm per man which would have been used for their initial march to the battlefield.[195] The only other possibility is that the units of Alexander's phalanx advanced on Issus in their normal intermediate order, but then a process other than 'doubling' (one not referred to in the narratives) was used to deploy the army to a half depth.

Furthermore, Polybius must be referring to the amount of ground required to accommodate Alexander's whole army rather than just the pike-phalanx. Alexander had 12,000 phalangites, divided amongst six *merarchiae* of 2,000 men each, at Issus. Within each of these *merarchiae* there would have been eight *syntagmae*. Each *syntagma*, when in its standard sixteen by sixteen arrangement in intermediate order of 96cm per man, possessed a frontage of around sixteen metres. If each of the *syntagmae* within the *merarchiae* were arranged side-by-side, each *merarchia* would then possess a frontage of 128m and Alexander's whole phalanx of six *merarchiae* a frontage of 768m (or four and a quarter *stades*). When deployed to a half depth of eight, but retaining the intermediate-order interval, Alexander's pike-phalanx would possess a total frontage of around 1,536m (or about eight and a half *stades*). Such a formation would easily fit within the fourteen *stade* plain of Issus, as Callisthenes describes it, if this was the only formation that had to be considered. However, at Issus, Alexander also employed his cavalry, *hypaspists* and light troops. Thus Polybius' issue concerns not only the arrangement of the pike-phalanx, but the disposition of every unit of Alexander's army as part of the larger battleline combined. Regardless, the important thing to note is that the entire breadth of the plain was used to accommodate Alexander's army and, depending upon how the accounts are viewed, a shallower depth may have been employed so that the army would stretch across it.

If the ground, troop numbers and tactical considerations permitted it, a shallow formation could also be used so that part of the army outflanked an opposing formation. Aelian states that reducing the depth of a formation (*leptysmos*) could be used to outflank an enemy on one wing (*hyperkerasis*) or on both wings (*hyperphalangisis*).[196] Frontinus tells us that Philip, when engaged against the Illyrians in 358BC, noticed that the front of the enemy's formation was strong but its flanks were weak. Philip accordingly placed his best men on his right so as to attack the enemy's left.[197] While the depth of Philip's formation is not stated, it is clear that his battleline stretched beyond that of his opponent. As such, the possibility that Philip had adopted a wider, yet shallower, formation to provide the opportunity for this flanking move cannot be ruled out.

Formations of varying depths could also be used to deceive an opposing army. A deep deployment could hide the size of an army.[198] Conversely a wide, shallow, deployment could make an army appear bigger than it really was. When Antigonus, who was greatly outnumbered, was fighting against Eumenes, he sent false information to Eumenes' camp which stated that his reinforcements had arrived. The next day, Antigonus drew up his phalanx to twice its usual width by reducing its depth. Polyaenus states that 'the enemy, having heard from the heralds about the presence of allies, and observing the extended length of the phalanx (although its depth was contemptible), was afraid to join battle and fled'.[199] However, while such deceptions were clearly beneficial if the opponent fell for the ruse, such modes of trickery did have their drawbacks if the enemy still chose to fight. Arrian observed that 'shallow formations are good for deception, but are useless in prolonged engagements'.[200]

Arrian's statement about the unsuitability of shallow formations for combat highlights the real purpose of the depth of the phalanx: density. In another part of his examination of the pike-phalanx, Arrian states that the phalanx 'is deployed deeper when [it needs to be denser], if it is necessary to repel enemies by density and force...or if it is necessary to repulse those who are charging...'.[201] Polybius states that the men in the more rearward ranks of the phalanx, 'by the very weight of their bodies pressing against those in front of them, add force to the assault'.[202] Similarly, Aelian states that 'by pressing forward with the weight of their bodies, [those at the back] increase the momentum of the phalanx and leave no possibility of seeking safety in flight to those in the forward ranks'.[203] Arrian, on the other hand, says that only the members of the sixth rank pushed those before them forward.[204]

Some scholars have seen the depth of the phalanx as being a means of providing the formation with rows of reserves behind those who could engage so that, should any of the combatants fall in action, a replacement could easily step forward into his place.[205] This would in part agree with Arrian's statement that shallow formations were of no use in protracted engagements. While the ability to field reserves would certainly be one benefit of any formation of relative depth, it should be noted that, due to the way only the pikes of the first five ranks could be presented for battle, even a deployment to a half depth of eight would still possess three rows of reserves who would be otherwise unengaged. The key to understanding the advantages of phalanxes of the standard depth of sixteen, or the double depth of thirty-two, seems to be the concept of density and momentum outlined by various ancient writers. The fact that this density and momentum is not attributed to shallower formations suggests that it was an important element of these deeper ones.

As noted by Aelian and Polybius one of the roles of those in the rear ranks was to help drive the formation forward. This push, or *othismos*, is a difficult concept to comprehend. Warry, for example, says that the phalanx was equally prepared to thrust with its pikes or to push with its shields.[206] There are a number of issues with such a conclusion. Firstly, only the members of the first rank of the phalanx, unimpeded due to no one being positioned in front of them, were capable of delivering a thrusting attack with the lengthy *sarissa*. Secondly, due to the serried array of pikes projecting forward of the formation, and the minimal gap between each set of weapons, it was almost impossible for an opponent to get close enough to the members of the pike-phalanx where they would have to 'push with their shields'. The writers of the military manuals all state that the weapons projecting ahead of the phalanx, even in its normal deployment, provided comfort and confidence to those within it – especially for those in the front rank.[207] Such statements would seem odd if opponents attacking the phalanx could regularly get close enough to the phalangites so that their lengthy pikes were basically useless and they were pushing against each other 'shield against shield'.

Warry seems to be applying the concept of the 'pushing' *othismos* of Classical Greek warfare to the pike-wielding formations of the Hellenistic Age. Yet even here the pushing of 'shield against shield' was only ever a part of very few battles in the Classical Period, and only when certain circumstances allowed for it.[208] It seems more likely that, for the pike-phalanx, any *othismos* was through the members of the pike-phalanx presenting their pikes to the enemy and 'pushing' forward with the whole formation to try and drive the enemy back and/or fracture their line.[209] Thus, at Pydna in 168BC, when the Macedonians pressed the tips of their *sarissae* into the shields of the opposing Romans, not only did this keep the enemy at bay as Polybius states, but the formation could also drive forward, pushing with their weapons, to try and force the Roman maniples back.

Such a method of offence sheds light on a number of other passages relating to the pike-phalanx in action. The references to the rearward ranks pressing the weight of their bodies against those in front, whether all of the rear ranks as per Asclepiodotus, Aelian and Polybius, or just those in the sixth rank as per Arrian, would add impetus to any push and drive forward of the formation. Importantly, any man in a rearward rank pushing forward would do so by pressing his shield into the butt-spike of the *sarissa* held by the man in front. This had a number of benefits. Not only would it drive the more forward man ahead, and prevent him from breaking and running, as the ancient commentators state, but pushing forward with the shield against the butt-spike would also brace the weapon in position and provide it with more stability with which to push an opponent. It would also prevent the butt-spike from accidentally injuring the man behind during the rigours of combat. Undoubtedly, this pushing with the shield would be done by every member of the phalanx other than those in the front rank – even by those in ranks two to five who have their *sarissae* lowered for battle. This would then provide both concerted impetus and force, as well as mutual protection and safety, to the formation as a whole.

Furthermore, this ability for the push of the rear ranks to brace more forward weapons in position demonstrates that Polybius' statement that the *sarissa* was held 4 cubits (196cm) from the rear end is incorrect. If the pike-phalanx was arranged in an intermediate-order interval of 96cm per man, the man behind each rank would not be able to brace the weapon in position as it would extend beyond the interval he was occupying (see page 85). Connolly, in his experiments with the pike-phalanx, found that if the front rank held their place (simulating contact with an enemy) while the rear ranks continued to advance, this resulted in a slight compression of the lines.[210]

Due to this ability for the rearward ranks to both brace weapons into position and to help drive the formation forward, one of the main benefits of a deep formation would be that it could add 'punch' to any offensive move. Importantly, this cannot have been a benefit of thirty-two deep, double depth, deployment. It is unlikely that anyone pushing from the back rank of such a formation would be able to exert any sort of pressure against those at the front. Any pushing made by the rear ranks would have simply been diffused throughout the formation. It is really only the first few ranks at the front of the formation which could have had any real effect on each other. This explains why Arrian states that it was only the men of the sixth rank who exerted this pressure without actually being able to engage. For the men of more rearward ranks, their 'pushing' would have been more figurative in nature, merely to help keep the formation together and to prevent those in the front from breaking as the ancient manuals state. The rear ranks could certainly help drive the formation forward by pressing into the backs of the men in front of them (who push those in front of them in turn), but their pressure could not directly affect the weapons presented for battle by the front ranks. This then suggests that other factors came into play when a decision to deploy to a double depth was made.

Plutarch states that, at Sellasia in 222BC, the advantage lay with the Macedonians due to their superior armour and the density of their formation.[211] This may be a reference to the Macedonian left wing which was deployed thirty-two deep for an attack up the narrow frontage of the hill of Olympus. It is interesting to note that, once engaged on the slopes of the hill, the Spartan charge was able to force this deep formation back. Consequently, it can only be assumed that the 'density' of the formation was advantageous in areas other than in the momentum of their uphill assault.

One possibility is that the advantage of a double-depth deployment was in the greater number of lines of reserves that this deep formation would have possessed. Another possibility is that a deep formation would have been almost impossible for an opposing line to break psychologically. Bennett and Roberts suggest that the back rows of the phalanx provided 'a crucial psychological feeling of depth and support' to those in front.[212] Undoubtedly, the members of a deep formation would find comfort in knowing that there was a vast amount of manpower behind them, all driving them forward or helping them hold their position, and the thought of engaging a formation that was unlikely to break must have greatly intimidated anyone facing one.[213]

The unlikelihood of a deep formation breaking also accounts for where such formations were used. Deep formations could be employed where a fight could be expected to be particularly hard fought, on ground that was somehow unfavourable, or when a concentration of force against a certain point in an enemy line was the tactic to be employed. In 335BC, Alexander the Great arranged his troops in a deep formation to direct his attack at the weakest point in the Taulantian line.[214] The concentration of force using a double-depth formation can also be seen at Cynoscephalae in 197BC where the thirty-two deep phalanx on the Macedonian right wing was used to advance down the hill and push back the attacking Romans. At Sellasia, the deep Macedonian phalanx would have been sent against the Spartans on Olympus not only because of the narrowness of the terrain, but also because the Spartans, if they chose to counter-attack, would be fighting with the benefit of advancing downhill and so a deep formation would have been needed to resist such an attack. It is interesting to note that, when the Spartans did counter-attack, the Macedonian line was forced back rather than it breaking or routing. This could only have been a result of the deep formation that had been adopted being somewhat self-supporting which prevented the Macedonian withdrawal from turning into a rout.

Gabriel suggests that one of the benefits of a square formation (presumably the sixteen-by-sixteen *syntagma*) is that it is a lot easier to control on the battlefield than an oblong formation is – especially when all it has to do is advance.[215] Gabriel also suggests that a square formation could change direction very easily while still presenting a strong front.[216] Indeed, a square formation like the *syntagma* could change its direction almost immediately due to its uniform shape. If an opposing force suddenly appeared to either side of an advancing *syntagma* (or to the rear for that matter) all that needed to occur to quickly meet this new threat would be for the formation to halt, have everyone raise their pikes vertically so as to not impede each other, have everyone turn to face the enemy, and then lower their pikes to engage. Turning to the left or right is a movement called a *klisis* (κλίσις) in the military manuals, while turning completely about is a manoeuvre called a *metabolē* (μεταβολή).[217] The fact that such movements are referred to as a regular part of phalangite drill suggests that pike-phalanxes could easily perform such movements if the situation required.

Due to the shape of a *syntagma* in its regular deployment, no matter which way the members of the unit faced, the formation would retain its square configuration. However, the one difference between simply turning to face a new enemy threat and wheeling the formation about to face it is that, in simply turning to face an unexpected threat, the file leaders in the front rank would no longer be facing the enemy. If a square formation simply turned to face to the left, for example, the file leaders would then make up the right-hand file of the unit. Similarly, if the unit faced to the right, the file leaders would be on the left. This was an arrangement of the phalanx known as a *paragogē* (παραγωγή).[218] Again, the reference to such arrangements suggests that pike-phalanxes could operate in such a manner. Bringing the file-leaders to bear against an enemy attacking from the side could only be achieved if the formation wheeled rather than turned, but this would take considerably more time and space which, due to the nature of the threat and the battlefield, the formation may not have.

The use of the semi-independent square *syntagmae* as the basic building blocks of the pike-phalanx raises the question of how each *merarchia* of a phalanx was organized if the army was deployed thirty-two deep. Unfortunately, both the narratives and the manuals do not provide any indication for how a double-depth phalanx was created – unlike how they outline the means of creating a half-depth deployment through the process of doubling. However, the structure of the units and sub-units of the pike-phalanx provides clues as to how a double-depth deployment could be accomplished. Each *chiliarchie* of the pike-phalanx, for example, was made up of four *syntagmae*. There are a number of ways in which these units could be arranged to create a formation thirty-two deep when each *syntagma* was arranged in its standard sixteen-by-sixteen layout. One possibility is that each of the four square *syntagmae* of one *chiliarchie* were themselves arranged in a square (Fig.52).

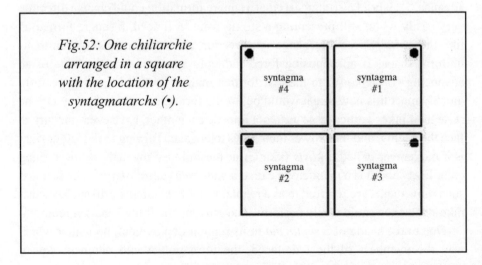

Fig.52: One chiliarchie arranged in a square with the location of the syntagmatarchs (•).

Within this arrangement, each of the four individual *syntagmae* retain their square structure and, if need be, can move independently of each other to attack or to reform the line if a different configuration is later needed. Additionally, because each of the *syntagmae* retain their square sixteen-by-sixteen arrangement, by deploying them in a larger square, the whole formation becomes a larger, double-depth square of thirty-two-by-thirty-two. Such an arrangement could easily be made out of a *chiliarchie* initially deployed in extended line by simply having the two most right-hand units (i.e. *syntagmae* #1 and #4) move forward and left to take up a new position ahead of the other two using a series of advances and changes of facing to ensure that they ended up facing forward. This would result in both lines of the new square formation having commanding *syntagmatarchs* leading the files on either side and the even distribution of command abilities of the senior officers remains intact and balanced (Fig.53).[219]

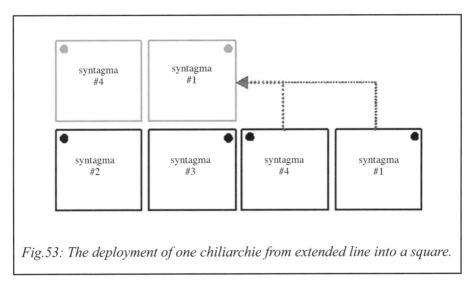

Fig.53: The deployment of one chiliarchie from extended line into a square.

Arranging a *chiliarchia* into a double-depth square in such a way would have all of the same benefits of movement and control associated with the square *syntagma*, just on a bigger scale. Alternatively, a double-depth square formation could be created by having every second file of each *syntagma* take up a position behind the file to its right – in effect making each *syntagma* a unit eight men across and thirty-two deep. When four such units were combined, each *chiliarchie* would then be thirty-two-by-thirty-two. However, in this manner of arrangement the individual *syntagmae* would not be in a square formation and would lose any benefits that such a configuration possessed – although this in itself is not a reason for dismissing such a possible configuration of the double-depth phalanx.

The use of square *syntagmae* and *chiliarchiae* also provides options for how each of the *mararchiae* of a pike army could be deployed. If each *chiliarchie* was arranged in a square, then both formations could simply be positioned beside one another to create a unit with a double depth of thirty-two – made up of two squares which could operate independently on the battlefield (Fig.54).

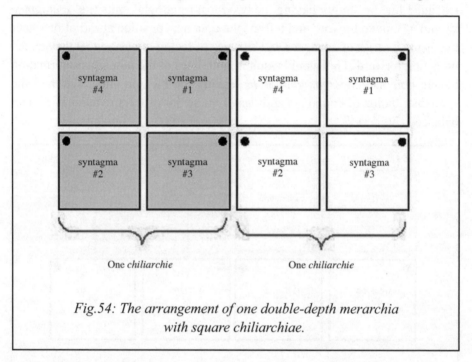

Fig.54: The arrangement of one double-depth merarchia with square chiliarchiae.

Alternatively, each of the two *chiliarchiae* of the *merarchia* could have been deployed in extended line, but with one of the *chiliarchiae* positioned behind the other. This would still result in a formation with a double depth of thirty-two (Fig.55).

While such a deployment loses any of the benefits of having the *chiliarchiae* arranged in semi-independent squares, the use of two parallel extended lines could be used to meet other tactical requirements.

At Gaugamela in 331BC, Alexander deployed his phalanx with a second line behind his front so that 'his phalanx faced both ways' (καὶ δευτέραν τάξιν ὡς εἶναι τὴν φάλαγγα ἀμφίστομον).[220] It has been suggested that this second line was made up of Greek allies and mercenaries not referred to by Arrian.[221] However, it cannot be discounted that Alexander's deployment

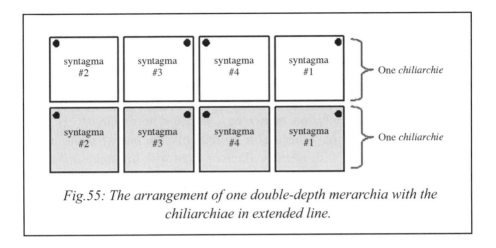

Fig.55: The arrangement of one double-depth merarchia with the chiliarchiae in extended line.

was the arrangement of the pike-phalanx to a double-depth with one *chiliarchie* of each *merarchia* positioned behind the other to create a double line.[222] Alexander had adopted just such a deployment, with his phalanx in two lines, for his advance to the Granicus in 334BC.[223] Arrian states that the commanders of these reserve units had orders to face about and receive any enemy attack if the position was encircled.[224] This provides a number of clues as to the nature of Alexander's deployment and the identification of these reserve units. Firstly, Arrian states that the reserve units were under the command of separate officers to those in the front line. While this does not discount the possibility of them being allies or mercenaries, it is also possible that they were the second *chiliarchiae* of each of the six *merarchiae* – the front *chiliarchiae* being under the command of the six *merarchs*, while the reserve lines would be commanded by *chiliarchs*.

Secondly, Arrian states that Alexander's second line at Gaugamela was expected to face about only if the rear of the Macedonian position was threatened. This suggests that the initial purpose of the second line of troops was to support those at the front. This would suggest an initial deployment of the pike-phalanx to a double depth of thirty-two with both *chiliarchiae* in parallel extended lines facing forward. The term that Arrian uses to describe Alexander's formation at Gaugamela (*amphistromon*) is used in the military manuals (Arrian's included) as the name for a pike-phalanx deployed in two parallel lines but with each line facing in opposite directions.[225] This finds close parallels with the stated purpose of Alexander's deployment at Gaugamela and the reverse facing of the rearward line can only be assumed to be something that the second line would do if required, rather than as part of its initial deployment. Aelian states that a phalanx with lines facing in two directions was very useful when engaged against enemies whose army was strong in cavalry – probably because of the likelihood of being encircled.[226] Alexander was greatly outnumbered at Gaugamela, and the Persians had large cavalry contingents which, unlike at Granicus and Issus, would be fighting on a level plain where they could manoeuvre and operate effectively. Consequently, no matter how shallow he deployed his line, Alexander would not have been able to outflank the enemy position, or prevent encirclement, with infantry alone or even with his combined army as a whole. Consequently, a double-depth deployment of the phalanx may have been adopted to provide density to his front which would bear the full brunt of the Persian opening charge but the arrangement was done in such a way that the rear half of the phalanx could easily turn about to face attacks from behind if needed. Both Arrian and Curtius state that to meet the threat of encirclement was the reason for the nature of Alexander's deployment.[227] Not only would the use of a deep formation to resist an enemy attack correlate with the exact purpose of such a formation as stated by Arrian, but this again suggests the deployment of the *chiliarchiae* in two parallel lines with the ability of the rear line to face about to meet an attack from the rear. If this is correct, it seems that Alexander recognized that the sheer weight of Persian numbers opposing him at Gaugamela would bear heavily on the front ranks of his phalanx and would afford the Persians the opportunity to encircle his smaller force. Consequently, it seems that Alexander had masterfully deployed his phalanx to effectively meet both scenarios.

The use of standardized units and sub-units within the pike-phalanx clearly provided commanders with a variety of possible ways to arrange their forces to meet any challenge and adapt to the changing nature of the battlefield. The use of shallow, standard or deep formations to meet any threat, conform to any tactic and operate on any terrain made the Hellenistic pike-phalanx a very flexible formation for commanders to work with and the ancient literary sources contain numerous accounts of engagements which highlight the sheer adaptability of the pike-phalanx in action. Thus while sixteen deep can be considered the 'standard' for the basic building block of the *syntagma*, one method of deployment for the larger army as a whole was just as likely as the other; and its arrangement was very much dependent upon the circumstances of the individual encounter. This, coupled with the greater tactics employed by many commanders, and the way that the individual phalangites fought, easily accounts for why the pike-phalanx was able to dominate the battlefields of the ancient world for nearly two centuries.

PHALANX TACTICS: THE HAMMER AND THE ANVIL

How was the pike-phalanx used to gain such a position of dominance in the ancient world? Evolving from the developments that the Classical Greek way of war had undergone at the end of the fifth century BC, warfare in the Hellenistic Age was a time of the effective and co-ordinated use of the various arms of an army to create opportunities where a decisive blow against an enemy formation could be delivered. For many scholars the two dominant arms of the Hellenistic army are the pike-phalanx and the cavalry. Both arms played different roles on the Hellenistic battlefield. In many analyses of the conflicts of the time, the tactics of the Hellenistic Age are often referred to as those of the 'hammer' and the 'anvil'.

The concept of the 'hammer and anvil' tactic of the Hellenistic Age is almost universally accepted and is thought to have been in use from the very beginning of the time period. Fuller, for example, says that Philip II 'decided to make his cavalry his decisive arm; that is, it would replace the phalanx as the instrument of shock, while the phalanx...would constitute the base of cavalry action. Instead of assaulting, normally the phalanx would threaten to do so, and through the terror its advance always instilled it would immobilize the enemy and morally prepare the way for the decisive charge.'[228] Thus for Fuller the purpose of the pike-phalanx was not to begin any offensive action directly, but to merely advance against an enemy, forcing them to remain in place in order to face it, while the cavalry swept around the flanks to attack the immobilised enemy from the sides. In this way the pike-phalanx became the 'anvil' for the cavalry's 'hammer'. Many scholars have accepted this analogy and used it as the basis for their own examinations of Hellenistic warfare.[229] Fuller additionally likens the warfare of the Hellenistic Age, following Clausewitz, to a struggle between two boxers – with the phalanx in the centre forming the body and the cavalry on both wings the arms which are used to deliver blows.[230] English, one of the few who argues against the hammer and anvil analogy, states that it is the pike-phalanx that is the strike weapon of Hellenistic warfare, rather than it being a defensive platform, and that the lengthy *sarissa* was similarly an offensive weapon rather than a defensive one.[231] This conclusion seems unlikely as it does not correlate with the accounts of many of the battles of the Hellenistic Period (see following).

Indeed, one need look no further than an examination of the four main engagements fought by Alexander the Great to see the 'hammer' and the 'anvil' in effect. In each of these confrontations (Granicus (334BC), Issus

(333BC), Gaugamela (331BC) and the Hydaspes (326BC)) Alexander's tactics were to advance the pike-phalanx to pin the opposing line in place while the right wing of the cavalry charged ahead to either knock out or nullify the enemy's left wing so that it could then turn inwards to deliver the decisive blow against the enemy centre from the side. There is a clear reason why Alexander's cavalry charge was always directed against the opponent's left wing. If, once their left wing had been routed by Alexander's horsemen, the enemy infantry units in the centre turned to face this new mounted threat, they would have to expose their right, unshielded side to the serried array of pikes presented by the pike-phalanx advancing against them from what used to be their front. If, on the other hand, the enemy centre chose to remain in place and engage the phalanx, they would then simply be mowed down by the cavalry attacking from the side. Had Alexander chose to attack with his cavalry from the left on the other hand, the enemy could turn to meet them and still present their shields to the advancing phalanx and so be able to resist this attack better. Alexander, however, never did this and always attacked in a way that made the enemy vulnerable. Alexander's tactics were to not only dictate the tempo of the engagement, but to also create opportunities to make the right wing charge as effective as possible, thus leaving the enemy centre virtually in a no-win situation. Alexander clearly knew what worked, what worked well, and how to best employ the various sections of his army to their best effect. This accounts for why his tactics are almost identical at every major battle he fought with only slight variations to cater for the terrain and the size of the opposing forces.

Timing was critical when employing the hammer and anvil stratagem. If the flanking attack was committed too early, before the pike-phalanx was able to pin the opposing formation in place, the enemy could simply turn to meet this threat with little risk. If, on the other hand, the flanking attack was delayed, or committed late, the pike-phalanx had to be able to hold the enemy in position long enough for the flanking units to arrive. This, in part, accounts for the greater depth of the Hellenistic pike-phalanx. The necessity of timing also explains Philip's feigned withdrawal at Chaeronea: to draw the Athenian left wing forward, rout it with a counter-attack, and provide enough time for the rest of the phalanx, which had been deployed obliquely to the left-rear of Philip's position, to advance and pin the opposing line in place. The timing of the hammer and anvil tactic also explains why many of Alexander's flanking movements moved obliquely to the right: not only did this action create many of the gaps that his flanking units were able to exploit, but it also provided the pike units in the centre the necessary time to engage the opposing centre.

Modern combat doctrine follows the exact same hammer and anvil tactics employed by Philip, Alexander and the Successors – a tactic known as the '4F' principle in the US Army: Find the enemy; Fix them in place using part of your forces; Flank the enemy with the other part of your forces; and Finish off the enemy in a crossfire.[232] These tactics work at any level of operation; from small unit infantry tactics to those employing entire mechanized divisions.[233]

The use of the cavalry 'hammer' and phalanx 'anvil' by Macedonian armies carried over into the engagements of the Successor Period as well. Many of the battles of this time opened with a charge of mounted troops, either on horseback, on elephants, or both, from the wings with the pike-phalanx only coming into action later in the encounter. Such opening tactics can be seen in the accounts of the battles of Paraetacene (317BC), Gabiene (316BC), Gaza (312BC), Ipsus (301BC), Heraclea (280BC), Asculum (279BC), Raphia (217BC), Mantinea (207BC) and Magnesia (190BC).[234] Other battles of the Hellenistic age, such as Chaeronea in 338BC and Sellasia in 222BC, also followed a similar tactic of opening the fighting with an attack from the right flank.[235] However, the difference between these two battles and many of the others during the Hellenistic Period is that they were predominantly infantry engagements.

Even when all seemed lost the cavalry could still carry the day. Phocion is said to have been able to recall his scattered units of infantry and cavalry, reform them, and achieve victory through the use of his horsemen.[236] Similarly, when Eumenes' infantry had been routed in a fight against Neoptolemus in 321BC, he was able to secure victory by having his cavalry attack the enemy forces pursuing his fleeing troops.[237] The sheer number of battles which involve flanking charges by mounted troops or even infantry clearly demonstrates that the use of the 'hammer and anvil' was an integral part of Hellenistic tactical doctrine across the entire time period.

Chapter Thirteen

The Anvil
in Action

In order for the decisive hammer stroke of the flank attack to be successful, the pike-phalanx, acting as the anvil, had to be able to effectively engage the centre of an enemy formation and pin it in place. This was where the fighting style, deployment and equipment peculiar to the Hellenistic phalangite came into play. Polybius flatly states that the pike-phalanx was far superior to the formations used in Greece and Asia in the earlier parts of the Hellenistic Period.[1]

Depending upon the type of opponent the pike-phalanx was facing, the phalangites within the units of the formation could operate in a number of ways to achieve the objectives of the overall battle plan. If the opponents were heavily armoured, such as Greek hoplites or Roman legionaries, then all the phalanx had to do to effectively pin them in place was to hold them at bay with the tips of their *sarissae*. This is exactly what the Macedonians did at the battle of Pydna in 168BC according to Plutarch.[2] In this position the phalangites were relatively secure. Greek hoplites had a *combat reach* of around 2.4m with their spears.[3] This is far shorter than even the amount that the shortest *sarissa* projects ahead of the man wielding it. Consequently, Greek hoplites could not even reach an opposing phalangite with their weapons unless they managed to somehow get 'inside' the first few rows of projected pikes. Yet Polybius says that this was all but impossible even for a Roman legionary carrying a shield that was thinner than the Greek *aspis* and wielding the small *gladius* instead of a lengthy spear.[4] Furthermore, when a phalangite and an opposing hoplite are in this position, the phalangite in the front rank has not even made an attacking move with his weapon and has another 48cm of *effective reach* with which to try and dispatch his opponent if an opportunity to deliver a killing blow presented itself (Fig.56).

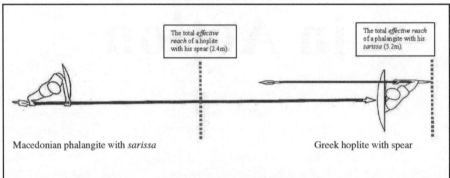

The total *effective reach* of a hoplite with his spear (2.4m).

The total *effective reach* of a phalangite with his *sarissa* (5.2m).

Macedonian phalangite with *sarissa*   Greek hoplite with spear

Fig. 56: The difference in effective reach between a phalangite and a hoplite.

This brings into question Polybius' statement (18.29) that the *sarissa* was held by the last 4 cubits of its length. If the purpose of such a lengthy weapon was to provide greater reach so that enemies could be engaged at a safe distance, it would seem illogical that it would be designed in a way that the weapon's point of balance meant that a third of its length was behind the person wielding it rather than projecting ahead of him. Most hand-held weapons such as swords, daggers and, in particular, the spear used by the Greek hoplite, are all configured with a point of balance towards the rearward end of the weapon where the hand grasps it so that the person wielding it can use the length of the weapon to its full advantage. The *sarissa*, developed from the hoplite *doru*, is unlikely to have been any different and Polybius' statement is most likely incorrect (see pages 83-88).

Greek hoplites, for the most part, also fought in the intermediate-order employed by the Macedonian pike-phalanx.[5] This meant that each hoplite across the front rank of a Greek formation would face five pikes levelled at him by the opposing Macedonians (Fig.57).

Even if the Greeks constructed a close-order shield wall using an interval of only 48cm per man, a formation commonly used by well trained troops like the Spartans, two Greeks would still face the five pikes presented by each file of the Macedonian phalanx. Additionally, the interlocked nature of the closer formation meant that, despite two hoplites facing each file of phalangites, individual hoplites would be unable to try and move between the sets of

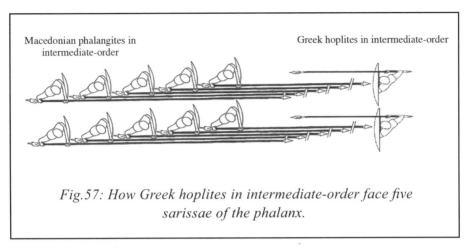

Macedonian phalangites in
   intermediate-order

Greek hoplites in intermediate-order

*Fig.57: How Greek hoplites in intermediate-order face five
sarissae of the phalanx.*

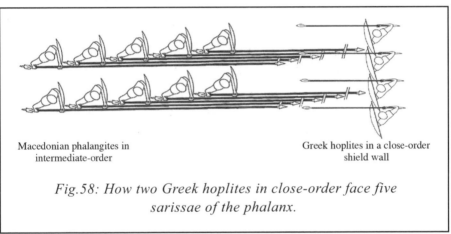

Macedonian phalangites in
   intermediate-order

Greek hoplites in a close-order
   shield wall

*Fig.58: How two Greek hoplites in close-order face five
sarissae of the phalanx.*

sarissae to engage the phalangites at closer quarters as they were locked into
the formation (Fig.58).

Thus no matter what order of interval an enemy formation of hoplites
adopted, the members of the pike-phalanx were well protected behind their
wall of serried pikes and were able to engage an opponent at a greater
distance than the enemy could. This made the pike-phalanx practically
unbeatable in a front-on confrontation – a sentiment specifically stated in
many of the ancient texts.

Despite such limitations, opponents like Greek hoplites could mount
a strong resistance to the Macedonian phalanx. The Theban Sacred Band,
for example, is said to have been able to hold their ground against the
Macedonians at the battle of Chaeronea in 338BC and that, following the
battle, Philip found them all dead to a man where they had stood.[6] At the

Granicus in 334BC, the hoplite mercenaries fighting on the Persian side are said to have inflicted great losses among the Macedonians as they were fighting at close quarters.[7] How are such accounts to be interpreted? How were hoplites able to resist an assault on their position which, as Diodorus states for Chaeronea, caused bodies to pile up until Alexander was finally able to break their line? There are several ways that such descriptions can be interpreted. Firstly, the members of the Sacred Band may have been able to withstand the attacks of the phalangites simply through the use of their larger shields. The Macedonians may have pressed the tips of their pikes into the Greek shields to keep them at bay and something of a stalemate could have ensued as the front rank phalangites sought to drive the Theban hoplites back or exploit targets of opportunity. It must be noted, however, that such a resistance by the Sacred Band would not include offensive actions against the Macedonians who would have been out of reach. Thus their resistance would have been wholly defensive and totally one sided. Alternatively, it must also be considered that such accounts may simply be literary motifs to both emphasize the ability of the hoplites and, by default, glorify the Macedonians who defeated them. Finally, it has also been suggested that Alexander led a unit of cavalry against the Thebans at Chaeronea rather than pike-wielding infantry.[8] Plutarch states that the Thebans had faced troops carrying the *sarissa*.[9] However, Plutarch may be referring to the long cavalry lance rather than the infantry weapon. If this is the case, then the Sacred Band did not face phalangites at all and may have been able to mount a stronger, and partially offensive, resistance to cavalry. If, on the other hand, the Sacred Band were engaged by pike infantry, then their deaths can only have been the result of thrusts delivered by the members of the front rank of the pike-phalanx. Due to the difficulty in committing such attacks against opponents with such large shields and strong defensive armour, this would account for the difficulty the Macedonians seem to have had in defeating the hoplites of the Sacred Band – with the lengthy *sarissa* eventually being the deciding factor.

Even if the Sacred Band itself did not face Macedonian phalangites at Chaeronea, the remainder of the Athenian and Theban line certainly did. Macedonian victory at other parts of the line seems to have been more easily obtained, partially through Philip's use of a feigned retreat on his right wing, than in the confrontation with the Sacred Band. This not only highlights the difference between the experienced members of the Sacred Band and the other hoplites of the part-time citizen militia of Athens and Thebes, but would also correlate with the account of the stand of the Sacred Band as a literary motif to emphasize the fact. Despite any strong stand against the Macedonians at Chaeronea, the Greek hoplite was greatly outclassed by phalangites wielding a lengthy pike when both sides engaged frontally. At the Granicus, Alexander's phalangites were able to easily overcome contingents of mercenary hoplites fighting for the Persians, while at Issus mercenary hoplites were only able to partially resist the advance of the pike-phalanx with the aid of an elevated position on the river bank.

Sekunda suggests that the shorter pike of the Iphicratean *peltast* was created to counter the spears of the Greek hoplite.[10] Such a development of military weaponry would certainly make sense. Prior to the Iphicratean reforms, the Greek hoplite had dominated the battlefields of the ancient world. The most effective way to overcome such troops, armed with a long thrusting spear, would have been to engage them with a longer one, and this is the essence of pike-based combat – to outreach your opponent. This, in part, explains why, when more engagements against the Macedonians were imminent during the wars of the Successors, many city states of Greece abandoned the use of hoplites in favour of the adoption of the *sarissa* and the rest of the phalangite panoply.

A need to outclass the Greek hoplite also explains the greater depth of the Macedonian phalanx. The traditional hoplite deployment was eight deep – although formations of greater and lesser depths were also used.[11] Thus even the ten deep *dekads* of the Macedonian army in the time of Philip II, and later the sixteen deep *syntagmae* of the army of Alexander, were deeper than, and in some cases double the depth of, a hoplite formation. Not only would this provide a considerable amount of reserves to the Macedonian line (even though, due to the advantages of the pike-phalanx, they may not have really been needed), but such a deep formation would impart a strong psychological deterrent to anyone facing it, was mutually supporting due to its sub-units, and the depth provided momentum and encouragement to those within it. Had the Macedonians regularly deployed in half-files of eight (or even of five), on the other hand, then any psychological bonus provided by the pike-phalanx would be lost and its advantage would solely lay with the length of its weapons.

Unlike the Greek hoplite, the Persian forces faced by Alexander in the late fourth century BC fought in a completely different manner. Whereas the Greek hoplite was designed for massed, frontal, hand-to-hand, combat, the Persian way of war relied more on masses of cavalry, and their infantry were much more lightly armoured (if at all), were more mobile, and often relied on skirmishing, hit-and-run tactics, and using missile weapons to hit their enemy from a distance while relying on weight of numbers.[12] Thus the more lightly-armoured Persians were at a distinct disadvantage when fighting the Macedonians. Africanus states that 'face-to-face with [Macedonians], no barbarian would be able to stand firm, no matter how he was fitted out'.[13] Following the Persian defeat at Issus, Darius increased the length of the swords and lances that the Persians used as he considered that the Macedonians held a distinct advantage in the reach over their weaponry.[14] This shows that Macedonian success was in part due to the length of the *sarissa*.

The configuration of the phalangite panoply provided other benefits to massed combat. The adoption of the hammer and anvil tactic to use the pike-phalanx to simply pin an opposing formation in place meant that a lowered pike, supported by the *ochane* and the oblique posture adopted to wield it, could be held in this position for a considerable length of time without causing the phalangite to become fatigued due to the minimal amount of thrusting offensive actions which would/could be undertaken. Thus the experienced officers in the front rank could maintain their position at the 'cutting edge' of the pike-phalanx and would not need to be rotated to the rear of the formation (a move that is almost impossible due to the mass of lowered weapons extending between the files) due to muscular exhaustion. The ability to hold back an opponent with a shorter reach weapon would have also reduced the risk of those in the front ranks sustaining injuries – another reason why such men would need to be moved to the rear, but one that would be difficult to accomplish once the phalangites in the front ranks had the tips of their weapons pressed into the shield of an opponent.

Despite its advantages, engaging lightly-armoured opponents with a pike-phalanx also presented potential problems – especially if that opponent charged against the line of phalangites *en masse*. At the battle of Gaugamela, the Persians are said to have attacked 'all along the line' which suggests just such a massed charge.[15] What then happened when the two lines met? The accounts of the battle of Gaugamela (or other similar encounters) unfortunately do not go into such detail. However, the possible outcomes can be worked out with the understanding of the dynamics of fighting in the pike-phalanx.

When a wall of serried pikes met a mass of lightly-armoured opponents charging at it, one of two things would have happened – mostly likely concurrently along different parts of the line. If, on the one hand, the shields carried by the lightly-armoured opponent were able to resist penetration by the presented *sarissae*, then the opponent would be brought to an abrupt halt, those advancing behind him would slam into his back as the lines compressed, and the man in the front rank would be held back by the sheer length of the weapon pressed into it. As such, the two sides would remain 'at pike length' from each other and the phalanx could continue to keep the enemy at bay until a flanking attack could be delivered.

On the other hand, if the shields carried by the lighter troops could not stand up to the points of the *sarissae* (or if they did not carry a shield at all), or if the impact of the more rearward ranks of the charging formation slamming into those in front of them as they were brought to an abrupt halt by the weapons of the phalanx, applied pressure to the front ranks and drove them forward, the very impetus of the charge could result in many of the lightly armoured opponents impaling themselves on the pikes of the phalanx. While this would have undoubtedly caused significant casualties among the front ranks of the charging enemy, it would have had dire consequences for the members of the pike-phalanx as well.

If the momentum of a charge impaled an opponent on the tip of a *sarissa* this would effectively take that weapon out of action. Due to the pressure being exerted from the compressing ranks of the charging enemy formation, and the bracing in place of the front rank pikes by the men of the second rank of the phalanx, any phalangite who found an enemy pierced by his pike would have no means of retracting that weapon to continue fighting. The impaled body, falling to the ground, would take the tip of the pike with it. This would open up a small, but potentially dangerous gap in the 'serried wall of pikes'. The next row of the charging enemy formation, carried forward by their own momentum over the bodies of those who had fallen, could also find themselves impaled on the *sarissae* of the second rank of the pike-phalanx with similar results. Thus, by the time the first three to four rows of the enemy had been slain, most of the pikes projecting ahead of the formation could have been rendered useless and, due to many of the members of the front ranks retaining their positions while the members of the more rearward ranks were unable to commit their weapons for action due to the depth of the file, the front-rank phalangites would have had little recourse but to drop their pikes and engage at much closer quarters with swords. When this occurred, the pike-phalanx would lose much of the advantage that the lengthy *sarissa* was designed to provide. This could have been one way of neutralizing the effectiveness of the phalanx – although, in terms of manpower, it would be a very costly one.

Such a compounding series of events would also occur if the *sarissa* was regularly used to thrust at an enemy. If the weapon became lodged in the body of an opponent due to the delivery of a killing blow, the pike would similarly become useless unless that killing blow was delivered in the furthest half of the weapon's *combat reach* which would then provide the phalangite with enough room to extract it and continue fighting.

Prior to the engagement at Gaugamela, Alexander is said to have instructed Parmenion to order his troops to shave, stating: 'do you not know that in battle there is nothing handier to grasp than a beard.'[16] Such passages suggest that Alexander recognized that his front lines may have become overwhelmed if the more lightly armoured Persians chose to charge his position and that the fighting could 'devolve' into something more 'in your face' (almost literally). Plutarch similarly claims that phalangite combat was fought 'amid shields, pikes, battle-cries and the clash of arms'.[17] This again suggests how frantic some pike engagements could become – although it does not necessarily mean that all of them resulted in the pikes becoming unserviceable. This suggests that the primary function of the *sarissa* was for it to be pressed into an opponent's shield in order to keep him at bay rather than to be used as an offensive weapon except when targets of opportunity presented themselves.

This primary use of the *sarissa* as a pushing weapon also correlates with the concept that the *sarissa* came in two parts. It needs to be considered, if the *sarissa* was in two sections, why an opponent simply did not pull the front end off and make the weapon ineffective? The answer is that the nature of phalanx combat rarely created the conditions where such a thing was possible. Most opponents facing a pike-phalanx would be carrying a shield in their left hand and a weapon in their right. Thus the only way an enemy could actually grasp the front end of the *sarissa* would be if they somehow freed their right hand. This would be difficult, although not impossible for a Greek hoplite wielding a long thrusting spear – who could raise the weapon vertically and transfer it to their left hand while their shield remained carried by its central armband. It would be somewhat easier for troops such as Roman legionaries armed with a sword which could be put back in its scabbard. However, this could only be done after the legionaries had cast the *pila* during the opening stages of the confrontations as, unlike the hoplite *aspis*, the Roman *scutum*, with its single grip and lack of central armband, was carried in a manner which prevented the left hand from being used for anything other than porting the shield. Plutarch's statement that the Romans tried to grasp the Macedonian pikes in their hands at the battle of Pydna not only suggests that they were trying to separate the front end of the weapon from the back, but also that their swords were sheathed and their javelins had been cast in order for this to even be attempted.[18] When engaged against another pike-phalanx, such as during the wars of the Successors, no one would have had the ability to grab onto an opponent's *sarissa* due to the way that both hands were required to carry the weighty weapon.

Even if an opponent managed to get himself into a position to grab onto the pike, it was not guaranteed that this would render the weapon ineffective. At the battle of Edessa (c.300BC), for example, the Spartan king Cleonymus ordered the first two ranks of his troops to deploy without weapons so that they could grab onto the opposing *sarissae* while others swept around them to carry on the fight.[19] This indicates that just grabbing the pike alone would not take it out of action. Indeed, the nature of massed combat would make it almost impossible for such an opponent to detach the front half of the pike. If the pike-phalanx was itself advancing, the forward momentum of this movement would drive the tip of the pike into an opponent (either his shield or body) with a result that anyone holding onto the pike would be simultaneously pushed back, the lines of his formation would compress, and he would not have sufficient room to step backwards and pull the front off the *sarissa*. Similarly, if the opposing formation was advancing against a static phalanx, once the tip of the pike was pressed into the opponent's shield, the forward impetus of his own lines would also compress the formation and limit his ability to separate the two halves of the *sarissa*. This suggests that, at Pydna, both sides were somewhat static or moving only slowly – despite Plutarch's statement that the initial Macedonian advance was conducted quickly.[20] Even if the two lines were only marginally separated, if a grasping opponent possessed enough room to step back and try to dislodge the front end of the *sarissa*, all the phalangite had to do was to thrust the weapon forward into the shield of the opponent, to negate this action and prevent the separation from happening.

Additionally, any force being brought to bear against the tip of the *sarissa* would force the two halves of the *sarissa* together within the connecting tube. This would make it even harder for someone to try and separate the two halves. Furthermore, in order to gain enough grip and leverage on the shaft of a *sarissa* that was pressed into his shield to pull it apart, an opponent who had grasped the weapon would have needed to turn his body and use his body mass behind his pull on the shaft. Doing so would result in the opponent exposing a side of his body to the other pikes of the phalanx. This would place the opponent at considerable risk and seems an unlikely practice except in the most extreme circumstances. This again means that the conditions where an opponent could attempt to pull the front end off a *sarissa* on the battlefield would have been rare.

The only time where the *sarissa* would have been at risk of separating would have been if the phalangite had delivered an offensive blow at some distance within the *effective range* of the weapon and the tip somehow became stuck in the opponent. If this occurred, when the phalangite tried to retract the weapon, the rear half which he was holding may dislodge from the forward half if it was securely lodged in the enemy's body and the grip of the connecting tube on the shaft was not sufficient. This partially explains why the diameter of the connecting tube is slightly smaller than the diameter of the shaft – the elasticity of the metal provided the tube with more grip on the weapon. This additionally suggests that the primary function of the *sarissa* was to be pressed into an opponent's shield, where it would not easily become lodged, in order to keep that opponent at a safe distance. In turn, the very nature of phalangite combat suggests that a *sarissa* that came in two sections was not only possible, but would have been just as effective as one with a single solid shaft.

When confronted by cavalry, chariots and even elephants, the pike-phalanx faced a similar dilemma. If the momentum of the beast did not simply snap the *sarissa* at the moment of impact, or force it backwards injuring the man in the rank behind, the tip of the pike could become lodged in the body of the animal. To counter such threats, the units of the phalanx may have moved out of their way rather than engage them directly. At Gaugamela, Darius sent hundreds of scythe-bearing chariots at Alexander's phalanx in an attempt to disrupt the formation.[21] Both Arrian and Diodorus state that to counter this attack the Macedonians opened their lines to allow the chariots to pass through into a 'killing zone' behind them.[22] Connolly suggests that this opening of the lines was conducted by having whole *syntagmae* move to the right into new positions behind adjacent units.[23] While this would create a sixteen metre wide opening for the chariots to charge into as Connolly notes, it seems unlikely that such a manoeuvre could be undertaken. In order to effectively draw the Persian chariots into these gaps, they would have to appear almost instantaneously otherwise the drivers would simply change their course and direct their vehicles at a different part of the Macedonian line where they could inflict casualties. Even if the members of the *syntagmae* were still holding their pikes vertically, rather than presenting them lowered to engage the chariots directly, it would still take considerable time to move a unit the size of the *syntagma* out of the way of a charging chariot and to a new position behind the front line.

This makes the moving of entire units to create gaps to counter a chariot attack unlikely.[24] The ancient texts do contain references to a pike-phalanx 'opening its files' to create gaps for lightly armed skirmishers to retreat through. However, this is more likely the moving of the members of alternate files into the intervals of the neighbouring file (in effect taking it from intermediate order to close order front-to-back). To undertake such a movement would also require the phalangites to have their pikes raised vertically. Appian describes how the pike-phalanx of Antiochus at Thermopylae in 191BC opened its files to allow skirmishers to withdraw, the formation then reformed into an intermediate order (*puknon*) and then, once in position, lowered their pikes for battle.[25] This shows that the normal deployment for battle was the intermediate order, that the opening of the files had to therefore be a move into a close order, and that the pikes were held in a vertical position until the intermediate order formation was re-established. 'Opening' into a close order by having every alternate file merge with the one to its right would create 48cm gaps between each file which skirmishers could pass down before the files resumed their original positions. However, these gaps would not be large enough to accommodate a charging chariot. Curtius, interestingly, makes no mention of this opening of the lines at Gaugamela and states that the pike-phalanx stood its ground looking 'like a rampart...creating an unbroken line of spears'.[26] This would suggest that the phalanx did not divide itself but engaged any chariots that made their way through the screening units of skirmishers positioned ahead of them.

Lucian's reference to the Thracians going down on one knee to brace a lengthy pike in position to engage cavalry raises the possibility that Alexander's front ranks did something similar at Gaugamela.[27] However, Curtius' statement that the phalanx created an 'unbroken line of spears' seems to more closely fit with their standard arrangement of standing upright with the pikes of the first five ranks lowered.[28] If this was how the pike-phalanx met Darius' chariots at Gaugamela, then it seems safe to assume that the collision of the horses and chariots with the line of infantry would have caused some *sarissae* to snap, others to be forced out of the hands of the men wielding them, and others to impale the charging horses. This would have caused great disruption to the line – an effective, but costly, tactic.

As if to confirm such events, the narratives state that on the left wing, Parmenion's troops suffered badly due to cavalry assaults.[29] We are not told whether the beleaguered troops in question were cavalry, pike-infantry, or both, but Parmenion was in overall command of the entire left wing, including units of the pike-phalanx, and the all-out attack by the Persians suggests that troops of both types were pressed hard – although Parmenion does save the situation with his Thessalian cavalry. If the Persian cavalry did attack the pike units on the Macedonian left then, as would have happened in the centre, much of the phalanx may have been disrupted by the mounted charge which greatly diminished the integrity of the line.

The other thing to consider is whether cavalry or chariot drivers would even attack a formation bristling with extended pikes. In 479BC at the battle of Plataea, a contingent of Phocian hoplites adopted a close-order formation to resist a charge by Persian cavalry. This formation would have also presented a wall of extending spears. Recognising the strength of this formation, Herodotus tells us the Persian cavalry turned and withdrew rather than engage.[30] Africanus also states that hoplites could easily repulse an attack of cavalry 'with their spears'.[31] It seems that, if the mounted Persian troops at Gaugamela had possessed a clear view of what they were charging at, they would have similarly not engaged a pike-phalanx with a frontal assault. Diodorus says that there was a great amount of dust being kicked up on the plain of Gaugamela and that Darius used this dust to hide his escape.[32] If this dust was blowing into the 'no man's land' between the lines at Gaugamela – i.e. the wind was blowing the dust towards the Persians and away from the Macedonians – as Diodorus' account of Darius using it to screen his movements would suggest, then it is possible that the Persian chariots and cavalry may not have even seen what they were charging at until it was too late to alter their course. Additionally, such impaired vision would make it unlikely that the units of the pike-phalanx would have had enough time to open their lines to create gaps to allow the chariots to pass through (as suggested by Arrian and Diodorus), and they would have more likely engaged the mounted troops directly as they emerged charging from the dust – which would then correlate with the account of Curtius.[33] Heckel, Willekes and Wrightson suggest that the 'lane' that the Persian cavalry charged into at Gaugamela was actually the gap that had formed in the phalanx by the separation of the units of Meleager and Simmias and that Arrian (or his sources) was using Xenophon's account of the battle of Cunaxa to account for how the Persians had managed to get their mounted troops through the phalanx.[34] If this is the case then the 'lanes' were not actually created but were 'accidental'. It also seems that, if the line could be maintained, the pike-phalanx could hold its own against a mounted charge and may have even been terrifying enough (if clearly visible) to force one to back down.

At the Hydaspes in 326BC, Alexander's army faced contingents of chariots as well as elephants – the first time they had fully engaged such beasts.[35] Unlike at Gaugamela, the terrain at the Hydaspes is reported to have been muddy due to a recent downpour and so there would have been little dust impeding anyone's vision.[36] In the opening engagement, Alexander's troops, after crossing the river, were attacked by advance units of the Indian army which included a contingent of chariots.[37] Supposedly recounting a record of the battle written by Alexander himself, Plutarch states that Alexander expected to be attacked by Indian cavalry.[38] Arrian states that one of Alexander's intentions was to fight a delaying action against this force with his own cavalry to allow his infantry to move up.[39] This would suggest that Alexander thought that his infantry would be able to effectively engage whatever units Porus sent against him. Yet most of this advance guard of the Indian army was taken out using cavalry and skirmishers rather than the pike-phalanx.[40]

For the main engagement, Porus had positioned both elephants and chariots across the front of his line.[41] The phalanx was ordered not to advance into action until the enemy wings had been disrupted (another example of the timing requirements of the use of the hammer and anvil tactic).[42] According to Curtius, despite the trumpeting of the elephants unsettling the members of the pike-phalanx, the beasts seem not to have troubled Alexander. The king is said to have claimed 'our spears are long and sturdy; they can never serve us better than against these elephants and their drivers – dislodge the drivers and slay the beasts'.[43] This would suggest that Alexander believed that the pike-phalanx could effectively engage such animals. Plutarch, on the other hand, states that Alexander recognized the threat of the Indian elephants and so chose to mount a flank attack against the enemy using his cavalry.[44]

Alexander's flank attacks are said to have taken out most of the Indian chariots and the phalanx then advanced in the centre.[45] Reports vary as to how, if at all, the pike-phalanx engaged the elephants. Diodorus says the phalangites only engaged the units of Indian infantry positioned between the beasts.[46] Other reports state that as the phalanx advanced, the elephants were the targets of attacks by skirmishers who showered the animals with missiles and, when they got close, hacked at their legs and trunks with axes and swords.[47] Curtius says that the whole phalanx (by which he means the entire infantry line) succeeded in punching through the Indian forces in a single advance – which must have involved some direct confrontation between the pike-phalanx and the elephants.[48] Regardless of which account is accepted, it seems clear that Alexander's pike-phalanx could, at least partially, hold its own against cavalry, chariots and even elephants.

For the most part, cavalry is used to either attack other cavalry or to deliver flank attacks against enemy infantry (as per the actions at Gaugamela and the Hydaspes). Rarely is the pike-phalanx used to counter an attack by mounted troops. Despite this, Aelian's military manual contains a number of chapters devoted to the different type of cavalry formations that could be used against the pike-phalanx and the various infantry formations that are best employed to counter them.[49] Importantly, all of the formations detailed by Aelian would require substantial time to adopt and could not have been done on the spur of the moment. Aelian also does not describe opening the files as a tactic that could be employed to counter mounted troops or chariots. This suggests that the chariot charge at Gaugamela was screened by the dust that had been kicked up on the battlefield and, therefore, the pike-phalanx probably did engage, and resist, them directly (as per Curtius) as both sides would have had little time to react.

The capacity for the pike-phalanx to easily resist a cavalry charge seems confirmed when the use of cavalry in the age of the Successors is examined. Snodgrass outlines how the ratio of infantry to cavalry greatly increased from the time of Alexander to the Age of the Successors, providing the following figures as examples (Table 19):

Period/Encounter	Ratio of Infantry to Cavalry
Age of Alexander	2:1
Battle of Sellasia (222BC)	Antigonus Doson - 8:1
Battle of Raphia (217BC)	Antiochus - 5:1 Ptolemy IV - 5-10:1
Battle of Cynoscephalae (197BC)	Philip V - 8:1

Table 19: Examples of the ratio of infantry to cavalry
in the times of Alexander and the Successors.[50]

The reduction in the reliance upon cavalry is a clear demonstration of changes in the type of opponent that was being fought from the time of Alexander into the time of the Successors. During Alexander's conquest of Asia, the Persian and Indian forces that his army primarily engaged were strong in cavalry and, while numerous in infantry, that infantry was not armed in a manner which made them effective against the pike-phalanx. Consequently, using the pike-phalanx to both engage an opponent and pin their centre in place while the strong contingents of Macedonian cavalry struck the flanks was a sound tactic.

However, by the time of the Successors, the main infantry opponent was another pike-phalanx rather than enemies wielding short reach weapons such as those Alexander had faced. Thus the fundamentals of strategy, and the composition of armies, seem to have changed. Had the pike-phalanx been unable to stand up to cavalry, one would expect to see a decrease, rather than an increase, in the ratio of infantry to cavalry within Hellenistic armies. Instead, the amount of cavalry declines. This suggests that cavalry was not considered to be an effective force for use against an opposing contingent of pikemen except when attacking them from the flank.

Rather than solely rely on cavalry as the shock arm of an army, warfare in the Age of the Successors saw the development of other technologies for war to give one side an advantage on the battlefield. This accounts for the increase in the length of the *sarissa* from 12 cubits (576cm) in the time of Alexander to 16 cubits (768cm) at the end of the fourth century BC. If two pike-phalanxes were engaged against each other, with both carrying weapons of the same length, then neither side would hold an advantage over the other in terms of the reach of their combatants. Two forces equipped in such a way would simply pin each other in place and the outcome of the battle would rely more on the morale of the two sides, the depth of opposing formations (as per Sellasia), the success of any flanking attack, or the simple luck of the day. If, on the other hand, one side wielded lengthier weapons, that side would be able to both pin their opponent in place, and deliver killing blows if the opportunities presented themselves, while remaining outside of the range of the shorter weapons of their enemy, much in the same way that Philip's phalanx held an advantage over Greek hoplites, or that of Alexander over the Persians.

Additionally, due to the way that both hands are employed to wield the *sarissa*, as the left arm extends as the weapon is thrust forward, the shield, strapped to the left forearm, is moved out of its defensive position across the front of the body. While this does not pose any threat when the phalangite is engaged against an opponent bearing a short-reach weapon like a sword or spear, it would place him at considerable risk when engaged against another pike-phalanx when both sides were bearing weapons of the same length. Such actions would have been very limited and only been undertaken when a certain killing blow could be delivered.

Furthermore, due to the way that the phalangite shield was securely mounted to the left forearm, and because both hands were employed to wield the *sarissa*, should a phalangite find his shield pinned in place by an opponent bearing a longer pike, that phalangite would have no way to thrust his weapon forward in an offensive manner even if an opportune target presented itself. In effect, using a longer pike and pressing its tip into the shield of an opposing phalangite rendered that enemy offensively impotent – incapable of any action like advancing or attacking – and would force him to simply stand his ground until he was slain either by a killing blow delivered by the longer weapons of the enemy, or mowed down by cavalry attacking from the flanks. This not only accounts for the increases in the length of the *sarissa*, but also highlights that the primary use of this weapon across the Hellenistic Age was to simply pin the opposing side in place.

Mounted troops were still a main strike weapon against pike-wielding phalangites during the wars of the Successors, as is demonstrated by the battles of Paraetacene, Gabiene, Gaza, Asculum, Raphia and Magnesia where cavalry were used to deliver flanking attacks against enemy formations. However, the changed nature of combat during this time meant that the opportunities where such attacks could be delivered were more limited, or had to be created in a different manner, to the ways they had been in the time of Alexander.

At battles such as Gaugamela and the Hydaspes, Alexander had used the oblique movements of his right wing of the cavalry to force part of the enemy (usually an opposing force of cavalry) to mirror his actions which resulted in the creation of an exploitable gap in the opposing line. The large, infantry-based armies of the Successor Period, on the other hand, were less likely to fragment in such a way due to their lesser mobility and, in many of the encounters of this time, the infantry line remains in place while cavalry are only used to initially engage other cavalry. Additionally, the traditional hammer and anvil tactics, and especially those employed by Alexander, would have been well known to most Hellenistic commanders of the age and any attempt to create a gap via an oblique advance of cavalry on either wing would have been anticipated. Consequently, in order to open gaps in the enemy line, other avenues of attack had to be employed. Thus there is an increase in the use of elephants in the Successor Period – whose charge was used to either smash an opposing wing and create a gap for cavalry to charge into (Gabiene, Raphia and Magnesia), to deliver sweeping attacks against the flank of an enemy line (Heraclea), or to make frontal charges against enemy infantry in an attempt to break the formation apart (Gaza, Ipsus and the 'elephant victory').

Julius Africans paints a vivid picture of the impact of a charge of elephants and the terror that this may cause:

...they make a terrifying spectacle at first sight: both to horses which are unaccustomed to them, and to men as well; and when [the enemy] equipped them with a tower, they considered them a source of fear, a sort of rampart advancing before the battle-line. Their trumpeting is piercing and their charge irresistible...It was...a multiple vision of military superiority: volleys of many missiles being cast from above by men with the advantage [of the tower], the area of the elephant's feet being impregnable, and the enemy even fleeing far away. The battle was not one fought on an equal footing; against the elephant a siege operation would have to be conducted. When the front ranks were broken [by the elephants], as always happened, the troops turned and their ranks, shattered by the enemy, were primed for total destruction...Who could stand up to an avalanche from the collapse of a cliff-side? An elephant in combat makes the impression of a mountain: it overturns, it throws down, it smashes, it slaughters, and it does not spurn anyone who is lying in its way...It grabs horse, man and chariot with its trunk, strikes them violently, turns them upside down, and drags them to its own feet. By leaning on them with its knees, it squashes them – aware of its own weight – made even heavier by the addition of towers.[51]

The use of elephants in such ways created a new shock arm for many Hellenistic armies. Diodorus states that a contingent of elephants in the army of Polyperchon in 318BC had 'a fighting spirit and a momentum of body that was irresistible'.[52] One of the reasons given for why Alexander's troops were reluctant to advance any further than the Hyphasis River in 326BC was the report that the kingdoms on the far side were said to have possessed thousands of elephants.[53] This demonstrates the impact that such beasts could have on the psyche of even seasoned professionals.

The threat of elephants thundering into infantry and breaking the formation apart seems to have been of considerable concern to some commanders who employed novel ways to counter the beasts. At Megalopolis in 318BC, Damis had wooden boards studded with sharp nails projecting upwards buried in shallow trenches near the entrance to the city. When the elephants of Polyperchon advanced into the breach, the animals stepped on the nails and, according to Diodorus, were in such pain that they could neither advance nor retreat while their drivers, and some of the beasts, were brought down by missiles fired from the flanks.[54] At Gaza in 312BC, Ptolemy deployed a series of spikes (χάραξ) chained together which were strung across the front of his right-wing infantry in the hope that these traps would foul any charge made by the opposing elephants of Demetrius.[55] This tactic worked. Diodorus describes how many of the beasts were wounded by treading on the spikes (he states that the softness of their feet is an elephant's one true liability) and caused the attack to collapse in disorder.[56] Africanus similarly states that iron caltrops, when stepped upon, bring the elephant to a halt by penetrating into the soles of their feet.[57] At Asculum in 279BC, the Roman consul Publius Cornelius Mus employed a different method to counter enemy elephants: 300 ox-drawn 'anti-elephant' wagons. Dionysius of Halicarnassus describes these contraptions as being equipped with:

> upright beams upon which were mounted movable traverse poles that could be swung round as quickly as thought in any direction that one may wish and, on the ends of these poles, there were either tridents or sword-like spikes or scythes – all of iron; or again they had cranes that hurled down heavy grappling-irons. Many of the poles had attached to them, projecting in front of the wagons, fire-bearing grapnels covered in tow that had been liberally smothered with pitch, which men standing on the wagons were to set fire to as soon as they came near the elephants and then rain blows with them upon the trunks and faces of the beasts. Furthermore, standing on the wagons, which were four-wheeled, were many light troops: bowmen, stone-throwers and slingers who threw iron caltrops...[58]

These wagons initially held back the on-rush of the elephants as the battle began until both the oxen and the men controlling the wagons were brought down by light troops.[59] The use of such devices to inhibit an attack that would inflict the kind of carnage that Africanus describes shows just how powerful the elephant arm of a Hellenistic army had become and also the dominant role that such animals played during the wars of the Successor Period.[60]

While cavalry seems to have been relegated to a lesser role than it had before, its use in flanking attacks to try and drive the enemy from their position, once an exploitable gap had been created, ensured that they still held a valuable position on the battlefield. Regardless, almost all commanders of the Hellenistic Age still relied on a massive pike-phalanx as the basis for their army – the only formation which had the ability to stand up to opposing infantry armed in a similar manner.

However, when the pike-phalanx began to confront the legions of Rome in the third century BC it, yet again, faced an enemy who fought in a completely different style. The legionaries of the Roman Republic were able to put up strong resistance to, and even gain victories over, the Macedonian phalanx at battles such as Heraclea, Asculum, Cynoscephalae and Pydna.[61] How this was accomplished was a matter of considerable examination even in ancient times. Polybius, for example, devotes a digressionary section of his histories to discussing how the Romans were able to overcome the pike-phalanx, while Africanus devotes a similarly sized section to comparing the armaments and fighting styles of the Greeks, Macedonians, Persians and Romans.

According to Polybius, the Romans formed their lines with an interval of three Greek feet (τρισί ποσί) of free space on either side of each man, and from front to back, in order to provide him with enough room to fight.[62] Thus each space occupied by the Roman legionary was six Greek feet across. Each Greek foot equated to about 32cm; meaning that an interval of six Greek feet equalled 192cm – the same as the open order for phalangite formations detailed in the military manuals. As if in confirmation of this, Polybius goes on to state that when adopting this order of spacing, each Roman faced two files of the pike-phalanx (which each used the intermediate-order interval of 96cm per man) and, as a result of this difference in interval each legionary in the front rank of the Roman formation would have to face ten *sarissae* (Fig.59).[63]

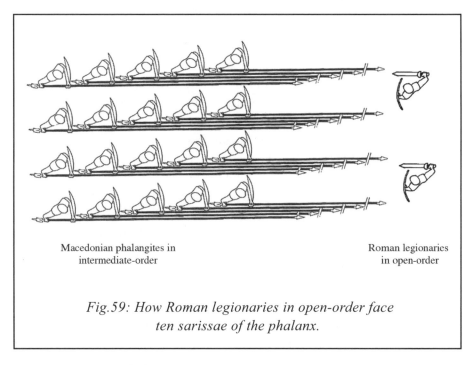

Macedonian phalangites in
intermediate-order

Roman legionaries
in open-order

*Fig.59: How Roman legionaries in open-order face
ten sarissae of the phalanx.*

The Roman legionary, armed with the short, *gladius*-type sword, had a far shorter *effective reach* due to the size of his weapon and, like both the Persians and the Greek hoplite, would have similarly been unable to reach an opposing phalangite when the *sarissa* was pressed into his shield – just as Plutarch says happened at the battle of Pydna.[64] How could such an open formation have ever stood a chance against the pike-phalanx?

At Heraclea and Asculum, Pyrrhus' pike-phalanx gained costly victories over the legions of Rome.[65] The battles of Cynoscephalae and Pydna resulted in Roman victories – although these were close-run encounters. Indeed, Polybius remarks that some found it incredible that the pike-phalanx could be defeated at all (accounting for which became the purpose of the digression in his histories).[66]

The advantage of the Roman legionary lay with his mobility and adaptability. According to Polybius, the Hellenistic phalangite was less suited to besieging enemy positions, being besieged, marching and encamping in open country and to meet unexpected attacks.[67] Similarly, Africanus states that legionaries:

> have an advantage in agility, ready for both attacks and withdrawals, and they are swifter both in taking high ground and in the use of the sword... And they themselves are also trained in every technique of close-combat so that, in knowledge, both they and the Greeks are evenly matched, but in the lightness of their equipment they have the advantage.[68]

Another area in which Polybius says the Roman legion was superior to the Hellenistic pike-phalanx was in its method of deployment. Polybius states that one way to defeat a pike-phalanx was to engage it with one part of your army while reserve units that had originally been posted behind the front lines could sweep around and either attack the phalanx from the side, or charge into gaps that had opened in the line.[69] This bears a close resemblance to the hammer and anvil tactics employed by Alexander and the Successors. The main difference is that in Polybius' strategy both the frontal assault and the flanking attack are delivered by infantry as opposed to the infantry anvil and cavalry hammer of Hellenistic warfare. It was here that the mobility of the Roman legionary really came into play.

Such manoeuvres could not have been conducted by pike-infantry alone due to their cumbersome equipment. This explains why it was more mobile cavalry, and later elephants, which delivered decisive blows against enemy formations in Hellenistic warfare. The Roman legionary, on the other hand, was an all-purpose soldier – capable, as Polybius and Africanus state, of rapid movement in most directions. Livy claims that, where the Macedonian pike-phalanx lacked mobility and formed a single line, the Roman army was more elastic – being made up of numerous contingents – which could act individually or in concert as required.[70] However, it must be noted that, due to the use of units and sub-units within the pike-phalanx, Livy's claim is not entirely accurate.

Despite this apparent elasticity, in a frontal clash, the Roman legionary was hopelessly outclassed by the pike-phalanx. The long *sarissae* could be pressed into his shield which would keep him at bay with little offensive recourse at his disposal. Africanus additionally states that the Roman legionary shield (the *scutum*) was less effective in closed ranks as the soldier carrying it was unable to push against it with his shoulder due to the way that the grip was configured.[71] Livy, on the other hand, states that the *scutum* provided better protection than the Macedonian *peltē*.[72] The inability to exert counter pressure with the Roman *scutum* as noted by Africanus partially explains why Plutarch states that, at the battle of Pydna, the Romans could not break the Macedonian line head on, and that the Romans resorted to various means to render the pikes facing them unserviceable – including hacking at them with their swords, trying to push them with their shields, or grasping them in their hands.[73]

What the Roman legionary needed was to get to closer quarters with his opponent, where his larger shield and short sword could be used to the most effect. Africanus' statement that the necks of the Greeks were particularly vulnerable to Roman swords suggests that close-quarters battle was where the Roman legionary was in his element and that they regularly got close enough to an enemy to inflict such injuries.[74] Similarly, Livy's reference to the Greek horror at the wounds inflicted by the *gladius* in 200BC highlight just how formidable a close-quarters weapon the Roman short sword was.[75]

In order to be in a position to inflict such injuries, when facing a pike-phalanx, legionaries needed to attempt the removal of the *sarissa* from the combative equation (as they tried to do at Pydna) in order to get closer to their opponent, or to attack from the flanks or into gaps in the opposing formation (as Polybius suggests). Polybius additionally states that one of the advantages of the legionary was that he was also effective when fighting as an individual.[76] Africanus, on the other hand, states that no Roman soldier fights on his own.[77] While this initially seems contradictory to the statement made by Polybius, both claims can be reconciled in the ability of the Roman legionary, in close-quarters combat, to fight somewhat singularly while, at the same time, still being part of a unit.

Additionally if, due to the variables of combat, the opposing Roman and Macedonian formations aligned so that each legionary was directly opposite each alternate file of the pike-phalanx (in effect only facing five *sarissae* rather than ten) this still poses little problem to the legionaries, but could cause considerable trouble for the phalanx. Even though every second file of the pike-phalanx in this scenario would not have an opponent standing directly in front of it, these unopposed files could not continue to advance lest they move forward, between the legionaries, and into a position where they could be attacked at close quarters from the sides with little chance of retaliation, and while compromising the integrity of the phalanx as a whole at the same time. Thus every second file of the phalanx is essentially nullified by the Roman open-order without having to actually engage it. The Macedonian phalanx, on the other hand, could not adopt a similar open-order deployment so that each file of phalangites faced a single file of legionaries as this would create avenues in the lines which the more mobile Romans would have been able to exploit. Thus, no matter how the enemy was deployed, the security and integrity of the pike-phalanx lay in the adoption and maintenance of an intermediate-order spacing.

Consequently, by having the Romans fight in an open order which the pike-phalanx was not able to exploit, fewer Romans were placed in a position where they could potentially be killed by pike thrusts delivered by the front rank of the phalanx. This, in turn, meant that Roman formations could be wider, and have more men placed in reserve (just as Polybius advises) as they were not needed to hold the phalanx in place. Furthermore, by standing further apart, the Roman legionaries also had enough room to use their shield to parry or deflect the pikes they were facing into the vacant gaps between their files. This could also provide many of them with the opportunity to move inside the reach of the front rank weapons and then, using the same procedure, move further ahead to where they could attack the phalangites from a close distance – again, with little chance of retaliation. It is unlikely that the Romans adopted such an open formation when engaging against other more mobile opponents like Gauls or Germans, and Polybius' reference to the open order of their deployment in his comparison to the Macedonian style of fighting must be something specifically undertaken by the Romans in order to negate many of the advantages of the pike-phalanx.

Another advantage that both Polybius and Africanus fail to comment on is that the Roman legionary also carried several javelins. Polybius himself, in an earlier section of his work, outlines how the members of the Roman infantry carried a light and heavy missile weapon (the *pilum*) some 3 cubits (144cm) in length.[78] Livy, who does mention the *pilum* in his comparison of the Roman and Macedonian fighting styles, claims that the javelin was 'a much more effective weapon than the spear'.[79] As two opposing sides closed with one another, volleys of javelins would be launched from the Roman lines. The room required to cast such missiles also partially accounts for the openness of the Roman order of battle as described by Polybius. These missiles would slay many members of the opposing front ranks of an enemy formation like a pike-phalanx and, as the phalanx continued to advance over the bodies of their fallen comrades, the phalanx would begin to break apart just prior to any closer contact.

Such weapons would have dire consequences for an advancing pike-phalanx. Polybius states that the rear ranks of the phalanx held their pikes at an angle over the heads of the men in front to protect them from incoming missile fire.[80] However, this would only work against missiles such as arrows, fired on a high curving trajectory from a large distance, which would rain down and then become entangled in the angled pikes. Javelins, on the other hand, were cast at a much shorter range and on a much flatter trajectory. Consequently, the front ranks of the pike-phalanx were only protected by their shields and armour and, due to the weight of the *sarissa* which they were carrying in both hands, could not quickly or easily raise a shield to receive or deflect an incoming *pilum*. This meant that the front ranks of an advancing phalanx were particularly vulnerable to the main Roman missile weapon.

Many years after Polybius, Julius Caesar would comment on the effectiveness of the Roman javelin by stating that '...several of [the enemy's] shields could be pierced and pinned together by a single *pilum*...'[81] If this was as true in the Hellenistic Age as it was in 58BC when Caesar wrote this passage, then the members of the front rank of an advancing pike-phalanx, unable to deflect an incoming missile with their shield due to the other elements of their panoply, would have been considerably susceptible to injury or death from the opening volleys of Roman javelins. The removal of many of the front rank of the phalanx (where all of its experienced officers were positioned), and the partial disruption of the line caused by volleys of javelins, would have therefore 'softened' the phalanx and aided the Roman legion engaging with it. However, due to the depth of the phalanx, these volleys of missiles would not break the formation or destroy it completely and, by the time the javelins had found their mark, a hand-to-hand encounter between phalanx and legion was all but inevitable.

Skarmintzos suggests that the triumph of Rome over the Hellenistic pike-phalanx was more the result of Rome's generals than the direct superiority of the legion and that Roman commanders adapted their tactics to overcome the advantages held by the phalangite.[82] While partially true, these adapted tactics would not have been possible without the mobility of the legion as described by the likes of Polybius and Africanus. Tarn, on the other hand, blames the failure of the phalanx on its negligence to protect its flanks – a practice he says the Macedonians began to neglect during their pike-phalanx versus pike-phalanx encounters with the Spartans at the end of the third century BC.[83] Again, while partially correct, it was really only the mobility of the Roman legion which could exploit this, and other, vulnerable areas of the pike-phalanx.

When strengthened by a united front, and with its wings protected, the phalanx was almost unstoppable so long as the cohesion of the line was retained. The use of semi-independent units and sub-units meant that, rather than being a solid, rigid line, the pike-phalanx possessed a level of tactical flexibility and adaptability which allowed it to perform on almost any terrain. The formation's weaknesses lie in its inability to turn and face new threats, the gaps that would invariably form in the line, and the inadequacy of the phalangite in individual combat should those gaps be exploited. It was this that ultimately led to the phalanx's downfall. The utilisation of a strong but more mobile form of fighting saw many of the advantages of the pike-phalanx lost and witnessed the demise of the pike-phalanx from its place of dominance on the battlefields of antiquity.

Chapter Fourteen

The Legacy of the Hellenistic Pike-Phalanx

Diodorus declared that 'Macedonian spears had conquered Asia and Europe'.[1] The organization of Hellenistic pike formations into units and sub-units, combined with an elaborate and symmetrically distributed command structure, made the pike-phalanx a very ordered instrument of war – one that was mutually supporting, offensively and defensively strong, and adaptable to the varied tactical requirements of the ancient battlefield. Phalangite formations, for example, could be deployed to different depths, with the most common being that of sixteen ranks deep, but could also be arranged in deeper or shallower configurations depending upon the terrain, the size of the opposing force, the decisions of those in command and the situation of the day.

Similarly, the interval that each man and file occupied within the pike-phalanx could also vary depending upon what action was to be performed. The 48cm close-order interval, on the one hand, was only adopted by a pike formation to undertake manoeuvres such as wheeling as the more compressed nature of this close interval meant that the formation had less distance to travel as it changed direction. Importantly, in this order, the pikes of the phalanx had to be held vertically due to its compact nature and, as such, the close-order formation could not be used offensively. The intermediate-order interval of 96cm per man, on the other hand, provided enough space for the weapons held by the first five men of each file to be lowered for combat. In contrast, the open-order interval of 192cm per man was too open to be effective in combat and would have only been used by armies on the march as the larger amount of room between each man facilitated ease of movement while carrying a panoply of equipment in excess of 21kg, including a long pike.

The length of the *sarissa* meant that the majority of combat in the Hellenistic Age (at least initially) was conducted at 'pike length' with the opposing sides separated by several metres. This was due to the way in which the *sarissa* was employed to hold an opponent at bay, and gave the Hellenistic pike-phalanx a great tactical advantage over opponents who carried weapons with a much shorter *effective reach*. If an opportune target did present itself, it was the members of the front ranks of the phalanx – the only ones with the freedom of movement to actually thrust their weapon forward – who would have been able to exploit it. For the other members of the phalanx, their role was to cover the gaps between each file, to engage prone opponents and to provide density and reserves to the formation as a whole while the enemy was held back.

The ability to engage an enemy from a safe distance, one where that opponent could not reciprocate offensively, allowed the pike-phalanx to effectively pin an enemy formation in place as part of the standard tactic of pike-phalanx combat – that of the hammer and anvil. Across the entire Hellenistic Period this tactic of using the pike-phalanx in the centre to hold an enemy formation (or part thereof) in position while other, more mobile, troops swept around to strike from the flanks remained unchanged. This in itself demonstrates the integral part that the pike-phalanx played in the conflicts of this time period.

While the functionality of the phalangite remained little changed across the Hellenistic Period, the tools and strategies of war constantly developed in attempts to overcome the advantages held by the pike-phalanx, particularly during the time of the Successors following the death of Alexander the Great in 323BC when pike-phalanx fought pike-phalanx and a decisive tactical advantage of some kind was required to secure victory. Consequently the *sarissa* itself became longer to outreach an enemy armed in a similar fashion, and beasts such as elephants were more regularly employed to try and smash an opposing formation apart. Many of these tactical 'experiments' met with mixed results; the pike could only be increased in length until a point was reached where it was almost impossible to wield, and contingents of elephants could be countered with other elephants, a solid wall of pikes, missile troops or other measures.

As a result, the phalangite, and the pike-phalanx, retained its position of supremacy on the battlefield.

Yet for all of its advantages, the pike-phalanx did have its limitations. This was mainly the potential for gaps to open in the line – due in part to the phalanx being comprised of semi-independent units and sub-units. If such gaps could be exploited by a more mobile opponent, one who could get inside the rows of projecting pikes, this was when the pike-phalanx was most vulnerable as the long *sarissa* made the phalangite a very ineffective individual combatant, and larger units were incapable of turning to meet threats from the sides once their pikes had been lowered. It was the ability of the Roman legionaries to take advantage of this inherent weakness in the Hellenistic pike-phalanx which ultimately led to the formation's demise.

Yet the defeat of Hellenistic pike-phalanxes by the Romans at Cynoscephalae in 197BC and then at Pydna in 168BC did not see the total removal of the Macedonian way of war from the battlefields of antiquity. In 86BC the legionaries of Rome fought against the pike-phalanx (and other troops) of the Pontic king Mithridates at the second battle of Chaeronea. Much like the phalangites of nearly three centuries earlier, Plutarch describes the Pontic phalangites as being equipped with bronze shields and pikes.[2] During the battle, and in actions reminiscent of the engagement at Pydna, the Romans are said to have cast their javelins, drawn their swords, and 'struggled to push aside the pikes so that they could get to close quarters as soon as possible'.[3] As if to confirm that very little change had taken place in both the use of the pike-phalanx and the Roman methods of countering it, the Pontic army was defeated by independent Roman units pressing their advantage along different parts of the line as the pike-phalanx began to break up. The second battle of Chaeronea was the last real offensive action that a Hellenistic-style pike-phalanx would undertake. As has been noted, by a strange quirk of fate, the location of this encounter was very close to the battlefield where, in 338BC, Philip II fought against the Greeks with the new Macedonian army and ushered in the Hellenistic Age and the rise to dominance of the pike-phalanx for almost the next century and a half.[4] Three hundred years after the first battle of Chaeronea things had now come full circle and the demise of this style of warfare would occur in almost the exact same place.

Even this defeat was not the last time the ancient world saw a Macedonian-style phalanx. In AD66 the Roman emperor Nero enrolled recruits into 'The Phalanx of Alexander the Great' for an intended campaign against the Parthians.[5] This expedition never occurred and the 'phalanx' does not seem to have seen any action. It is also uncertain exactly how these troops were armed. Bennett and Roberts, for example, suggest that they were actually equipped as standard Roman legionaries.[6] There are, however, clues as to their equipment which suggest otherwise. As well as a Parthian campaign, Nero had also planned expeditions to Greece and Alexandria.[7] Both of these locations had strong connections to Alexander and Nero may have been entertaining dreams of attempting to emulate the former conqueror.[8] Additionally, the Parthians were the descendants of the Persians and, if Nero was attempting to emulate Alexander, he may have thought that the best way to defeat the Parthians was with the very instrument that Alexander had used to defeat Persia: a pike-phalanx.

In AD217 the emperor Caracalla, also thinking himself to be the 'New Alexander', similarly created a phalanx for a Parthian campaign. Cassius Dio states:

> Caracalla was so enthusiastic about Alexander [the Great] that he used certain weapons and cups that he believed had once been his, and he also set up many likenesses of him in both army camps and in Rome. He organized a phalanx, composed entirely of Macedonians, 16,000 strong, named it 'Alexander's Phalanx', and equipped it with the arms that warriors had used in his day. These consisted of a helmet of raw oxhide, a three-ply linen cuirass, a bronze shield, long spear, short spear, high boots and sword. Not even this, however, satisfied him, but he had to call his hero 'the Augustus of the East'; and once he actually wrote to the senate that Alexander had come to life again in the person of the Augustus, and that he might live on once more in him...[9]

Herodian states:

> At times we saw ridiculous portraits [of Caracalla], statues with one body which had on each side of a single head the faces of Alexander and the emperor. Caracalla himself went about in Macedonian dress, especially the broad sun hat and short boots. He enrolled picked youths into a unit which he called his Macedonian phalanx; its officers bore the names of Alexander's generals.[10]

Cowan suggests that the short spear mentioned by Cassius Dio as part of the panoply for these new phalangites could have been a javelin.[11] This would then find parallels with the 'regular' Macedonian equipment carried by Horratus/Coragus in his heroic duel with the Athenian Dioxippus in 326BC as recounted by Curtius and Diodorus.[12] Such equipment would then supply a close connection between a description of one of Alexander's troops and those of Caracalla who was trying to emulate them. Furthermore, the number of men in Caracalla's 'phalanx' (16,000) bears close similarities with the size of the numerically perfect phalanx (16,348) that is outlined in military manuals such as that of Aelian which, although written a century before the reign of Caracalla, reported to be writing about the army of Alexander the Great.[13]

Cassius Dio reports seeing the training of Caracalla's phalangites in Nicomedia.[14] However it is still uncertain exactly how such troops were armed. An epitaph from Apamea, for example, refers to a trainee *'phalangarius'* from Legio II Parthica which had been raised by Septimius Severus in AD197.[15] However, the term 'phalangarius' has two meanings – both of which have military connotations. The first is a reference to a member of the phalanx. Yet many Greek writers of the middle empire use the term phalanx to refer to a Roman battle line without specifically meaning that the men within it were armed as pikemen.[16] The other definition refers to someone who carries equipment on a long pole.[17] This last definition bears many similarities to the 'Marian Mules' of the late Roman Republic who had to carry their personal equiment suspended from a long pole slung over their shoulder – an operational aspect of the Roman army that was kept in place at least into the second century AD.[18] Consequently, the *phalangarii* of Legio II Parthica could simply be troops within the standard battle line, or troops porting their own equipment (or both), but they may not have necessarily been armed as pikemen. Cowan suggests that these *phalangarii* may have been the cause for the brief resurgence of the spear-wielding *triarii* around AD300.[19] If so, then the *phalangarii* may have been spearmen rather than pikemen.

Additionally, reports of the Romans fighting in Parthia suggest that they may not have been using long pikes. Both Herodian and Africanus state that the Roman weapons were not long enough to counter the Parthian cavalry lances.[20] This again would suggest that the *phalangarii* may have been spearmen rather than pikemen.

It seems that the Macedonian-style pike-phalanx itself had disappeared from the battlefields of the Roman Empire, but its prestige, titulature and even some of the materials that its shields had been faced with remained. The achievements of Alexander the Great also remained as a benchmark which Roman emperors attempted to surpass. In the third century AD for example, Severus Alexander:

> made every effort to appear worthy of his name and even to surpass the Macedonian king, and he used to say that there should be a great difference between a Roman and a Macedonian Alexander. Finally, he provided himself with soldiers armed with silver shields and with golden [shields], and also a phalanx of 30,000 men, whom he ordered to be called phalangarii, and with these he won many victories in Persia. This phalanx, as a matter of fact, was formed from six legions, and was armed like the other troops, but after the Persian wars received higher pay.[21]

This is a clear demonstration that the soldiers of Severus' army, while imitating many aspects of the army of Alexander the Great, were armed in their standard fashion rather than as pikemen. As such, if the term *phalangarii* is not just an anachronistic appellation, it would seem that these troops may have been carting their own equipment which would then warrant such a title.

However, the Macedonian way of war also continued to provide inspiration following the fall of Rome. The Byzantine emperor Leo IV (reigned AD866-912), for example, incorporated much of the content of Aelian's manual on the Hellenistic pike-phalanx into his own version of the *Tactica* in the late ninth century AD. An Arabic version of Aelian was also published around 1350.[22] While the Byzantines and the Muslims did not put the principles of the Hellenistic pike-phalanx into widespread practice, the fact that these texts were copied so extensively in Europe and the Middle East attests to the pride of place that an understanding of the Macedonian way of war had in the development of later military institutions.

More than three centuries later, the large European pike and musket armies of the sixteenth and seventeenth centuries would be heavily modelled on the Macedonian system and the works of Asclepiodotus, Aelian and Arrian became popular once again as instructional manuals for the reigning monarchs of the time. This would be the last time an army armed and organized (if only in part) in the Macedonian manner would fight on the world stage.

With the defeat of the Pontic army of Mithridates by the Romans at the second battle of Chaeronea in 86BC the use of the pike-phalanx, at least one that would have seemed familiar to Philip, Alexander and the

Successors, came to an end. The pike-phalanx was defeated once and for all by an opponent who employed and transmitted a very different way of fighting – one that, as with all developments in the arts and technologies of war, outclassed that which had come before it. Yet in an ironic turn of fate, the pike-phalanx was eclipsed by a culture which still held the principles of the army of Alexander the Great in high regard and which, through the continued emulation and study of it, left the ideals of the Hellenistic pike-phalanx as a lasting legacy which directly influenced western warfare for centuries to come. This, more than anything else, highlights the important part that the 'invincible beast' of the Macedonian pike-phalanx has had on shaping the world as we know it.

Notes

Preface

1   Livy (37.42) goes out of his way to specifically state that this was the name of the long Macedonian pike.
2   *Excerpta Polyaeni*, 18.4
3   Livy, 44.41
4   Plut. *Alex*. 33
5   Plut. *Aem*. 19
6   Diod. Sic. 17.4.4
7   M.M. Markle, 'The Macedonian *Sarissa*, Spear and Related Armour' *AJA* 81.3 (1977) 323
8   J. Pietrykowski, *Great Battles of the Hellenistic World* (Pen & Sword, Barnsley, 2009) viii
9   Xen. *Cyr*. 2.3.9-10
10 Polyb. 12.25e
11 A.M. Snodgrass, *Arms and Armour of the Greeks* (Johns Hopkins University Press, Baltimore, 1999) 114
12 N. Sekunda, 'Land Forces' in P. Sabin, H. Van Wees and M. Whitby (eds.), *The Cambridge History of Greek and Roman Warfare Vol. I: Greece, the Hellenistic World and the Rise of Rome* (Cambridge University Press, 2007) 330
13 A. B. Bosworth, *Conquest and Empire: The Reign of Alexander the Great* (Cambridge University Press, 2008) 10
14 C.G. Thomas, 'What you Seek is Here: Alexander the Great' *Journal of the Historical Society* 7.1 (2007) 61
15 Bosworth, *Conquest and Empire*, 10
16 Griffith (*Mercenaries of the Hellenistic World* (Cambridge University Press, 1935) 322) outlines one of the other problems that researchers encounter when using the archaeological record when he states that 'there is a certain body of archaeological evidence throwing some light upon the arms and accoutrements of Hellenistic soldiers, but it is both difficult to collect (much of it is unpublished), and of a highly specialised interest'.
17 S. English (*The Field Campaigns of Alexander the Great* (Pen & Sword, Barnsley, 2011) 16) states that 'history is very far from an exact science and we can ultimately do no better than make our best efforts in any ancient reconstruction'. However physical re-creation takes us a step closer (almost literally in many regards) to achieving a fuller understanding of an ancient form of combat by placing us in the shoes (or sandals) of the warriors of the past. Such processes then fill many of the gaps left in models based solely on analysis of the traditional source material.
18 For other examples of the use of physical re-creation as a means of investigating aspects of ancient warfare see: C.A. Matthew, *A Storm of Spears: Understanding the Greek Hoplite at War* (Pen & Sword, Barnsley, 2012); Markle, 'The Macedonian *Sarissa*', 323-339; R.E. Dickinson, 'Length Isn't Everything – Use of the Macedonian *Sarissa* in the Time of Alexander the Great' *JBT* 3:3 (2000) 51-62; W. Donlan and J. Thompson, 'The Charge at Marathon: Herodotus 6.112' *Classical Journal* 71:4 (1976) 339-343; P. Connolly 'Experiments with the *Sarissa* – the Macedonian Pike and Cavalry Lance – a Functional View' *JRMES* 11 (2000) 103-112; W.B. Griffiths, 'Re-enactment as Research: Towards a Set of Guidelines for Re-enactors and Academics' *JRMES* 11 (2000) 135-139; N.G.L Hammond, 'Training in the Use of a *Sarissa* and its Effects in Battle 359-333 B.C.' *Antichthon* 14 (1980) 53
19 Ael. Tact. *Tact*. 2, 8
20 Asclep. *Tact*. 2.7
21 Hoplites: see: Polyb. 18.29; phalangites: Polyb. 4.12, 18.28; *peltasts*: Polyb. 5.23; *sarissaphoroi*; see: Polyb. 12.20; Arr. *Anab*. 1.14.1; Livy 36.18
22 Diod. Sic. 18.2.3
23 Plut. *Mor*. 197c

24 πεζοί : Diod. Sic. 19.14.5-8; πυκνότητα: Diod. Sic. 16.3.2
25 Much like the word *peltast* used by many Greek writers, Livy seems to use the term *caetrati* interchangeably to mean either 'skirmisher' or 'pikeman'. Indeed Livy (31.36) himself defines the *caetrati* as 'what the Greeks call *peltasts*'. At 35.27 Livy describes *caetrati* as clearly being skirmishers armed with 'slings, darts and other light arms' (*cum fundis et iaculis et alio levi genere armaturae*). Yet in other places the actions of the *caetrati* seem to be the actions of pikemen (for example see: Livy 33.15-16, 37.40, 44.41; Plut. *Aem.* 19; see also: Pietrykowski, *Great Battles* 230, n.169; Snodgrass, *Arms and Armour*, 123; F.W. Walbank, *Philip V of Macedon* (Cambridge University Press, 1940) 291-293; M.B. Hatzopoulos, *L'organization de l'armee macédonienne sous les Antigonides* (Kentron Hellēnikēs kai Rōmaïkēs Archaiotētos, Ethnikon Hidryma Ereunōn, Athens, 2001) 71-72; E. Foulon, 'Hypaspistes, *peltastes*, chrysaspides, argyraspides, chalcaspides' REA 98 (1996) 53-62; E. Foulon, 'La garde à pied, corps d'élite de la phalange hellénistique' BAGB 1 (1996) 17-31. The *caetrati* are often associated with the pike-phalanx. Yet Polybius also comments on the inflexibility of the pike-phalanx (18.31-32) and how the soldier in the phalanx is of no use in either small detachments or individually (18.32). This then suggests that the *caetrati* were armed in a manner that was better suited to massed formation fighting – i.e. as pikemen. Walbank (Philip V, 292-293) suggests that while these *peltasts*/*caetrati* were pikemen, they were more lightly armed than the regular phalangites. Sekunda (*Macedonian Armies after Alexander 323-168BC* (Osprey, Oxford, 2012) 33) suggests that they may have been carrying shorter *sarissae*.
26 This also occurs in many modern works, which contain generalized statements such as 'the phalanx presented a serried array of spear points', without considering that the men of the phalanx did not actually carry spears.

Who Invented the Pike Phalanx?

1 Marathon: Hdt. 6.117; Thermopylae: Hdt. 8.24; Plataea: Hdt. 9.70
2 For example see the accounts of the battle of Coronea (394BC): Xen. *Hell.* 4.3.17-19; Plut. *Ages.* 18.
3 For example see the accounts of the defeat of the Spartans by Athenian light troops at Sphacteria (425BC – Thuc. 4.33-40) and Lechaeum (390BC – Xen. *Hell.* 4.5.11-17) where a whole *morai* of Spartans was defeated by light troops; for light troops being able to engage hoplites at a distance see: Xen. *Hell.* 4.8.33-39; Xen. *Anab.* 6.3.7-9; for light troops being able to easily flee from hoplites see: Thuc. 4.34; Xen. *Anab.* 6.3.4.
4 Just. *Epit.* 8.1.1
5 Harpocration, Lexicon s.v. *pezhetairoi* (Anaximenes – Jacoby *FrGrHist.* 72 F4) – Ἀναξιμένες ἐν ᾇ Φιλιππικῶν περὶ Ἀλεξάνδρου λέγων φησίν. ἔπειτα τοὺς μὲν ἐνδοξοτάτους ἱππεύειν συνεθίσας ἑταίρους προσηγόρευσε, τοὺς δὲ πλείστους καὶ τοὺς πέζους εἰς λόχους καὶ δεκάδας καὶ τὰς ἄλλας ἀρχὰς δειλὼν πεζεταίρους ὠνόμασεν, ὅπως ἑκάτεροι μετέχοντες τῆς βασιλικῆς ἑταιρίας προθυμότατοι διατελῶσιν ὄντες; on Anaximenes' history of Philip and Alexander and his other interactions with them see; Diog. Laert. 5.11; Paus. 6.18.2
6 Ael. Tact. *Tact.* 3
7 Dem. 2.17
8 Theopompus – *FrGrHist* 115 F348 - Θεόπομπός φησιν ὅτι ἐκ πάντων τῶν Μακεδόνων ἐπίλεκτοι οἱ μέγιστοι καὶ ἰσχυρότατοι ἐδορυφόρουν τὸν βασιλέα καὶ ἐκαλοῦντο πεζέταιροι; see also: Photius s.v. πεζέταιροι; Etym. Magn. 699.50-51 s.v. Πεζέταιρος; Gabriel (*Philip II of Macedonia – Greater than Alexander* (Washington, Potomac, 2010) 107) suggests that a passage of Frontinus (Strat. 2.3.2) also refers to Philip and the *pezhetairoi*. However, Frontinus only refers to Philip's 'strongest men' (*fortissimus suorum*) and not the *pezhetairoi* specifically by name.
9 P. Connolly, *Greece and Rome at War* (Greenhill Books, London, 1998) 69; see also: N. Sekunda, *The Army of Alexander the Great* (Osprey, Oxford, 1999) 28; M. Thompson, *Granicus 334BC: Alexander's First Persian Victory* (Oxford, Osprey, 2007) 25; A.M. Snodgrass, *Arms and Armour of the Greeks* (Johns Hopkins University Press, Baltimore, 1999) 115; J. Pietrykowski, *Great Battles of the Hellenistic World* (Pen & Sword, Barnsley, 2009) 32; B. Bennett and M. Roberts, *The Wars of Alexander's Successors 323-281BC Vol.II: Battles and Tactics* (Pen & Sword, Barnsley, 2009) 4; A.B. Bosworth, *Conquest and Empire: The Reign of Alexander the Great* (Cambridge University Press, 2008) 259
10 J. Warry, Alexander 334-323BC (Osprey, Oxford, 1991) 14; see also: A. Erskine, 'The πεζέταιροι of Philip II and Alexander III' *Historia* 38.4 (1989) 386
11 W. Heckel, *The Wars of Alexander the Great 336-323BC* (Osprey, Oxford, 2002) 24; see also: W. Heckel and

R. Jones, *Macedonian Warrior: Alexander's Elite Infantryman* (Osprey, Oxford, 2006) 31; W. Heckel, *The Conquests of Alexander the Great* (Cambridge University Press, 2008) 25; E.M. Anson, 'The Hypaspists: Macedonia's Professional Citizen-Soldiers' *Historia* (1985) 247

12 Gabriel, *Philip II*, 54; it is uncertain whether Gabriel then assumes that part of the army was first called the *pezhetairoi* by the undesignated Alexander of Anaximenes and the term was then applied to the whole phalanx by Philip, or whether, contrary to Anaximenes, Gabriel believes that the term *pezhetairoi* was first used by Philip. Tarn (*Alexander the Great Vol.II* (Chicago, Ares, 1981) 141) similarly suggests that it was Philip II who bestowed the name *pezhetairoi* on the infantry of the phalanx.

13 Erskine, 'πεζέταιροι', 392-393

14 Erskine, 'πεζέταιροι', 393; Griffith ('Philip as a General and the Macedonian Army' in M.B. Hatzopoulos and L.D. Loukopoulos (eds.), *Philip of Macedon* (Ekdotike Athenon, Athens, 1980) 58) suggests that Alexander II created the *hypaspists* around 370BC. Similarly, Gabriel (Philip II, 71) says that the bodyguard of Philip were also the *hypaspists*. See also: Anson, 'Hypaspists', 246. The validity of the passage of Anaximenes' which refers to the mysterious 'Alexander' as the creator of the *pezhetairoi* has been the cause of contention for a considerable period of time, as has been the value of Anaximenes himself as a reliable source. As early as the first century BC Dionysius of Halicarnassus (Is 19) described Anaximenes as 'feeble and unconvincing...good at many things but master of none'. Pausanias (6.18.5) also levelled criticisims at Anaximenes' credibility. Such comments have carried on to the present day. For a discussion of the nature of Anaximenes' passage and the creation of the *pezhetairoi* see: A.M. Anson, 'Philip II and the Creation of the *Pezhetairoi*' in P. Wheatley and R. Hannah (eds.), *Alexander and His Successors – Essays from the Antipodes* (Claremont, Regina Books, 2009) 88-98

15 Arr. *Anab.* 1.28.3; for a commentary on this passage see: A.B. Bosworth, *A Historical Commentary on Arrian's History of Alexander* (Clarendon Books, Oxford, 1998) 170-171. Bosworth, in this work at least, also suggests that Anaximenes is referring to Alexander I in the role of mythologised lawgiver. In a later work ('The Argeads and the Phalanx' in E. Carney and D. Ogden (eds.), *Philip II and Alexander the Great: Father and Son, Lives and Afterlives* (Oxford University Press, 2010) 99), Bosworth alternatively offers that the most likely candidate for the Alexander of Anaximenes is Alexander II.

16 Arr. *Anab.* 1.28.7

17 Arr. *Anab.* 7.2.1; for a brief commentary on this passage see: Erskine, 'πεζέταιροι', 392-393

18 Arr. *Anab.* 7.11.3

19 For the convincing argument that the *pezhetairoi* were infantry units raised in Lower Macedonia while the *asthetairoi* were units from Upper Macedonia see: A.B. Bosworth, 'ΑΣΘΕΤΑΙΡΟΙ' *CQ* New Series 23.2 (1973) 245-253

20 Arrian (*Anab.* 3.11.8) describes the arrangement of the Companion Cavalry at Gaugamela as: '[The] right wing was held by the Companion Cavalry, the Royal Squadron in front; commanded by Cleitus, son of Dropides. In successive order came those of Glaucias, Aristo, Sopolis...Heraclides...Demetrius, Meleager and lastly that led by Hegelochus...The Companion Cavalry as a whole was commanded by Philotus, son of Parmenion.'

21 P.A. Brunt, 'Anaximenes and King Alexander I of Macedon' *JHS* 96 (1976) 151

22 Brunt, 'Anaximenes and King Alexander I', 151

23 Brunt, 'Anaximenes and King Alexander I', 151

24 Thuc. 2.100

25 N.G.L. Hammond, 'Training in the Use of a *Sarissa* and its Effects in Battle 359-333BC' *Antichthon* 14 (1980) 54

26 S. English, *The Army of Alexander the Great* (Pen & Sword, Barnsley, 2009) 4

27 Hdt. 6.44-45

28 Thuc. 1.62

29 Thuc. 2.100

30 Xen. *Hell.* 5.2.38

31 Polyb. 29.21

32 Thuc. 2.100

33 Just. *Epit.* 8.4.2

34 For example see: Brunt, 'Anaximenes and King Alexander I', 153; Bosworth, *Commentary on Arrian's History*, 170

35 R. Develin, 'Anaximenes ("F Gr Hist" 72) F 4' *Historia* 34.4 (1985) 494

36 Brunt, 'Anaximenes and King Alexander I', 151

37 Diodorus (16.2.5) says that Perdiccas III lost 4,000 men in battle with the Illyrians in 359BC while the remainder, an amount unfortunately not numbered, lost all heart for further confrontation. This reference to surviving members of the army shows that, at this time, the Macedonian military was larger than the 4,000 men who were lost. This further suggests that the loss was considerable, possibly half of Macedon's forces, but was not 'total'. Later, Philip II was able to field 10,000 men (Diod. Sic. 16.4.3) – a number which Fuller suggests may have been the size of the Macedonian army as early as 364BC; see: J.F.C. Fuller, *The Generalship of Alexander the Great* (Wordsworth, Hertfordshire, 1998) 26. English (*Army of Alexander*, 72), on the other hand, suggests that Macedon was militarily weak prior to the time of Philip II. How an army possibly 10,000 strong could be considered 'weak' is not explained.

38 Brunt, 'Anaximenes and King Alexander I' 153; English, *Army of Alexander*, 4

39 N.G.L. Hammond, *Alexander the Great: King, Commander and Statesman* (Duckworth, London, 1980) 26; N.G.L. Hammond, 'The Various Guards of Philip II and Alexander III' *Historia* 40.4 (1991) 404

40 Gabriel, *Philip II*, 61

41 A. Erskine, 'The πεζέταιροι of Philip II and Alexander III' *Historia* 38.4 (1989) 391

42 Plut. Pelop. 25

43 Develin, 'Anaximenes' 494; Hammond, *King, Commander and Statesman*, 25

44 These reforms were the adoption of the *aquila* (eagle) as the dominant standard of the Roman military, the adoption of the cohort as the basic tactical unit of the legions, and the creation of 'Marius Mules' by which each legionary was required to carry his own equipment and rations. These reforms followed on from the 'head count' reform of 107BC in which Marius opened up enlistment in the legions to volunteers. See: C.A. Matthew, *On the Wings of Eagles: The Reforms of Gaius Marius and the Creation of Rome's First Professional Soldiers* (Cambridge Scholars, Newcastle upon Tyne, 2010)

45 Bosworth, 'The Argeads' 97-98

46 Diodorus and Cornelius Nepos are the only writers to give detailed accounts of the Iphicratean reforms. Xenophon's *Hellenica* omits any reference to the year 374BC at all. Book six, chapter two of the *Hellenica* begins with the Peace of 375BC and then moves straight on to Iphicrates' campaign against Corcyra in 373BC. As such, the details of Iphicrates' reforms are not covered. Additionally, Xenophon's *Hellenica* is noticeably light on coverage of the events in Macedon at this time. Despite covering the history of the Greek world down to 362BC, there is no mention of the reign of Amyntas in the *Hellenica* after the events of 382BC, Amyntas' death in 370BC is not mentioned, nor is the entire reign of Alexander II or the rise of Perdiccas III. Consequently, Xenophon's *Hellenica* is of limited value for the examination of the rise of the Macedonian state in the mid-fourth century BC. Both Polyaenus and Frontinus, despite devoting a considerable number of passages to Iphicrates, do not mention his reforms either. However, it must be considered that both of these writers were compiling works devoted to the art of strategy and so a discussion of the invention of a particular type of fighting may not have been seen as warranted. Webster ('In Search of the Iphikratean Peltast/Hoplite' *Slingshot* 287 (2013) 5-6) dates the reforms to 373BC or later despite Diodorus clearly dating them to 374BC.

47 Nepos, *Iphicrates*, 1.3-4.

48 Diod. Sic. 15.44.1-4; Ueda-Sarson ('The Evolution of Hellenistic Infantry Part 2: Infantry of the Successors' *Slingshot* 223 (2002) 24) suggests that the shield of the reformed Iphicratean *peltast* was oval. This may come from how Diodorus describes the new shield as 'symmetrical' (συμμέτρους). While an oval shield is symmetrical, so too is a perfectly round one. Sekunda ('Land Forces' in P. Sabin, H. van Wees and M. Whitby, *The Cambridge History of Greek and Roman Warfare Vol. I: Greece, the Hellenistic world and the rise of Rome* (Cambridge University Press, 2007) 327) suggests that συμμέτρους means 'of the same size' and that the *peltē* was the same size as the hoplite *aspis* but without its distinctive offset rim. This seems unlikely as the bearing of a shield 90cm in diameter would not allow the *sarissa* (or the Iphicratean pike) to be held in both hands.

49 L. Ueda-Sarson, 'The Evolution of Hellenistic Infantry Part 1: The Reforms of Iphicrates' *Slingshot* 222 (2002) 30-31

50 Fuller, *Generalship of Alexander*, 42

51 In the years following his reforms, Iphicrates embarked on a naval campaign against the island of Corcyra (see: Xen. *Hell.* 6.2.10-39). It may be the maritime nature of this campaign that has led Ueda-Sarson to conclude that the reforms altered the equipment of the Athenian marines.

52 C.A. Matthew, 'When Push comes to Shove: What was the *Othismos* of Hoplite Combat?' *Historia* 58.4 (2009) 400.

53 Champion (*Pyrrhus of Epirus* (Pen & Sword, Barnsley, 2009) 24) says that Iphicrates increased the spear

carried by 'light troops' (rather than hoplites as is specifically stated in the ancient texts), and that it was only increased from a length of 3m (300cm – which is too big for a hoplite spear) to 4m (400cm – which is too short for a weapon that has been doubled in length – or even increased by half (see n.54)). Such a conclusion clearly does not correlate with the details given by Nepos and Diodorus.

54 Arr. *Tact.* 12.7; Markle ('Use of the *Sarissa* by Philip and Alexander of Macedon' *AJA* 82.4 (1978) 487) says the spear was only lengthened to 12 feet (3.6m); see also J. Warry, *Warfare in the Classical World*, (University of Oklahoma Press, 1995) 67. Similarly, Heckel and Jones (*Macedonian Warrior*, 10) and Sekunda ('Military Forces' 327) also states that the spear was only increased in length to 1½ times that of the hoplite spear. Ueda-Sarson ('Reforms of Iphicrates' 30) suggests that Nepos says the spear was doubled in length while Diodorus says it was only increased by half and that the difference may have been due to the two writers using different source material. Such claims may be, in part, due to the terminology used by Diodorus. The word used by Diodorus to describe the size of the new weapon of the Iphicratean reforms is ἡμιολίῳ (from ἡμιόλιος) meaning 'one and a half' or 'half as much' which, following Warry's, Markle's, Sekunda's and Ueda-Sarson's line of thought, would result in a new weapon 382cm in length. However, the term ἡμιολίῳ can also mean 'as large again' (i.e. doubled) which would then reconcile what Diodorus says with the description of the Iphicratean reforms given by Nepos.

55 M.M. Markle, 'The Macedonian *Sarissa*, Spear and Related Armour' *AJA* 81.3 (1977) 323.

56 For example see: Arr. *Anab.* 1.27.8, 1.28.4

57 As with many things relating to the warfare of this time, there are many different scholarly opinions about how both the Thracian and Iphicratean *peltasts* were armed as well. Gabriel (*Philip II*, 183), for example, states that Thracian *peltasts* (rather than Iphicratean *peltasts*) carried the small *peltē* and a spear longer than that of traditional Greek hoplites. It is uncertain if this conclusion is a confused interpretation of the description of the armament of the Iphicratean *peltast* given by Diodorus and Nepos – although Lucian (*Dial. mort.* 699.1-3) does describe a Thracian armed with a *peltē* and *sarissa*. Parke (*Greek Mercenary Soldiers* (Ares, Chicago, 1981) 79-80) suggests that Iphicratean *peltasts* still functioned as hoplites. English (*Army of Alexander*, 24, 69-70) and Pietrykowski (*Great Battles*, 15) imply that the Iphicratean *peltasts* were armed with spears or javelins, that they wore little armour and that they still operated as light troops but with standardized equipment. English then goes on to say (p.135) that both the *hypaspists* and the *pezhetairoi* were essentially *peltasts* with little armour and a *peltē*. Many aspects of these interpretations clearly go against the descriptions of the Iphicratean *peltast*, the *hypaspists* and the *pezhetairoi* that are given in the ancient sources. English (p.69) also criticises the works of Diodorus by stating that he shows a 'serious lack of understanding of the military situation of the day' because, according to English, Diodorus fails to recognize the existence of *peltasts* prior to the time of Iphicrates. Similarly, Everson (*Warfare in Ancient Greece* (Sutton Publishing, Stroud, 2004) 172) suggests that passages such as those of Diodorus are full of errors because 'the first clear misconception is that hoplites became *peltasts*' and that what Diodorus was referring to were skirmishers. Despite the transformation of hoplites into *peltasts* being exactly what Diodorus and Nepos state, such conclusions are clearly mistaken. *Peltasts*, in the role of a skirmisher, had existed for some time prior to Iphicrates. What Diodorus is commenting on, and what makes the reforms of Iphicrates so noteworthy, is that the soldiers resultant from the reform were not skirmishers but pikemen who were also armed with a small shield. Despite the errors in their conclusions (see n.53 and n.54 above) Champion (*Pyrrhus*, 24) still calls the Iphicratean *peltast* the 'prototype of the Macedonian phalangite', while Everson (*Warfare*, 173) says that the lengthened spear may have been the forerunner to the *sarissa*.

58 Sekunda, *Army of Alexander*, 27; R. Sheppard (ed.), *Alexander the Great at War* (Osprey, Oxford, 2008) 54; D. Head, 'The Thracian *Sarissa*' *Slingshot* 214 (2001) 10-13

59 Didymus, *On Dem.* 13.3-7; see also: Just. *Epit.* 9.3.1-3; Dem. 18.67; Plutarch (*Mor.* 331B) calls the weapon a *longche* which may be another name for the *sarissa* (see page 59).

60 Lucian, *Dial. mort.* 439-440 - ...ὑποστὰς δὲ ὁ Θρᾷξ τῇ πέλτῃ μὲν ὑποδὺς ἀποσείεται τοῦ Ἀρσάκου τὸν κοντόν, ὑποθεὶς δὲ τὴν σάρισαν αὐτόν τε διαπείρει καὶ τὸν ἵππον.

61 J.G.P. Best, *Thracian Peltasts and their Influence on Greek Warfare* (Wolters-Noordhoff, Groningen, 1969) 3-16, 102-110, 139-142, 144

62 Xen. *Anab.* 4.7.15-16

63 See the Dillery's note to Xenophon's passage in the Loeb edition of the *Anabasis*.

64 See Warner's note to Xenophon's passage in the Penguin edition of the *Anabasis* (called *The Persian Expedition*).

65 Ueda-Sarson, 'The Reforms of Iphicrates', 31

66 On the Thracians using long spears see: Lucian, *Dial. mort.* 439-440; Plut. *Mor.* 331b; on the arms and

armour of the Egyptians see: Hdt. 2.182, 3.47, 7.89; Xen. *Cyr.* 6.2.10, 7.1.33.

67 Ael. Tact. *Tact.* 2; Ueda-Sarson ('Infantry of the Successors', 24) suggests that the reason why the deployment and operation of *peltasts* is not covered as a separate issue in all of the military manuals is that they were functioning as phalangites – for which the manuals go into great detail about their drill, functions and formations.

68 Polyb. 5.23, 10.31, 10.49; Ueda-Sarson ('Infantry of the Successors', 24) connects this unit of 10,000 *peltasts* with the 'Silver Shields' – a unit of phalangites.

69 Ael. Tact. *Tact.* 2; Tarn (*Hellenistic Military and Naval Developments* (Cambridge University Press, 1930) 6) seems to interpret Aelian as referring to traditional Greek hoplites and *peltast* skirmishers rather than to hoplites and phalangites.

70 Aldrete, Bartell and Aldrete (*Reconstructing Ancient Linen Armour – Unraveling the Linothorax Debate* (The Johns Hopkins University Press, Baltimore, 2013) 22, 46-47) note that natural linen varies in colour from dirty grey to brown, but that one method of bleaching the fabric was to rub it with natron or potash. This may also be another origin of the term *argilos*. For a detailed discussion of the variety of colours and designs found on Greek examples of linen armour see: Aldrete, Bartell and Aldrete, *Reconstructing Ancient Linen Armour*, 47-56. The weight of a set of linen armour would depend upon a number of factors including how thickly it was made, whether it was constructed entirely of linen or was a 'composite' of a leather core with a linen facing and backing, whether the armour possesses flaps (*pteruges*) to protect the groin, if the armour was reinforced with metal scales and so on. Replica *linothorakes* made by Aldrete, Bartell and Aldrete for their experiments, constructed solely from layered linen glued together, made to fit a person slightly larger than the average ancient Greek weighed around 4kg (see: Aldrete, Bartell and Aldrete, *Reconstructing Ancient Linen Armour*, 146).

71 Bardunias ('The Linothorax Debate: A Response' *Ancient Warfare* 5.1 (2011) 4-5) notes how a covering of clay would actually absorb water and so various gums and glues would be needed to be used as a sealant to make the armour water resistant. It should also be noted that Theophrastus (*Lap.* 62-64) states how certain kaolins, such as Melian, Tymphaic or Samian earth, were used in the bleaching of textiles. Consequently, the term *argilos* may also be a reference to the materials used to bleach the cloth used in the construction of the *linothorax* (see also the commentary on page 208-213 of Caley and Richard's edition of Theophrastus). Bardunias further states that armour would be much more effective if, rather than applying a layer of clay to the outside, it was instead worked into the weave which would then make the armour both pliable to slow impacts, but able to present a resistant surface to fast impacts such as would be received from a stabbing weapon like a spear or sword.

72 Diod. Sic. 16.93.4

73 For the two one-handed combative postures that the Greek hoplite employed to wield their spear see: Matthew, '*Othismos*' 400-405; Everson (*Warfare*, 174) states that, even if the spear was increased in length so that it had to have been carried with both hands, he finds it unlikely that Iphicrates' *peltasts* would have used it in this manner as it would have reduced their mobility. What Everson has failed to appreciate is that the troops being reformed were not highly mobile skirmishers but heavily armed pikemen and that the creation of a pike-phalanx did, in fact, reduce the mobility of the formation as Everson suggests. How he thinks the Iphicratean *peltast* did use a lengthened spear, if not in a two-handed manner, Everson does not elaborate on other than suggesting (p.174) that the sources may be incorrect in their description and that the spear was only lengthened by half, to about 12ft in length, which then could have possibly been wielded in one hand. However, such an interpretation of the source material seems unlikely (see n.54 above).

74 Plutarch (*Cleom.* 11) states how the Spartan king Cleomenes taught his men to 'use a long pike, held in both hands, instead of a short spear'. Tarn (*Military and Naval Developments*, 28) suggests that both hands were only needed to use the pikes of the second century BC when they had increased to 14 cubits (or 672cm) in length. Clearly, this statement cannot be considered accurate.

75 This was partially dependent upon the type of wood that was used to fashion the core of the shield. Pliny the Elder (*HN* 16.209) states that the best woods for making shields were poplar and willow. Krentz ('Hoplite Hell: How Hoplites Fought' in D. Kagan and G.F. Viggiano, *Men of Bronze – Hoplite Warfare in Ancient Greece* (Princeton University Press, 2013) 136) notes that a shield made from poplar or willow would weigh half as much as one made of oak, and two-thirds to three-quarters the weight of a shield made of pine.

76 See: Matthew, *Storm of Spears*, 39-44

77 See: Matthew, *Storm of Spears*, 39-44

78 Ueda-Sarson ('The Reforms of Iphicrates', 30) suggests that the lengthy spear of the Iphicratean reforms was used to partially offset the reduction in protection brought about via the adoption of the *peltē*.

79 This is similar to (albeit a two-handed version of) the 'low position' – one of several combative postures adopted by the Classical Greek hoplite to fight – which similarly placed little muscular stress on the muscles of the arm. See: Matthew, *A Storm of Spears*, 113-130

80 D. Featherstone, *Warriors and Warfare in Ancient and Medieval Times* (Constable, London, 1997) 57; Ueda-Sarson, 'The Reforms of Iphicrates', 34

81 Asclep. *Tact.* 5.1; Theophr. *Caus.* pl. 3.12; Arr. *Tact.* 12.7

82 Best (*Thracian Peltasts*, 139) suggests that it was the Thracian *peltast*, and not the Iphicratean *peltast*, that bore all of the hallmarks of the Macedonian phalangite. Bests states '[Thracian] equipment is characterized by the long thrusting spear (*sarissa*) and the *pelte*.' This, at least, is how Lucian (*Dial. mort.* 439-440) describes the Thracians. However, for the other forms of Thracian *peltast*, while they did carry a small shield, their spear was not unlike that of the classical Greek hoplite, or even smaller javelins for use in skirmishing, rather than the long pike of the Iphicratean *peltast* or the later Macedonian phalangite. Best goes on to suggest (p.104) that the reforms of Iphicrates, as described by Diodorus and Nepos, never actually took place and that this 'proves that both authors had no idea of the military situation in Iphicrates' time'. Such a conclusion seems unlikely when the connection between the Iphicratean *peltast* and the Macedonian phalangite is considered. Ueda-Sarson ('The Reforms of Iphicrates', 33) also makes a possible connection between the Iphicratean *peltast* and the Macedonian phalangite. However, because he believes that the reformed spear was only lengthened by half (see n.53 and n.54), rather than being enlarged into a pike, he subsequently dismisses this conclusion.

83 Tarn, *Military and Naval Developments*, 43

84 Aeschin. 2.26-28

85 Aeschin. 2.29; Nepos, *Iphicrates*, 3.2; when Alexander II was assassinated by Ptolemy in 368BC, the crown passed to Philip's other brother, Perdiccas III. However, Perdiccas was too young to rule and Ptolemy acted as his regent until 365BC when Perdiccas had him executed and assumed sole rule. In 359BC Perdiccas attempted to conquer Upper Macedonia from the Illyrians, but the expedition was a disaster and Perdiccas was killed along with 4,000 of his troops (see: Diod. Sic. 16.2.4-5). The crown then passed to Perdiccas' son, Amyntas, but he too was too young to rule alone. Perdiccas' last surviving brother, Philip, acted as Amyntas' tutor and regent for only a short time before declaring himself king of Macedonia in the same year (359BC). Amyntas was not considered to be any threat to Philip and he was tied to Philip through a marriage to Philip's daughter Cynane. Hammond (*Alexander the Great King, Commander and Statesman*, 24-25) suggests that Amyntas, being from a more senior line of the Macedonian royal house, may have even been considered the possible successor of Philip. However, this all changed when Alexander III came to the throne in 336BC and promptly had Amyntas (who was not only his cousin but also his stepbrother) executed.

86 Markle, 'Use of the *Sarissa*' 486

87 Bosworth, 'The Argeads' 99; Bosworth also suggests (101-102) that the Macedonians under Alexander II may have fought as hoplites rather than pikemen. This would seem odd when the pike-phalanx of Iphicrates seems to have been adopted by Macedon around the same time, if not earlier. For Pelopidas in Macedon see: Diod. Sic. 15.67.4

88 Aldrete, Bartell and Aldrete, *Reconstructing Ancient Linen Armour*, 15

89 Bennett and Roberts, *The Wars of Alexander's Successors Vol.II*, 3; see also: Sekunda 'Land Forces', 329; J.K. Anderson, *Military Theory and Practice in the Age of Xenophon* (University of California Press, Berkeley, 1970) 131, 306; Bennett and Roberts also state that 'such closeness to the tactical reformers [as Philip had with Iphicrates] is required to explain the radical military thinking [of] Philip'. A closeness to a tactical reform would certainly help understand where the Macedonian pike-phalanx had come from, but this does not necessarily pertain to only Philip when many other Macedonian rulers are just as likely.

90 English, *Army of Alexander*, 69

91 Webster ('Iphikratean Peltast/Hoplite', 6) suggests that there was a now lost biography of Iphicrates which outlined his reforms in relation to drill, training and discipline as well as crediting Iphicrates with the reforms initiated by Philip of Macedon – a detail which then carried over into the works of Diodorus and Nepos who could have used this lost text as source material. This seems highly speculative and simply accepts that Philip was the creator of the pike-phalanx without considering much of the other available evidence.

92 Sekunda, 'Land Forces' 329; see also: Pietrykowski, *Great Battles*, 15; A.R. Burn, 'The Generalship of Alexander' *Greece & Rome* Second Series 12.2 (1965) 141; G. Cascarino, *Tecnica della Falange* (Rome, Il Cerchio, 2011) 21; B.A. Noguera (L'évolution de la phalange macédonienne: le cas de la sarisse' *Ancient*

Macedonia 6.2 (1999) 849-850 (839-850)) suggests that, for the Macedonians, the word 'sarisa' applied to all spears while, for the Greeks, it referred specifically to the Macedonian pike. The tenth century *Suda* (s.v. Σὰρισσα) merely calls the *sarissa* 'a long Macedonian spear' (δόρυ μακρὸν παρὰ Μακεδόσι) without providing any details of its origins or where the name of the weapon came from.

93 P.A. Rahe, 'The Annihilation of the Sacred Band at Chaeronea' *AJA* 85.1 (1981) 87; see also: Sheppard, *Alexander the Great at War*, 54; see also: Ueda-Sarson, 'The Reforms of Iphicrates', 33l; Rahe also incorrectly connects Diodorus' reference (16.3.1-2) to Philip using a formation in close-order with this adoption of the pike-phalanx.

94 Champion, *Pyrrhus*, 23-24

95 Diod. Sic. 15.44.3

96 Markle, 'Use of the *Sarissa*', 483-497

97 Heckel and Jones, *Macedonian Warrior*, 31; see also: Heckel, *The Wars of Alexander*, 24

98 Develin ('Anaximenes', 496) outlines how it is suggested that the *pezhetairoi*, who had acted as a bodyguard for Philip II according to Diodorus and Theopompus, were replaced by the *hypaspists* under Alexander III while the term *pezhetairoi* may have become synonymous with the rest of the phalanx (or at least part of it). Erskine ('πεζέταιροι', 390-391, 393-394), who suggests that the passage of Anaximenes could be consistent with a small unit of troops as much as it could be with the entire Macedonian phalanx, offers that the change occurred because, upon ascending to the throne, Alexander needed to replace Philip's bodyguard with a different unit because both he and his mother Olympias were implicated in Philip's assassination (see Just. *Epit*. 9.7.1-14) and, as such, his hold on power was somewhat tenuous and so Philip's *pezhetairoi* bodyguard was replaced with Alexander's *hypaspists*. Suspicion also seems to have fallen on Philip's guards and his closest associates (*P. Oxy*. 15.1798) and this would also account for Alexander needing to form a new contingent of guards.

99 English, *Army of Alexander*, 5; Strangely, English later states (p.27) that Alexander III introduced the term *pezhetairoi* which goes against his earlier statement.

100 Polyb. 12.20

101 Markle, 'Use of the *Sarissa*', 487

102 G.T. Griffith, 'Philip as a General and the Macedonian Army' in M.B. Hatzopoulos and L.D. Loukopoulos (eds.), *Philip of Macedon* (Heinemann, London, 1980) 58. Griffith goes on to suggest (p.59) that 'it was Philip presumably (although we are not told when) who introduced the *sarissa*, the very long pike which became eventually the standard weapon of the Macedonian infantry', yet also concedes (p.62) that 'our record of Philip's battles is a travesty, which does not allow us to see precisely in what ways the arrival of the *sarissa* changed the tactics of the infantry phalanx'.

103 Polyaenus, 2.38.2; see also: Diod. Sic. 16.35.2

104 Markle, 'Use of the *Sarissa*', 487; see also: Champion (Pyrrhus 24-25) who suggests they are all skirmishers. Hammond ('Training in the Use of a *Sarissa* and its Effects in Battle 359-333 B.C.' *Antichthon* 14 (1980) 60), on the other hand, states that Markle's belief that the phalangites were using missile weapons is 'irreconcilable'.

105 Arr. *Anab*, 3.18.3; Diod. Sic. 17.68.2-3; Curt. 5.3.17-23

106 Arr. *Anab*. 3.18.2

107 Arr. *Anab*. 1.1.9; Alexander had his men lay down and cover themselves with their shields so that the wagons would simply roll over the top of them. English (*The Field Campaigns of Alexander the Great* (Pen & Sword, Barnsley, 2011) 25) suggests that this may indicate that Alexander's phalangites had actually borrowed the larger hoplite *aspis* from the allied Greek and mercenary troops. This seems unlikely. Sheppard (*Alexander the Great at War*, 73), on the other hand, suggests that these men were *hypaspists* carrying the larger hoplite *aspis*.

108 Gabriel (*Philip II*, 129) suggests that Polyaenus does not use the term 'phalanx' in his description of the battle with Onomarchus but uses the term 'battle-line' instead. Gabriel is clearly mistaken.

109 Arr. *Anab*. 1.28.3

110 Arr. *Anab*. 3.12.3

111 Markle, 'Use of the *Sarissa*', 487

112 M. Andronicos, '*Sarissa*' *BCH* 94.1 (1970) 91-107

113 Warry (*Warfare*, 68), for example, states that 'Philip of Macedon was a man of many-sided genius...He created an army on a new model and used it in war with brilliant strategic and tactical ability'. Donvito ('Clad in Gold and Silver' *Ancient Warfare* V.6 (2012) 10) calls Philip one of the 'fathers' of the Hellenistic army in a clear assumption that he created this new style of fighting (the other named 'father' is Alexander

the Great). Hammond ('What May Philip Have Learnt as a Hostage in Thebes?' *GRBS* 38.4 (1997) 366) flatly states that 'the infantry were taught to use a new weapon and to fight in a new formation, both of which were Philip's inventions'. Connolly (*Greece and Rome*, 64) states that 'the rise of Macedon was due almost entirely to one man, Philip II. On his accession to the throne in 359BC he set about building up the most formidable fighting machine the world had yet seen.' For other examples of models which offer Philip as the creator of the pike-phalanx see: Tarn, Alexander the Great Vol.II, 140-141; J. Kromayer and G. Veith *Heerwesen und Kriegführung der Geichen und Römer* (C.H. Beck, Munich, 1928) 98; Gabriel, *Philip II*, 64; Champion, *Pyrrhus*, 24, 60; I. Worthington (ed.), *Alexander the Great – A Reader* (Routledge, London, 2012) 18; Thomson, *Granicus*, 24; Hammond, 'Training', 54-57; G.T. Griffith, 'Philip as a General' in Hatzopoulos and Loukopoulos (eds.), *Philip of Macedon*, 58; J. Pietrykowski, 'In the School of Alexander – Armies and tactics in the Age of the Successors' *AncWar* 3.2 (2009) 21; Sheppard, A*lexander the Great at War*, 54; Pietrykowski, *Great Battles*, x; Fuller, *The Generalship of Alexander*, 24, 26, 35, 49, 50; Bennett and Roberts, *Wars of Alexander's Successors II*, xvi, 3; Bosworth, *Conquest and Empire*, 10; I. Worthington, *Philip II of Macedonia* (Yale University Press, New Haven, 2008) 27; Heckel, *Conquests of Alexander*, 16

114 Connolly, *Greece and Rome at War*, 64, 68-69; Champion (*Pyrrhus*, 24) dates the reforms to between 349BC and 338BC, but goes on to say (page 25) that it is easier to accept the simpler concept of a single period of reform earlier in Philip's reign than it is to accept the idea of one lot of reforms early in his reign, and more reforms a decade later.

115 G. Cawkwell, *Philip of Macedon* (Faber & Faber, London, 1978) 150; see also: Gabriel, *Philip II*, 61, 64-65, 245

116 Diod. Sic. 16.1.5

117 English, *Army of Alexander*, 4-5

118 N.G.L Hammond and G.T. Griffith, *A History of Macedonia Vol.II* (Oxford University Press, 1979) 213, 407; see also: E.N. Borza, *In the Shadow of Olympus: The Emergence of Macedonia* (Princeton University Press, 1990) 202; Worthington, *Philip II*, 26

119 Bennett and Roberts, *Wars of Alexander's Successors II*, 3

120 Diod. Sic. 16.2.5

121 Philip and the Phocians: Polyaenus, 2.38.2; see also: Diod. Sic. 16.35.2; Alexander at the Persian Gates: Arr. *Anab*, 3.18.3; Diod. Sic. 17.68.2-3; Curt. 5.3.17-23

122 Markle, 'The Use of the *Sarissa*', 483

123 R.E. Dickinson 'Length Isn't Everything – Use of the Macedonian *Sarissa* in the Time of Alexander the Great' *JBT* 3:3 (2000) 51

124 Cawkwell, *Philip of Macedon*, 157; for other examples of the use of backwards extrapolation to try and understand the warfare of the age of Philip see: Heckel and Jones, *Macedonian Warrior*, 7; Fuller, *Generalship of Alexander*, 49

125 Cawkwell, *Philip of Macedon*, 157

126 Everson, *Warfare*, 174; Daniel ('The Taxeis of Alexander and the change to Chiliarch, the Companion Cavalry and the change to Hipparchies: A Brief Assessment' *AncW* 23.2 (1992) 45) also suggests that Philip may have been 'impressed' by the Theban use of a long pike. However, the Thebans did not use a pike at this time (see following).

127 Gabriel, *Philip II*, 61; Worthington (*Philip II*, 22) similarly sums up the state of Macedon at the time of Philip's rise to the throne by stating that 'Philip's inheritance was the worst that any new king could face, and lesser men would have failed miserably in their attempts to save Macedonia.'

128 Diod. Sic. 16.2.4-5

129 Gabriel, *Philip II*, 61

130 Diod. Sic. 16.4.3; Hammond ('Casualties and Reinforcements of Citizen Soldiers in Greece and Macedonia' *JHS* 109 (1989) 63) states that after the time of Philip the strength of the Macedonian military could have been 20,000 infantry and 600 cavalry. Fuller (*Generalship of Alexander*, 26) suggests that soon after he came to the throne, Philip 'reorganized the Macedonian army and recruited it to a force of 600 cavalry and 10,000 infantry'. Fuller uses this as evidence for the claim that Philip was a great military reformer. However, what Fuller seems to have failed to consider is that Perdiccas III may have possessed an army of this size back in 359BC and that Philip was merely replacing the losses that had been incurred in the conflict with the Illyrians rather than reforming the nature of the Macedonian military.

131 Diod. Sic. 16.85.5

132 Alexander the Great took 12,000 Macedonian infantry (including the *hypaspists*) to Asia in 334BC (Diod. Sic. 17.17.3), another 12,000 were left in Macedonia under the command of Antipater (Diod. Sic.

17.17.4-5), and several thousand had already been sent to Asia as part of an expeditionary force in 336BC (Diod. Sic. 17.7.10).

133 Diodorus (16.1.3-6, 16.3.1-2) provides an overview of the reign and accomplishments of Philip. Interestingly, the creation of the pike-phalanx is not mentioned.

134 Arr. *Anab.* 7.9.2

135 Diod. Sic. 16.3.1; Just. *Epit.* 8.5.6-7; see also: Gabriel Philip II, 34, 37, 50, 71, 82; N.G.L Hammond, *The Macedonian State: Origins, Institutions and History* (Clarendon Press, Oxford, 1989) 151; Bosworth convincingly argues that the term *pezhetairoi* (πεζέταιροι), meaning 'foot companions', could be either a generic reference to the pike-phalanx as a whole or a specific reference to infantry units raised only in Lower Macedonia. Bosworth additionally argues that the term *asthetairoi* (ἀσθέταιροι), a term found seven times in the works of Arrian (*Anab.* 1.28.3, 2.23.2, 4.23.1, 5.22.6, 6.6.1, 6.21.3, 7.11.3), is a reference only to infantry units recruited from Upper Macedonia following Philip II's annexation of that region; see: A.B. Bosworth, 'ΑΣΘΕΤΑΙΡΟΙ' *CQ* New Series 23.2 (1973) 245-253.

136 Arr. *Anab.* 7.9.4;

137 Arrian (*Anab.* 1.2.5) does mention specific cavalry contingents that have come from Upper Macedonia in Alexander's army, while Diodorus (17.57.2) mentions infantry contingents from the regions of Elymoitis, Orestis, Lyncestis and Tymphaea (see also: Curt. 4.13.28).

138 Hammond, *King, Commander and Statesman*, 27; Fuller (*Generalship of Alexander*, 47) similarly suggests that Philip created the professional Macedonian army. See also: Gabriel, *Philip II*, 46

139 Snodgrass, *Arms and Armour*, 114-115

140 Harpocration, *Lexicon* s.v. *pezhetairoi* (Anaximenes - Jacoby *FrGrHist.* 72 F4); Ael. Tact. *Tact.* 3

141 Ael. Tact. *Tact.* 34

142 Front. *Strat.* 4.1.6

143 Just. *Epit.* 11.6.4-5

144 F.E. Adcock, *The Greek and Macedonian Art of War* (University of California Press, Berkeley, 1957) 25; see also: Gabriel, *Philip II*, 24; Featherstone, *Warriors and Warfare*, 57; Warry, *Warfare*, 69; Snodgrass, *Arms and Armour*, 116; Sheppard, *Alexander the Great at War*, 51-52; Cascarino, *Tecnica della Falange*, 18; Hammond ('Royal Pages, Personal Pages, and Boys Trained in the Macedonian Manner during the Period of the Temenid Monarchy' *Historia* 39.3 (1990) 275) states that Philip's 'new weapon was the pike' despite citing, on previous pages, many of the sources which suggest that the Macedonian military was professional (or at least semi-professional) prior to his reign.

145 Plut. *Pel.* 26; Diodorus presents a different, and conflicting version of events. In Book 15 of his *History*, Diodorus simply states (15.67.4) that in the year 369/368BC Pelopidas 'made an alliance with Alexander, the Macedonian king, and he took from him as a hostage his brother Philip whom he sent to Thebes'. However, in the following book (16.2.2), Diodorus elaborates on his earlier tale by stating that Amyntas, the father of both Alexander II and Philip had been defeated by the Illyrians in 383BC (the year that Philip had been born) and that the Illyrians had subsequently taken Philip as a hostage only to later place him in the care of the Thebans. In another version of events, Justin (*Epit.* 7.5.1) says that Philip was ransomed by Alexander II and then sent to Thebes. However, regardless of how he actually got there, it is clear that Philip was taken to Thebes as a hostage when he was about 13 or 14 years old.

146 Cawkwell, *Philip of Macedon*, 150; see also: Snodgrass (*Arms & Armour*, 116) who says the same thing.

147 Warry, *Warfare*, 69

148 Connolly, *Greece and Rome*, 64, 69; Connolly also suggests (page 69) that it was Philip II who called the men of his phalanx 'Infantry Companions' in order to raise their dignity. This conclusion seems to ignore the passage of Anaximenes which states that it was Alexander II, and not Philip II, who first referred to the pike-infantry as *pezhetairoi* (see page 4).

149 Fuller, *Generalship of Alexander*, 24; see also: Worthington, *Philip II*, 18

150 Plut. *Pelop.* 26 - ἐκ δὲ τούτου καὶ ζηλωτὴς γεγονέναι ἔδοξεν Ἐπαμεινώνδου, τὸ περὶ τοὺς πολέμους καὶ τὰς στρατηγίας δραστήριον ἴσως κατανοήσας

151 App. *Syr.* 41

152 Plut. *Pelop.* 26; Diodorus (16.2.2) similarly states that Philip was entrusted to the care of Epaminondas' father. In his *Moralia* (178c), Plutarch states that the young Philip lived with Philon rather than Pammenes.

153 For accounts of the battle see: Xen. *Hell.* 6.4.1-15; Plut. *Pel.* 23; Diod. Sic. 15.55.1-15.56.4

154 Thuc. 4.93

155 Xen. *Hell.* 6.4.2; Bennett and Roberts (*Wars of Alexander's Successors II*, 4) suggest that the Macedonian pike-phalanx was not as deep as the weighty, 48-deep, Theban phalanx. However, there is no reference to

the Thebans ever using a formation of this depth.

156 Piraeus: Xen. *Hell.* 2.4.12; Mantinea: Xen. *Hell.* 6.5.19

157 Front. *Strat.* 4.1.6

158 For example see: Polyb. 6.25; Arr. *Tact.* 42.1; Asclep. *Tact.* 2.1-2; Ael. *Tact.* 5.1-2

159 Diod. Sic. 17.11.3, 17.13.4

160 Featherstone, *Warriors and Warfare*, 57

161 S. Skarmintzos, 'Phalanx versus Legion: Greco-Roman Conflict in the Hellenistic Era' *AncWar* 2.2 (2008) 30; see also Hammond (*The Genius of Alexander the Great* (Duckworth, London, 1997) 13) who makes a similar suggestion.

162 *FrGrHist* 4.356 F1; see also: Athenaeus 11.506f

163 E.M. Anson, 'The Introduction of the Sarisa in Macedonian Warfare' *AncSoc* 40 (2010) 58-59

164 Worthington, *Philip II*, 26

165 Just. *Epit.* 7.2.6

166 Diod. Sic. 15.18.2-3; Xen. *Hell.* 5.2.11, 5.2.36, 5.3.8

167 Ueda-Sarson, 'The Reforms of Iphicrates', 33

168 Ueda-Sarson, 'The Reforms of Iphicrates', 33

169 Ueda-Sarson, 'The Reforms of Iphicrates', 33

170 Thuc. 4.124

171 Hom. *Il.* 6.319; see also: T.D. Seymour, *Life in the Homeric Age* (Biblo & Tannen, New York, 1963) 665

172 Gabriel, *Philip II*, 65

173 A. Aymard, 'Philippe de Macédoine, otage á Thèbes' REA 56 (1954) 21

174 For example see: J.R. Ellis, *Philip II and Macedonian Imperialism* (Thames & Hudson, London, 1976) 43; M.B. Hatzopoulos, 'La Béotie et la Macédoine à l'époque de l'hégémonie Thébaine' in *La Béotie antique: Lyon-Saint-Étienne – 16-20 Mai 1983* (Editions du Centre national de la recherche scientifique, Paris, 1985) 252; Gabriel, *Philip II*, 3-4, 24-27

175 Diod. Sic. 15.67.4,

176 Just. *Epit.* 7.5.1-4

177 Plut. *Alex.* 7

178 Plut. *Pel.* 26

179 For example see: F. Geyer, *Makedonien bis zur Thronbesteigung Philipps II* (R. Oldenbourg, Munich, 1939) 133; Hammond and Griffith, *A History of Macedonia II*, 186, 205; Bennett and Roberts, *Wars of Alexander's Successors II*, 3. Worthington (*Philip II*, 17), in something of a mid-point position, suggests a period of detention for Philip between 368BC and 365BC – see also: Gabriel (*Philip II*, 3-4, 24) who suggests that Philip was sent to Thebes in 369/368BC and then returned in 365BC.

180 Constant campaigning by the Spartans against the Thebans is said to have taught the Thebans how to fight better (Plut. *Ages.* 26; Plut. *Pel.* 15; Plut. *Mor.* 189f, 213f, 217e; see also: Polyaenus, *Strat.* 1.16.2; E*xcerpta Polyaeni*, 13.1); as Bosworth points out (*Conquest and Empire*, 11), after Philip II annexed Thessaly in 352BC, he made no attempt to convert the Thessalians into an effective military force based upon the Macedonian model. This may have been partially due to a reluctance to teach the people of a newly conquered region a more effective way of fighting in case they should rebel.

181 *IG* VII 2418; for the details of the this decree see: P. Roesch, 'Un Décret inédit de la Ligue Thébaine et la flotte d'Epaminindas' *REG* 97 (1984) (pp.45-60) 45-50

182 Hammond and Griffith, *A History of Macedonia II*, 186, 205; see also Hammond, *Macedonian State* 74

183 Justin. *Epit.* 7.5.2

184 Worthington, *Philip II*, 19; on page 17 Worthington suggests that Philip was detained in Thebes until 365BC which would then correlate with the treaty between Thebes and Macedon.

185 Plut. *Pel.* 27

186 Just. *Epit.* 7.5.1-4

187 On the death of Alexander II see: Diod. Sic. 15.71.1

188 In an earlier passage (15.67.4), Diodorus also has Alexander II send the young Philip to Thebes as per Justin's account. *Aeschines* (2.26) places Philip in the court of Ptolemy of Alorus in 369BC at which time Ptolemy ascends the throne of Macedon following the death of Alexander II.

189 Diod. Sic. 16.2.3

190 Just. *Epit.* 7.5.3

191 Diod. Sic. 16.3.2

192 Champion, *Pyrrhus*, 24

193 M.M. Markle, 'A Shield Monument from Veria and the Chronology of Macedonian Shield Types' *Hesperia* 68.2 (1999) 243

194 Thompson, *Granicus*, 24

195 Gabriel, *Philip II*, 61; see also: English, *Army of Alexander*, 4; Hammond, 'Training', 53-58; Rahe, 'The Annihilation of the Sacred Band', 87; Sheppard, *Alexander the Great at War*, 52; Heckel and Jones, *Macedonian Warrior*, 6

196 P. McDonnell-Staff 'Hypaspists to Peltasts: The Elite Guard Infantry of the Antigonid Macedonian Army' *Ancient Warfare* 5.6 (2012) 22; Warry (*Warfare*, 73) similarly says that open-order formations were 180cm per man (presumably based upon the Attic cubit of 45cm), close-order (by which he means intermediate-order), which was used for battle, was 90cm per man, and the locked shield formation used for defensive purposes was 50cm per man.

197 Asclep. 4.1 – τὸ πυκνότατον, καθ' ὃ συνησπικὼς ἕκαστος ἀπὸ τῶν ἄλλων πανταχόθεν διέστηκεν πηχυαῖον διάστημα, τό τε μέσον, ὃ καὶ πύκνωσιν ἐπονομάζουσιν, ᾧ διεστήκασι πανταχόθεν δύο πήχεις ἀπ' ἀλλήλων, τό τε ἀραιότατον, καθ' ὃ ἀλλήλων ἀπέχουσι κατά τε μῆκος καὶ βάθος ἕκαστοι πήχεις τέσσαρας; for the changes in the size of the cubit from the Classical to the Hellenistic period see the following section of the length of the *sarissa*.

198 Ael. Tact. *Tact.* 14; Polyb. 18.29; Arr. *Tact.* 11.3

199 It is interesting to note that Diodorus seems to confuse the formation he is referring to by calling it both the *synaspismos*, the 48cm interval order of Asclepiodotus, and a 'compact order', the alternate name for the 96cm intermediate-order of Asclepiodotus that was the closest combative deployment phalangites could adopt. However, as Diodorus specifically states that the formation adopted by Philip was able to interlock shields, it seems that what he is referring to is the smaller interval of 48cm per man. It is also interesting to note that the term used by Diodorus to describe the infantry is 'fighters' (πυκνότητα) - a term used by later writers to mean either hoplites or phalangites.

200 Ael. Tact. *Tact.* 32; see also: Asclep. *Tact.* 12.9; Mcdonnell-Staff ('Hypaspists to Peltasts', 23) suggests that it was impossible for a formation in close-order to change direction. This goes completely against the procedure of 'wheeling' outlined by Aelian and Asclepiodotus which specifically employs a close-order to change the direction the phalanx is facing.

201 Ael. Tact. *Tact.* 32

202 Polyb. 18.26; 18.30; in an interesting passage, Arrian (*Anab.* 5.17.7) states that, at the end of the battle of the Hydaspes, Alexander ordered his troops, who were encircling the Indian position, 'to lock shields, concentrate in intermediate-order, and advance the phalanx' (τοὺς πεζοὺς δὲ ξυνασπίσαντας ὡς ἐς πυκνοτάτην ξυγκλεισιν ἐπάγειν τὴν φάλαγγα ἐσήμηνε). This passage initially seems somewhat confusing and has resulted in a variety of translations and interpretations which have as their general premise the connotation of the troops locking shields, forming the most compact order as possible, and then advancing. While this interpretation would make some sense, it would require an understanding that the troops had adopted a close-order formation in order to execute the wheel of their formation to accomplish the encirclement, but that they remained in this non-offensive configuration while they faced the remaining Indian forces. This would then mean that the Macedonians had no capability to engage the Indians and were simply surrounding them in the hope that they surrendered. Another way of reading the text however would be for the troops to adopt a close-order formation in order to execute the wheel and then, once in their new position, to open back out into an intermediate-order which would allow them to lower their pikes into a combative position. In either way the Indians were still surrounded, but in this alternate interpretation, the Macedonians would still possess an offensive capability. This seems to be confirmed by Arrian who says that, following the encirclement, the Indians were being killed from all sides.

203 P. Connolly, 'Experiments with the *Sarissa* – The Macedonian Pike and Cavalry Lance – A Functional View' *JRMES* 11 (2000) 111

204 See: Matthew, *Storm of Spears*, 179-197

205 See: Matthew, *Storm of Spears*, 179-197; it is also interesting to note that the term for the close-order formation (*synaspismos*) has its root in the term for the hoplite shield (*aspis*) rather than the shield of the phalangite (*peltē*). This is yet another indication that the term is to be associated primarily with hoplites rather than phalangites. See also: Markle 'The Use of the *Sarissa*', 484.

206 For examples of the 'elite' status and operations of the *Hypaspists* and The Royal Guard see: Diod. Sic. 14.6.7, 17.45.6, 17.57.2, 17.61.3, 17.99.4, 17.110.1; Arr. *Anab.* 1.5.2, 1.5.10, 1.6.5-6, 1.6.9, 1.8.3-4, 1.14.2, 1.28.3, 2.4.3-4, 2.8.3, 2.23.2-2.24.4, 2.27.1, 3.11.9, 3.17.2, 3.18.5, 4.26.6, 4.29.1, 4.30.3, 5.13.1, 5.23.7, 6.2.2, 7.11.2-3. For the commemoration of the *hypaspist* units by Alexander on some of his coinage in

recognition of their service see: C.A. Matthew, 'For Valour: The 'Shield Coins' of Alexander and the Successors' *JNAA* 20 (2009/2010) 15-35.

207 Tarn (*Alexander the Great Vol.II*, 140) states that the term *hypaspist* probably had 'something to do with [the unit's] armament, though what it may have been is unknown'. The armament of the *hypaspists* is another area of Hellenistic warfare which has roused considerable scholarly debate. Gabriel (*Philip II*, 71), for example, says that the *hypaspists* were armed as hoplites under Philip but were armed with *sarissae* for the battle of Chaeronea in 338BC (p.271). English (*Army of Alexander*, 29-30) and Milns ('Philip II and the Hypaspists' *Historia* 16.4 (1967) 509-510; 'The Hypaspists of Alexander III – Some Problems' *Historia* 20 (1971) 181) say that Philip invented the *hypaspists* but that they were always armed with *sarissae*. Bosworth (*Conquest and Empire*, 260) also suggests that the *hypaspists* were pikemen. Anson ('Introduction of the Sarisa' 61-63) suggests that the *hypaspists* were a unit of professional infantry guardsmen under both Philip and Alexander III, modelled on the Theban Sacred Band, and that they were armed as both hoplites and phalangites. Hammond ('The Battle of the Granicus River' *JHS* 100 (1980) 81-83, 86) states that it is clear that the *hypaspists* were armed with pikes from Polyaenus' account (4.2.2) of Philip's actions at Chaeronea and Arrian's account of Alexander's actions at Pelium (*Anab.* 1.6.2). Strangely, neither of these passages actually mention the *hypaspists* so it is uncertain upon what this conclusion is based. Everson (*Warfare*, 177) suggests that the *hypaspists* were only lightly armed but, as their name suggests, they may have carried substantial shields. Sekunda ('Land Forces' 333) similarly suggests that the *hypaspists* were 'more lightly equipped than the main body of the phalanx'. Heckel (*Conquests of Alexander*, 26) suggests that the *hypaspists* were armed as hoplites, but with the lighter *linothorax*. Park ('The Fight for Asia – The Battle of Gabiene' *AncWar* 3.2 (2009) 33-35) says that the 'Silver Shields', Alexander's former *hypaspists* who later fought with Eumenes, were armed with the *sarissa* in 316BC. Sekunda ('Land Forces', 339), on the other hand, suggests that Eumenes' Silver Shields were carrying the *aspis* – which would then mean that they were operating as hoplites rather than pikemen. Bennett and Roberts (*Wars of Alexander's Successors II*, 9) suggest that the Silver Shields were phalangites fighting in the heart of the pike-phalanx by the time of the Successors. Featherstone (*Warriors and Warfare*, 58), on the other hand, claims that the *hypaspists* were created by Alexander the Great and were armed as hoplites. Donvito ('Clad in Gold and Silver' 8) says that the *hypaspists* were elite troops trained to fight 'in all fashions available at that time: as hoplites, phalangites and light troops'. Cawkwell (*Philip of Macedon*, 33) says that by the time of Alexander the Great the *hypaspists* were armed as skirmishers so as to act as a more mobile hinge between the phalanx and the cavalry. Considering the nature of the engagements that the *hypaspists* were involved in at this time (sometimes operating on their own) this last conclusion seems unlikely. The very nature of the engagements that the *hypaspists* were involved in required men who were both relatively mobile, yet still very well protected. This suggests that they were armed as hoplites. Sekunda (*Army of Alexander*, 27) suggests that the way in which Arrian (1.1.9) describes Alexander's troops as laying down and covering themselves with their shields to let wagons that were being rolled down upon them to harmlessly pass over them indicates that they were carrying the larger *aspis* rather than the smaller Macedonian shield. However, it must be considered that to undertake such a manoeuvre troops would be unlikely to be wielding their weapons, especially the *sarissa*, at the same time and, subsequently, they may have been covering themselves with the smaller shields. Alternatively, it is possible that this passage is referring to troops armed in the manner of the Classical hoplite. For further discussions on the *hypaspists* see: Heckel and Jones, *Macedonian Warrior*, 17-18; Snodgrass, *Arms and Armour*, 115; Tarn, *Alexander the Great Vol.II* 140; Pietrykowski, 'In the School of Alexander', 24; Warry, *Warfare,* 72; Sheppard, Alexander the Great at War, 84-85; Pietrykowski, *Great Battles*, 16; Fuller, *Generalship of Alexander*, 49-50; J.R. Ellis, 'Alexander's Hypaspists Again' *Historia* 24.4 (1975) 617-618; R.A. Lock, 'The Origins of the Argyraspids' *Historia* 26.3 (1977) 373-378; E.M. Anson, 'Alexander's Hypaspists and the Argyraspids' *Historia* 30.1 (1981) 117-120

208 Ellis, *Philip II*, 53, 58

209 Brunt, 'Anaximenes', 152; however, both Archaic and Classical Greek formations were a lot closer than Brunt would suggest. Xenophon (*Cyr.* 6.3.25; *Mem.* 3.1.7-8), for example, describes the deployment of the phalanx as being like the structure of a house; with the strongest elements making up the roof and the foundations (i.e. the front and rear ranks respectively) in a reference to the positioning of officers across the front and back of the line. In a passage which suggests the use of massed, phalanx-like, formations in the Archaic Period, Nestor is described as deploying his forces in a similar fashion by Homer in the *Iliad* (4.297-300). Homer also describes the Trojans as advancing with 'rank following rank' (13.800) and the Greeks fighting in native contingents (13.308-314, 16.168-199)

210 Anson ('Citizen-Soldiers', 248) suggests that Philip placed more mobile *hypaspists* on the right wing of

his formation (the traditional location for Alexander's *hypaspists*) when engaging the Illyrians in 358BC.

211 Thuc. 4.126; Ennius (*Ann.* Frg.540) describes the Illyrians as fighting with long spears and curved swords. While these bear many similarities to the *doru* and *macharia* carried by the Greek hoplite, the use of a standard formation (or not) by the Illyrians is not mentioned by Ennius.

212 Diod. Sic. 15.13.2

213 Polyaenus, *Strat.* 4.2.10

214 Champion, *Pyrrhus*, 24

215 Markle, 'Use of the *Sarissa*', 491; Markle then argues that the *sarissa* was not used again until 331BC.

216 Diod. Sic. 17.2.3

217 Markle, 'Use of the *Sarissa*', 486

218 Polyaenus, *Strat.* 3.9.35; Nepos, *Iphicrates*, 2.1-2

219 Polyaenus, *Strat.* 4.2.10; regular training, including accustomizing troops to privations, fostering physical fitness and weapon handling skills, training with wooden practice weapons and engaging in mock battles and exercises was a part of the Macedonina military system that continued from the time of Philip well into the age of the Successors. For Philip see: Diod. Sic. 16.3.1-2; Front. *Strat.* 2.1.9; Polyaenus, *Strat.* 4.2.1, 4.2.10; Alexander: Arr. *Anab.* 7.6.1, 7.12.2; Diod. Sic. 17.2.3, 17.1008.2, 17.110.3, Plut. *Alex.* 4, 40, 47; Successors: Ployb. 5.63-5.64; Livy 40.6.6-7

220 This did not stop the Athenian state from awarding panoplies of hoplite equipment to war orphans (*Aeschines* 3.154) or issuing elements of the hoplite panoply to the citizens following the *ephebic* reform of 335BC (Arist. *Ath. Pol.* 42.3-4). For an examination of the issue of this equipment see: B. Bertosa, 'The Supply of Hoplite Equipment by the Athenian State Down to the Lamian War' *JMH* 67.2 (2003) 361-379. Ueda-Sarson ('The Reforms of Iphicrates', 32) suggests that Diodorus is referring to troops from Phocis armed in the manner of the Iphicratean reforms in 355BC when he describes (16.24.2) how Philomenus selected 'a thousand men from Phocis; whom he called *peltasts*' (τῶν Φωκέων ἐπέλεξε χιλίους, οὓς ὠνόμασε πελταστάς). Ueda-Sarson argues that, had these troops been traditional skirmishers, Diodorus would have just used a term such as 'a thousand Phocian *peltasts*', and that the emphasis placed upon what they were called suggests that they were something out of the ordinary. If this is the case, then it would seem that some of the city states of Greece may have taken the principles of the Iphicratean reforms on board within two decades of their initial implementation. It is possible that some of the states of Greece had reformed their army in response to similar events which had already taken place in Macedon. Another point to consider is that someone would have paid for Iphicrates to undertake his military 'experiment'. Unfortunately, the sources do not outline how this reform was funded or by whom (for a discussion on who may, or may not, have funded the equipping of mercenary troops see: P. McKechnie, 'Greek Mercenary Troops and their Equipment' *Historia* 43.3 (1994) 297-305). Regardless of who funded the reform, the fact that both Diodorus and Nepos place considerable emphasis on its details suggests that it was not done on a small scale and, as such, it can only be assumed that the expense incurred to create the Iphicratean *peltast* was considerable. Possibly due to the expense of creating a pike-phalanx, other states took much longer to adopt this mode of fighting. Both the Spartans and the Achaeans, for example, did not use the pike-phalanx until the late third century BC (Plut. *Cleom.* 11; Polyaenus, Strat. 6.4.3). There are three main sources for military reforms for the Achaeans undertaken by Philopoemen in 207BC (Polybius (11.9) only mentions the reforms in passing) but their interpretation has aroused a certain level of scholarly debate. Plutarch (*Philopoemen*, 9) states that 'they [i.e. the Achaeans] used *thureoi* which were easily carried because they were so light, and yet were too narrow to protect the body, and spears which were much shorter than the Macedonian *sarissa*...Philopoemen... persuaded them to adopt the *sarissa* and *aspis* instead of the *doru* and *thureos*'. Pausanias (8.50.1) states that 'whereas they [i.e. the Achaeans] used to carry small light spears and long shields like the Celtic thureoi or the Persian wicker shields, he persuaded them to put on breastplates and greaves, and to additionally use Argive shields (*aspides*) and long spears'. Polyaenus (*Stratagems*, 6.4.3) states that 'Philopoemen taught the Achaeans to use the *aspis* and *sarissa* instead of the large oblong *thureos* and *doru*, to protect themselves with helmets, breastplates and greaves...'. These passages have aroused considerable scholarly debate. Some have seen this as a change from fighting in a Homeric fashion, with the large tower shields of the Archaic Age, to the style of hoplites – no doubt based upon the use of the term aspis to describe the new shield that the Achaeans used (for example see: H.L. Lorimer, *Homer and the Monuments* (Macmillan, London, 1950) 195; A.M. Snodgrass, *Early Greek Armour and Weapons* (Edinburgh University Press, 1964) 61, 184, 203). Others have seen this as a change from fighting with a shield not unlike the Roman scutum, as per one definition of the term *thureos*, to fighting as hoplites (for example see: J.K. Anderson, 'Philopoemen's Reform of the Achaean Army'

CP 62.2 (1967) 104-106). While others see it as a change to fighting as Hellenistic Phalangites (for example see: M.F. Williams, 'Philopoemen's Special Forces: *Peltasts* and a New Kind of Greek Light-Armed Warfare (Livy 35.27)' *Historia* 53.3 (2004) 257-277). This last interpretation seems to more closely fit with the use of the term *sarissa* to describe the new weapon that the Achaeans employed. Such being the case, it can only be concluded that the use of the term *aspis* for the shield that they used following the reform has to be a generic expression to simply mean 'shield' and cannot be a reference to the use of the large shield of the Classical hoplite (despite Pausanias' explicit reference to 'Archive shields') as the large diameter of the hoplite *aspis* would not allow the left hand to be used to help carry the pike. Conversely, Anderson ('Philopoemen's Reform', 104) suggests that the use of the term *sarissa* in the ancient sources may be a generic term for 'long spear' – which could then allow for the use of the larger *aspis* depending upon just how long these new 'long spears' were.

221 N. Sekunda, *The Persian Army 560-330BC* (Osprey, Oxford, 2005) 27
222 Diod. Sic. 16.3.1
223 For a discussion of the continuance of the issue of equipment and other customs by the state see: N.G.L. Hammond, 'The Continuity of Macedonian Institutions and the Macedonian Kingdoms of the Hellenistic Era' *Historia* 49.2 (2000) 141-160
224 Justin 8.6.3; Diodorus (16.2.4) similarly states that Macedonia was 'in a bad way' when Philip came to the throne. Plutarch (*Mor.* 327D) additionally states that not only were Philip's treasuries empty, but the state also had an outstanding debt of 200 talents. By the time Alexander launched his campaign into Asia, he only had cash reserves of 70 talents which, according to Duris, was only enough to fund the expedition for 30 days (Plut. *Alex.* 15; Plut. *Mor.* 327e).
225 The mines provided Philip with approximately 1,000 talents of gold per year (Diod. Sic. 16.8.7; Dem. 9.50) while he also collected tribute from the Thracians (Diod. Sic. 16.71.2) and taxes from the Thessalians (Dem. 6.22, 19.89).
226 Diod. Sic. 16.4.5-6
227 Sheppard, *Alexander the Great at War*, 53; see also: Heckel and Jones, *Macedonian Warrior*, 9-10. Hammond ('Training' 58) says that the use of the *sarissa* was one of the reasons for Philip's victory. While this may be so, his victory using this weapon does not necessarily mean he invented it. Diodorus (16.8.6-7) states that the increased output from the mines of Philippi meant that, once Philip had captured them, he was able to hire large numbers of mercenaries. These troops would have undoubtedly not been using the *sarissa*.
228 Anson, 'Introduction of the Sarisa', 64; see also: Heckel, *Conquests of Alexander*, 16
229 For an analysis of the comparative costs involved in making both linen and metal armour see: Aldrete, Bartell, and Aldrete, *Reconstructing Ancient Linen Body Armor*, 149-160
230 Gabriel, *Philip II*, 61
231 This would then agree with Bosworth's conclusion that the term *pezhetairoi* refers to the infantry units raised in Lower Macedonia while the term *asthetairoi* refers to units recruited from Upper Macedonia following Philip II's annexation of that region; see: Bosworth, 'ΑΣΘΕΤΑΙΡΟΙ', 245-253
232 Diodorus (16.4.6) states that when Philip confronted the Illyrian forces of Bardylis in 358BC, it was the use of his cavalry to strike the flanks of the Illyrian position which contributed greatly to the victory.
233 See: Anderson, *Military Theory*, 160
234 Bennett and Roberts (*Wars of Alexander's Successors II*, 3), somewhat oddly, suggest that while in Thebes the young Philip must have gained an appreciation for deeply weighted phalanxes such as those that Epaminondas had used at Leuctra in 371BC and Mantinea in 362BC. The odd thing is that Philip was not in Thebes in 362BC and so it would have been impossible for any information on the battle of Mantinea to have had an influence on him while he was there between 369-367BC.
235 For example see: Gabriel, *Philip II*, 108-109; Hammond, 'Philip as Hostage' 357-365; see also: N.G.L. Hammond, 'The Battle between Philip and Bardylis' *Antichthon* 23 (1989) 1-9
236 Diod. Sic. 15.55.1; Plut. *Pelop.* 23
237 It is also interesting to note that, despite the contribution of the slanting phalanx to the Theban victory at Leuctra in 371BC, it is not used at Mantinea in 362BC, nor is it used against Philip at Chaeronea in 338BC.
238 Diod. Sic. 16.4.6; Frontin. *Str.* 2.3.2; for an examination of this confrontation see: Hammond, 'Philip and Bardylis', 1-9
239 Polyaenus, *Strat.* 4.2.2; Diod. Sic. 16.86.1-6; Fuller (*Generalship of Alexander*, 35) suggests that Philip's tactics at Chaeronea 'resembled those of Leuctra'. However, other than in the most vague of terms, they clearly did not.

240 Arrian (*Anab*. 3.13.1) says that the Macedonian line advanced against the Persians with a move 'to the right'. Similarly, Diodorus (17.57.6) and Curtius (4.15.1) also say that the Macedonian advance was at an oblique angle. Whether the whole army was initially deployed in this manner is far from certain. However, from the ancient texts it seems clear that this oblique advance was only conducted by the right wing of the army. This would have resulted in the line adopting an oblique angle regardless of whether it had been originally deployed in that way or not. For a discussion on this battle, and the Macedonian advance in particular, see: Fuller, *Generalship of Alexander*, 163-182; A.M. Devine, 'Grand Tactics at Gaugamela' *Phoenix* 29.4 (1975) 374-385

241 Gabriel, *Philip*, 26; Heckel and Jones (*Macedonian Warrior*, 10) suggest that Philip learnt the value of attacking a point in an enemy line in depth while he was a hostage in Thebes.

242 Plut. *Pelop*. 23; Xen. *Hell*. 6.4.12; Polyaenus, *Strat*. 2.3.15; *Excerpta Polyaeni*, 14.7; the numbers of both sides at Leuctra may have been close to equal despite the ancient texts providing varying figures (see: Polyaenus, *Strat*. 2.3.8, 2.3.12; Frontin. *Str*. 4.2.6; Diod. Sic. 15.52.5; see also: Anderson, *Military Theory*, 197-198). The Thebans were deployed fifty ranks deep while the Spartans were arrayed twelve deep. This means that, while the Theban contingent was four times deeper than the Spartan one, it was also four times narrower across its frontage.

243 Cleombrotus: Xen. *Hell*. 6.4.13; Spartan losses: Xen. *Hell*. 6.4.15; Paus. 9.3.14; Plut. *Mor*. 193b; Plutarch (*Pelop*. 23) states that the entire Boeotian phalanx, not just the Thebans, advanced obliquely against the Spartans – although not all of the Boeotian allies would have engaged the Spartans directly but must have engaged some of the allies on the Spartan left wing. Diodorus (15.55.2) states that the Boeotian right was ordered to slowly withdraw to try and draw the Spartan left wing forward.

244 Diod. Sic. 16.4.5-6; Frontin. *Str*. 2.3.2; see also: Just. *Epit*. 7.6.7

245 Plut. *Pel*. 18; see also: Diod. Sic. 16.86.1-6

246 Arr. *Anab*. 1.6.10

247 E. Badian, 'The Battle of the Granicus: A New Look' *Ancient Macedonia II* (1977) 292

248 Gabriel, *Philip II*, 24

249 Gabriel, *Philip II*, 24; Hammond, 'Philip as Hostage', 357

250 For the presence of the Boeotian cavalry at Leuctra see: Xen. *Hell*. 6.4.13; Kromayer and Veith, in their topographical reconstruction of the battle and the deployment of the various contingents involved (Map 28 – Leuktra 371 v. Chr. Schlachtkarte) show the Theban infantry striking the Spartan line between the first and second *morai* while their left flank is protected by their cavalry. See: Kromayer, J. and Veith, G., *Schlachten-Atlas zur Antiken Kriegsgeschichte* (revised by R. Gabriel (ed.) and re-released as *The Battle Atlas of Ancient Military History* (Canadian Defence Academy Press, Ontario, 2008) 32. Cary ('Thebes' in J.B. Bury, S.A. Cook and F.E. Adcock (eds.), *The Cambridge Ancient History Vol. 6 – Macedon 401–301BC* (Cambridge University Press, 1933) 82) suggests that the novel tactics used at Leuctra for which Epaminondas became famous were not the deployment of a deep phalanx nor the arrangement of an oblique line (both of which had been used in the past, particularly by the Thebans) but was the co-ordinated use of infantry and cavalry; an innovation that later became the trademark tactic of the Macedonians. Tudela ('The Hoplite's Second Shield – Defensive Roles of Greek Cavalry', *AncWar* 2:4 (2008) 36, 38) calls the cavalry 'the hoplite's second shield' and states that one of the defensive roles of a contingent of cavalry was to protect the flanks (particularly the left) of an advancing hoplite phalanx; see also: I.G. Spence, T*he Cavalry of Classical Greece* (Clarendon Press, Oxford, 1993) 151–162.

251 Diod. Sic. 16.4.5-6; Frontin. *Str*. 2.3.2; see also: Just. *Epit*. 7.6.7

252 Ael. Tact. *Tact*. 18; in chapter 39 Aelian states that Philip invented the cavalry wedge. However, this may be more a case of its adoption into the Greek world rather than its actual creation.

The *Sarissa*

1   Author's illustrations

2   M. Andronicos, '*Sarissa*' BCH 94:1 (1970) 96-98

3   Andronicos, '*Sarissa*', 98; Sheppard (*Alexander the Great at War* (Osprey, Oxford, 2008) 54) says the butt is bronze. This is clearly incorrect.

4   Andronicos, '*Sarissa*' 96

5   N. Sekunda, *The Army of Alexander the Great* (Osprey, Oxford, 1999) 28; N. Sekunda, *Macedonian Armies after Alexander 323-168BC* (Osprey, Oxford, 2012) 15

6   Sekunda, *Army of Alexander*, 28

7   Everson (*Warfare in Ancient Greece* (Sutton Publishing, Stroud, 2004) 177) says that this butt is from a
      spear rather than a *sarissa*. However, the weight of the butt would make it much more suitable to a pike.
8   Everson, *Warfare*, 177
9   Sekunda, *Army of Alexander*, 28
10   For details of finds of the hoplite *sauroter* from Olympia see: H. Baitinger, *Die Angriffswaffen aus Olympia*
      (Berlin, Walter de Gruyter, 2000) 224–231, pl.59–62. For the average characteristics of the common
      sauroter of the hoplite spear see: C.A. Matthew, C.A., *A Storm of Spears: Understanding the Greek Hoplite
      at War* (Pen & Sword, Barnsley, 2012) 4-5
11   Sekunda, *Army of Alexander*, 28
12   Andronicos, '*Sarissa*', 98
13   Andronicos, '*Sarissa*', 98
14   Andronicos, '*Sarissa*', 101
15   Dickinson ('Length Isn't Everything – Use of the Macedonian *Sarissa* in the Time of Alexander the Great'
      JBT 3:3 (2000) 51-62), for example, uses a weapon equipped with the large head in his examination of the
      forces required to balance and reposition the weapon. See also: N.G.L. Hammond, *Alexander the Great:
      King, Commander and Statesman* (Duckworth, London, 1980) 32; S. English *The Army of Alexander the
      Great* (Pen & Sword, Barnsley, 2009) 20; R. Gabriel, *Philip II of Macedonia – Greater than Alexander*
      (Potomac, Washington, 2010) 65; J.R. Mixter, 'The Length of the Macedonian *Sarissa* During the Reigns
      of Philip II and Alexander the Great' *AncW* 23.2 (1992) 23; W. Heckel and R. Jones, *Macedonian Warrior
      – Alexander's Elite Infantryman* (Osprey, Oxford, 2006) 6, 13; A.M. Snodgrass, *Arms and Armour of the
      Greeks* (Johns Hopkins University Press, Baltimore, 1999) 140; G.T. Griffith, 'Philip as a General and
      the Macedonian Army' in M.B. Hatzopoulos and L.D. Loukopoulos (eds.), *Philip of Macedon* (Ekdotike
      Athenon, Athens, 1980) 65; R. Sheppard (ed.), *Alexander the Great at War* (Oxford, Osprey, 2008) 54;
      G. Cascarino, *Tecnica della Falange* (Il Cerchio, Rome, 2011) 113; B. Bennett and M. Roberts, *The Wars
      of Alexander's Successors 323-281BC Vol.II: Battles and Tactics* (Pen & Sword, Barnsley, 2009) 4; A.B.
      Bosworth, *Conquest and Empire: The Reign of Alexander the Great* (Cambridge University Press, 2008)
      260; W. Heckel, *The Conquests of Alexander the Great* (Cambridge University Press, 2008) 16
16   M.M. Markle, 'The Macedonian *Sarissa*, Spear and Related Armour' *AJA* 81.3 (1977) 325
17   Markle, 'Spear', 325
18   Markle, 'Spear', 325; see also: P.M. Petsas, 'Ἀνασκοφὴ ἀρχαίου νεκροταφείου Βεργίνης [1960-1961]'
      Archaeologikon Deltion 17A (1961-1962) 230 pl.146a
19   M.M. Markle, 'Use of the *Sarissa* by Philip and Alexander of Macedon' *AJA* 82.4 (1978) 487
20   Markle, 'Use of the *Sarissa*', 488
21   Markle, 'Spear', 325; Petsas,'Ἀνασκοφὴ', 235-236
22   Markle, 'Spear', 325; Interestingly, Markle does not state why he believes that a 34cm head is too big for a
      hoplite spear but a 31cm head is sufficient.
23   G. Sotiriades, 'Das Schlachtfeld von Chäronea und der Grabhügel der Makedonen' *AthMitt* 28 (1903) 309
24   Markle, 'Spear', 325-326; interestingly, in a later article ('Use of the *Sarissa*', 490-491) Markle suggests
      that the Macedonian infantry were not using the *sarissa* at Chaeronea which would then dismiss his own
      conclusion about the head discovered there. Conversely, Sotiriades ('Das Schlachtfeld von Chäronea',
      301-330) argued that the heads found at Chaeronea were too big to be from hoplite spears but were more
      suitable for *sarissa*e.
25 J. Champion, *Pyrrhus of Epirus* (Pen & Sword, Barnsley, 2009) 24; Champion therefore concludes that, as
      Olynthus was sacked in 349BC and the battle of Chaeronea, at which the *sarissa* was used, took place in
      338BC, the *sarissa* had to have been invented at a point in time between 349BC and 338BC during the
      reign of Philip II.
26   M.M. Markle, 'Macedonian Arms and Tactics Under Alexander the Great' in B. Bar-Sharrar and E.N.
      Borza (eds.), *Macedonia and Greece in Late Classical and Early Hellenistic Times* (National Gallery of
      Art, Washington, 1982) 89; see also: P. Connolly, 'Experiments with the *Sarissa* – the Macedonian Pike and
      Cavalry Lance – a Functional View' *JRMES* 11 (2000) 103
27   Connolly, 'Experiments', 107
28   Connolly, 'Experiments', 107
29   A.M. Snodgrass, *Early Greek Armour and Weapons* (Edinburgh University Press, 1964) 116-133
30   Snodgrass, *Early Greek*, 123
31   Snodgrass, *Early Greek*, 123-126
32   See: Matthew, *Storm of Spears*, 2-4

33 Grattius, *Cynegetieon*, 117-120 – (*Macetum immensos libeat si dicere contos. Quam longa exigui spicant hastilia dentes*). Sekunda ('Land Forces' in P. Sabin, H. Van Wees and M. Whitby (eds.), *The Cambridge History of Greek and Roman Warfare Vol. I: Greece, the Hellenistic World and the Rise of Rome* (Cambridge University Press, 2007) 329) similarly states that the *sarissa* possessed a 'small iron head' but does not outline any other details.

34 Sekunda (*Macedonian Armies*, 14) also dismisses the idea that the large head found at Vergina has come from a *sarissa* but suggests that it may have come from the spear of a member of the Royal bodyguard and that its size was a mark of distinction.

35 N.G.L. Hammond, The Genius of Alexander the Great (Duckworth, London, 2004) 13

36 See also: Connolly ('Experiments', 106) who reaches this same conclusion.

37 Pliny (E), *HN*, 35.110; see also: M. Bieber, 'The Portraits of Alexander the Great' *PAPS* 93.5 (1949) 387

38 For a discussion of the fresco in Kinch's tomb see: Kinch, 'Le tombeau de Naiusta', 281-288

39 C. Nylander, 'The Standard of the Great King: A Problem with the Alexander Mosaic' *OpRom* 14.2 (1983) 33

40 Koepfer ('The *Sarissa*', *AncWar* 3.2 (2009) 37) states that the weight of the head was 970g. This seems to be a typographical error for the weight of the smaller head found at Vergina (97g).

41 Connolly, 'Experiments', 107

42 A 'butt-cap' is suggested by the diminutive noun *styrakion* (στυράκιον) used by Thucydides (2.4.3) to distinguish it from the more elongated *styrax* (στύραξ) – another name for the larger spike on the end of the hoplite spear.

43 Markle, 'Spear', 325

44 Andronicos, '*Sarissa*', 98

45 For example see: Markle 'Use of the *Sarissa*', 323, Connolly, 'Experiments', 103; Andronicos, '*Sarissa*', 105-107; Gabriel, *Philip II*, 65; Hammond, *King Commander and Statesman*, 32; Heckel and Jones, *Macedonian Warrior*, 13; P. Connolly, *Greece and Rome at War* (Greenhill Books, London, 1998) 69; Koepfer, 'The *Sarissa*', 37; Sheppard, *Alexander the Great at War*, 54, 81; Cascarino, *Tecnica della Falange*, 113; Bennett and Roberts, *Wars of Alexander's Successors II*, 4

46 Dickinson, 'Length Isn't Everything', 52-53

47 English (*Army of Alexander*, 19) dismisses this idea that the tube is for connecting two halves of a *sarissa* together by suggesting that a join could not be maintained in such a heavy piece of equipment. However, as is shown throughout this work, a *sarissa* made in two parts is easily held together quite effectively through the use of a small connecting tube based upon the find made at Vergina by Andronicos (see Plates 13, 15-17).

48 This idea is also favoured by some scholars. For example see: Heckel and Jones, *Macedonian Warrior*, 14; Sheppard, *Alexander the Great at War*, 54

49 For modern accounts of attempts to retrace the route of Alexander's campaign and the difficult nature of the terrain, for example at Aornus (Pir-Sar) in Pakistan, see: A. Stein, *On Alexander's Track to the Indus* (Asian Publications, Lucknow, 1985) 120-134; M. Wood, *In the Footsteps of Alexander the Great* (BBC Worldwide, London, 2001) 178-182

50 Plut. *Pyrr.* 25; Dion. Hal. 20.10-12

51 Markle, 'Use of the *Sarissa*', 492; see also: Griffith, 'Philip as a General' in Hatzopoulos and Loukopoulos (eds.), *Philip of Macedon*, 59

52 S. English, *The Field Campaigns of Alexander the Great* (Pen & Sword, Barnsley, 2011) 29

53 See: Matthew, *Storm of Spears*, 12-14

54 For the reason why a head with a weight of 174g was used in this calculation, and for the formula used to calculate the point of balance, see the following section on 'The Balance of the *Sarissa*' (page 81).

55 B. Cotterrell and J. Kamminga, *Mechanics of Pre-Industrial Technology: An Introduction to the Mechanics of Ancient and Traditional Material* (Cambridge University Press, Melbourne, 1990) 163-175

56 Arr. *Anab.* 3.26.3

57 Curt. 7.1.9

58 Curt. 9.7.19

59 Diod. Sic. 17.100.6

60 Polyaenus, *Strat.* 2.38.2

61 Diod. Sic. 17.100.2-7

62 P. McDonnell-Staff, 'Hypaspists to Peltasts: The Elite Guard Infantry of the Antigonid Macedonian Army' *Ancient Warfare* 5.6 (2012) 22; Bosworth (*Conquest and Empire*, 260-261) also suggests that phalangites

carried a pike and a spear based upon the descriptions of the equipment carried by Coragus in his 'heroic duel' (Curt. 9.7.19; Diod. Sic. 17.100.2) and that ascribed to the 'Macedonian phalanx' of the Roman emperor Caracalla (Dio. 78.7.1-2).

63 Plut. *Mor.* 197c

64 For example see: Hdt 1.52, 7.69; Soph. *Tr.* 512, 856; Pindar, *Nemean* 10.60

65 Plut. *Mor.* 331b

66 Herodotus (1.56) suggests that the original 'Macedonons' were not even indigenous to Greece, but were part of the migrations of Dorian peoples who, after moving southward from central Europe in the tenth century BC, later settled in the Peloponnese. Similarly, Demosthenes' *Third Philippic* (9.31) states that Philip II 'is not only no Greek, nor [is he] related to the Greeks, but [he is] not even a barbarian from any place that can be named with honor, but [he is] an annoying scoundrel from Macedonia - where it has never been possible to buy a decent slave'. Demetrius of Phaleron, a late fourth century writer whose work is cited by Polybius (29.21), states that around 400BC the name of the Macedonians had hardly even been heard of due to the obscurity of their kingdom.

67 Arr. *Anab.* 4.8.8-9

68 For Hammond's discussion of the armament of the Guards see: N.G.L. Hammond, 'The Various Guards of Philip II and Alexander III' *Historia* 40.4 (1991) 396-418

69 Death of Cleitus: Curt. 8.1.45; execution of Philotas: Curt. 7.1.9

70 Plut. *Alex.* 51

71 *IG* IV2 1.121-122; see: E.J. Edelstein and L. Edelstein, *Asclepius – A Collection and Interpretation of the Testimonies* (The Johns Hopkins Press, Baltimore, 1945) no.423

72 English (*Army of Alexander*, 19-20) says that the *sarissae* belonging to the troops under the command of Alexander the Great were carried in wagons as part of the army's baggage train, and that the phalangites carried normal hoplite spears while they were on the march through difficult terrain – although he does concede (p.24) that there is no reference to either Philip or Alexander requisitioning large amounts of hoplite weaponry. Following Markle, English also suggests (p.22) that the Macedonians may have only used the *sarissa* at the battle of Gaugamela in 331BC as the terrain at other engagements would have made wielding it difficult. It is unlikely that an army as professionally organized as the Macedonians after the time of Philip would have two sets of arms for each man due to the logistical and operational problems of such an exercise. It is more likely that a *sarissa* that could be divided into two halves was used, as this would allow the weapons to be easily transported by each man and provide a certain level of tactical flexibility which could be utilized depending upon the nature of the terrain and the engagement. Furthermore, the lengthy *sarissa* is actually much better suited for engagements such as river crossings (for example: Alexander's battles at Granicus and Issus) than a shorter spear is (see the chapter The Anvil in Action).

73 Livy 35.27-30

74 Livy 35.27

75 M.F. Williams, 'Philopoemen's Special Forces: Peltasts and a New Kind of Greek Light-Armed Warfare (Livy 35.27)' *Historia* 53.3 (2004) 260

76 Plut. *Mor.* 183a

77 The name for the cavalry lance (*xyston* – ξυστόν) derives from the word ξέω meaning to 'scrape' or 'carve'. Sekunda ('Land Forces', 345) translates the term as 'whittled down'. It seems that the cavalry lance was a shortened version of the infantry pike – with additional differences in the configuration of the butt. Despite these differences, the artistic representations of this weapon show that it was still lengthy.

78 Arr. *Anab.* 1.15.5

79 Greek lances: Xen. *Eq.* 12.12; hoplite spears: *AP* 6.122, 6.123; other weapons: Strabo 12.7.3; Hdt. 7.92, Xen. *Hell.* 3.4.14, Xen. *Cyr.* 7.1.2

80 Theophr. *Caus.* pl. 3.12; quite a number of scholars see this passage of Theophrastus as a reference to the material that the *sarissa* was made from or simply state that the shaft was made from this wood (for example see: N.G.L. Hammond, 'What May Philip Have Learnt as a Hostage in Thebes?' *GRBS* 38.4 (1997) 367; Hammond, *King, Commander and Statesman*, 32-33; Gabriel, Philip II, 64; English, A*rmy of Alexander*, 17; M. Thompson, *Granicus 334BC: Alexander's First Persian Victory* (Osprey, Oxford, 2007) 24; Heckel and Jones, *Macedonian Warrior*, 13; Connolly, *Greece and Rome*, 69; Sheppard, *Alexander the Great at War*, 54; I. Worthington, *Philip II of Macedonia* (Yale University Press, New Haven, 2008) 27; P. Manti, 'The *Sarissa* of the Macedonian Infantry' *AncW* 23.2 (1992) 32. However, this is clearly not the case.

81 Manti, 'The *Sarissa*', 34; *English, Army of Alexander*, 18

82 Manti, 'The *Sarissa*', 33-34; see also English, *Army of Alexander*, 18

83 Manti, 'The *Sarissa*', 34, 39

84 Manti, 'The *Sarissa*', 37-39

85 English (*Army of Alexander*, 21) says that the foreshaft guard is 2 cubits long. This is clearly incorrect. Even in the smaller system of measurements used by Manti which uses a smaller (33cm) cubit, a tube 17cm in size can in no way be considered 2 cubits long. If the depictions of the socket of the head seen on the Alexander mosaic (which is cited by Manti and English as proof of a foreshaft guard on the *sarissa*) are taken as 2 cubits in length as English suggests, then the blade of the head would be more than 60cm in length – a size for which there is no archaeological proof and which would greatly alter the balance of the weapon.

86 This is calculated by deducting the thickness of the metal twice (equating to each side of the socket's diameter) from the overall diameter to determine what the inner size of the socket is.

87 This side of the tube found at Vergina is very rarely shown in images of it.

88 For example see: Markle, 'Spear', 324; Heckel and Jones, *Macedonian Warrior*, 13; Koepfer, 'The *Sarissa*', 37; Sheppard (*Alexander the Great at War*, 54) simply offers that the shaft was between 3-4cm in diameter.

89 For example see: Sekunda, *Macedonian Armies*, 16

90 Snodgrass, *Arms and Armour*, 118-119

91 Tyrt. 1; Hom. *Il.* 4.47, 19.390; Koepfer ('The *Sarissa*', 37) suggests that the shaft was made from either Cornelian Cherry or Ash, rather than specifically from one or the other. Park ('The Fight for Asia – the Battle of Gabiene' *AncWar* 3.2 (2009) 33) and Sekunda ('Land Forces' 329), on the other hand, say that the shaft of the *sarissa* was only made of Ash. Everson (*Warfare in Ancient Greece*, 175) says that the shaft of the *sarissa* may have originally been made from Cornelian Cherry, but that the longer weapons of the later Hellenistic Period were probably made of Ash.

92 Stat. *Theb.* 7.269; Theo. *Caus. Pl.* 3.11.3-4

93 Pliny (E) *HN* 16.228; See also: H. Lumpkin, 'The Weapons and Armour of the Macedonian Phalanx' *JAAS* 58.3 (1975) 197

94 Biton 52 – see E.W. Marsden, *Greek and Roman Artillery Vol.II – Technical Treatises* (Clarendon Press, Oxford, 1971) 71

95 Theo. *Caus. Pl.* 3.16.1-3, 5.6.2, 5.7.6

96 How a sufficient number of straight pieces of timber to adequately fashion the hundreds or thousands of spear and pike shafts used by their forces were procured by states like ancient Macedon is far from certain. There was a clear preference as to the type of wood used in shield construction (for example see: Theophr. *Caus. pl.* 5.3.4; Plin. (E) *HN* 16.77) but whether this wood was cultivated specifically for this purpose or merely obtained by other means is not outlined. Nor is there clear evidence for the cultivation of trees specifically for weapon manufacture. One method of obtaining mass quantities of shafts would have been by splitting larger logs lengthways into sections which could then be shaped into a shaft – most likely with a taper (see: Sekunda, *Macedonian Armies*, 16). Another method would have been by simply using the trunk of a young sapling to fashion a single shaft (as the spoke-shave bearing troops described by Xenophon are likely to have done in the field). This would suggest that some states in ancient Greece had access to large numbers of cultivated trees which may have been specifically grown for the purpose of weapon making, and would indicate that some sort of specialised industry, possibly on a massed scale, for the manufacture of weapons was in effect. Yet another way of obtaining this timber would have been through the process of 'coppicing' existing trees. Coppicing is the process whereby numerous shoots of new growth grow from the trunk of an existing tree when it is felled. Depending upon the species of tree, these shoots may grow long and straight. In some instances (again depending upon the species of tree and for how long the new shoots are left to grow before being cut themselves) the shoot may have a uniform thickness along its entire length or a slight taper. The growth of such shoots from the trunks of felled trees was a phenomenon known to the Greeks (see: Theophr. *Caus. pl.* 2.7.2, 3.7.1-2, 4.16.1-4). However, there is no reference to this (or any other) kind of specialised agricultural technique being used on a mass scale for the production of weaponry. In his study of the use of timber by ancient armies, Meiggs only concentrates on the use of wood in the construction of fleets, fortifications, buildings and siege equipment; with the omission of the use of timber in weapon manufacture. See R. Meiggs, *Trees and Timber in the Ancient Mediterranean World* (Clarendon Press, Oxford, 1982) 154-217. As such, any suggested means for the procurement of wood by the state for the manufacture of spears, pikes and other weapons can only be considered speculative.

97 Xen. *Cyr.* 6.2.32

98 Xenophon (*Cyr.* 6.2.32) states that the fashioning of shafted weapons was an acquired skill. Xenophon (*Lac.* 11.2) also refers to specialised craftsmen who accompanied Spartan armies in the field. Whether any of these craftsmen were spear makers is not stated.

99 It is interesting to note that the diameter of the tube is not uniform along its length but is smaller at one end. It would seem likely that this smaller end was attached to the forward end of a segmented *sarissa* – with the elasticity of the metal allowing it to grip onto the shaft. The larger diameter of the other end of the tube could then simply be slotted onto the rearward half of the *sarissa* when the weapon was assembled.

100 I would like to express my gratitude to the editors of the journal *Antichthon* for allowing me to reproduce material that had been released in an earlier article (C.A. Matthew, 'The Length of the *Sarissa*' *Antichthon* 46 (2012) 79-100) in parts of this section.

101 Asclep. *Tact.* 5.1 - δόρυ δὲ αὖ οὐκ ἔλαττον δεκαπήχεος, ὥστε τὸ προπῖπτον αὐτοῦ εἶναι οὐκ ἔλαττον ἢ ὀκτάπηκυ, οὐ μὴν οὐδὲ μεῖζον ἐτώλεσαν δύο καὶ δέκα πηχέων, ὥστε τὴν πρόπτωσιν εἶναι δεκάπηχυν.

102 Ael. Tact. *Tact.* 12 δόρυ δὲ μὴ ἔλαττον ὀκταπήχους, τὸ δὲ μήκιστον μέχρι τοῦ δύνασθαι ἄνδρα κρατοῦντα χρῆσθαι εὐμαρῶς. In chapter 1 of his work on tactics, Aelian details many of the sources that he has used as the basis for his research when putting his work together. The sources cited by Aelian include specific works on tactics by Stratocles, Frontinus, Aeneas, Cyneas the Thessalian, Pyrrhus the Epriote and his son Alexander, Clearchus, Pausanias, Evangeleus, Polybius, Eupolemus, Iphicrates, Poseidonius and Brion. Sadly, many of these works have not survived the passing of the centuries and are only briefly mentioned in the works of other writers like Aelian. One interesting omission from the list of sources provided by Aelian is the *Tactica* of Asclepiodotus. This poses two possibilities. Either a) Aelian borrowed heavily from the earlier work of Asclepiodotus, with which it shares numerous similarities, but did not want to draw attention to the fact that he had done so, or b), as it has been theorised, Asclepiodotus had simply released a work on tactics written by Poseidonius, of whom he was a pupil and whom Aelian does list as a source, under his own name. The question of the true authorship of Asclepiodotus' work, and therefore Aelian's potential use of it as a source, will probably never be satisfactorily addressed.

103 A.M. Devine, 'The Short Sarissa: Tactical reality or Scribal Error?' *AHB* 8:4 (1994) 132

104 Theophr. *Caus. pl.* 3.12; obviously the height of each Cornellian Cherry tree would vary so Theophrastus must only be making a generalization. However, such a generalization would not work unless the tree was compared to something of a standard length – such as a weapon. Indeed, the whole effectiveness of ancient combat using men in a massed formation like the Hellenistic pike-phalanx revolved having weapons that were relatively standardized across all members of a unit or army as this would then influence how those troops could be employed and which tactics and strategies could be used in battle (see following). It is thus not surprising that we find numerous references to both weapons and armour of specific sizes and shapes, for the Greeks, Macedonians and Romans, in the works of the ancient military writers and historians.

105 Polyaenus, *Strat.* 2.29.2; N.G.L. Hammond and F.W. Walbank (*A History of Macedonia III 336-167 BC* (Oxford University Press, 1988) 262) date this battle to 274 BC; see also: Sekunda, *Macedonian Armies*, 13

106 Polyb. 18.29 – τὸ δὲ τῶν σαρισῶν μέγεθός ἐστι κατὰ μὲν τὴν ἐξ ἀρχῆς ὑπόθεσιν ἑκκαίδεκα πηχῶν, κατὰ δὲ τὴν ἁρμογὴν τὴν πρὸς τὴν ἀλήθειαν δεκατεττάρων.

107 Ael. Tact. *Tact.* 14

108 Sekunda, *Army of Alexander*, 27

109 Heckel and Jones, *Macedonian Warrior*, 13

110 Markle, 'Spear', 323

111 Everson, *Warfare*, 175

112 Snodgrass, *Arms and Armour* 119

113 Ael. Tact. *Tact.* Praef.; the dedication to Hadrian appears in Robertello's 1552 edition and Arcerius' 1613 edition of Aelian's work. The dedication is then carried into Bingham's 1616 edition and Augustus' 1814 edition. The Köchly and Rüstow 1885 edition, on the other hand, has an alternate version of the text where the work is dedicated to Trajan (αὐτόκρατορ Καῖσαρ υἱὲ θεοῦ Τραϊανὲ σεβαστέ). It seems that this edition is of an earlier version of the work, initially begun under Nerva, for Trajan but which was, as Aelian admits, abandoned only to be later taken up again. Consequently, the later version of the text had its dedication altered to the new emperor, Hadrian.

114 Sekunda (*Army of Alexander*, 27) suggests that the *sarissa* in the time of Alexander may have actually been shorter than the length Theophrastus provides even though he is writing only a year after Alexander's

death. What Sekunda does not outline is what, if anything, occurred during this one year period to warrant the pike being lengthened. It seems more likely that Theophrastus is referring to a weapon that was used both during the time of Alexander and the period immediately following when Theophrastus was writing.

115 Hdt. 1.60; AP 12.50

116 Hdt. 2.149 - τῶν ποδῶν μὲν τετραπαλάστων ἐόντων, τοῦ δὲ πήχεος ἑξαπαλάστου; see also: W.F. Richardson, *Numbering and Measuring in the Classical World* (Bristol Phoenix Press, Bristol, 2004) 29-32

117 Richardson, *Numbering and Measuring*, 29-32; for examples see: Champion, *Pyrrhus*, 24; Gabriel, *Philip II*, 65; English, *Army of Alexander*, 17, 23-24; W.W. Tarn, *Hellenistic Military and Naval Developments* (Cambridge University Press, 1930) 14, 28; Heckel and Jones, *Macedonian Warrior* 6, 13; Snodgrass, *Arms and Armour*, 118; Connolly, *Greece and Rome*, 69; J. Warry, *Warfare in the Classical World* (University of Oklahoma Press, Norman, 1995) 72-73; Sheppard, *Alexander the Great at War*, 54, 82; Sekunda (*Macedonian Armies*, 13) suggests that a 16 cubit weapon measured 7.92m. This would make the cubit equal to 49.5cm.

118 I. Dekoulakou-Sideris, 'A Metrological Relief from Salamis' *AJA* 94:3 (1990) 445-451; the relief is now in the Piraeus Museum (#5352).

119 That is: 2.0cm x 4 = 8cm per 'palm', 4 x 8cm = 32cm per 'foot' and 6 x 8 cm = 48cm per cubit.

120 Dekoulakou-Sideris, *Metrological Relief* (n.17) 449

121 For the Olympic/Peloponnesian foot see: O. Broneer, *Isthmia Vol.I: Temple of Poseidon* (American School of Classical Studies at Athens, Princeton, 1971) 175-180.

122 Connolly ('Experiments', 106-107) bases his replica weapons on an Attic *cubit* of 48.7cm.

123 Broneer, *Isthmia*, 175-180

124 A. Michaelis, 'The Metrological Relief at Oxford' *JHS* 4 (1883) 339

125 W.B. Dinsmoor, 'The Basis of Greek Temple Design in Asia Minor, Greece and Italy' *Atti VII Congresso Internazionale di Archologia Classica I* (L'Erma di Bretschneider, Rome, 1961) 358-361

126 IG I³ 1453 (*M&L* 45) – clause 12

127 See: J.V.A. Fine, *The Ancient Greeks: A Critical History* (Cambridge University Press, 1983) 367; H.B. Mattingly, 'The Athenian Coinage Decree' *Historia* 10 (1961) 148-188; H.B. Mattingly, 'Epigraphy and the Athenian Empire' *Historia* 41 (1992) 129-138; H.B. Mattingly, 'New Light on the Athenian Standards Decree' *Klio* 75 (1993) 99-102

128 F. Hultsch, *Griechische und Römische Metrologie* – 2nd Ed. (Weidmann, Berlin, 1882) 30-34, 697; W. Dörpfeld, 'Beiträge zur antiken Metrologie 1: Das solonisch-attische System' *Ath. Mitt. VII* (Athens, 1882) 277; see also: F. Lammert, 'Sarisse' *RE Second Series IA* (Stuttgart, 1920) col. 2516; Marsden, *Greek and Roman Artillery*, ix

129 Tarn, *Military and Naval Developments*, 15; W.W. Tarn, *Alexander the Great Vol.II* (Ares, Chicago, 1981) 169-171

130 Arr. *Anab.* 5.4.4

131 Arr. *Anab.* 5.19.1

132 Diod. Sic. 17.88.4; Plut. *Alex.* 40; Diodorus (17.91.7) also refers to another Indian king, Sopeithes, whose height he says was 'exceeding (ὑπεράγων) 4 *cubits*'.

133 Tarn, *Alexander the Great Vol.II*, 170; this was a view shared by Manti ('The Sarissa', 40).

134 Tarn, *Military and Naval Developments*, 15

135 J.R. Mixter, 'The Length of the Macedonian Sarissa During the Reigns of Philip II and Alexander the Great' *AncW* 23:2 (1992) 22.

136 Manti, 'The Sarissa', 39-41; Strangely, Manti claims (p.40) that part of the proof of the existence of the 33cm *cubit* is that the large head, supposedly from a *sarissa*, found by Andronicos at Vergina measures exactly 1 ½ Macedonian *cubits*. However, 51cm (the size of the head) is not 1.5 times a small 'Macedonian' *cubit* of 33 cm, nor does it conform with a system incorporating a larger *cubit* of 45cm (unless it is assumed that the head measures 1 *cubit* and 2 *daktyloi*), so the size of the head found by Andronicos cannot be used to definitively support either position in the debate over the size of the *cubit*. Manti also says that the size of the tube found by Andronicos (17cm) is equal to ½ a *bematist's cubit* and that the size of the butt-spike that was found (44.5cm) equals 1 1/3 Macedonian *cubits*. None of these calculations work out correctly in any system of measurements. For other views on this debate see: Markle, 'Spear', 323; Dickinson, 'Length Isn't Everything', 51-52; R. Lane Fox, *Alexander the Great* (Penguin, London, 1973) 511.

137 Manti, 'The *Sarissa*', 41-42; Koepfer ('The *Sarissa*', 37) suggests that Theophrastus, Polybius and Asclepiodotus all used different *cubits* in their measurements of the *sarissa*.

138 Markle, 'Spear', 323; Lane Fox, *Alexander*, 511; Dickinson, 'Length isn't Everything', 51-62

139 Markle, 'Spear', 323; Lane Fox, *Alexander*, 511

140 Ath. 10.442b; Pliny (E), *HN* 6.61-62

141 See Pliny (E), *HN* 6.61-62; Strabo 11.8.9; a good tabulated summary of these recorded measurements can be found in D.W. Engels, *Alexander the Great and the Logistics of the Macedonian Army* (University of California Press, Berkeley, 1978) 157.

142 Engels, *Logistics*, 158; Heron, *Dioptra*, 34

143 Strabo 11.8.9

144 In this system 1 *daktylos* = 1.37cm, 1 palm = 5.5cm, 1ft = 22cm, a *cubit* = 33cm and a *stade* (600ft) = 13,200cm or 132m.

145 In this system 1 *daktylos* = 1.875cm, 1 palm = 7.5cm, 1ft = 30cm, a *cubit* = 45cm and a *stade* (600ft) = 18,000cm or 180m.

146 In this system 1 *daktylos* = 2.0cm, 1 palm = 8.0cm, 1ft = 32cm, a *cubit* = 48cm and a *stade* (600ft) = 19,200cm or 192m.

147 In his examination of Alexander's Bactrian campaign, English (*Field Campaigns*, 167) shows Alexander's army taking an indirect route around the mountains separating these two points which, according to the scale on the accompanying map, comes to only 180km. In a direct line, the distance works out to 160km. This is clearly incorrect regardless of which unit of measure is being used.

148 The largest margin of error is Strabo's calculation of the distance from Prophthasia (Juwain) to Arachoti Polis (Kelat-i-Ghilzai) which he gives as 4,120 *stades* or 791km in the larger 'late-Attic' system. The actual distance is 845km – a difference of around 9%.

149 Tarn, *Alexander the Great Vol.II*, 16

150 Manti, 'The *Sarissa*', 40

151 Hdt. 1.68

152 Hdt. 6.117

153 Plut. *Thes*. 36

154 Tarn, *Alexander the Great Vol.II*, 170

155 Email correspondence with J. Stroszeck of the Deutsches Archäologisches Institut 1-4 March 2008; see also: J. Stroszeck, 'Lakonisch-rotfigurige Keramik aus den Lakedaimoniergräbern am Kerameikos von Athen' AA 2 (2006) 101-120; H. van Wees, *Greek Warfare – Myths and Realities* (Duckworth, London, 2004) 146-147; C.F. Salazar, *The Treatment of War Wounds in Greco-Roman Antiquity* (Brill, Leiden, 2000) 233-234; W.K. Pritchett, *The Greek State at War – Part IV* (University of California Press, Berkeley, 1985) 133-134; L. van Hook, 'On the Lacedaemonians Buried in the Kerameikos' *AJA* 36.3 (1932) 290-292.

156 A. Schwartz, 'Large Weapons, Small Greeks: The Practical Limitations of Hoplite Weapons and Equipment' in D. Kagan and G.F. Viggiano, *Men of Bronze – Hoplite Warfare in Ancient Greece* (Princeton University Press, 2013) 165-167

157 Comparably, Vegetius (*Mil.* 1.5) states that both Roman cavalrymen and front line legionaries of the fourth century AD should be between 172cm and 177cm in height. This further suggests that an average height of around 170cm for the Greeks of the earlier Classical period is not implausible.

158 Asclep. *Tact*. 5.1

159 Ael. Tact. *Tact*. 12

160 K. Liampi, *Der makedonische Schild* (Rudolf Habelt, Bonn, 1998) 52-55, pl.1; P. Adam-Veleni, 'Χάλκινη ασπίδα από τή Βεγόρα τής Φλωρίνας' *Ancient Macedonia 5.1 – Papers Read at the Fifth International Symposium, Thessaloniki* (1993) 17-28; see also: Sekunda, 'Land Forces', 337; Sekunda, Macedonian Armies, 18; for finds of other shields (or parts thereof) which have been given similar estimates of size see: D. Pandermalis, 'Basile[ōs Dēmētr]iou' Myrtos: *Mnēme Ioulias Vokotopoulou* (Thessaloniki, 2000) xxi; P. Juhel, 'Fragments de "boucliers macédoniens" au nom de Roi Démétrios trouvés à Staro Bonce (République de Macédoine)' *Zeitschrift für Papyrologie und Epigrafik* 162 (2007) 165-180

161 Liampi, *Schild*, 53; K. Liampi, 'Der makedonische Schild als propagandistiches Mittel' *Meletemata* 10 (1990) 157-171; see also: Adam-Veleni, ('Χάλκινη ασπίδα', 17-28) who estimates the diameter of the Vegora shield as 73.6cm and may be the source for Liampi's later figure.

162 N.G.L. Hammond, 'A Macedonian Shield and Macedonian Measures' ABSA 91 (1996) 365

163 Hammond, 'Macedonian Shield', 365; Hammond also points out ([pg. 365] n. 6) that the military

fortifications at Dion and Thessaloniki, and the heroön at Yiannitsa, were all built using the 48cm cubit standard. This shows that the Macedonians were using a system of measurements incorporating a cubit much larger than most previous scholars have suggested. Markle, in his examination of a monument found in Macedonia containing representations of various shield types, cites several other monuments, reliefs and tomb paintings where small round shields are depicted with a diameter of around 75cm and larger round shields are depicted with a diameter of around 95cm. Markle interprets the smaller shields as being the Macedonian *peltē* and the larger ones being representations of the hoplite *aspis*. In both cases Markle states that the shields are depicted life size. However, while the size of the *aspis* could range from 80-122cm, the archaeological evidence indicates that the *peltē* was around 64-66cm. Thus the representations of shields that Markle cites are actually slightly bigger than life size – but both to the same degree. See: M.M. Markle, 'A Shield Monument from Veria and the Chronology of Macedonian Shield Types' *Hespeira* 68.2 (1999) 219-254.

164 For a discussion of the mould see: E.M. Moormann, *Ancient Sculpture in the Allard Pierson Museum Amsterdam* (Allard Pierson Series, Amsterdam, 2000) 187-188; see also: Liampi, *Schild*, 59-61, pl.5; interestingly, a metrological relief now in the Museum of Lepcis Magna contains a representation of a so called 'Ptolemaic cubit' which measures 52.5cm. Such a system of measures would incorporate a 'foot' of 45cm. However, it is clear from the shield mould that the Ptolemies, at least at the time when the mould was in use, were either not using this larger standard (or else the mould would have to be 2 'Ptolemaic feet' or 90cm in diameter) or that they were making shields with a diameter less than the 2 'feet' detailed by Asclepiodotus.

165 See: U. Peltz, 'Der Makedonische Schild aus Pergamon der Antikensammlung Berlin', *Jahrbuch der Berliner Museen*, Bd. 43 (2001) 331-343.

166 Everson, *Warfare*, 178

167 For example see: Heckel and Jones, *Macedonian Warrior*, 14; Connolly, *Greece and Rome*, 79; English, *Army of Alexander*, 24; M. Launey, *Recherches sur les armées hellénistiques: Bibliothèque des écoles françaises d'Athènes et de Rome* (Paris, de Boccard, 1949) 354; J. Kromayer and G. Veith, *Heerwesen und Kriegführung der Griechen und Römer* (C.H. Beck, Munich, 1928) 108; Worthington, *Philip II*, 27.

168 That is, 65cm / 32 *daktyloi* = 2.03cm per *daktylos*.

169 If the average diameter of the phalangite shield was greater than 64cm, this would mean that the units of measure used to represent its size were even larger than the 32cm foot.

170 Asclep. *Tact.* 4.1; close-order: τὸ πυκνότατον, καθ' ὃ συνησπικὼς ἕκαστος ἀπὸ τῶν ἄλλων πανταχόθεν διέστηκεν πηχυαῖον διάστημα; intermediate-order: τό τε μέσον, ὃ καὶ πύκνωσιν ἐπονομάζουσιν, ᾧ διεστήκασι πανταχόθεν δύο πήχεις ἀπ' ἀλλήλων.

171 Ael. Tact. *Tact.* 14; Polyb. 18.29; Arr. *Tact.* 11.3

172 For how the hoplite armies of the Classical age conformed to an interval of 45-50cm when deployed in close-order see: C.A. Matthew, 'When Push Comes to Shove: What was the *Othismos* of Hoplite Combat?' *Historia* 58:4 (2009) 406-407; numerous ancient writers state that the pikes held by the front ranks of the Hellenistic phalanx projected between the files and beyond the front of the line (see: Arr. *Tact.* 12; Polybius 18.29; Ael. Tact. *Tact.* 14; Asclep. *Tact.* 5.1). The positioning of these weapons between the files make it impossible for phalangites to stand in the 48cm interval of the close-order formation and create a line with 'interlocked shields' (or 'with shields brought together' as the term *synaspismos* can be translated) while keeping the shield in a protective position across the front of the body and the weapons poised for combat. It appears that phalangites only adopted a close-order to undertake such manoeuvres as 'wheeling' which required their pikes to be carried vertically (see: Ael. Tact. *Tact.* 31). Ignoring this statement by Aelian, some scholars have come up with different theories and models for how a file of phalangites may have stood in order to fit into a close-order interval and still engage in combat; with the men standing almost side-on (for example see: J. Warry, *Warfare in the Classical World* (University of Oklahoma Press, Norman, 1980) 72-73; Connolly, *Greece and Rome*, 109). Such models seem incorrect as they do not conform with the terminology (in such models the shields of the phalangites are neither 'interlocked' nor are they 'brought together'), the shield is removed from its protective position across the front of the body, and/or the phalangite is contorted into a position which would make the effective use of his weapon all but impossible. This suggests that the close-order formation was only used by troops armed as Classical hoplites to create a close-order 'shield wall' or by phalangites who were holding their pikes vertically while undertaking particular drill movements, as Aelian states.

173 Arr. *Tact.* 12.7 - τὸ δὲ μέγεθος τῶν σαρισῶν πόδας ἐπεῖχεν ἑκκαίδεκα.

174 For the details of the reforms of Iphicrates see pages 11-12.

175 *Excerpta Polyaeni*, 18.8
176 See: Asclep. *Tact*. 5.1; Ael. Tact. *Tact*. 13.3; Arr. *Tact*. 12.3; Polyb. 18.29
177 Plut. *Aem*. 19
178 Snodgrass, *Arms and Armour*, 119; Hammond, *King, Commander and Statesman*, 32
179 Hammond, 'Macedonian Shield', 366. This is additionally strange considering Hammond is one of the scholars who suggests that an even-fronted phalanx may have been used.
180 Such a formation was tested using volunteers armed with staves by Delbrück (*Geschichte der Kriegskunst im Rahmen der politischen Geschichte I* (G. Stilke, Berlin, 1900) 404) who seems to have had no trouble using it. However, he does not elaborate on whether he tested how the formation would work if members in the front ranks started to fall as detailed above.
181 Markle, 'Spear', 324, 326; Mixter ('Macedonian *Sarissa*', 24) clearly states that the butt was 'required as a counterweight for the heavy *sarissa* head'.
182 Asclep. *Tact*. 5.1: δόρυ δὲ αὖ οὐκ ἔλαττον δεκαπήχεος, ὥστε τὸ προπῖπτοναὑτοῦ εἶναι οὐκ ἔλαττον ἢ ὀκταπηχυ, οὐ μὴν οὐδε μεῖζον ἐτέλεσαν δύο καὶ δέκα πηχέων, ὥστε τὴν πρόπτωσιν εἶναι δεκάπηχυν
183 Ael. Tact. *Tact*. 14: τὸ δὲ τῶν σαρισσῶν μέγεθός ἐστι κατὰ μὲν τὴν ἐξ ἀρχῆς ὑπόθεσιν ἐκκαίδεκα πηχῶν, κατὰ δὲ τὴν ἀλήθειαν δεκατεσσάρων. τούτων δὲ δύο πήχεις ἀφαιρεῖται τὸ μεταξὺ τοῖν χεροῖν διάστημα τῆς προβολῆς. αἱ δὲ λοιπαὶ δέκαδυο πήχεις προπίπτουσι πρὸ τῶν σωμάτων. Similarly, Arrian (*Tact*. 12) says that the pike of the Iphicratean *peltast* 'approached' 16 Greek feet (512cm or just less than 11 *cubits*) in length. Of this length, Arrian says that four Greek feet (128cm or just under 3 *cubits*) were taken up with the grip so that the weapon extended 12ft (384cm or 8 *cubits*) beyond the bearer (τῶν σαρισῶν πόδας ἐπεῖχεν ἐκκαίδεκα καὶ τούτων οἱ μὲν τέσσαρες ἐς τὴν χεῖρά τε τοῦκατέχοντος καὶ τὸ ἄλλο σῶμα ἀπετείνοντο, οἱ δώδκα δὲ προεῖχον πρὸ τῶν σωμάτων ἐκάστου τῶν πρωτοστατῶν). This provides yet another correlation between the Iphicratean *peltast* and the subsequent Hellenistic phalangite.
184 Polyb. 18.29
185 For example see, Mixter, 'Macedonian *Sarissa*', 24; Snodgrass, *Arms and Armour*, 118; English, *Army of Alexander*, 21; A. Smith, 'The Anatomy of Battle – Testing Polybius' Formations' *AncWar* 5.5 (2011) 41-45; Sheppard, Alexander the Great, 54, 83; Cascarino, *Tecnica della Falange*, 62-63; Hammond, *Genius of Alexander*, 14; Sekunda, *Macedonian Armies*, 16
186 Connolly, 'Experiments', 103; I am fortunate enough to be in possession of one of the *sarissae* that Markle had made for the filming of a Japanese documentary in 2002. These mock pikes had no butt-spike and a fibreglass replica of the large 'head' found at Vergina (which gives the weapon a relatively central point of balance), and a 5.1m shaft 35-38mm thick. The shaft alone weighs 4.8kg and, if a metallic head and butt was attached, the complete weapon would weigh over 6kg – in the press release about the filming of the documentary it was stated that the pike weighed 8kg (see: http://www.une.edu.au/news/releases2002/September/110-02.html). Due to the sheer length and weight of the pike (at either 6kg or 8kg), and the absence of a heavy butt (which would shift the point of balance towards the rear of the weapon), it is very difficult to deploy this weapon in a combative stance and keep the tip off the ground, while holding it at the correct point of balance 2 cubits from the rearward end, and maintain it in this position for any more than a few minutes. Thus, in agreement with Connolly, I can only state that a weapon of this weight and configuration seems to be particularly unsuitable for a combat situation and, as such, it seems more likely that the *sarissa* was lighter and more manageable. For further discussion on the replicas used by Markle in his experiments see: P. Manti, 'The Macedonian *Sarissa*, Again' *AncW* 25.2 (1994) 77-91.
187 Connolly, 'Experiments', 103; see also: Connolly, *Greece and Rome*, 77; Connolly goes on to suggest that if Markle and Andronicos had considered this, their conclusions might have been very different. Interestingly, Connolly also states ('Experiments', 109) that a replica *sarissa* that he made for experimentation (based upon the small (97g) head and butt found at Vergina, and with a tapered shaft 5.84m in length) had a point of balance 35cm in front of the 4 cubit mark given by Polybius (i.e. about 227cm from the rear end). In his earlier book (p.77) he stated that this same configuration created a point of balance right on Polybius' 4 cubit mark. Regardless, this cannot be correct. Even a weapon with a uniform shaft using these attachments will have a point of balance only 66cm from the rear end rather than Connolly's 227cm. Additionally, Connolly states ('Experiments', 109) that a similar weapon, but with the 235g head described by Markle, will have a point of balance 48cm ahead of Polybius' 4 cubit mark (i.e. 240cm from the rear tip). Again, this cannot be correct as even a weapon with a uniform shaft configured in this manner would have a point of balance 117cm from the rear end. With additional weight distributed to the back due to a tapering of the shaft, the point of balance would shift even further

rearward. Sekunda (*Army of Alexander*, 28) says it is uncertain whether the *sarissa* possessed a butt or not. However, the one thing that definitively proves that it did, is that a large weight on the rear end of the shaft is required to give the weapon the correct point of balance.

188 English, *Army of Alexander*, 16

189 English, *Army of Alexander*, 20

190 English (*Army of Alexander*, 20) says that the head of the *sarissa* was 51cm long and 1.22kg in weight. This is clearly taken from the find made by Andronicos. English also says that the butt of the *sarissa* was 45cm long and weighed 1.04kg. While this is slightly inconsistent with the data given by Andronicos, English's characteristics are clearly taken from this find. Heckel and Jones (*Macedonian Warrior*, 13) also use the large head and butt found at Vergina in their examinations. However, even a cursory glance at these figures shows that the head is heavier than the butt and, therefore, any weapon using both of these attachments would not have a rearward point of balance. Clearly such implications have not been considered by these scholars.

191 Plutarch (*Aem.* 20) specifically states that the *sarissa* was wielded in both hands.

192 Polyb. 18.30; Ael. Tact. *Tact.* 14

193 Dickinson ('Length Isn't Everything', 52) says that the hands were separated by only 50cm. This seems unlikely regardless of which model for the position of the grip is being considered. To maintain a solid grip on the weapon, the hands would need to be spaced at least 70cm apart. This distance, plus the amount of the butt beyond the socket, and the thickness of the shield, equals the 96cm of a 2 *cubit* spacing for the grip.

194 See: P.J. Callaghan, 'On the Date of the Great Altar of Zeus at Pergamon' BCIS 28.1 (1981) 117; Markle, 'Shield Monument', 248-249; Liampi (*Schild*, 81-82) also dates this plaque to the second century BC.

195 Some modern works contain illustrations of how some scholars perceive the phalanx deployed while wielding the pikes in the Polybian manner. In all of these illustrations, the last 2 cubits of the weapon project into the interval occupied by the man behind. For example see: Cascarino, *Tecnica della Falange*, 62; a deployment with weapons wielded in this manner poses considerable problems with the proper functioning of the phalanx (see following).

196 Note: for ease of representation the weapon held by the man in the second rank has been removed from this image.

197 Note: for ease of representation any part of the weapon held by the man in the second rank that extends forward of his shield has been removed from this image.

198 Polyb. 18:30

199 Snodgrass, *Arms and Armour*, 119

200 Sekunda (*Army of Alexander*, 27) suggests that the *sarissa* possessed a head smaller than a hoplite spear so that it could more easily penetrate armour. However, not only was a head with a weight similar to that of a hoplite spear required to give the *sarissa* its correct point of balance, but the hoplite spear could easily penetrate the bronze plate armour of the day (see: Matthew, *Storm of Spears*, 130-146).

201 Sheppard, *Alexander the Great*, 54; Bosworth, *Conquest and Empire*, 260

202 Markle, 'Spear', 324

203 Markle, 'Spear', 334

204 Markle, 'Spear', 324

205 Gabriel (*Philip II*, 65) says the weight of the shaft was 4.25kg which is not overly different even though his calculations are based upon a 12 cubit weapon but using a smaller cubit of 45cm in length.

206 Tarn, (*Military and Naval Developments*, 12) suggests that the pike, while longer, was probably lighter than the hoplite spear. This is clearly wrong. Not only was the shaft of the *sarissa* thicker and longer than that of the hoplite spear (35mm in diameter and just over 5m long for the *sarissa* compared to 25mm in diameter and just over 2.5m long for the hoplite spear), the butt of the *sarissa* (1,070g for the Vergina find) was also considerably heavier than that of the hoplite spear (average 329g). Just the shaft of the *sarissa* (4.07g) was nearly triple the weight of the average hoplite spear which weighed just over 1.4kg. How Tarn came to such a conclusion is not outlined. Mixter ('Macedonian *Sarissa*', 24-25) says that the *sarissa* weighed 6.5kg which he states is seven times more than the longest spear. This is also clearly incorrect. Heckel and Jones (*Macedonian Warrior*, 13) suggest that the *sarissa* weighed between 6.3kg and 6.8kg. However, they were basing their calculations on the larger head found at Vergina which was not part of the *sarissa*.

207 Connolly ('Experiments', 103) notes how the *sarissa* re-created by Markle, using the large head found at Vergina, weighed 6.58kg (see also: Markle 'Spear', 324). This figure is also given by Gabriel (*Philip*

II, 65) and Mixter ('Macedonian *Sarissa*', 25). However, as has been shown, this larger head is unlikely to have come from a *sarissa*, and therefore Markle's calculations need to be revised. If the larger head (1,235g) is replaced with one weighing 174g, the overall weight of Markle's reconstruction would have been 5.5kg – the same weight as given above.

The Phalangite Panoply

1   A.M. Snodgrass, *Arms and Armour of the Greeks* (Johns Hopkins University Press, Baltimore, 1999) 114

2   Asclep. *Tact.* 5.1; τῶν δὲ φάλαγγος ἀσπίδων ἡ Μακεδονικη χαλκῆ ὀκτωπάλαιστος, οὐ κοίλη; Anderson ('Shields of Eight Palms' Width' CSCA 9 (1976) 1-6) suggests that Asclepiodotus' reference to the best shield being 'not too hollow' (as he interprets it) is actually advice that the *peltē* should be considerably more concave than the hoplite *aspis*. As proof of this he cites various artistic images which depict a shield (either *peltē* or *aspis*) which are distinctly 'bowl-like'. However, such depictions go against the archaeological evidence which suggest that the *peltē* was relatively flat and it can only be assumed that the depiction of shields with a more pronounced curvature is the result of the artistic medium employed and the skill of the artisan. Similarly Markle 'A Shield Monument from Veria and the Chronology of Macedonian Shield Types' *Hesperia* 68.2 (1999) 247-249) uses imagery from Hellenistic coins to suggest that two different types of shield were used – one with a pronounced bowl and the other that was relatively flat. However, due to the small size of the coins cited, the detail of the shield being depicted can hardly be considered totally accurate and, as noted, a deeply concave shield goes against the archaeological evidence.

3   Ael. Tact. *Tact.* 12; ἀσπὶς μὲν οὖν ἐστιν [ἡ] ἀρίστη χαλκῆ, Μακεδονική, οὐ λίαν κοίλη, ὀκταπάλαιστος.

4   The previous scholarly contention over the size of the cubit and other associated units of measure in the Hellenistic world has led to earlier scholars suggesting a smaller size of 60cm for the Macedonian shield based upon the descriptions given by Asclepiodotus and Aelian and the use of a 'foot' of 30cm (for example see: M.M. Markle, 'Use of the *Sarissa* by Philip and Alexander of Macedon' *AJA* 82.4 (1978) 326; R. Gabriel, *Philip II of Macedonia – Greater than Alexander* (Potomac, Washington, 2010) 64; J.R. Mixter, 'The Length of the Macedonian *Sarissa* During the Reigns of Philip II and Alexander the Great' *AncW* 23.2 (1992) 24; N.G.L. Hammond, *Alexander the Great: King, Commander and Statesman* (Duckworth, London, 1980) 32; W. Heckel and R. Jones, *Macedonian Warrior: Alexander's Elite Infantryman* (Osprey, Oxford, 2006) 14; P. Connolly, *Greece and Rome at War* (Greenhill Books, London, 1998) 70, 79; S. English, *The Army of Alexander the Great* (Pen & Sword, Barnsley, 2009) 24; G. Cascarino, *Tecnica della Falange* (Rome, Il Cerchio, 2011) 21, 60, 105-110. Other scholars simply offer a range for the size of the Macedonian shield between 60-75cm (for example see: T. Everson, *Warfare in Ancient Greece* (Sutton Publishing, Stroud, 2004) 178; P. McDonnell-Staff 'Hypaspists to *Peltasts*: The Elite Guard Infantry of the Antigonid Macedonian Army' *Ancient Warfare* 5.6 (2012) 22; A. Smith 'The Anatomy of Battle – Testing Polybius' Formations' *AncWar* 5.5 (2011) 43; R. Sheppard (ed.), *Alexander the Great at War* (Osprey, Oxford, 2008) 54, 84). Hatzoupoulos and Juhel ('Four Hellenic Funerary Stelae from Graphyra, Macedonia' *AJA* 113.3 (2009) 426), on the other hand, state that the shield was 64cm in diameter. Similarly, Connolly ('Experiments with the *Sarissa* – the Macedonian Pike and Cavalry Lance – a Functional View' *JRMES* 11 (2000) 109), created a replica shield 63cm in diameter for his experiments. While both of these statements may be based upon the physical evidence (Connolly, for example, based his replica on the size of the Pergamum shield), a standard unit of measure incorporating a foot of 32cm and a cubit of 48cm for the Hellenistic period does not seem to have been considered. Anderson ('Shields of Eight Palms', 1-6) suggests that Macedonian shields were actually larger than this. Anderson bases his conclusion on such things as the size of the depiction of the Macedonian shield in the Tomb of Lyson and Kallikles relative to other parts of the panoply, particularly armour, depicted within the same image. From this analysis he suggests that the Macedonian shield was closer in size to the hoplite *aspis* (somewhere around 80cm across). Sekunda (*Mcedonian Armies after Alexander 323-168BC* (Osprey, Oxford, 2012) 5) similarly suggests that the shields depicted on the Pergamon Plaque are of a larger variety than the phalangite *peltē*. However, such conclusions assume that all of the elements of the panoply in the images are drawn to the same scale. Anderson further cites the depiction of shields on the Monument of Aemilius Paullus at Delphi as proof that the Macedonian shield was larger than 64cm; see also: A.J. Reinach, 'La frise du monument de Paul-Émile à Delphes' *BCH* 34 (1910) 444-446. However, these shields are depictions of the *aspis* rather than the *peltē*. Strangely, Anderson even admits that 'the only round shield on the monument whose inner surface is visible...has a...*porpax* and...*antilabe* near the rim, exactly like the traditional Greek hoplite shield'. Anderson also

states that 'Hellenistic artists naturally prefer [to depict] the traditional Greek hoplite *aspis*' in their art. This would then throw a question mark over all of Anderson's conclusions. Pritchett (*Ancient Greek Military Practices: Part I* (Berkeley, University of California Press, 1971) 144-145) suggests that it is safer to follow base conclusions upon archaeological remains rather than an interpretation of passages like that of Asclepiodotus. While Pritchett believed that the archaeological record provided a size of around 80cm for the *peltē*, both the archaological and literary records more closely correlate to a size of 64cm.

5   N.G.L. Hammond, 'What May Philip Have Learnt as a Hostage in Thebes?' *GRBS* 38.4 (1997) 367; in another work (*King, Commander and Statesman*, 32), Hammond describes the *peltē* as a small metal shield; see also: Sekunda, *Macedonian Armies*, 18. These are presumably references to the shield's covering rather than suggesting that it was only made of metal.

6   B. Bennett and M. Roberts, *The Wars of Alexander's Successors 323-281BC Vol.II: Battles and Tactics* (Pen & Sword, Barnsley, 2009) 4

7   See Perrin's translation of Plutarch's passage (*Aem.* 20) in the Loeb Classical Library edition.

8   *Porpax*: Eur. *Hel.* 1376; Eur. *Phoen.* 1127; Eur. *Tro.* 1196; Soph. *Aj.* 576; Strabo 3.3.6; Plut. *Mor.* 193e; *antilabe*: Thuc. 7.65; Strabo 3.3.6

9   E. Kunze, *Bericht Über die Ausgrabungen in Olympia – Vol.V* (Berlin, Verlag Walter de Gruyter and Co., 1956) 35-68, pl.11-33; E. Kunze, *Bericht Über die Ausgrabungen in Olympia – Vol.VI* (Verlag Walter de Gruyter and Co., Berlin, 1958) 74-117, pl.13-32; M. Andronicos, *Olympia* (Ekdotike Athenon, Athens, 1999) 31-32, 70, 76; M.T. Homolle, *Fouilles de Delphes – Tome V* (Ancienne Librairie Thorin et Fils, Paris, 1908) 103-106; Connolly, *Greece and Rome*, 53; Everson, *Warfare*, 121-122, 161; P. Cartledge, 'Hoplites and Heroes: Sparta's Contribution to the Technique of Ancient Warfare' *JHS* 97 (1977) 13; N. Sekunda, *Greek Hoplite 480-323 BC* (Osprey, Oxford, 2004) 10, 50; C.A. Matthew, *A Storm of Spears: Understanding the Greek Hoplite at War* (Pen & Sword, Barnsley, 2012) 40.

10   Pliny (E), *HN* 16.209

11   P. Krentz, 'Hoplite Hell: How Hoplites Fought' in D. Kagan and G.F. Viggiano, *Men of Bronze – Hoplite Warfare in Ancient Greece* (Princeton University Press, 2013) 136.

12   Connolly, in his reconstructions of the pike-phalanx, had his replica *peltē* made from a dished oak core with a diameter of 63cm (see: Connolly, 'Experiments', 109). Connolly states that his replica weighed 8.06kg. This seems far too heavy when it is considered that a larger hoplite *aspis* made of pine weighs 6-8kg. Strangely, in an earlier work (*Greece and Rome*, 79) Connolly states that a replica shield that he had made weighed 5kg. Gabriel (*Philip II*, 64), suggests that a shield 61cm in diameter weighed 5.4kg. These lighter weights seem more likely, as the shield made for this research, fashioned from a pine core 64cm in diameter and approximately 1cm thick, and with a central armband and other fittings, weighs 4.6kg.

13   Mixter, 'Macedonian *Sarissa*', 24; Hammond, 'Philip as Hostage', 367; English, *Army of Alexander*, 23; S. English, 'Hoplite or Peltast? – Macedonian 'Heavy' Infantry', Ancient Warfare 2:1 (2008) 35; Heckel and Jones, *Macedonian Warrior*, 14; Everson, *Warfare*, 178; B. Bennett and M. Roberts, *The Wars of Alexander's Successors 323-281BC Vol.II: Battles and Tactics* (Pen & Sword, Barnsley, 2009) 4; Snodgrass, *Arms and Armour*, 118; J. Pietrykowski, *Great Battles of the Hellenistic World* (Pen & Sword, Barnsley, 2009) 15; J. Warry, *Warfare in the Classical World* (University of Oklahoma Press, Norman, 1995) 68; Markle, 'Shield Monument', 220; I. Worthington, *Philip II of Macedonia* (Yale University Press, New Haven, 2008) 27; Anderson ('Shields of Eight Palms', 1-6) suggests that larger shields, closer in size to the hoplite *aspis*, could also be carried in this manner. Sekunda ('Land Forces' in P. Sabin, H. Van Wees, and M. Whitby, (eds.), *The Cambridge History of Greek and Roman Warfare Vol. I: Greece, the Hellenistic World and the Rise of Rome* (Cambridge University Press, 2007) 338) suggests that the *ochane* is a hand grip similar to the *antilabe* of the hoplite *aspis*. This is clearly incorrect.

14   Hom. *Il.* 2.388, 5.796, 11.38, 12.401, 14.404, 16.803, 17.290, 18.480; Hdt. 1.171; see also: Matthew, *Storm of Spears*, 255-256

15   Plut. *Cleom.* 11

16   Plut. *Aem.* 19

17   M.M. Markle, 'The Macedonian *Sarissa*, Spear and Related Armour' *AJA* 81.3 (1977) 326

18   Some scholars suggest that both an armband and shoulder-strap was used to carry the *peltē*; for example see: M. Thompson, *Granicus 334BC: Alexander's First Persian Victory* (Osprey, Oxford, 2007) 24; Snodgrass, Arms and Armour, 117-118; Sheppard, *Alexander the Great at War*, 54

19   For the details of some surviving examples of the *porpax* see: P.C. Bol, *Argivische Schilde* (De Gruyter,

Berlin, 1989) 126-164

20 Heckel and Jones, *Macedonian Warrior*, 14

21 For example see: Connolly, 'Experiments', 109-111; Connolly, *Greece and Rome*, 79; Everson, *Warfare*, 178; Sheppard, *Alexander the Great at War*, 81

22 Many modern illustrations of the inside of the *peltē* show the *ochane* mounted towards the top of the shield (for example see: Connolly, *Greece and Rome*, 78; Warry, *Warfare*, 73). It was found in the construction of the phalangite panoplies used in this research that a shoulder-strap mounted in this position did not provide good support for the shield due to the angle of the arm when wielding the *sarissa*. It seems much more likely that the shoulder-strap was mounted in such a way that it would not only support the shield in a protective position, but also partially take the weight of the heavy *sarissa* as well. Connolly, in his reconstructive tests, used a *peltē* with this exact same configuration.

23 Everson, *Warfare*, 178; see also: Connolly, *Greece and Rome*, 79

24 Connolly ('Experiments', 109) states that the replica used in his experiments had this configuration. Cawkwell (*Philip of Macedon* (Faber & Faber, London, 1978) 158) simply states that the *peltē* was carried on the forearm. This seems unlikely.

25 Strangely, the illustration accompanying Thompson's examination of Alexander the Great's battle at the Granicus River in 334BC shows the entire phalanx fighting with their shield slung across their back. This cannot be correct. There would be no point in carrying a piece of defensive armament like a shield if it was not going to be placed in a protective position across the front of the body for combat; see: Thompson, *Granicus*, 74-75.

26 Sekunda (*The Persian Army 560-330BC* (Osprey, Oxford, 2005) 27) interprets Diodorus' reference to the reforms of Iphicrates (15.44.2) as saying that the *peltē* and the *aspis* were the same size. This is clearly an incorrect reading of the passage as Diodorus specifically states that 'the Greeks were using shields that were large and difficult to handle. These he discarded and replaced with small round ones of moderate size.'

27 Arr. *Anab.* 7.11.3; Curt. 4.13

28 Diod. Sic. 17.57-59, 18.63, 19.12-48; Plut. *Eum.* 13-19;

29 Livy, 37.40

30 Bronze Shields serving with Antigonus at Sellasia (222BC): Polyb. 2.66; with Perseus at Pydna (168BC): Plut. *Aem.* 18; Livy 44.41; Diod. Sic. 31.8.10; Gold Shields: Plut. *Eum.* 14; White Shields serving with Antigonus (222BC): Plut. *Cleom.* 23; at Pydna (168BC): Livy 44.41; Diod. Sic. 31.8.10; for a discussion on whether the 'Gold Shields' ever existed or not see: N. Sekunda, *The Seleucid Army* (Montvert, Stockport, 1994) 14-15; Sekunda, *Macedonian Armies*, 35-37

31 Polyb. 30.25

32 1 Macc. 6.39

33 Everson, 178

34 *SEG* 40.524; the whole decree reads: 'those not carrying the weapons appropriate to them are to be fined according to the regulations: for the armour (*kotthybos*), two obols, the same amount for the helmet (*konos*), three obols for the *sarissa*, the same for the sword (*machaira*), for the greaves (*knemides*) two obols, for the *aspis* a drachma. In the case of officers (*hegemons*), double [shall be the fine] for the arms mentioned, two drachmas for the breastplate (*thorax*), a drachma for the half-breastplate (*hemithorakion*). The secretaries (*grammateis*) and the chief assistants (*archyperetai*) shall extract the penalty, after identifying the transgressors to the King (*basileus*)' - τοὺς μὴ φέροντας τι τῶν καθηκόντων αὐτοῖς ὅπλων ζημιούτωσαν κατά τα γεγραμμένα· κοτθύβου ὀβολοὺς δύο, κώνου τὸ ἴσον, σαρίσης ὀβολοὺς τρεῖς, μαχαίρας τὸ ἴσον, κνημίδων ὀβολοὺς δύο, ἀσπίδος δραχμήν. Ἐπὶ δὲ τῶν ἡγεμόνων τῶν τε δεδηλωμένων ὅπλων τὸ διπλοῦν καὶ θώρακος δραχμὰς δύο, ἡμιθωρακίου δραχμήν. Λαμβανέτωσαν δὲ τὴν ζημίαν οἱ γραμματεῖς καὶ οἱ ἀρχυ[πηρέτ]αι, παραδείξαντες τῶι βασιλεῖ τοὺς ἠθετηκότας.

35 McDonnell-Staff, 'Hypaspists to Peltasts', 22

36 Plut. *Cleom.* 23

37 Hes. *Op.* 129-234

38 Livy 44.41

39 Liampi's work (*Der makedonische Schild* (Rudolf Habelt, Bonn, 1998)) is the most comprehensive study of the designs found on Macedonian shields.

40 For a discussion of the developmental stages of these designs, and their influence of pottery decorations, see: P.J. Callaghan, 'Macedonian Shields, 'Shield Bowls' and Corinth: A Fixed Point in Hellenistic Ceramic Chronology?' *AAA* 11 (1978) 53-60

41 For an examination of the various designs found of hoplite shields see: G.H. Chase, *The Shield Devices of the Greeks in Art and Literature* (Ares, Chicago, 1979); Sekunda (*Macedonian Armies*, 20) suggests that the central design changed with the rise of a new monarch or to commemorate certain events. While this may have been the case in the time of the Successors, it seems that in the time of Alexander the Great, the central motifs were to identify specific units within the army (see following).

42 These symbols are: on coins issued in Macedonia c.335BC - the lightning bolt (see: M.J. Price, T*he Coinage in the Name of Alexander the Great and Philip Arrrhidaeus Vol.I and II* (British Museum Press, London, 1991) 397-416; C.A. Matthew 'For Valour: The Shield Coins of Alexander and the Successors' *JNAA* 20 (2009/2010) 23 (15-34)); on coins issued in Sardes c.334BC - a bust of Herakles or a caduceus (Price, *Coinage in the Name of Alexander*, 2546-2614; Matthew, 'For Valour' 23); on coins issued in Miletus c.334-333BC – a Gorgon head, a pellet or an axe (Price, *Coinage in the Name of Alexander*, 2063-2072; Matthew, 'For Valour' 23-24); on coins issued in Asia Minor c. 333-332BC – a bust of Herakles (Price, *Coinage in the Name of Alexander*, 2801-2808a; Matthew, 'For Valour' 24); on coins issued from Salamis c.332BC – a Gorgon head (Price, *Coinage in the Name of Alexander*, 3157-3162a; Matthew, 'For Valour' 24).

43 Hephaestion's cavalry unit carried an image of him after he died (Arr. *Anab.* 7.14.10). The word Arrian uses (*semeion* – σημεῖον) can be interpreted as a banner of some kind (as per Thuc. 4.42), or as a design carried on the shield (as per Hdt. 1.171; Eur. *Ph.* 1114).

44 Heckel and Jones, *Macedonian Warrior*, 14-15

45 Polyaenus, Strat. 4.2.10

46 Aen. *Tact.* 29.4

47 *SEG* 40.524

48 Plut. *Aem.* 32

49 P. Dintsis (*Hellenistische Helme* (Giorgio Bretschneider Editore, Rome, 1986) 169-177) also examines what is known as 'Egyptian limestone helmets' (*Die Ägyptischen Kalksteinhelme*). These are basically stone busts, with very few (if any) details of the face, but with a much higher level of detail of a helmet that is being worn.

50 Dintsis, *Hellenistische Helme*, 1-23, supplement 1, cards 1-2

51 An example of the mould used to create such helmets is now housed in the Allard Pierson Museum, Amsterdam. It has been suggested that this, and other examples of stone helmet 'templates' may actually be store models, put on display in the armour maker's workshop, so that customers would be able to get an idea of what a finished product may look like. For a discussion of the different theories relating to these 'templates' see: E.M. Moormann, *Ancient Sculpture in the Allard Pierson Museum Amsterdam* (Allard Pierson Series, Amsterdam, 2000), 187, 189-191, pl.88-90.

52 Dintsis, *Hellenistische Helme*, 23-57, supplement 2, cards 3-4

53 Dintsis, *Hellenistische Helme*, 57-75, supplement 3, cards 5-6

54 Dintsis, *Hellenistische Helme*, 75-77, supplement 4, cards 7-8

55 Dintsis, *Hellenistische Helme*, 77-87, supplement 5, card 9

56 Dintsis, *Hellenistische Helme*, 87-97, supplement 6, cards 10-11

57 Dintsis, *Hellenistische Helme*, 135-143, supplement 10, card 17-18

58 Dintsis, *Hellenistische Helme*, 105-113, supplement 8, cards 13-14

59 A fourth century grave near Salonika contained a Chalchidian helmet (now in the British Museum) which suggests that this type of helmet was still being worn in some areas at this time. See: Snodgrass, *Arms and Armour*, 116.

60 Dintsis, *Hellenistische Helme*, 113-132, supplement 9, card 15

61 Dintsis, *Hellenistische Helme*, 97-105, supplement 7, card 12

62 Dintsis, *Hellenistische Helme*, 143-149, supplement 11, card 19

63 Dintsis, *Hellenistische Helme*, 149-169, supplement 12, card 20

64 Dintsis, *Hellenistische Helme*, 173-183, supplement 13, card 21

65 Dintsis, *Hellenistische Helme*, 183-199, card 22

66 B.M. Kingsley, 'The Cap that Survived Alexander' *AJA* 85 (1981) 39-46

67 Warry, *Warfare*, 89; the common Corinthian style helmet of the Greek Classical Age, did provide a good level of vision and did not need to evolve in this manner as Warry suggests (see. Matthew, *Storm of Spears*, 96-101). However, the development of more open-style helmets during the later fifth century BC, and the incorporation of openings for the ears, did improve both the ventilation and the hearing. However, the protection of such helmets was not absolute (as indeed it was not for the Corinthian either).

Philip II lost an eye during the siege of Methone in 354BC (Diod. Sic. 16.34.5), and Plutarch (*Mor.* 339c) details how Tarrias was struck in the eye by a missile during Philip's siege of Perinthus in 340BC.

68 Afric. *Cest.* 7.1.53-55

69 For example see: N. Sekunda, *The Army of Alexander the Great* (Osprey, Oxford, 1999) 27, pl.D-H; J. Warry, *Alexander 334-323BC* (Osprey, Oxford, 1991) 82; Warry, *Warfare*, 89; Thompson, *Granicus*, 74-75; Heckel and Jones, *Macedonian Warrior*, 16-17, Pl. A-E; Connolly, *Greece and Rome*, 79-80

70 Warry, *Warfare*, 89

71 Dintsis, *Hellenistische Helme*, 113-132, supp. 9, card 15

72 Dintsis, *Hellenistische Helme*, 113-132, supplement 9, card 15. Warry (*Warfare*, 89) uses the term 'Thracian' for both the ridged helmet and for the type with the bulbous extension which is otherwise known as the 'Phrygian'.

73 Dintsis (*Hellenistische Helme*, 23-55, 218-233) lists six helmets of this type in the collections of various museums around the world. All are made of bronze and are dated between the fourth century BC and the third century BC. However, the depiction of this style of helmet on coinage, in monumental reliefs, and on grave *stele* continued well into the second century BC.

74 The large horse hair crests used in the Classical period, and on other helmets used throughout the Hellenistic Period, achieved this same result. For an object falling down upon the wearer, such as an arrow fired in a lob, the easiest way for the missile to penetrate the plate metal of the helmet is to hit the relatively flat surface of the crown of the helmet almost perpendicular so that the largest amount of its impact energy is used to pierce the plate. However, the addition of a crest, ridge or bulbous extension ensures that there are no flat surfaces on the top of the head which an incoming arrow might strike.

75 For the Alexander Sarcophagus see: Dintsis, *Hellenistische Helme*, 221-222; for the Pergamum frieze see: Dintsis, *Hellenistische Helme*, 230-231; Dintsis (*Hellenistische Helme*, 289-290) lists the helmets from the tomb of Lyson and Kallikles in his category of Glockenhelmes ('Bell' Helmets) rather than under the Thracian.

76 See: M. Mitchiner, *Indo-Greek and Indo-Scythian Coinage Vol.I – The Early Indo-Greeks and their Antecedents* (Hawkins Publications, Sanderstead, 1975) 21; Dintsis, *Hellenistische Helme*, 215

77 For an excellent photo of this helmet see: Sekunda, *Army of Alexander*, 26

78 Everson, *Warfare*, 181; Connolly, *Greece and Rome*, 80

79 For a comprehensive commentary and overview of the decades old debate over whether Tomb II is that of Philip II or not which, at the time of writing at least, seems to favour that it is, in fact, the tomb of Alexander the Great's half-brother Philip III, see: M.B. Hatzopoulos, 'The Burial of the Dead (at Vergina) or The Unending Controversy on the Identity of the Occupants of Tomb II' *Tekmeria* 9 (2008) 91-118; for discussions and counter-discussions of the potential occupants of other tombs at Vergina and the possible owners of some of the items found therein see: E.N. Borza, 'Royal Macedonian Tombs and the Paraphernalia of Alexander the Great' *Phoenix* 41.2 (1987) 105-121; N.G.L. Hammond, 'Arms of the King: The Insignia of Alexander the Great' *Phoenix* 43.3 (1989) 217-224

80 Everson, *Warfare*, 184

81 Sekunda, *Army of Alexander*, 27, Pl. F-G plus commentaries; Warry (*Warfare*, 89) also suggests that the painting of helmets, or decorating them with elaborate enamel inlays (such as a mixture of sulphur and copper) was a fairly common practice in the Hellenistic period.

82 Everson, *Warfare*, 184

83 Everson, *Warfare*, 183

84 Everson, *Warfare*, 183

85 Afric. *Cest.* 1.1.45-50; Ὀλίγα δὲ τούτων παρεποίησαν οἱ ἐπίγονοι Μακεδόνες διὰ τὸ τῶν πολέμων ποικίλον, κοινὴν καὶ κατὰ Βαρβάρων καὶ πρὸς αὐτοὺς τὴν ὅπλισιν ἐπισκευάσαντες. σημεῖον δὲ [τὸ] ἐλευθέρας τῶν μαχομένων τὰς ὄψεις ὑπὸ πίλῳ Λακωνικῷ ἐν τῇ Μακεδονικῇ γεγενῆσθαι. καλοῦσι δὲ χρῆμα καὶ ἐπιτήδευμα [τοῦτο] τὸ τοῦ στρατιώτου βασιλέως.

86 Hatzopoulos and Juhel, 'Four Hellenic Funerary Stelae', 426-428

87 Men. *Per.* 174

88 Snodgrass, *Arms & Armour*, 125

89 Plut. *Mor.* 180b; Afric. *Cest.* 1.1.45-48

90 Antipater, *Garland of Philip*, 41 - καυσίη ἡ τὸ πάροιθε Μακηδόσιν εὔκολον ὅπλον καὶ σκέπας ἐν νιφετῶι καὶ κόρυς ἐν πτολέμωι ἱδρῶ διψήσασα πιεῖν τεόν

91 *SEG* 36.1221

92 Ph.Byz. *Mech. Syn.* 5.77-78

93 For example see the images and discussion of the facade of the Aghios Athanasios tomb in: M. Tsimbidou-Avloniti, *Makedonikoi Taphoi ston Phoinika kai ston Aghio Athanasio Thessalonikes* (Ekdose tou Tameiou Archaiologikon Poron kai Apallotrioseon, Athens, 2005); M. Tsimbidou-Avloniti, 'Excavating a painted Macedonian tomb near Thessaloniki. An astonishing discovery' in M. Stamatopoulou, and M. Yeroulanou (eds.), *Excavating Classical Greek Culture: Recent Archaeological Discoveries in Greece* (Beazley Archive and Archaeopress, Oxford, 2002), 91-97; M. Tsimbidou-Avloniti, 'Les peintures funeraires d'Aghios Athanassios' in S. Descamps-Lequime (ed.), *Peinture et couleur dans le monde grec antique* (Musée du Louvre, Paris, 2007), 57-67; M. Tsimbidou-Avloniti, "La tombe macedoine d'Hagios Athanasios pres de Thessalonique" in A.M. Guimier-Sorbets, M.B. Hatzopoulos, and Y. Morizot (eds.), *Rois, cites, necropoles: Institutions, rites et monuments en Macedoine* (Meletemata 45) (Research Centre for Greek and Roman Antiquity, Athens, 2006), 321-331; O. Palagia, "Hellenistic Art" in R. Lane Fox, (ed.), *Brill's Companion to Ancient Macedon: Studies in the Archaeology and History of Macedon, 650 BC-300 AD* (Brill, Leiden, 2011), 477-493; interestingly, the figures wearing the *kausia* on this facade are not wearing any other armour whereas those who are wearing armour also wear a metal helmet.

94 C. Saatsoglou-Paliadeli, 'Aspects of Ancient Macedonian Costume' *JHS* 113 (1993) 123

95 Heckel and Jones, *Macedonian Warrior*, 15-16; Sheppard, *Alexander the Great at War*, 81

96 Afric. *Cest.* 1.1.25-26

97 Afric. *Cest.* 1.1.25-28 - ὅ τε πῖλος περὶ τῇ κεφαλῇ κυνῆν [ἔχων] ἐτέραν δὲ ἐπιθήκην χαλκοῦ καὶ ἄλλην ἐπὶ ἄλλη περικεφαλαίαν τυγξάνει πρὸς τὰ ἀπὸ σφευδόνης βλήματα περιθλωμένης μὲν τῆς ἔξω λεπίδος καὶ συνεικούσης ὡς μὴ ἐφικέσθαι τὸ πεμφθὲν τοῦ ἐνδοτέρω τῆσ κεφαλῆς ἐπιβλήματος

98 For a discussion of what the *kausia* was made from, its shape and other aspects of its use see: Saatsoglou-Paliadeli, 'Ancient Macedonian Costume', 123-143

99 Eustathius, *Il.* 255 – καυσία, ἥ τισ ἢν κάλυμμα κεφαλῆς Μακεδονικὸν ἐκ πίλου, ὡς τιάρα, σκέπουσά τε ἀπὸ καύσωνος καὶ ὡς εἰς περικεφαλαίαν συντελοῦσά τι

100 Polyaenus, *Strat.* 5.44.5

101 G.T. Griffith, 'MAKEDONIKA: The Macedonians Under Philip and Alexander' *Proceedings of the Cambridge Philological Society* (1956-1957) 3-10; English, *Army of Alexander*, 25; S. English, *The Field Campaigns of Alexander the Great* (Pen & Sword, Barnsley, 2011) 90, 140.

102 Cawkwell, *Philip*, 158; Snodgrass, *Arms & Armour*, 117; J. Pietrykowski, 'In the School of Alexander – Armies and Tactics in the Age of the Successors' *AncWar* 3.2 (2009) 23; Mixter ('Macedonian *Sarissa*', 24) says they did not wear cumbersome metal breastplates; Heckel and Jones (*Macedonian Warrior*, 15-16) and Sheppard (*Alexander the Great at War*, 81) suggests not only that armour may not have been worn by members of the rear ranks of the phalanx, but that they did not wear helmets either – only the *kausia*. Pietrykowski (*Great Battles*, 15) confusingly suggests that the armour worn by the front ranks of the phalanx was lightened, and that the rear ranks may have worn no armour at all, but then goes on to describe the phalanx as 'row upon gleaming row of spearheads projecting from a huddled mass of armour and shields'. Worthington (*Philip II of Macedonia* (New Haven, Yale University Press, 2008) 27) suggests that phalangites did not wear any armour as the densely arrayed series of pikes of the phalanx was protection enough. D. Karunsnithy ('Of Ox-Hide Helmets and Three-Ply Armour: The Equipment of Macedonian Phalangites as Described through a Roman Source' *Slingshot* 213 (2001) 33-40) suggests that men in the rear ranks may have been bare-legged, carried a cheap, mass-produced, wicker shield and a helmet made of leather. See also: D. Lush, 'Body Armour in the Phalanx of Alexander's Army' *AncW* 38.1 (2007) 27.

103 Hom. *Il.* 2.529

104 Snodgrass (*Arms and Armour*, 123) says that the trend from the fourth century onwards was to reduce the weight of infantry armour. From the perspective of the Iphicratean reforms of 374BC, this is certainly the case.

105 The barrel-like shape created by wrapping this panel around the body has also led to this style of armour being referred to as a 'tube and yoke' or 'box' corslet. For further discussions of the evidence for, and construction of, the *linothorax* see: E. Borza, *In the Shadow of Olympus: The Emergence of Macedonia* (Princeton University Press, 1990) 298-299; G.S. Aldrete, S. Bartell and A. Aldrete, *Reconstructing Ancient Linen Body Armour: Unraveling the Linothorax Mystery* (The Johns Hopkins University Press, Baltimore, 2013) 22-30, 31-57, 58-89

106 P. Bardunias, 'Don't Stick to Glued Linen – The Linothorax Debate' *AncWar* 4.3 (2011) 50

107 Sheppard (*Alexander the Great at War*, 82) suggests that the *linothorax* weighed between 5.0-6.3kg. The *linothorax* made for this research, which was a composite armour with a double layer of *pteruges* made

from 5mm thick ox-hide covered on both sides with 2 layers of linen which were then painted with a white-wash, decorated, sealed, and had the edges reinforced with stitched cloth, resulting in a material 7-9mm thick, weighs 5.1kg.

108 For some of the discussions relating to this armour see: Bardunias, '*Linothorax* Debate', 48-53; Aldrete, Bartell and Aldrete, *Reconstructing Ancient Linen Body Armour*, 57-68.

109 Bardunias, '*Linothorax* Debate', 51

110 Aldrete, Bartell and Aldrete, *Reconstructing Ancient Linen Body Armour*, 149-159

111 Thompson, *Granicus*, 24

112 See: Mitchiner, *Indo-Greek and Indo-Scythian Coinage*, 21

113 Plut. *Alex*. 32

114 McDonnell-Staff, 'Hypaspists to Peltasts', 24

115 Snodgrass (*Arms and Armour*, 122) suggests that only cavalry wore metal cuirasses.

116 Paus. 10.26.2

117 Connolly, *Greece & Rome*, 54-55; see also: Everson, *Warfare*, 142

118 Snodgrass, *Arms & Armour*, 50; Connolly, *Greece & Rome*, 54-59; Everson, *Warfare*, 142

119 This is based upon the weight of a replica muscled cuirass that was used in the research on the warfare of the Classical Age discussed in Matthew, *A Storm of Spears*.

120 English, *Army of Alexander*, 25

121 Connolly, *Greece and Rome*, 70, 80; Sekunda (*Macedonian Armies*, 3) similarly suggests that the depiction of the muscled cuirass on the *Stele* of Nikolaos identifies him as a file-leader.

122 Snodgrass, *Arms and Armour*, 117

123 Warry, *Alexander*, 82

124 Arr. *Anab*. 1.28.7

125 Diod. Sic. 17.44.2 – διὰ γὰρ τῶν θωράκων λαὶ τῶν ὑποδυτῶν παρεισπίπτουσα ἡ ἄμμος καὶ διὰ τὴν ὑπερβολὴν τῆς θερμασίας λυμαινομένη. See also: Curt. 4.3.26. Lush ('Body Armour', 20) suggests that the similarities between the accounts of this incident found in Diodorus and Curtius may be indicative of the use of a common source – possibly an eye-witness to the event such as Cleitarchus.

126 Lush ('Body Armour', 19) suggests that while there are no specific references to which infantry units were involved in this event, it is clear that the troops were Macedonians rather than mercenaries, and the lack of references to the *hypaspists* would suggest that they were members of the pike-phalanx.

127 Nepos, *Iphicrates*, 1.3-4

128 Plut. *Alex*. 60

129 Diod. Sic. 17.95.4 – ἐκομίσθησαν δὲ καὶ πανοπλίαι διαπρεπεῖς πεζοῖς μὲν δισμυρίοις καὶ πεντακισξιλίοις; Curt. 9.3.21 – *adduxerat armaque xxv milibus auro et argento caelata pertulerat*. It is interesting to note that Diodorus uses the word 'panoplies' (πανοπλίαι) while Curtius uses the word 'arms' (*armaque*). As such, none of them specifically say that what is received is armour. Consequently, anything combustible that conforms with this description, such as shields or pike-shafts could be the items (or even part thereof) that were replaced while the older items were burned. Lush ('Body Armour', 20) argues that other items, such as ox-hide helmets or non-metallic greaves, may have also been burned.

130 Curt. 9.3.21-22

131 McDonnell-Staff ('Hypaspists to Peltasts' 23) suggests that the *linothorax* was just made of leather (see also: Sekunda, *Army of Alexander*, 27; Lush, 'Body Armour', 23). While it is possible that some sets were made in this manner, descriptions of this type of armour being specifically made from linen (such as Aeneas Tacticus' use of the term *thorakes lineoi* (29.4)) suggest that it was either made wholly of layers of cloth glued together, or that it may have been made of leather covered with layers of linen. Everson (*Warfare*, 192) basing his conclusion of the find of an iron cuirass fashioned in the likeness of a *linothorax* found in the so-called 'Tomb of Philip II' in Vergina, suggests that the armour may have been made from metal plates covered with linen or some other cloth.

132 Ael. Tact. *Tact*. 2, 48; the term that Aelian uses to describe the additional protective layer is *kassides*. The translation of *kassides* is something of a problem. Some have seen this word as a reference to a covering of leather of some kind – possibly scales or even an outer layer of armour, while others interpret it as meaning the attachment of tin plates or scales to the armour, while yet others see it as a reference to metal armour (like chain-mail) – see also n.131 above for the theory that the *linothorax* was possibly made from metal plates covered with cloth. It is clear from the passage that Aelian is suggesting that the *peltasts* have some form of additional protection as part of their defensive armament. Unfortunately, there is no clear way of distinguishing what that was by this passage alone.

133 *SEG* 40.524

134 Heckel and Jones (*Macedonian Warrior*, 15) suggest that the Amphipolis Decree makes no mention of armour except for officers without detailing what they consider the *kotthybos* to be and that, by default, the regular phalangite may have worn the unmentioned *linothorax*. Anderson ('Shields of Eight Palms', 2) similarly suggests that the Amphipolis decree only refers to the armour worn by officers. Conversely, Connolly (*Greece and Rome*, 80) readily identifies the *kotthybos* with the *linothorax*. Everson (*Warfare*, 190), on the other hand, suggests that the *kotthybos* was a padded tunic worn under the armour. If this is the case then the wording of the Amphipolis decree would suggest that the regular pikeman only wore a padded tunic and nothing else while the officers wore either a breastplate or half breastplate. It seems more likely that the *kotthybos* was the *linothorax* or some other form of lighter armour while the *lorica* and *hemithorakia* were plate metal cuirasses. The phalangite would most likely have worn some form of padding under his armour – possibly of wool, sheepskin or leather – to prevent chafing, to provide greater protection and to make the armour more comfortable to wear. However, this should not be confused with an actual padded tunic. Even if the *kotthybos* is interpreted as a padded tunic of some description, the phalangite was still armoured to a certain extent contrary to what some scholars suggest (see also: Lush, 'Body Armour', 18).

135 *SEG* 40.524

136 Hatzopoulos and Juhel 'Four Hellenic Funerary Stelae', 425-428

137 Polyaenus, *Strat*, 4.3.13

138 Diod. Sic. 20.83.2

139 Plut. *Aem*. 18, 21; Frontinus (*Strat*. 3.2.11) describes Charmades, a Ptolemaic general (c.280BC), as also wearing a cloak.

140 The stylised musculature on the front panel of the bronze muscled cuirass served a number of purposes: it was a clear display of wealth and status as this type of armour would have been expensive; it made the wearer look more imposing to an opponent in battle; and the numerous curves on the armour meant that there were very few flat surfaces on the front. This provided a greater level of protection to the wearer as any weapon impact which struck a curved section of the armour would require more energy to penetrate it and weaker blows would simply be deflected by the rounded muscles on the metal plate (see The Penetration Power of the *Sarissa* from page 225). An almost complete example of a Hellenistic muscled cuirass was found at Prodromi in Epirus. This armour has been dated to 330BC – making it contemporary with the reign of Alexander the Great. Interestingly, this cuirass is made of iron rather than bronze and has gold fittings. While iron provided a better level of protection than bronze, as a material, it is much harder to work with and much more difficult to fashion into a thin plate which demonstrates the level of craftsmanship attained in this area by this time. The lower rim of the cuirass (now heavily damaged) seems to be quite flared which suggests that this was for a cavalry officer. Regardless, both the material and the gold fittings indicate this as the armour of a wealthy noble (see: A. Choremis, 'Metallimos Oplismos apo tou tapho sto Prodromi tes Thesprotias' *Athens Annals of Archaeology* 13 (1980) 10-12); Everson, *Warfare*, 187-189).

141 Alexander's wounds: Arr. *Anab*. 6.10.1; Curt. 9.5.9; Plut. *Alex*. 63; Plut. *Mor*. 341c, 344c, 345a; Demetrius' armour: Dido. Sic. 19.81.4; 20.52.1

142 Plut. *Eum*. 8

143 M. Andronikos, 'Vergina: The Royal Graves in the Great Tumulus' *Athens Annals of Archaeology* 10 (1977) 26-27; Hatzopoulos and Loukopoulos, *Philip of Macedon*, 225, pl.127; Connolly, *Greece and Rome*, 58

144 Everson (*Warfare*, 192) suggests that this armour may have even been able to withstand impacts from bolts fired from catapults as per Plutarch (*Dem*. 21).

145 Everson, *Warfare*, 192-193

146 There is little reason to doubt that, should a noble wish it, he could have worn a *linothorax* instead of a metal breastplate unless it is assumed that the wearing of a muscled cuirass was a compulsory distinguishing badge of rank. However, it seems unlikely that a choice of what armour to wear was available to the lower ranks who, being poorer, would have been restricted to their issued kit. If certain officers were able to wear a metal cuirass as a mark of rank and/or distinction, there is no reference anywhere to what rank (if any) and above this choice was reserved for.

147 The *hypaspists*, armed as hoplites, would have most likely also worn the classical muscled cuirass. However, the *hypaspists* were not part of the pike-phalanx.

148 Xen. *Eq*. 12.1

149 E. Jarva, *Archaiologia on Archaic Greek Body Armour* (Studia Archaeologica Septentrionalia 3), (Pohjois-Suomen Historiallinen Yhdistys, Societas Historica Finlandiae Septentrionalis, Rovaniemi, 1995) 20

150 Diod. Sic. 14.41.4-5

151 Ducrey (*Warfare in Ancient Greece* (Schocken Books, New York, 1986) 54) suggests that one of the roles of the tunic was to prevent chafing. Pausanias (5.5.2) states that *Byssos* (βύσσος) grew exclusively in Elis. Jones and Ormerod translate the word *Byssos* in this passage as meaning 'fine flax' (see their translation for the Loeb Classical Library, 1966). Levi translates the same passage as 'fine linen' (which is made from flax) but suggests in a footnote that what Pausanias may be describing is actually cotton (see his translation for Penguin Books, 1971 – page 206 n.32). Wild flax/linen (*Linum bienne*) was quite common in early Europe and Pausanias may be distinguishing the difference between this plant and cultivated linen/flax (*Linum usitatissium*). Edmonds translates Theocritus' use of *Byssos* (2.73) to mean 'silk'. Jones similarly states that Strabo (15.1.20) is referring to silk in his use of the term *Byssos* in his translation. Conybeare says that Philostratus (VA 2.20) is referring to cotton. Wright translates Empedocles' use of *Byssos* (93) as simply 'linen'. The word *linon* (λίνον) is used by many other ancient authors to describe articles of clothing and is usually also translated as 'linen'; see: Aesch. *Supp.* 121, 132; Ar. *Ran.* 1347; *AP* 6.231; Hom. *Il.* 9.661; Philostr. *VA* 2.20. Regardless of what plant the word *Byssos* is actually referring to, it is clear that the Greeks had access to at least one cultivatable plant (and possibly more) which would produce fibres which could be spun into cloth for the production of clothing.

152 This was experienced when wearing the replica muscled cuirass that was used in the research on the warfare of the Classical Age discussed in Matthew, *A Storm of Spears*. During the testing undertaken for that project a bib of sheepskin was worn to pad the shoulders.

153 Tomb paintings show Hellenistic phalangites wearing both tunics and cloaks in a variety of colours. Plutarch (*Aem.* 18), on the other hand, states that the Macedonians at Pydna were wearing red tunics. This may be a reference to a type of uniform for the Macedonian military. Conversely, Heckel and Jones (*Macedonian Warrior*, 16) state that uniformity of equipment, other than for the Romans, is a relatively modern concept. However, if the phalangite carried a *sarissa* of a standardized length, carried a shield that was not only of a similar size right across the Hellenistic period but may have also been indicative of his unit, then this would suggest the basic elements of a Macedonian uniform regardless of whether a tunic of the same colour was worn or not. If a standard tunic was worn, then this would have added to the sense of a uniform. Standardisation of equipment added to *esprit de corps* and aided unit bonding – things that are very beneficial to a professional army. Plutarch (*Eum.* 8) also says that one of the special awards that could be handed out by Macedonian royalty was the distribution of purple cloaks and caps to distinguished soldiers. This suggests that these awards were of a standard colour as well.

154 Everson, *Warfare*, p.142

155 Some vase illustrations of hoplites from the Classical period show them wearing tunics with a cross-hatch pattern on them. Some have interpreted this as a depiction of a quilted tunic. See: Bardunias, 'Linothorax Debate', 52-53; Warry (*Warfare*, 67) has the Iphicratean *peltast* wearing padded armour as well.

156 Polyaenus, *Strat.* 4.2.10

157 *SEG* 40.524; Aen. *Tact.* 29.4

158 Everson, *Warfare*, 195

159 See: P. Jaeckel, 'Pergamenische Waffenreliefs' *Jarbuch Waffen und Kostumkunde* (Dt. Kunstverl, Munich, 1965) fig. 40

160 Plut. *Aem.* 34

161 Everson (*Warfare*, 195) suggests that these men are dismounted cavalry.

162 For example see: R. Post, 'Alexandria's Colourful Tombstones – Ptolemaic Soldiers Reconstructed' *AncWar* 1.1 (2007) 38-43

163 It must also be considered that it gets quite cold in northern Greece and Macedonia during the winter. It seems unlikely that the people of these regions would have walked around in open-toed sandals or barefoot at times when the region could be under a metre or more of snow. Plutarch (*Mor.* 340e) describes how, during his campaign, Alexander the Great 'dug through nations buried in deep snow'. Undoubtedly things like socks, cloaks and tunics were a common part of the everyday apparel for the people of these areas and it seems unlikely that such articles of clothing were not used by the Macedonian military as well. For a brief discussion of the regularity of the Macedonian boot see: Saatsoglou-Paliadeli, 'Macedonian Costume', 145

164 Heckel and Jones, *Macedonian Warrior*, 16; see also: Connolly, *Greece and Rome*, 80; Sheppard, *Alexander the Great at War*, 83-84

165 Everson, *Warfare*, 195

166 Plut. *Alex.* 16

167 N.G.L. Hammond, 'The Battle of the Granicus River' *JHS* 100 (1980) 78, 80

168 Nepos, *Iphicrates*, 1.3-4; Diod. Sic. 15.44.1-4

169 See: Everson, *Warfare*, 125-126, 163-164

170 Everson, *Warfare*, 177; Choremis, 'Metallimos Oplismos', 15-16

171 Snodgrass, *Arms & Armour*, 119; interestingly, Snodgrass also notes the depiction of the hacking *macharia* on the Lion Hunt mosaic from Pella but does not seem to associate this with an infantry weapon.

172 Connolly, *Greece and Rome*, 77

173 Everson, *Warfare*, 177

174 Plut. *Mor.* 191e, 216c, 217e; Plut. *Dion* 58; Plut. *Lyc.* 19; in typical Laconic fashion, the Spartans apparently rebutted any disparaging comment about the length of their swords by stating that, while their swords were short, they still got close enough to the enemy to use them.

175 Sheppard, *Alexander the Great at War*, 82

176 *SEG* 40.524

177 Hatzopoulos and Juhel 'Four Hellenic Funerary Stelae', 425-428

178 Plut. *Aem.* 20

179 Plut. *Mor.* 344d

180 Heckel and Jones, *Macedonian Warrior*, 17

181 W.W. Tarn, *Hellenistic Military and Naval Developments* (Cambridge University Press, 1930) 12

182 Plut. *Aem.* 20

183 Anderson, 'Shields of Eight Palms", 2

184 This was the basis for the adoption by the Romans of the straight-edged *gladius* which, according to writers like Vegetius (*Mil.* 1.12), the Roman legionaries were taught to thrust with rather than swing in a slashing motion. Polybius (18.30), on the other hand, says that the Romans used their swords for both cutting and thrusting which, he says, required more room and a greater interval between each man.

185 Heckel and Jones, *Macedonian Warrior*, 17

186 Polyb. 6.23

187 Gabriel (*Philip II*, 64), one of the few scholars who does provide an estimated weight for the phalangite panoply, gives a total of 18kg. This is only slightly less than the figure given above for the equipment carried by the Iphicratean *peltast* or a phalangite in the age of Alexander the Great (see following).

188 Not only would the weight of a longer shaft have to be considered, but the longer pikes would have also required a heavier butt in order to offset the added weight of the longer shaft so as to provide the weapon with the correct point of balance.

189 Ael. Tact. *Tact.* 2

190 English, *Army of Alexander*, 31

191 Diod. Sic. 19.84.7

192 Polyb. 18.1

193 Polyaenus, *Strat.* 4.2.10

194 Front. *Strat.* 4.1.6

195 D.W. Engels, *Alexander the Great and the Logistics of the Macedonian Army* (University of California Press, Berkeley, 1978) 18

196 Diod. Sic. 19.96.4

197 Diod. Sic. 20.73.3; the large amount of these personally carried rations was most likely due to the scarcity of food and water to be found in the arid Sinai Peninsula.

198 Plut. *Eum.* 9

Bearing the Phalangite Panoply

1 Hammond 'Training in the Use of a *Sarissa* and its Effects in Battle 359-333 B.C.' *Antichthon* 14 (1980) 62) goes as far as to call the *sarissa* the 'weapon par excellence' of the Macedonian phalangite.

2 M.M. Markle, 'The Macedonian *Sarissa*, Spear and Related Armour' *AJA* 81.3 (1977) 323

3 For example see: P. Connolly, 'Experiments with the *Sarissa* – the Macedonian Pike and Cavalry Lance – a Functional View' *JRMES* 11 (2000) 109; P. McDonnell-Staff 'Hypaspists to Peltasts: The Elite

Guard Infantry of the Antigonid Macedonian Army' *AncWar* 5.6 (2012) 22-23; J. Warry, *Warfare in the Classical World* (University of Oklahoma Press, Norman, 1995) 73; Connolly not only suggests the use of a side-on stance, but also bases his model on Polybius' description of holding the *sarissa* 4 *cubits* from the end which, as outlined previously, seems to be incorrect (see pages 83-88).

4   See: Connolly, 'Experiments', 109

5   For example see: Warry, *Warfare*, 73; strangely, despite having an illustration depicting the use of a side-on posture, Connolly ('Experiments', 110), wields the pike in a manner more akin to the oblique posture in a photograph of himself using the replica *sarissa* that he had made for his own experiments. The phalangite is also depicted in more of an oblique posture in another of Connolly's works (see: *Greece and Rome at War* (Greenhill Books, London, 1998) 78). Whether the depiction of a completely side-on stance is an error or not, and why this diagram does not correlate with the stance used in the photograph, is not explained.

6   Polyaenus, *Strat.* 4.2.2

7   Plut. *Alex.* 33

8   Plut. *Aem.* 18

9   Any style of open-faced helmet like the Boeotian or the *pilos* would not encounter any limitation to the movement of the head due to a lack of elongated cheek pieces. On styles like the Attic and the Chalcidian, the cheek pieces are not elongated and merely cover the sides of the face. These too would pose no restriction to the rotation of the head.

10 Tyrt. 10, 11

11 Veg. *Mil.* 1.20

12 Diodorus (16.86.3) describes how the bodies of the fallen piled up at Chaeronea in 338BC. Similarly, Appian (*Syr.* 36) describes the battlefield of Magnesia (190BC) as being littered with the bodies of men, horses and elephants. Plutarch (*Aem.* 21) says a great mound of dead and equipment was strewn across the battlefield of Pydna (168BC). Plutarch (*Pyrr.* 28) additionally describes how freshly-turned earth did not provide a firm footing for combat.

13 Arr. *Tact.* 16.13

14 Diod. Sic. 17.55.4

15 Arrian (*Anab.* 1.14.7) says that Alexander's cavalry at the Granicus were carried downstream 'in the direction that the river took them' which suggests that the current was quite strong. If the current was strong enough to push horses and riders downstream, then it would have had a major effect on infantry who did not have a secure posture and stable footing.

16 In the diagram accompanying his examination of the phalanx, Connolly ('Experiments', 109) has the right leg/foot extending far to the right, beyond being in line with the shoulder and to the rear of the interval that the phalangite is occupying. This would separate the feet by a considerable distance and, while not physically impossible, the further the feet are apart beyond shoulder width, the less stable the posture becomes. Interestingly, in the photograph on the following page of Connolly wielding his replica panoply, the right leg is much closer in – in line with the right shoulder.

17 Asclep. 4.1; for the changes in the size of the cubit from the Classical to the Hellenistic period see the earlier section on 'the length of the *sarissa*'

18 Asclep. *Tact.* 4.3; Ael. Tact. *Tact.* 11

19 Arr. *Tact.* 12.4; Polyb. 18.29; Asclep. *Tact.* 4.3; Ael. Tact. *Tact.* 11; despite such evidence, Sekunda (*The Seleucid Army* (Montvert, Stockport, 1994) 9) interprets Asclepiodotus as meaning that the 'natural interval on the battlefield was six feet'. This equates to the open-order of 4 cubits per man and it is unlikely that such a deployment was ever used by the pike-phalanx to engage in combat.

20 Asclep. *Tact.* 4.3; Ael. Tact. *Tact.*11; Arr. *Tact.* 11.4

21 Ael. Tact. *Tact.* 32

22 Asclep. *Tact.* 12.9; Ael. Tact. *Tact.* 32

23 McDonnell-Staff 'Hypaspists to Peltasts', 22-23

24 Warry, *Warfare,* 73

25 A.M. Snodgrass, *Arms and Armour of the Greeks* (Johns Hopkins University Press, Baltimore, 1999) 121

26 P. Sabin, *Lost Battles* (Continuum, London, 2009) 48

27 Front. *Strat.* 2.2.1

28 See: Arr. *Tact.* 12.3; Ael. Tact. *Tact.*13.3; Polyb. 18.29; Asclep. *Tact.* 5.1; Sheppard (*Alexander the Great at War* (Osprey, Oxford, 2008) 54) says the weapons of only the first 3-4 ranks projected ahead of the line. This goes against what most ancient writers tell us about the pike-phalanx.

29 This is basic trigonometry. The person creates a triangle with their body by forming a diagonal line across their interval. The line from shoulder to shoulder represents the hypotenuse of the triangle, while the angle between the hypotenuse and the base of the triangle (basically the rear of the interval) is 45° (the angle of the upper body). Cos45° x 45cm = 32cm

30 A. Smith, 'The Anatomy of Battle – Testing Polybius' Formations' *AncWar* 5.5 (2011) 43 (41-45)

31 Smith, 'Anatomy of Battle' 43, 45; strangely, Smith also states (p.45) that when each man occupied a space 48cm across with his body in an oblique posture, that there was 24cm of 'fighting space' to either side of him. While there is clearly a 24cm projection of shield to the phalangite's left, there cannot be another 24cm of space to his right if, as claimed in the earlier statement, the lateral space the individual occupied was between 78cm and 92cm.

32 See: C.A. Matthew, *A Storm of Spears: Understanding the Greek Hoplite at War* (Pen & Sword, Barnsley, 2012) 182-197

33 In reality, the pike held by the foremost phalangite should be on his right (at the front of the image) rather than crossing over his body to be deployed behind his shield and in between the man next to him. Such 'errors' were a common feature of depictions of the classical hoplite in the vase paintings of the fifth century BC as well. Whether this was the result of a lack of understanding or skill on the part of the artist, or was a conscious design element meant to show off more of the human physique in the image with the least amount of obstruction, is unknown.

34 Plut. *Aem.* 19-20

35 Even this amount of space would leave the hands of the phalangite vulnerable to being squashed between the weapons of the phalanx – especially as the formation advanced and the weapons that had been lowered for combat moved and flexed. It is possible, although there is no direct evidence for it, that phalangites wore some form of protection on their hands – something in the way of gloves (χειρίς), the leather thonging wrapped around the hands of boxers (ἱμάντες), or some other form of protective padding. Such coverings would have protected the phalangite's hands within the massed confines of the pike-phalanx.

36 Connolly, ('Experiments', 112) found that the phalanx that he re-created using replica shields and pikes was impossible to move, let alone charge, if each man occupied a space of 69cm and suggested that, if a wider interval was adopted, movement would have been possible, but the men in the front rank ran the risk of being speared by the men in the fifth rank. However, according to the ancient sources, the pike held by the fifth man extended beyond the front of the formation by a distance of 2 cubits. Thus even with a spacing of 96cm per man for the intermediate order, the tip on the weapon held by the man in the fifth rank was ahead of the man in the front by nearly a metre and would have also been located off to the front man's right-hand side by about 20cm.

37 Arr. *Anab.* 5.17.7; Brunt translates this passage literally, suggesting that Alexander's phalanx created a formation 'with locked shields'.

38 For example see: Plut. *Aem.* 19-20

39 For example see: J. Pietrykowski, *Great Battles of the Hellenistic World* (Pen & Sword, Barnsley, 2009) 83, 233, who, following Arrian's and Plutarch's descriptions, offers that pike-phalanxes fought with interlocked shields at the Hydaspes in 326BC and at Pydna in 168BC.

40 S. English, *The Field Campaigns of Alexander the Great* (Barnsley, Pen & Sword, 2011) 195, 208

41 W.W. Tarn, *Alexander the Great Vol.II* (Ares, Chicago, 1981) 96, n.3; for a comprehensive discussion of the use of Greek terminology by Roman writers, and Roman terms for the description of Greek formations, for the 'phalanx' and a battle-line or a dense formation see: E.L. Wheeler, 'The Legion as Phalanx in the late Empire (I)' in Y. le Bohec and C. Wolff (eds.), L'*Armée Romaine de Dioclétien à Valentinien Ier* (De Boccard, Paris, 2004) 309-358; E.L. Wheeler, 'The Legion as Phalanx in the Late Empire (II)' RÉMA 1 (2004) 147-175

42 See: Plut. *Crass.* 24; Front. *Strat.* 2.3.15

43 Both Caesar (*B Gall.* 2.6.2, 7.85.5) and Livy (10.29.26, 10.41.4) use the term *testudo* to describe both formations of Roman infantry and formations of Gallic infantry in what are again likely to be the uses of similes (especially in the case of the Gallic troops) to describe compact-order formations.

44 It is interesting to note that the root of the term *synaspismos* used to describe the close-order formation is the word *aspis*, the name of the hoplite shield, rather than the word *peltē* – further suggesting its correlation with troops armed as hoplites. Aelian, in his description of the different intervals used by infantry formations, states that these were applicable to hoplites (*Tact.* 11). This is a clear differentiation between the Classical hoplite and the pikeman which is referred to as a *peltast* by Aelian.

The lexicographer Photius (*sv.* ὑπασπιστής) has the term *hypaspists* (ὑπασπισταί) synonymous with 'spear bearer' (δορυφόροι) – again suggesting that these troops were hoplites rather than pikemen. For further doubts about phalangites employing the close-order formations see: J. Kromayer, and G. Veith, *Heerwesen und Kriegführung der Geichen und Römer* (C.H. Beck, Munich, 1928) 358.

45 N.G.L. Hammond, 'What May Philip Have Learnt as a Hostage in Thebes?' GRBS 38.4 (1997) 367; see also: A. Blumberg, 'Inspired by the Bard: Philip II, Alexander the Great and the Homeric Ethos' *AncWar* 3.3 (2009) 20; interestingly, in another work (*Alexander the Great: King, Commander and Statesman* (Duckworth, London, 1980) 32) Hammond says that each man in the front rank occupied a space of 1m across. This is the intermediate-order interval which, not only prevents phalangites from interlocking their shields, even if no weapon is carried, but is also twice the size of the interval used by Greek hoplites to create a 'shield wall'.

46 The diagram accompanying Connolly's examination of the phalanx ('Experiments', 109) shows the phalangites covering this exact same distance with their bodies via the use of a scale graduated in cubits.

47 Connolly, 'Experiments', 109

48 S. English, *The Army of Alexander the Great* (Pen & Sword, Barnsley, 2009) 21

49 Hammond, 'Philip as Hostage', 367; see also McDonnell-Staff ('Hypaspists to Peltasts', 22-23) who makes a similar claim. McDonnell-Staff further suggests (p.22) that the normal (or open) order of the phalanx was 2m per man and that this was how the phalanx deployed sixteen deep. He then suggests that, as they approached the enemy, the phalanx moved into a 'close order' of 1m per man (which is actually the intermediate-order of the manuals) by 'doubling' the files to create a formation eight men deep. However, there is no reference in the ancient literature to this manoeuvre being undertaken in such a manner and it is especially unlikely that such movements, if ever carried out, would have been done as the phalanx was about to engage as Mcdonnell-Staff suggests.

50 Polyb. 18.29-30; Worthington (*Philip II of Macedonia* (Yale University Press, New Haven, 2008) 27) calls the *pyknosis* order of deployment 'close order'. This is a misinterpretation of the ancient literature as Polybius and the writers of the manuals use the term *pyknosis* to describe the intermediate-order interval for the phalanx.

51 Sheppard (*Alexander the Great at War*, 80) also states that the close-order formation, of about 45cm per man, was used to receive an enemy charge.

52 Bennett and Roberts (*The Wars of Alexander's Successors 323-281BC Vol.II: Battles and Tactics* (Pen & Sword, Barnsley, 2009) 4) suggest that close-order was a defensive formation. However, they do not outline what sort of attack they think this tight formation was used to resist.

53 The hoplite shield wall was the most dominant formation of the battlefields of the ancient world prior to the creation of the pike-phalanx. In every encounter where a close-order formation was employed, the side that had adopted it was always victorious regardless of whether the opponent was infantry or cavalry (see: Matthew, *Storm of Spears*, 205-217)

54 The passage of Diodorus cited by English (17.57.5) does not actually say anything about deployment or order and it can only be assumed that this is some form of typographical error.

55 J. Champion, *Pyrrhus of Epirus* (Pen & Sword, Barnsley, 2009) 87; Similarly, Morgan ('Sellasia Revisited' *AJA* 85.3 (1981) 328) uses an intermediate-order interval of three feet per man to calculate the frontage of a formation of 10,000 phalangites at the battle of Sellasia, drawn up in a double depth of 32 men per file. McDonnell-Staff ('Sparta's Last Hurrah – The Battle of Sellasia (222BC)' *AncWar* 2.2 (2008) 27-28) suggests that these troops were deployed in a double-depth close-order of three feet per man with interlocked shields. However, there are several issues with this. Firstly, three feet per man is the intermediate-order outlined in the manuals, not the close-order. Secondly, an interval of three feet per man would prevent men carrying the *peltē* from interlocking them. Polybius gives clues as to the manner of deployment when he says that it made 'a hedge of pikes' (συμφράξαντες τὰς σαρίσας). Thus these troops cannot be in a close-order shield wall which required their weapons to be held vertically. Lastly, McDonnell-staff interprets the word that Polybius uses (2.66) to describe the deployment (*epallallos phalaggos* - ἐπαλλήλου φάλαγγος) as meaning 'interlocked shields'. However, the term simply means 'united phalanx' and is not an indication of its order or interval, merely that the troops are in formation.

56 Warry, *Warfare*, 73

57 English, *Army of Alexander*, 21; in a similarly confused passage, McDonnell-Staff ('Hypaspists to Peltasts', 22) states that the normal (or open) order of the phalanx was 2m per man, and that this was

how the phalanx deployed sixteen deep. McDonnell-Staff goes on to suggest that, as they approached an enemy, the phalanx moved into 'close-order' (*pyknosis*) of 1m per man by 'doubling' the files to create a formation eight deep. However, *pyknosis* is the term used for the intermediate-order of 96cm per man by the ancient writers, rather than for the 'close-order' of 48cm per man as McDonnell-Staff states (even though he says the formation used a 1m interval). Additionally, the use of an 8 deep file seems to have been an improvised formation rather than a standard deployment, which further confuses McDonnell-Staff's statements. Similarly Sheppard (*Alexander the Great at War*, 80) states that the close-order *pyknosis* formation had an interval of around 91cm per man (based upon the smaller Attic cubit of 45cm) which, again, is actually the intermediate-order of the manuals.

58 R. Gabriel, *Philip II of Macedonia – Greater than Alexander* (Potomac, Washington, 2010) 66

59 Strangely, on a previous page (p.64) Gabriel had suggested that when moving into the 'compact' intermediate-order, the phalangites swung their shields across their backs. This would go against the point of carrying them for battle and also would not allow them to interlock into a close-order 'shield wall'. Despite the fact that phalangites were not physically able to adopt such a formation, any suggestion that they could do so while their shield was slung across their back clearly goes against the terminology used to describe it. See also: Cascarino (*Tecnica della Falange* (Il Cerchio, Rome, 2011) 61) who also bases his examination of the interval of the phalanx on the wrong size for the cubit.

60 Gabriel, *Philip II*, 130

61 Smith ('The Anatomy of Battle', 41-45) used the Polybian model to try and analyse the spatial requirements of the men within the phalanx. However, many aspects of this examination are confusing and do not correlate with what Polybius himself states. For example, he states (p.42) that 2 cubits equates to a distance of 48cm and that, based upon Polybius, each man occupied a space of 48cm. While a distance of 48cm is correct for a single cubit in the Hellenistic standard (see: The Length of the *Sarissa*), a single cubit could not have been only 24cm in size as Smith suggests. Strangely, Smith later states that10 cubits equated to 480cm which must assume a single cubit of 48cm - correctly following the Hellenistic standard. Initially this would seem to be a typographical error. However, Smith goes on to state, following the Polybian model for carrying the *sarissa*, that the rearward section of the weapon projected behind the bearer by only 48cm – which must be based upon a single cubit of 24cm. This incorrect conclusion then influences much of the remainder of the analysis (albeit in a somewhat contradictory fashion). In the diagram accompanying the article (p.42), the phalangites are shown to be occupying a space 96cm across. This would then correlate with the use of a 48cm cubit and the adoption of an intermediate-order spacing of 2 cubits per man. However, the phalangites are only depicted with the pikes held by the first two ranks projecting between the files, contrary to Polybius' statement that the weapons of the first five ranks extended ahead of the line, with no additional space left for the remaining three pikes. As the shield, with its diameter of 64cm, is portrayed as being positioned across the front of the interval that each man occupies, this leaves the impression that each man's left fist (holding the *sarissa*) is over 20cm in size. Additionally, as the pike is shown to only extend behind the bearer by 48cm, the resultant formation has the butt-spike of the forward ranks wedged into the groin of the man behind. Such a deployment seems highly unlikely due to its dangerous nature. Furthermore, when the pike is wielded at waist height, the butt of the *sarissa* points at the shield of the man behind and so could not be wedged into his groin unless it is assumed that it was somehow positioned to pass under the shield (again unlikely). Such a depiction is even more confusing when, on page 45, Smith states that the Hellenistic phalangite wielded the *sarissa* in the same manner as the pikemen of sixteenth and seventeenth century Europe with the weapon held at shoulder height. It is only at the end that Smith arrives at the true spatial requirements of the phalangite by stating that each man occupied a space of about 48cm with his body while 24cm of extra 'fighting space' was on either side – bringing the total to the 96cm of the intermediate-order. This is fundamentally correct. As well as the interval occupied by the individual in an oblique body posture, the shield projected to the left by about 24cm and 24cm on the right would be taken up by the pikes belonging to the first five ranks all levelled beside on another.

62 W. Heckel and R. Jones, *Macedonian Warrior: Alexander's Elite Infantryman* (Osprey, Oxford, 2006) 16)

63 For example see: Warry, *Warfare*, 73; Cascarino, *Tecnica della Falange*, 62

64 Connolly, 'Experiments', 111

65 Connolly, 'Experiments', 112

Phalangite Drill

1   For example, Hammond (*Alexander the Great: King, Commander and Statesman* (Duckworth, London, 1980) 32) says: 'strict discipline and precise drill were essential, and the best training of all was provided by experience in battle' but does not examine this concept of essential drill any further. English (*The Army of Alexander the Great* (Pen & Sword, Barnsley, 2009) 22), on the other hand, merely says that the weight and size of the *sarissa* would make it difficult to handle without any further discussion.

2   Ael. Tact. *Tact*. 53

3   Asclep. *Tact*. 12.11; Arr. *Tact*. 32

4   Ael. Tact. *Tact*. 32

5   Arr. *Anab*. 1.6.1-3 - καὶ τὰ μὲν πρῶτα ἐσήμηνεν ὀρθὰ ἀνατεῖναι τὰ δόρατα τοὺς ὁπλίτας, ἔπειτα ἀπὸ ξυνθήματος ἀποτεῖναι ἐς προβολήν, καὶ νῦν μὲν ἐς τὸ δεξιὸν ἐγκλῖναι τῶν δοράτων τὴν σύγκλεισιν, αὖθις δὲ ἐπὶ τὰ ἀριστερά. καὶ αὐτὴν δὲ τὴν φάλαγγα ἔς τε τὸ πρόσω ὀξέως ἐκίνησε καὶ ἐπὶ τὰ κέρατα ἄλλοτε ἄλλη παρήγαγε.

6   For the military definitions of these two terms see: Ael. Tact. *Tact*. 23, 35; Asclep. *Tact*. 10.1, 11.2-7; Arr. *Tact*. 20, 28; Rooke's translation of Arrian's *Anabasis* from 1814 is somewhat abbreviated and presents an even more confusing picture: '...ordered the armed soldiers to advance first, with their spears erect; and upon a signal given, to reverse them, and sometimes to direct them to the right, and then to the left, as occasion required'.

7   Polyb. 2.69

8   Arr. *Anab*. 1.6.4

9   Diod. Sic. 17.57.6

10   Plut. *Eum*. 14

11   M.M. Markle, 'Use of the *Sarissa* by Philip and Alexander of Macedon' *AJA* 82.4 (1978) 492

12   Arr. *Anab*. 1.4.1 – ὑπὸ δὲ τὴν ἔω Ἀλέξανδρος διὰ τοῦ ληίου ἦγε, παραγγείλας τοῖς πεζοῖς πλαγίαις ταῖς σαρίσσαις ἐπικλίνοντας τὸν σῖτον οὕτω προάγειν ἐς τὰ οὐκ ἐργάσιμα. Markle gets around this apparently contradictory material by suggesting that these pikes were not intended to be used as combat weapons, but were carried with the army only to be used to flatten grain fields. This seems highly unlikely as there would be no point in an army the size of Alexander's carrying a different weapon for each man if it was not going to be used in battle and was only meant to flatten crops – an activity for which there would be much better implements. English (*The Field Campaigns of Alexander the Great* (Pen & Sword, Barnsley, 2011) 29) says that Arrian actually used the word 'spears' rather than *sarissa* and that this may suggest that the troops being described were actually hoplites. This is clearly an incorrect reading of the passage.

13   Polyb. 18.30

14   Polyaenus, *Strat*. 4.2.10

15   English, *Field Campaigns of Alexander*, 175; somewhat confusingly, in his examination of the battle of the Hydaspes in 326BC (pp.207-208), English dismisses Arrian's use of the term 'locked shields' by stating that this would be impossible 'if the heavy infantry were equipped as we might expect them to have been, with the *sarissa* and *pelta*'. A few lines later, he then states that 'I am not convinced that the *sarissa* was used at the Hydaspes.'

16   Curt. 8.14.16

17   Arr. *Anab*. 7.6.9; Diod. Sic. 17.108.1-2; Plut. *Alex*. 47; Curt. 8.5.1

18   Arr. *Anab*. 7.12.2; Diod. Sic. 17.110.3; for a discussion of Alexander's training of local youths to become front-line infantry see: N.G.L. Hammond, 'Royal Pages, Personal Pages, and Boys Trained in the Macedonian Manner during the Period of the Temenid Monarchy' *Historia* 39.3 (1990) 261-290. Hammond calculates (p.279) that the number of these trained youths could have been as high as 125,000 across the timeframe of Alexander's campaign.

19   See: M. Mitchiner, *Indo-Greek and Indo-Scythian Coinage Vol.I – The Early Indo-Greeks and their Antecedents* (Hawkins Publications, Sanderstead, 1975) 21. English (*Field Campaigns of Alexander*, 208) dismisses this imagery as iconic rather than realistic. See also: A.B. Bosworth, *A Historical Commentary on Arrian's History of Alexander* Vol.II (Oxford University Press, 1995), 301. It is further interesting to note that, if the figure is Alexander, then the weapon in question is more likely to be a cavalry lance than an infantry pike.

20   For the characteristics of the hoplite spear see: C.A. Matthew, *A Storm of Spears: Understanding the Greek Hoplite at War* (Pen & Sword, Barnsley, 2012) 1-15

21   For a comprehensive discussion of the evolution of the sixteenth century training manual for pike and

musket armies see: D.R. Lawrence, *The Complete Soldier – Military Books and Military Culture in Early Stuart England, 1603-1645* (Brill, Leiden, 2009) 135-156; the Hellenistic pike-phalanx was not the only ancient mode of warfare which influenced the pike and musket armies of the sixteenth and seventeenth centuries. Chapter one of Edward Cooke's *The Character of Warre* (Tho. Purfoot, London, 1626) comments on the value of studying the Roman military writer Vegetius, but uses examples from Greek, Macedonian and Roman history, from the fourth century BC to the time of Vegetius, throughout his text.

22 See: J. Smythe, *Instructions, Observations and Orders Mylitarie*, (R. Johnes, London, 1595) 23; according to Blaise de Monluc (*Military Memoirs: Blaise de Monluc: The Valois-Habsburg Wars and the French Wars of Religion* (I. Roy (ed.)) (Longman, London, 1971) 188) the Germans were very adept at this style of fighting while the Swiss held their pikes in the middle of the shaft. Many of the woodblock prints of battles from this time period show pikemen holding their weapons either in this elevated position or at waist level in the Macedonian manner.

23 A. Smith, 'The Anatomy of Battle – Testing Polybius' Formations' *AncWar* 5.5 (2011) 45

24 N.G.L. Hammond, *The Genius of Alexander the Great* (Duckworth, London, 1997) 14; the same image is found in I. Worthington, *Philip II of Macedonia* (Yale University Press, New Haven, 2008) 28; this crouched position was one of the ways that the Classical Hoplite could wield his shorter spear with one hand (see: Matthew, *Storm of Spears*, 15-18).

25 A description of the pike is a common element in the military manuals of sixteenth and seventeenth century Europe. An anonymous pamphlet from the early seventeenth century (*The Exercise of the English, in the Militia of the Kingdome of England* (London, 1642) 1-2), for example, states that '... the pike of Ashen-wood for the Steale, and at the upper end an iron head, of about a handful long, with cheeks about the length of two foote, and at the butt-end a round strong socket of iron ending in a Pike, that is blunt, yet sharpe enough to fixe to the ground'.

26 See: Diod. Sic. 17.34.1; Asclep. *Tact.* 3.1-6; Ael. Tact. *Tact.* 10

27 Polyb. 2.69

28 For example see: Ptolemy's phalangites at Raphia (217BC): Polyb. 5.85; the Spartans at Mantinea (207BC): Polyb. 11.15; the Achaeans at Mantinea (207BC): Polyb. 11.16; the Macedonians at Cynoscephalae (197BC): Polyb. 18.24

The Reach and Trajectory of Attacks made with the *Sarissa*

1 R.E. Dickinson, 'Length Isn't Everything – Use of the Macedonian *Sarissa* in the Time of Alexander the Great' *JBT* 3.3 (2000) 51

2 J. Warry, *Warfare in the Classical World* (University of Oklahoma Press, Norman, 1995) 68

3 R.D. Milns, 'The Hypaspists of Alexander III – Some Problems' *Historia* 20 (1971) 188

4 J. Pietrykowski, *Great Battles of the Hellenistic World* (Pen & Sword, Barnsley, 2009) 230

5 I. Worthington, *Philip II of Macedonia* (Yale University Press, New Haven, 2008) 27

6 R. Gabriel, *Philip II of Macedonia – Greater than Alexander* (Potomac, Washington, 2010) 66

7 A. Smith, 'The Anatomy of Battle – Testing Polybius' Formations' *AncWar* 5.5 (2011) 43

8 See: Diod. Sic. 17.34.1; Asclep. *Tact.* 3.1-6; Ael. Tact. *Tact.* 10

9 Ael. Tact. *Tact.* 13; Arr. *Tact.* 12

10 See. C. A. Matthew, *A Storm of Spears: Understanding the Greek Hoplite at War* (Pen & Sword, Barnsley, 2012) 83-86

11 J. Chananie, 'The Physics of Karate Strikes' *Journal of How Things Work* 1 (1999) 3

12 Chananie, 'The Physics of Karate Strikes' 3

13 Sheppard (*Alexander the Great at War* (Osprey, Oxford, 2008) 54) suggests that the connecting tube used to join a segmented *sarissa* decreases the bend in the shaft and adds to the overall sturdiness of the weapon. Regardless of how true such an assumption is, both weapons with a single shaft and those with joined shafts possess a certain level of flex due to the length of the weapon and the weight of the amount of shaft that is forward of the left hand – compare Plates 1 and 15.

14 P. Connolly, 'Experiments with the *Sarissa* – the Macedonian Pike and Cavalry Lance – a Functional View' *JRMES* 11 (2000) 111

15 The only one of the statements issued by modern scholars cited at the beginning of this chapter that does not hold true is that of Warry who suggests that the phalanx was equally prepared to thrust with its pikes or push with its shields. Due to the rows of levelled pikes projecting ahead of the phalanx, it is unlikely that opponents carrying large shields like the hoplite *aspis* or the Roman *scutum* were regularly able to

move inside this array of presented weapons and physically collide with the shields of the front rank phalangites (See: The Phalanx in Battle).

16 Dickinson, 'Length Isn't Everything' 57; for the duel between Dioxippus and Coragus/Horratus see: Diod. Sic. 17.11.2; Curt. 9.7.19

The 'Kill Shot' of Phalangite Combat

1 C.A. Matthew, *A Storm of Spears: Understanding the Greek Hoplite at War* (Pen & Sword, Barnsley, 2012) 94)

2 The Corinthian style helmet also gave the wearer a natural range of vision (see: Matthew, *Storm of Spears*, 96-101) so this was not an improvement that had resulted from developments in defensive headgear in the fourth century BC.

3 Alc. *Frag.* 19

4 Matthew, *Storm of Spears*, 71-92

5 Plut. *Aem.* 20

6 Afric. *Cest.* 1.1.28-29 – γυμνὸν δὲ τὸ πρόσωπον καὶ αὐχὴν ἐλεύθεπος ἀκώλυτον τὴν πανταχοῦ περίσκεψιν χωρωῖ

7 Afric. *Cest.* 1.1.38-39 – σημεῖον δὲ [τὸ] ἐλευθέρας τῶν μαχομένων τὰς ὄψεις ὑπὸ πῖλῳ Λακωνικῷ ἐν τῇ Μακεδονικῇ γεγενῆσθαι

8 According to Diodorus (17.61.1) the dust kicked up at the battle of Gaugamela (331BC) was used to mask the escape of the Persian king Darius. As if to confirm how limited the vision was, Curtius (4.15.32) says that at Gaugamela the Macedonians 'wandered around like people in the dark; coming together only when they recognized a voice or heard a signal'. Plutarch (*Eum.* 16) states how the fine white dust kicked up at the battle of Gabiene (316BC) obscured vision (see also Diod. Sic. 19.42.1). Similarly, Appian (Syr. 33) describes how mist on the battlefield obscured vision during the battle of Magnesia in 190BC. Livy (37.41) says that the light rain (rather than mist) at Magnesia made it impossible for either end of the phalanx to be seen from the centre.

9 See: Matthew, *Storm of Spears*, 103-104

10 E. Kastorchis, 'ΠΕΡΙ ΤΟΥ ΕΝ ΧΑΙΡΩΝΕΙΑ ΑΕΟΝΤΟΣ' *Athenaion* 8 (1879) 486-491

11 C.F. Salazar, *The Treatment of War Wounds in Greco-Roman Antiquity* (Brill, Leiden, 2000) 233-234; W.K. Pritchett, *The Greek State at War – Part IV* (University of California Press, Berkeley, 1985) 136; Kastorchis, 'ΧΑΙΡΩΝΕΙΑ' 486-491; L. Phytalis, 'ΕΡΕΥΝΑΙ ΕΝ ΤΩ ΠΟΛΥΑΝΔΡΙΩ ΧΑΙΡΩΝΕΙΑΣ' *Athenaion* 9 (1880) 347-352, plate 1

12 T. Everson, *Warfare in Ancient Greece* (Sutton Publishing, Stroud, 2004) 129

13 Everson, *Warfare,* 129

14 Plut. *Aem.* 19; Polybius (18.24) similarly says that the nature of the Macedonian weapons gave them a distinct advantage over the Romans at the battle of Cynoscephalae in 197BC.

15 Plut. *Aem.* 20

16 Livy 44.42

17 Diod. Sic. 17.84.4 – οἱ γὰρ Μακεδόνες ταῖς σαρίσαις ἀναρρήσσοντες τὰς τῶν βαρβάρων πέλτας τὰς ἀκμὰς τοῦ σιδήρου τοῖς πνεύμοσιν ἐνήρειδον

18 Diod. Sic. 18.34.2

19 Livy, 36.18

20 Polyb. 5.84

21 IG IV2 1.121-122 (# 12 & 40); See also: E.D. Edelstein and L. Edelstein, A*sclepius: A Collection and Interpretation of the Testimonies* (The Johns Hopkins Press, Baltimore, 1945), #423

22 Curt. 3.11.6

23 Curt. 3.11.5

24 Just. *Epit.* 9.3.10

25 Diod. Sic. 17.25.4

26 Diod. Sic. 17.63.2-4

27 Diod. Sic. 19.32.1

28 Just. *Epit.* 28.4.5

29 Plut. *Aem.* 21

30 Magnesia: Livy 37.44; Pydna: Livy 44.42

31 N. Mashiro, *Black Medicine: The Dark Art of Death – The Vital Points of the Human Body in Close Combat* (Paladin Press, Boulder, 1978) 16

32 Mashiro, *Black Medicine*, 28-29

33 Mashiro, *Black Medicine*, 32

34 Mashiro, *Black Medicine*, 32

35 Hom. *Il.* 22.325; Xenophon (*Eq.* 12) also comments on the vulnerability of this area, and its need for adequate protection, among mounted troops.

36 For a discussion of the casualties suffered in phalangite contests see: P. Sabin, 'Land Battles' in P. Sabin, H. van Wees and M. Whitby (eds.), *The Cambridge History of Greek and Roman Warfare Vol. I: Greece, the Hellenistic World and the Rise of Rome* (Cambridge University Press, 2007) 414-416

37 Diod. Sic. 16.86.3

38 Diod. Sic. 19.30.5

39 Plut. *Aem.* 21

40 In many battles, the majority of casualties for the losing side were suffered at the hands of cavalry or skirmishers – either during a rout or from a flanking assault on the sides of their formation. For example, while Polybius (5.86) states that Ptolemy's victory at Raphia in 217BC was a result of the actions of the phalanx, he also states that the majority of the losses suffered by Antiochus were the result of a pursuit by his cavalry and mercenaries, rather than by being slain by the phalangites. Similarly, Appian (*Syr.* 36) states that most of the Roman infantry killed at Magnesia in 190BC were slain by cavalry rather than phalangites.

41 Diod. Sic. 17.34.8

42 Twisting the weapon also causes more damage to organs and soft tissue before the bayonet is withdrawn.

43 Veg. *Mil.* 1.12

44 Livy 31.34; Diodorus (28.8.1) has Philip make the off-hand remark that the dead probably do not care what size their wounds are.

45 Livy, 31.34

46 Warfare in Classical Greece seems to have similarly involved strikes directed towards the shield/chest area of an opponent with killing strikes directed against the head when the opportunity presented itself. See: Matthew, *Storm of Spears*, 93-112

47 It is possible that the ends of the two halves which slotted into the connecting tube were cut at an angle so that they would then slide into each other when the *sarissa* was assembled. If this was the case, then the complete weapon would be much more rigid and the pike as a whole could be twisted (albeit still with some difficulty) to aid extraction.

Accuracy and Endurance when Fighting with the *Sarissa*

1   Hom. *Il.* 13.785, 19.160; such sentiments are echoed in the later writings of Vegetius (*Mil.* 3.11) who states that 'when a man who is tired, or sweating, or has been running, enters battle against a man who is rested, alert and has been standing at his post, the two fight on unequal terms.'

2   Paus. 4.21.8-9; Diod. Sic. 14.105.1-2

3   Diod. Sic. 16.86.2, 17.11.5

4   Diod. Sic. 17.63.2

5   Plut. *Alex.* 33; see also: Curt. 4.13.17-25, 4.16.18

6   Diod. Sic. 19.30.5, 19.31.1

7   Plut. *Pyrr.* 21

8   Plut. *Aem.* 22

9   For a modern examination of the effects of fatigue and environmental conditions on performance see: J. Ramsey, 'Heat and Cold' in R. Hockey (ed.), *Stress and Fatigue in Human Performance* (Wiley & Sons, Chichester, 1983) 33-57

10 Front. *Strat.* 2.1.9; see also: Polyaenus, *Strat.* 4.2.7; Just. *Epit.* 9.3.9

11 Curt. 4.16.18

12 D. Holding, 'Fatigue' in Hockey (ed.), *Stress and Fatigue*, 145-164; N. Forestier and V. Nougier, 'The Effects of Muscular Fatigue on the Co-Ordination of Multijoint Movement in Humans' *Neuroscience Letters* 252.3 (1998) 187-190; K. Royal, D. Farrow, I. Mujika, S. Hanson, D. Pyne and B. Abernethy, 'The Effects of Fatigue on Decision Making and Shooting Skill in Water Polo Players' *Journal of Sports Sciences* 24.8 (2006) 807-815; P.R. Davey, R.D. Thorpe and C. Williams, 'Fatigue Decreases Tennis Performance' *Journal of Sports Sciences* 20.4 (2002) 311-318

13 G.P. Krueger, 'Sustained Work, Fatigue, Sleep Loss and Performance: A Review of the Issues' *Work and Stress* 3:2 (1989) 129-141

14 Leonis Imp. *Strat.* 9.4
15 Arist. *Eth. Nic.* 3.8.7-9
16 Just. *Epit.* 11.6.4-6
17 Front. *Strat.* 4.2.4
18 Technical skill/courage: Veg. *Mil.* 1.1, 1.8, 1.13, 2.23-24, 3.9-10; Vegetius also comments in several places on the value of fitness and recommends that Roman legionaries should be kept in peak physical condition, see: Veg. Mil. 1.1, 1.5-6, 1.9, 1.13, 1.19, 1.27, 2.23, 3.11.
19 Veg. *Mil.* 1.3
20 Veg. *Mil.* 3.9
21 Diod. Sic. 16.3.1
22 Front. *Strat.* 4.1.6; Polyaenus, *Strat.* 4.2.10
23 Dem. 9.49-50; 18.235
24 App. *Syr.* 3, 7
25 For example see: Curt. 4.9.19-21; for discussions on the march rates of Alexander's army see: R.D. Milns, 'Alexander's Pursuit of Darius through Iran' *Historia* 15 (1966) 256; C. Neumann, 'A Note on Alexander's March-Rates' *Historia* 20 (1971) 196-198; D.W. Engels, *Alexander the Great and the Logistics of the Macedonian Army* (University of California Press, Berkeley, 1980) 153-156; F.L. Holt, 'Imperium Macedonicum and the East: The Problem of Logistics' *Ancient Macedonia V* (1989) 585-592
26 Demetrius: Diod. Sic. 19.96.4; Antigonus: Diod. Sic. 20.73.3; Eumenes: Plut. *Eum.* 9
27 Diod. Sic. 19.41.2
28 Dod. Sic. 17.2.3
29 Diod. Sic. 29.2.1
30 App. *Syr.* 3, see also: App. *Syr.* 19
31 For the non-professional nature of the hoplite forces of Athens and other states in Greece see: Arist. *Pol.* 1256a7; Plut. *Mor.* 214; Plut. Ages. 26; Polyaenus, *Strat.* 2.1.7 The one stark contrast to these citizen militias was the army of the Spartans. The culture of the Spartans, with the harsh *agoge* system of education, simple diet, prohibition on art, luxury and commerce, indoctrinated the boys of Sparta into a military lifestyle from an early age and Spartan society was geared towards only one thing – war (see: Xen. Lac. 2.1-4.7; Plut. *Lyc.* 13, 16-25). Xenophon (*Hell.* 6.1.5), in an obvious comparison to the Spartan way of training, states that many citizen-based armies contain many who have either passed, or not yet reached, their prime and that there are few in the city who keep themselves in good physical condition. Vegetius, in comparisons of Athens and Sparta, (*Mil.* 3.0, 3.10) states that the Athenians cultivated arts other than the practice of war within their state, whereas the only concern of the Spartans was to prepare themselves for conflict. According to Aelian (*VH* 13.38-37), Alcibaides stated that it was not unexpected that the Spartans died so fearlessly in battle as they used death as a way of escaping from the harsh lifestyle they had left behind in Sparta. Some city-states of Greece fielded units of professional troops to augment their citizen forces. The Theban Sacred Band could be considered 'professional' due to their extensive training. A contingent of 1,000 select Argives were also given extensive military training at state expense, and these troops could also be considered 'professional' hoplites. For the details of other elite or professional units in ancient Greece see: W.K. Pritchett, *The Greek State at War – Vol.II* (University of California Press, Berkeley, 1974) 221-224.
32 Thuc. 2.39
33 P. Ducrey, *Warfare in Ancient Greece* (Schocken Books, New York, 1986) 67-70; interestingly, Curtius (4.7.31) says that the Macedonians, despite living under the rule of a king rather than in a democracy, and having a professional army as opposed to a citizen militia, 'lived in the shadow of liberty more than other races'. This was most likely because, under the stability of the monarchy, Macedonia had not witnessed any of the political turmoil – with its ostracisms, factions, confiscations and executions – which had run rampant through the 'democracy' of Athens for decades.
34 Arist. *Eth. Nic.* 3.8.13
35 Pindar (*Frag.* 110) states that war is sweet to those who are not familiar with it, but is frightening to those who have experienced it. Vegetius (*Mil.* 3.12) states that fear in battle is an ordinary reaction for the individual combatant. Yet this fear (and the reactions to it) could manifest itself in varying degrees. For a broad discussion of the forms and effects of different types of fear, courage and experience in battle, see: Arist. *Eth. Nic.* 3.7.10-3.8.16. In his book *On Killing*, Grossman outlines how some modern soldiers, unconditioned to the turmoils of war, may demonstrate a resistance to killing another human being in combat while those who have been 'conditioned' are more effective killers (see: D. Grossman,

On Killing (Back Bay Books, New York, 1996) 3, 13-15, 67-73). If this psychology holds as true for the warriors of the ancient world as it does for the modern, the professional Macedonian phalangite who had been 'conditioned' for war through his extensive training may have been more accepting to the notion of killing an enemy he faced than the members of many of the armies that Macedonian troops of the early Hellenistic Period fought against such as Greek hoplite militias or Persian conscripts.

36 C. Idzikowski and A. Baddeley, 'Fear and Dangerous Environments' in Hockey (ed.), *Stress and Fatigue*, 123-141

37 Idzikowski and Baddeley, 'Fear and Dangerous Environments' in Hockey (ed.), *Stress and Fatigue*, 127

38 Idzikowski and Baddeley, 'Fear and Dangerous Environments' in Hockey (ed.), *Stress and Fatigue*, 125

39 Excerpta Polyaeni, 18.4; Plut. *Alex.* 33

The Penetration Power of the *Sarissa*

1 J. Warry, *Warfare in the Classical World* (University of Oklahoma Press, Norman, 1995) 68

2 W. Heckel, *The Conquests of Alexander the Great* (Cambridge University Press, 2008) 16

3 R. Gabriel, *Philip II of Macedonia – Greater than Alexander* (Potomac, Washington, 2010) 65

4 N. Sekunda, *The Army of Alexander the Great* (Osprey, Oxford, 1999) 27

5 I. Worthington, *Philip II of Macedonia* (Yale University Press, New Haven, 2008) 27

6 Diod. Sic. 17.25.5

7 Diod. Sic. 17.84.4

8 Paus. 1.21.7; Borza ('The Royal Macedonian Tombs and the Paraphernalia of Alexander the Great' Phoenix 41.2 (1987) 112), most likely following the statement of Pausanias, suggests that 'the linen thorax cannot be considered as a serious and effective device.' While this may be true of resistance to strong thrusts with weapons like spears or pikes as Pausanias suggests, the *linothorax* did provide excellent protection against missile fire (see: C.A. Matthew, 'Testing Herodotus – Using Re-creation to Understand the Battle of Marathon' *AncWar* 5.4 (2011) 41-46). Aldrete, Bartell and Aldrete (*Reconstructing Ancient Linen Armour – Unraveling the Linothorax Debate* (The Johns Hopkins University Press, Baltimore, 2013) 92) suggest that as the length of the *sarissa* could be used to keep an opponent at a safe distance, the wearing of linen armour by the members of the pike-phalanx would have been predominantly for protection from weapons that the *sarissa* could not defend against such as arrows, javelins and sling bullets.

9 Halicarnassus: Diod. Sic. 17.25.4; Megalopolis: Diod. Sic. 17.63.2-4; Paraetacene: Diod. Sic. 19.32.1; Magnesia: Livy 37.44; Pydna: Livy 44.42

10 R. Gabriel and K. Metz, *From Sumer to Rome – The Military Capabilities of Ancient Armies* (Greenwood Press, Connecticut, 1991) xix

11 Gabriel and Metz, *Sumer to Rome*, xix

12 Gabriel and Metz, *Sumer to Rome*, xix

13 Gabriel and Metz, *Sumer to Rome*, xix-xx; brass targets were used as the properties of brass and bronze are almost identical.

14 Gabriel and Metz, *Sumer to Rome*, xx; the two inch 'killing depth' was based upon Vegetius (Mil. 12) who states that a wound penetrating to this depth, inflicted by the Roman gladius, is generally fatal. In a similar series of tests, Blyth (*The Effectiveness of Greek Armour against Arrows in the Persian Wars (490-479B.C.): An Interdisciplinary Enquiry* (British Library Lending Division [unpublished thesis – University of Reading], London, 1977) 24) suggests that to cause serious damage to the head, a weapon would only have to penetrate up to 3cm but would have to penetrate deeper to seriously damage the chest.

15 Gabriel and Metz, *Sumer to Rome*, xx

16 Gabriel and Metz, *Sumer to Rome*, xix, 57, 59-60, 63, 95; for calculations of the penetrative abilities of other pieces of ancient weaponry see: Blyth, *The Effectiveness of Greek Armour* 15-18, 81-85; C.A. Matthew, *A Storm of Spears: Understanding the Greek Hoplite at War* (Pen & Sword, Barnsley, 2012) 130-145; Aldrete, Bartell and Aldrete, *Reconstructing Ancient Linen Armour*, 91-128

17 Gabriel and Metz, *Sumer to Rome*, 59

18 See: Matthew, *Storm of Spears*, 2-4

19 Blyth (*Effectiveness of Greek Armour*, 71) tabulates the thickness of different parts of the helmet for nine different examples in the collection of Olympia.

20 Matthews (*The Battle of Thermopylae – A Campaign in Context* (History Press, Stroud, 2008) 55) states that the linen cuirass worn by the Classical Greeks was lighter and provided less protection than

the bronze corslet. However, it has been suggested by other scholars that the linen composite cuirass afforded no better protection or advantage in weight than its bronze equivalent (for example see: E. Jarva, *Archaiologia on Archaic Greek Body Armour* (Studia Archaeologica Septentrionalia 3), (Pohjois-Suomen Historiallinen Yhdistys, Societas Historica Finlandiae Septentrionalis, Rovaniemi, 1995) 135-143; T. Everson, *Warfare in Ancient Greece* (Sutton Publishing, Stroud, 2004) 111, 147). Some re-created examples of the *linothorax* certainly agree with this conclusion, at least in terms of weight. Nepos (*Iphicrates* 1.4) says that the linen corslet was lighter but gave the same protection as bronze. Pausanias (1.21.8), on the other hand, states that the linen corslet would not protect against spears and was only suitable for hunting – although Anderson (*Military Theory and Practice in the Age of Xenophon* (University of California Press, Berkeley, 1970) 23) suggests that Pausanias may have had no actual experience with this type of armour. Conversely, Aldrete, Bartell and Aldrete (*Reconstructing Ancient Linen Armour*, 127-128), based upon their own practical experimentations, conclude that for cloth armour, which deforms with impacts and is made of many individual layers which all offer resistance to penetration, a thickness of 11mm was equivalent to wearing plate metal 1.8mm thick.

21　Taken from A. Williams, *The Knight and the Blast Furnace – A History of the Metallurgy of Armour in the Middle Ages and the Early Modern Period* (Brill, Leiden, 2003) 928-929, 936

22　Confirmed through personal correspondence with R. Gabriel, 01 June 07 – 06 July 07

23　Williams, *The Knight and the Blast Furnace*, 929-930; for example, if a weapon hits a curved chest-plate 'square-on' to the centre of a barrel-shaped torso, it will hit roughly perpendicular to the surface of the armour. However, if the strike lands further around to either side of the centre, the further around the strike lands, the more of the curvature of the armour the thrust would have to overcome.

24　Taken from Williams, *The Knight and the Blast Furnace*, 937

25　Things such as imperfections in the armour and/or the brittleness of the metal it was made from would also have an impact on how easily a weapon might be able to penetrate it. See: Matthew, *Storm of Spears*, 140-141

26　Curt. 3.11.5

27　Diod. Sic. 17.84.4

28　Blyth, *Effectiveness of Greek Armour*, 14

29　A leaning attack would not have a very high velocity and this would reduce the overall amount of energy delivered. If, as an example, a leaning attack was made by a 220lb (100kg) phalangite who could get a quarter of his body weight (55lb/25kg) behind an attack made with an 11.9lb pike at a velocity of only 3.2ft/1m per second, the amount of energy delivered through the tip of the weapon would be approximately 10fpds/13.5j. This would not even be enough to penetrate armour with a thickness of 1mm if the strike was delivered perfectly perpendicular to the surface of the armour. However, it could still knock the person on the receiving end over. In an impromptu 'test' done with members of a sixteenth century pike and musket re-enactment group in 2009, I was asked to simply hold onto the extended tip of pike when it was lowered for action and wielded at shoulder height. The wielder of this weapon then simply lent forward and, due to the amount of energy being transferred down the length of the weapon, I was knocked backwards by several feet.

30　Plut. *Aem.* 20

31　Lucian, *Dial. mort.* 27

The Use of the Butt-Spike in Phalangite Combat

1　Lucian, *Dial. mort.* 27; Pietrykowski (*Great Battles of the Hellenistic World* (Pen & Sword, Barnsley, 2009) 81) suggests that Alexander's troops adopted this same position to counter Porus' elephants at the Hydaspes River in 326BC. However, there is no reference to such an undertaking in any of the accounts of the engagement.

2　Lucian, *Dial. mort.* 27; 20 *cubits*, or about 9.6m, seems incredibly big for a cavalry lance when it is considered that the longest infantry pikes only got to 16 cubits or just over 7.6m. However, a long lance would have given the Thracian more opportunity to parry the point and then set the *sarissa* as Lucian describes.

3　M.M. Markle, 'The Macedonian *Sarissa*, Spear and Related Armour' *AJA* 81.3 (1977) 324; W. Heckel and R. Jones, *Macedonian Warrior: Alexander's Elite Infantryman* (Osprey, Oxford, 2006) 14; M. Park, 'The Fight for Asia – The Battle of Gabiene' *AncWar* 3.2 (2009) 35; B. Bennett and M. Roberts, *The Wars of Alexander's Successors 323-281BC Vol.II: Battles and Tactics* (Pen & Sword, Barnsley, 2009) 4

4　C.A. Matthew, *A Storm of Spears: Understanding the Greek Hoplite at War* (Pen & Sword, Barnsley,

2012) 146-164

5   Polyb. 11.18; Plutarch (*Phil.* 10), recounting the same event, states that Philopoemen killed Machanidas with a blow delivered with the front tip of the weapon rather than with the butt.

6   Arr. *Anab.* 1.15.6; Plut. *Alex.* 16

7   Diod. Sic. 17.20.4-5; Diodorus also states that, even with the momentum of a charging horse, Alexander's lance was unable to penetrate the shield carried by the Persian he was attacking and this was why the head was broken off.

8   Xenophon (*Eq.* 8.10) details the use of the butt, rather than the head, in mock cavalry battles. This suggests that, regardless of its shape, the butt of the Greek lance could not inflict serious injury. It is also possible that the butt of the some Greek lances was more akin to the *sauroter* of the hoplite spear, and thus considerably more blunt than the head, rather than the large, 'head-like' butt of the Macedonian lance.

9   Front. *Strat.* 2.3.20

10   Diod. Sic. 17.100.6-7

11   Plut. *Aem.* 20

12   Plut. *Pyrrh.* 21

13   Hdt. 9.62; see also: Matthew, *Storm of Spears*, 150-155

14   Plut. *Aem.* 20 - οἱ μὲν γὰρ ἐκκρούειν τε τοῖς ξίφεσι τὰς σαρίσας ἐπειρῶντο

15   R. Sheppard (ed.), *Alexander the Great at War* (Osprey, Oxford, 2008) 82

16   Matthew, *Storm of Spears*, 151-153

17   Eur. *Phoen.* 1396-1399

18   Hdt. 7.224

19   Xen. *Ages.* 2.12-14

20   Diod. Sic. 15.87.1

21   P.H. Blyth, *The Effectiveness of Greek Armour against Arrows in the Persian Wars (490-479B.C.): An Interdisciplinary Approach* (British Library Lending Division [unpublished thesis – University of Reading], London, 1977)) 22-22e

22   Blyth, *Effectiveness of Greek Armour*, 22-22e

23   H. Baitinger, *Die Angriffswaffen aus Olympia* (de Gruyter, Berlin, 2000) plates 14, 17, 19, 20, 40, 41, 42

24   The exact opposite thing happens if the first metre of a hoplite *doru* is broken off in the same manner, and the point of balance actually shifts closer to the butt rather than away from it as it does with the *sarissa* (see: Matthew, *Storm of Spears*, 155-156). This is mainly due to the size, length and weights of the shafts of the two weapons involved. For example, with the *doru*, the butt weighs 329g, the head 153g and the 200cm shaft about 850g. Thus for an intact weapon, on either side of the mid-point (127cm forward of the tip of the butt), the rearward half weighs 754g (329g butt + 425g of shaft) and the forward half weighs 578g (153g head + 425g of shaft) – a difference of 176g and which gives the *doru* a point of balance 89cm from the rear tip. When the head and 70cm of shaft are broken off the front of the *doru*, the remaining weapon (155cm in length) would have, from its mid-point, a rearward half weighing 721g (329g butt + 52cm of shaft weighing 392g) and a forward half weighing 329g (77.5cm of shaft), or a difference of 392g. The increase in the difference between the forward and rear ends of an intact spear (179g) to that of a broken spear (392g) means that, on a broken *doru* the point of balance moves even further back towards the butt. However, it is different for the *sarissa*. For an intact *sarissa* 12 cubits (576cm) long, the rearward half weighs 3,105g (1,070g butt + 2,035g of shaft) while the forward half weighs 2,209g (174g head + 2,035g of shaft) – a difference of 896g. If this weapon then loses its tip and about 1m of shaft, the rearward half would weigh 2,617g (1,070g butt + 192cm of shaft weighing 1,547g), while the forward end would weight 1,918g (238cm of shaft) – a difference of 699g. Thus, unlike for the broken *doru*, the difference between the rear and forward halves of a broken *sarissa* decreases in relation to that of an intact weapon (896g (intact) compared to 699g (broken)). This then causes the point of balance for a broken *sarissa* to move away from the rearward tip and towards the centre – the opposite of what happens to the hoplite spear.

25   Connolly ('Experiments with the *Sarissa* – the Macedonian Pike and Cavalry Lance – a Functional View' *JRMES* 11 [2000] 109-112) experienced similar problems when trying to wield a replica *sarissa* which had a copy of the large 'head' found at Vergina attached to it.

26   C. Koepfer, 'The *Sarissa*' *AncWar* 3.2 (2009) 37; Park, 'The Fight for Asia', 35

27   Matthew, *Storm of Spears*, 160-164

28   See: Plut. *Aem.* 20

29   Polyb. 18.30

The Phalanx

1 Tarn (*Alexander the Great Vol.II* (Ares, Chicago, 1981) 142) states: 'there was no such formation in Alexander's army as 'the phalanx'; in both Greek and English it is only a convenient expression for the sum total of the battalions of the πεζέταιροι, the heavy infantry of the line...'

2 The Greek word for 'discipline' (*eutaxia* – εὐταξία) literally translates as 'arranged well' and can be seen as another variant of the generic use of the term taxis to refer to organization. The term 'phalanx' itself is also used by some ancient writers to simply refer to a battle formation which could include a variety of troops such as mercenaries and cavalry, and even to Persian formations, rather than just specifically to the Macedonian pike-phalanx. For example see: Arr. *Anab.* 2.8.2, 2.8.6, 2.8.8, 2.8.10, 2.9.1, 2.9.3-5, 2.10.4-6, 2.11.2, 2.11.7.

3 For the claim that his work analyses the pike-armies of Alexander the Great see: Ael. Tact. *Tact.* Praef.; despite such claims, it is clear that Aelian's work contains details relating to Macedonian armies from across the Hellenistic Period such as the varying lengths of the *sarissa* (chapters 12 and 14) and the use of war elephants (chapters 2 and 23) which did not become a common feature of Hellenistic warfare until the age of the Successors.

4 Ael. Tact. *Tact.* 1; the work of Stratocles is no longer extant. Frontinus' work on stratagems survives, but a specific work on tactics does not. Aeneas is mentioned by Polybius (10.44) where he discusses the signals that can be made by a fire-beacon in the case of an enemy attack. This work on 'How to Withstand a Siege' by Aeneas is still extant. However, a specific book on tactics or generalship (Polybius calls it 'A Commentary on the Office of General', and Aelian calls it 'Book on the Office of a General') has not survived. Cyneas is mentioned by Plutarch in his biography of Pyrrhus (*Pyrr.* 14) and by Cicero (*Ad. Fam.* 9.25.1). The writings of Pyrrhus are mentioned by both Cicero (*Ad. Fam.* 9.25.1) and Plutarch (*Pyrr.* 8) and Pyrrhus was thought to have been one of the greatest generals of all time (see: Plut. *Pyrr.* 8). Pyrrhus had a son called Alexander (See: Plut. Pyrrh. 1, 6, 9; Just. *Epit.* 18.1-3; Ath. 3.73) who is the one mentioned by Aelian. A reference to a book on tactics written by Alexander is found nowhere else other than in Aelian. According to Plutarch (*Phil.* 4), a work on tactics by Evangeleus was one of the texts studied by the Archaean general Philopoemen. The Polybius mentioned by Aelian is the same whose history is, for the most part, still extant. A specific work of his on tactics has not survived. However, there are numerous passages scattered throughout his history which demonstrate a firm understanding of tactics and strategy. The *Iphicrates* referred to by Aelian was a great Athenian commander and military reformer of the fourth century BC (see: Nepos, *Iphicrates*, 1.1-4; Diod. Sic. 15.44.1-4; Arr. *Anab.* 2.15) who had compared the structure of an army to that of a human body (Plut. *Pelop.* 2). Posidonius was a philosopher, scientist and strategist (see: Cic *Tusc.* 2.25.61; Plin. (E) *HN* 7.30). He is credited with inventing a 'sphere', or orrery, a mechanical model which correctly calculated the motions of the planets in the 1st century BC (see: Cic. *Nat.D.* 88). His work on tactics may have been the source of the slightly later work by Asclepiodotus. Sadly few, if any, of these works survive, and little is known about the other authors, or their works, that are mentioned by Aelian. The similar work of Asclepiodotus contains no such beginning overview of source material. The beginning of Arrian's *Tactics* is fragmentary, but does cite works by Pyrrhus, his son Alexander, Clearchus, Pausanias, Evangeleus, Polybius, Eupolemus, Iphicrates and Poseidonius. It is interesting to note that the sources cited by Arrian are listed in the same order as in the preceding work written by Aelian which suggests that Arrian used Aelian's work as a source. It is also suggested that Aelian was the true author of Arrian's *Tactics*, a revised version of his earlier work, and that Arrian had simply released it under his own name (see: T.F. Didbin, A*n Introduction to the Knowledge of Rare and Valuable Editions of the Greek and Latin Classics* (W. Dwyer, London, 1804) 5

5 Asclepiodotus (*Tact.* 2.1) says that the *lochos* was formally called a line (στίχος), a *synomoty* (συνωμοτία), or a *decury* (δεκανία). Aelian (*Tact.* 4-5) calls the file a *lochos*, but states that it is also called a *dekad* or an *enomotia*. Arrian (*Tact.* 6) says the *lochos* was called a line (στίχος) by some and a *decury* (δεκανία) by those who base their examinations on units of ten. Arrian then goes on to detail the possibility that another name for the file was an *enomotia*.

6 Arr. *Tact.* 5

7 For example see: J. Warry, *Warfare in the Classical World* (University of Oklahoma Press, Norman, 1995) 68; N.G.L. Hammond, *Alexander the Great: King, Commander and Statesman* (Duckworth, London, 1980) 32; Cascarino (*Tecnica della Falange* (Il Cerchio, Rome, 2011) 59) suggest that the file was simply ten deep under Philip II and then later increased to sixteen deep under Alexander the Great.

8   Polyb. 12.19, 12.21
9   Curt. 3.9.12
10   Harpocration, *Lexicon* s.v. *pezhetairoi* (Anaximenes – Jacoby *FrGrHist*. 72 F4)
11   D. Head, *Armies of the Macedonian and Punic Wars 359 BC to 146 BC* (Wargames Research Group, Sussex, 1982) 13
12   Arr. *Anab*. 2.8.4
13   On the process of 'doubling' see: Asclep. *Tact*. 10.17-20; Ael. Tact. *Tact*. 28; Arr. *Tact*. 25
14   Polyb. 12.21
15   P. McDonnell-Staff, 'Hypaspists to Peltasts: The Elite Guard Infantry of the Antigonid Macedonian Army' *Ancient Warfare* 5.6 (2012) 22
16   N.G.L. Hammond, *The Genius of Alexander the Great* (Duckworth, London, 1997) 13-14, 66
17   English (*The Field Campaigns of Alexander the Great* (Pen & Sword, Barnsley, 2011) 196) suggests that Alexander the Great had adopted an eight deep formation for the battle of the Hydaspes in 326BC in order to allow his line to extend beyond that of his Indian opponents. While possible, none of the narrative accounts of the battle state that this was the case.
18   In his translation, Melville Jones (*Testimonia Numaria Vol.II* (Spink and Sons, London, 2007) 18) says that the pay of the 'ten-stater man' was 'more than that of the *dimoirites*, but more than that of those who did not have any supplement'. This is most likely a typographical error as, later in his commentary, Melville Jones correctly follows the Greek by stating that the pay of the 'ten-stater man' was 'less than that of the man on double pay'.
19   Arr. *Anab*. 7.23.3-4; κατέλεγεν αὐτοὺς ἐς τὰς Μακεδονικὰς τάξας, δεκαδάρχην μὲν τῆς δεκάδος ἡγεῖσθαι Μακεδόνα καὶ ἐπὶ τούτῳ διμοιρίτην Μακεδόνα καὶ δεκαστάτηρον, οὕτως ὀνομαζόμενον ἀπὸ τῆς μισθοφορᾶς, ἥντινα μείονα μὲν τοῦ διμοιρίτου, πλείονα δέ τῶν οὐκ ἐν στρατευομένων ἔφερεν. ἐπὶ τούτοις δὲ δώδεκα Πέρσας καὶ τελευταῖον τῆς Μακεδόνα, δεκαστάτηρον καὶ τοῦτον. ὥστε ἐν τῇ δεκάδι τέσσαρας μὲν εἶναι Μακεδόνας. τοὺς μὲν τρεῖς τῇ μισθοφορᾷ προΰχοντας, τὸν δὲ τῇ ἀρχῇ τῆς δεκάδος, δώδεκα δὲ Πέρσας.
20   A.J. Heisserer, *Alexander the Great and the Greeks: the Epigraphic Evidence* (University of Oklahoma Press, Norman, 1980) 21-22
21   In other passages Arrian also seems to have retained earlier titles and terms. At 6.27.6, for example, Arrian describes how pack animals were distributed among the infantry by units of 100 (*hekatostyes*) and by file (ἑκατοστύας, τοῖς κατὰ λόχους). Bosworth ('The Argeads and the Phalanx' in E. Carney and D. Ogden (eds.), *Philip II and Alexander the Great: Father and Son, Lives and Afterlives* [Oxford University Press, Oxford, 2010] 94-97) convincingly argues that the use of the term *hekatostyes* is anachronistic, based upon an earlier decimal hierarchy of the phalanx, but applied to an army based upon files of sixteen as Alexander had at this time.
22   For Abreas see: Arr. *Anab*. 6.9.3
23   In his commentary on the passage, Melville Jones (*Testimonia Numaria Vol.II*, 18) says the base unit was seventeen men. However, the numbers involved clearly add up to sixteen.
24   Ael. Tact. *Tact*. 5
25   Arr. *Anab*. 7.22.3; Arr. *Tact*. 6.2; see also: Diod. Sic. 17.34.1; Asclepiodotus (*Tact*. 2.2) calls the half-file a *hemilochion*, under the command of a *hemilochite* (if the phalanx is arranged sixteen deep) or a *dimoiria*, under the commander of a *dimoirites* (if the phalanx is arranged twelve deep). See also: Ael. Tact. *Tact*. 9
26   Arr. *Tact*. 5.4; Asclep. *Tact*. 2.2; see also: Ael. Tact. *Tact*. 5
27   For example see: P. Connolly, *Greece and Rome at War* (Greenhill Books, London, 1998) 69; see also: S. English, *The Army of Alexander the Great* (Pen & Sword, Barnsley, 2009) 22, 25; N. Sekunda, 'Land Forces' in P. Sabin, H. Van Wees, and M. Whitby, (eds.), *The Cambridge History of Greek and Roman Warfare Vol. I: Greece, the Hellenistic World and the Rise of Rome* (Cambridge University Press, 2007) 333; Bosworth (*Conquest and Empire: The Reign of Alexander the Great* (Cambridge University Press, 2008) 270); states that the four Macedonian officers were positioned 'at the front and rear' of the file. It is unclear if Bosworth means at the front and rear of each half-file (which would then correlate with the structure outlined in the manuals), or if he means that the Macedonians occupy positions 1, 2, 15, and 16 with the twelve Persians in between. In another work ('The Argeads', 96) Bosworth positions Macedonians in the first three positions in the file with the *ouragos* at the rear. One of the few who offer an alternative placement for the officers is Cascarino (*Tecnica della Falange*, 50-51) who has officers positioned at the front and rear of each half-file as is described in the following pages. Arrian

continues his description of the new units by stating that the Persians were still armed with bows and javelins. This would seem unlikely as it does not readily create a pike-phalanx. It is more likely that the Persians were armed as phalangites. In 327BC, Alexander had 30,000 Persian youths trained as phalangites to replace parts of his phalanx (Arr. *Anab.* 8.6.1; Plut. *Alex.* 71; Diod. Sic. 17.108.1-3; Curt. 7.5.1). This would further suggest that the Persians incorporated into the phalanx in 324BC were armed as phalangites unless it is assumed that the use of mixed-arms within the new units was some form of tactical experiment.

28 See: Diod. Sic. 17.34.1; Asclep. *Tact.* 3.1-6; Ael. Tact. *Tact.* 10; Arr. *Anab.* 7.22.3; Arr. *Tact.* 6.2

29 Same principle applied to the formations of the earlier Classical hoplite and go back as far as the Archaic Age.

30 Arr. *Anab.* 7.22.3; Arr. *Tact.* 6.2; see also: Diod. Sic. 17.34.1

31 Connolly, *Greece and Rome*, 69

32 Asclep. *Tact.* 2.3

33 Ael. Tact. *Tact.* 5; Arr. *Tact.* 6-7

34 Asclep. *Tact.* 2.3

35 Cascernio, *Tecnica della Falange*, 55

36 Morgan 'Sellasia Revisited' *AJA* 85.3 (1981) 328) says that the 'double depth deployment' at Sellasia in 222BC referred to by Polybius (2.66) is a reference to a formation thirty-two deep. This is similar to Alexander the Great's initial deployment at Issus in 333BC and suggests a standard file size of sixteen (see also: J. Pietrykowski, *Great Battles of the Hellenistic World* (Pen & Sword, Barnsley, 2009) 172). Gabriel (*Philip II of Macedonia – Greater than Alexander* (Potomac, Washington, 2010) 67) suggests that the file was commonly sixteen deep under Alexander and deeper under the Successors.

37 R. Sheppard (ed.), *Alexander the Great at War* (Osprey, Oxford, 2008) 54, 79

38 Front. *Strat.* 4.1.6

39 Hammond ('The Battle of the Granicus River' *JHS* 100 (1980) 81, 83-84 (73-88)) suggests that in 334BC the standard file of Alexander's army was sixteen deep but, because Alexander needed to ensure that his smaller force could not be outflanked at the Granicus, he deployed in half-files of eight to give his formation a wider frontage equal to that of the Persians – the same as the manner in which he deployed for the later battle of Issus.

40 Warry (*Warfare,* 68) says the depth of the pike-phalanx was originally eight deep and later expanded to sixteen deep. This would also correlate with what some of the manual writers have to say about the early phalanx. See also: Hammond (*King, Commander and Statesman,* 32) who says that the phalanx was only 8 deep.

41 For standard hoplite deployments of the Classical Age see: C.A. Matthew, *A Storm of Spears: Understanding the Greek Hoplite at War* (Pen & Sword, Barnsley, 2012) 172-179

42 Asclep. *Tact.* 2.1-2

43 Xen. *Hell.* 6.4.12

44 Cascarino, *Tecnica della Falange*, 55-58

45 Connolly, *Greece and Rome*, 77

46 N. Sekunda, *The Seleucid Army* (Stockport, Montvert, 1994) 5-6

47 Paraetacene: Diod. Sic. 19.29.1-6; Gabiene: Diod. Sic. 19.27.6

48 Raphia: Polyb. 5.63-65; Pydna: Livy 44.40-43; Plut. *Aem.* 18-23

49 Bennett and Roberts (*The Wars of Alexander's Successors 323-281BC Vol.II: Battles and Tactics* (Pen & Sword, Barnsley, 2009) xv) suggest, due to the detail in Diodorus' account of the battle of Paraetacene, that Diodorus' source (Hieronymous of Cardia) may have not only seen these formations as they were drawn up on the field of battle, but may have also been present at the pre-battle council where the nature of the deployment was first decided.

50 See: Polyb. 18.18-27; Livy, 33.3-10; Plut. *Flam.* 8

51 App. *Syr.* 32; see also: Livy 37.40

52 Cass. Dio 78.7.1-2

53 Harpocration, *Lexicon* s.v. *pezhetairoi* (Anaximenes – Jacoby *FrGrHist.* 72 F4)

54 T. Daniel, 'The *Taxeis* of Alexander and the change to Chiliarch, the Companion Cavalry and the change to Hipparchies: A Brief Assessment' *AncW* 23.2 (1992) 44

55 Arr. *Anab.* 1.6

56 Daniel, 'The Taxeis of Alexander' 44; Daniel goes on to suggest (p.45) that the regular deployment for a *lochos* of 120 men was in fifteen files of eight men each – a configuration for which there is no evidence.

57 Sheppard, *Alexander the Great at War*, 79; Cascarino, *Tecnica della Falange*, 51-52

58 English, *Army of Alexander*, 111; N. Sekunda, *The Army of Alexander the Great* (Osprey, Oxford, 1999) 25; See also: English, *Field Campaigns of Alexander*, 10;

59 Cascarino, *Tecnica della Falange*, 52

60 For example see: Arr. *Anab*. 3.9.6 where each file commander (*lochagos*) encourages the men in his file (*lochos*), or *Anab*. 7.24.4 where Alexander distributes wine and sacrificial victims by file (*lochos*) and by a larger unit of around 100 men known as a *ekatostos*. Like many passages in ancient literature, Arrian's description of the units to which these distributions are made begins with the smallest and then moves to the biggest. Cascarino (*Tecnica della Falange*, 51-52) however, and despite citing this passage, suggests that in the time of Alexander the *ekatostos* was a unit of 128 men (similar to the *taxis* of the later manuals) while the *lochos* was bigger at 256 men (the same as the later *syntagma*). This interpretation would seem to be incorrect.

61 Unfortunately, many of the models cited are not referenced so it is difficult to determine the author's line of argument.

62 Sheppard, *Alexander the Great at War*, 79

63 For example: Connolly, *Greece and Rome*, 69; M.G. Carey, *Operational Art in Classical Warfare: The Campaigns of Alexander the Great* (US Army Command and General Staff College, Fort Leavenworth, 1997) 14; W.W. Tarn, *Alexander the Great Vol.II* (Ares, Chicago, 1981) 142-148; M. Thompson, *Granicus 334BC: Alexander's First Persian Victory* (Osprey, Oxford, 2007) 24; Daniel, 'The Taxeis of Alexander', 43; W. Heckel and R. Jones, *Macedonian Warrior: Alexander's Elite Infantryman* (Osprey, Oxford, 2006) 30; R.D. Milns, 'Arrian's Accuracy in Troop Details: A Note' *Historia* 27:2 (1975) 375; Sekunda, *Army of Alexander*, 28; N. Sekunda, *Macedonian Armies after Alexander 323-168BC* (Osprey, Oxford, 2012) 21; J.F.C. Fuller, *The Generalship of Alexander the Great* (Wordsworth, Hertfordshire, 1998) 50; W. Heckel, *The Conquests of Alexander the Great* (Cambridge University Press, 2008) 26

64 Cascarino, *Tecnica della Falange* 52-53

65 Diod. Sic. 17.17.3

66 For example see the entries for note 63.

67 For example see: A.M. Devine, 'Demythologizing the Battle of the Granicus' *Phoenix* 40.3 (1986) 268; Hammond, 'The Battle of the Granicus', 81-83, 86; it is interesting to note that, while Diodorus is the only source to provide a breakdown of the units under Alexander's overall command in 334BC, the sum total of the infantry forces whom he says accompanied Alexander into Asia does not agree with many of the other available accounts. Diodorus (17.17.34) states that the 12,000 Macedonians were part of the 32,000 infantry who marched into Asia with Alexander – with the breakdown being given as: 12,000 Macedonians; 7,000 allies; 5,000 mercenaries; 7,000 Odrysians, Triballians and Illyrians; and 1,000 archers and Agrianians. This figure to also given by Justin (*Epit*. 11.6.2) in his abridgement of the history of Pompeius Trogus. Arrian (*Anab*. 1.11.3), on the other hand, gives a lesser figure of 30,000 infantry – due to his use of Ptolemy and Aristobulos (who were not used by Diodorus) as source material who both provide the same figure. Anaximenes and Callisthenes, on the other hand, provide the differing figures of 43,000 and 40,000 respectively for the size of Alexander's infantry which may possibly be including the 10,000 men whom Philip had sent to Asia prior to 336BC to secure a bridgehead (see: Polyaenus, Strat. 5.44.4; Diod. Sic. 16.91.2; Just. *Epit*. 9.5.8-9). Plutarch (*Alex*. 15) merely provides a range of between 30,000 and 43,000 for the size of Alexander's infantry – no doubt using various texts with conflicting troop numbers as his source material. The varying size of Alexander's army as given in the different texts raises a problem with the division of Alexander's pike-phalanx into units of 1,500. The main issue is which overall figure for the size of Alexander's infantry is the correct one and what impact would the dismissal of the other figures have on the even division of his pikemen into units of a set size. If, for example, Arrian's total figure of 30,000 is taken as the more accurate, there is then a discrepancy of 2,000 men between this figure and that provided by Diodorus. But where should this 2,000 be subtracted from Diodorus' breakdown of contingents in order to rectify the balance? If it is taken from the number for the pike-phalanx, then instead of being accompanied by 9,000 phalangites as some scholars assert, Alexander may have only had 7,000 with him when he entered Asia. This figure would then not divide by the six named commanders into even units of 1,500 each. Alternatively, if any of the larger totals is taken as accurate, where should the extra men be allocated to Diodorus' numbers to make up the balance and how does this then impact the divisions of the pike-phalanx under the six named commanders? It is interesting to note that scholars who offer that Alexander's phalanx was divided into six units of 1,500 each base their conclusions on

the troop numbers given by Diodorus and the number of officers provided by Arrian even when the total troop numbers given in both sources are at odds with each other.

68 At the battle of the Granicus (334BC) Arrian (*Anab.* 1.14.2-3) mentions units under the command of (L-R) Craterus, Meleager, Philip, Amyntas, Coenus and Perdiccas arranged across the front of the line.

69 Tarn (*Alexander the Great Vol.II,* 136) suggests that Arrian not only does not always use the same word to describe something, but that his use of the word taxis is commonly used in its generic form to mean 'formation'. Tarn offers that Arrian's use of various words to mean the same thing was part of his literary style 'to avoid a jingle of sound...The worst jingle was to use the same word twice running, which he avoided like most...Greek writers'.

70 Cascarino, *Tecnica della Falange*, 52-53; see also Tarn, *Alexander the Great Vol.II*, 136

71 *Strategos*: for example see Arr. *Anab.* 1.24.1 where both Coenus and Meleager, the commanders of a 'taxis' at Granicus, are referred to as *straegoi*. However, this should not be seen as an indication that these two men had been promoted; for *taxiarch* see: Arr. *Anab.* 3.9.6

72 For example: cavalry – Arr. *Anab.* 1.15.4; mounted javelineers – Arr. *Anab.* 3.24.2; skirmishers – Arr. *Anab.* 4.24.10; archers – Arr. *Anab.* 5.23.7; Latin writers similarly used the word *ordo* to mean either a 'unit' or a 'formation' – for example see Livy 8.8.4, 8.8.7-9

73 For example see: Diod. Sic. 20.53.1

74 Plut. *Aem.* 17

75 Arr. *Anab.* 1.29.4

76 Arr. *Anab.* 2.8.4

77 According to Polybius (12.19), Callisthenes reported that Alexander received another 5,000 reinforcements just prior to the army's entry into Cilicia. Bosworth (*Conquest and Empire*, 267) suggests that these troops, who are not reported in any other source, were pikemen and these troops, combined with the 3,000 received at Gordium, increased the size of the pike-phalanx by more than half. However, this does not correlate with the fact that at the battle of Issus not long afterwards there are still only six named senior phalanx commanders. This leaves several possibilities: a) that Polybius' reporting of Callisthenes is incorrect, b) Callisthenes himself was incorrect (possibly placing the reinforcements from Gordium, numbered differently, into a differnt context, c) that these troops, if ever received, were not pike-men but Greeks and/ or mercenaries, d) that there were six senior phalanx commanders each in charge of units of around 3,000 men each – a fifty per cent increase on the size of the *merarchia*. This last possibility seems unlikely and we are only left to speculate who these troops were (if they existed at all).

78 Bennett and Roberts, *Wars of Alexander's Successors II*, 6

79 Polyaenus, *Strat.* 5.44.4; see also: Diod. Sic. 16.91.2; Just. *Epit.* 9.5.8-9

80 Arr. *Anab.* 1.6.1

81 Daniel, *'The Taxeis* of Alexander', 44

82 N.G.L. Hammond, 'Alexander's Campaign in Illyria' *JHS* 94 (1974) 82

83 Arr. *Anab.* 1.5.12

84 Arrian's use of Ptolemy as a source: Arr. *Anab.* 1.2.7; Arrian's use of Aristobulos: Hammond, 'Illyria' 77

85 Tarn (*Alexander the Great Vol.II*, 137) suggests that Arrian, and possibly his source Ptolemy, regularly rounded troop numbers down rather than up. This would then correlate with a unit arranged in 128 ranks being described as '120 deep'.

86 Ael. Tact. *Tact.* 36

87 Arr. *Anab.* 3.11.10

88 Curt. 5.2.2-3; For a discussion of Alexander's adoption of the office of 'court *chiliarch*', a ceremonial court position based upon the Persian *hazarapatis*, see: A.W. Collins, 'The Office of Chiliarch under Alexander and the Successors' Phoenix 55.3-4 (2001) 259-283

89 As noted earlier (see page 19), Develin ('Anaximenes ("F Gr Hist" 72) F 4' *Historia* 34.4 (1985) 496) argues that what Anaximenes is describing are these changes to the structure of the Macedonian army under Alexander the Great rather than the invention of something new under an earlier Alexander. Featherstone (*Warriors and Warfare in Ancient and Medieval Times* (Constable, London, 1997) 58) says that Alexander broke up the phalanx into smaller, more manoeuverable, units. Such a statement has to assume that either the pre-Alexander phalanx was a single rigid line (which goes against much of the available evidence) or that this is a reference to Curtius' statement regarding the creation of the *chiliarchiae*. Strangely, the creation of the *chiliarchiae* is actually an increase in the size of the units (from 500 to 1,000 according to Curtius) so it is not clear what Featherstone's conclusion is based upon.

90 Arr. *Anab.* 5.11.3

91 Fuller, *Generalship of Alexander*, 191-192

92 Arr. *Anab.* 5.12.2; Perdiccas seems to have retained this more senior command as later in the campaign he leads both cavalry and infantry together (see: Arr. *Anab.* 5.22.6). Coenus seems to have similarly held both a cavalry and infantry command at the Hydaspes and Arrian states that Antigenes is only put in charge of Coenus' infantry unit when Coenus leads his cavalry forward (Arr. *Anab.* 5.16.3). Antigenes' infantry unit was supported by a contingent of *hypaspists* under Seleucus (5.13.4) and archers under Tauron (5.14.1).

93 The figures for Alexander's losses in battle, his garrisons and the numbers of reinforcements received are notoriously conflicted among the ancient texts. Many of the earlier writers do not differentiate between the phalangites and the other types of infantry in Alexander's army when they provide figures. Furthermore, even the figures provided do not agree among the various writers. However, in many cases, the stated losses are so small that, even if these are assumed to be all phalangites the impact on the overall number of the army is not significant. If these figures are added/subtracted from the most likely figures for other events (when more than one ancient author provides conflicting figures) then an approximate number of Alexander's phalangites following the battle of Gaugamela in 331BC can be calculated. For example, Alexander crossed into Asia with 12,000 'Macedonians' – 9,000 of whom were pikemen (Diod. Sic. 17.17.3); thirty were lost at the Granicus in 334BC (Arr. *Anab.* 1.16.4; see also Just 11.6.12; Plut. *Alex.* 16 who give the figure as nine); fifty-six were lost at Halicarnassus (Arr. *Anab.* 1.20.10, 1.22.7); 3,000 reinforcements were received at Gordium (Arr. *Anab.* 1.29.4); 1,000 were lost at Issus (P. *Oxy.* 1798; see also: Curt. 3.11.27 (32 men); Arr. *Anab.* 2.10.7 (120 men); Diod. Sic. 17.36.6 (300 men); Just. 11.9.10 (130 men)); 400 were lost at Tyre (Arr. *Anab.* 2.24.4); 4,000 were left as a garrison in Egypt (Curt. 4.8.4); and 1,000 were lost at Gaugamela (P. *Oxy.* 1798). This would leave Alexander with a total of 5,514 men following the battle. A little later, Amyntas arrived with 6,000 Macedonian infantry reinforcements among others (Arr. *Anab.* 3.16.10; Curt. 5.1.40; Diod. Sic. 17.65.1). Arrian (3.16.11) states that these troops were distributed amongst the existing units. Another 3,000 were later received from Antipater while the army was in Artacana (Curt. 6.6.35). This gives a total of 14,514 men in the pike-phalanx by the time Alexander reaches India.

94 Curt. 5.2.5; Bosworth (*Conquest and Empire*, 268) suggests that the creation of these new positions was to create a sub-level of command which was occupied by men who had been dependent upon royal favour for their appointment – thus becoming totally loyal to Alexander.

95 Bosworth ('The Argeads and the Phalanx' in E. Carney and D. Ogden (eds.), *Philip II and Alexander the Great: Father and Son, Lives and Afterlives* (Oxford University Press, 2010) 92-93) suggests that there were actually nine recipients of the new title of *chiliarch* – with the last name being lost through manuscript corruption. This is based upon the possible use of the Latin word *novum* (nine) at the end of section 5.2.2 of Curtius which describes the contest that Alexander initiated – making it read 'he appointed judges and offered to those who wished to enter a contest of military valour prizes of nine'. As has been conjectured, this is most likely a corruption of the word *nova* (new) which would make the passage read 'he appointed judges and offered to those who wished to enter a contest of military valour prizes of a new kind'. This alternate reading would then reconcile this passage with the eight named recipients of the new titles. Bosworth also suggests that the 'nine' named recipients of the title of *chiliarch* were only those who were awarded this position by the acclamation of the troops and that Alexander himself may have appointed another nine (making a total of eighteen new *chiliarchs*). Bosworth suggests that these new officers were then distributed amongst the new *chiliarchiae* of the pike-phalanx, which he claims may have numbered 18,000 at this time. However, Bosworth has failed to consider that in a phalanx of such size, every second *chiliarchia* would actually be commanded by a *merarch* and so there is no need to assume that there are further recipients of this title other than those whom Curtius details.

96 Nicanor commanding the *hypaspists* at Granicus: Arr. *Anab.* 1.14.2; at Issus: Arr. *Anab.* 2.8.3; Curt. 3.9.7; at Gaugamela: Arr. *Anab.* 3.11.9; Curt. 4.13.27

97 Diod. Sic. 17.110.1

98 Arr. *Anab.* 17.11.3; Arrian also mentions a Persian *agema*. The *agema* was normally the bodyguard of Alexander. Both Diodorus (18.27.1) and Justin (*Epit.* 12.12.4) refer to both Persian and Macedonian contingents of bodyguards in 322BC. Bosworth (*A Historical Commentary on Arrian's History of Alexander* Vol II (Clarendon Books, Oxford, (1998) 9), on the other hand, argues that the Guards were always only Macedonians.

99 W. Heckel, *Who's Who in the Age of Alexander the Great* (West Sussex, Wiley-Blackwell, 2009) *sv. Atarrihias, Antigenes, Philotas* [3], *Amyntas* [6], *Antigonus* [3], *Amyntas* [7], *Theodotus, Hellanicus*

100 Heckel suggests that Atarrihias, Antigenes, Philotas the Augaean and Hellanicus were made *chiliarchs*, Antigonus and Theodotus were made *pentacosiarchs*, and Amyntas and Amyntas Lyncestes were made either *chiliarchs* or *pentacosiarchs*. If it is assumed that both Amyntas and Amyntas Lyncestes were *pentacosiarchs*, this would then make four *chiliarchs* and four *pentacosiarchs*. While this number of officers would allocate across an even *phalangarchia* of 4,000, the *hypaspists* did not reach this number until 324BC. Furthermore, such an allocation would ignore the fact that at least one of the units would be under command of a *phalangarch* who was the overall leader of the *hypaspists* contingent. Heckel's theory can only be seen as a manipulation of the available evidence to conform with a preconceived idea that all of the men were assigned to the *hypaspists*.

101 Arr. *Anab.* 5.16.3; Curt. 8.14.16

102 Antigenes was later discharged from service at Opis in 324BC (see Just. *Epit.* 12.12.8). In a note to their translation of Curtius (Penguin p.282), Yardley and Heckel suggest that Antigenes was a member of the *hypaspists*. This seems unlikely as it goes against his command of pike-infantry at the Hydaspes (see following). Berve (*Das Alexanderreich auf prosopographischer Grundlage Vol.II* (Beck, München, 1926) 41-42) suggests that Antigenes was given command of a phalanx unit and this more closely fits with the available evidence.

103 Arr. *Anab.* 3.29.1; an otherwise unidentified Philotas is also mentioned as an infantry commander earlier in Arrian's text during the siege of Halicarnassus (1.21.5) but the ambiguity of the text makes it difficult to determine exactly who is being referred to. Brunt, in a note to his translation of the passage, suggests that this is the son of Parmenion rather than Philotas the Augaean.

104 Arr. *Anab.* 1.21.5

105 Curt. 5.2.5, 8.1.36

106 Arr. *Anab.* 1.22.7; Arrian rarely uses the terms *chiliarch* and *chiliarchie* in his text and, when he does, it is usually in relation to either *hypaspists* or light troops (for example see: 3.29.7, 4.24.10, 4.30.5-6, 5.23.7, 7.14.10, 7.25.6). However, Adaeus is listed (1.22.4) as commanding a *taxis* (i.e. unit) which was 'supported by light troops'. This suggests that Adaeus was a *chiliarch* of 1,000 regularly infantry.

107 Arr. *Anab.* 1.22.4; in this passage Arrian uses the term taxis in its generic form to mean 'unit'.

108 Arr. *Anab.* 1.22.7

109 Tarn, *Alexander the Great Vol. II*, 148

110 Berve, *Das Alexanderreich*, 12

111 R.D. Milns, 'The Hypaspists of Alexander III – Some Problems' *Historia* 20.2-3 (1971) 189-190

112 Bosworth ('The Argeads', 93-94) also dismisses the idea that Adaeus was commanding mercenaries and suggests that Arrian was being 'inadvertently misleading' in his passage. Bosworth (as per Berve) also suggests that Adaeus was in command of two *pentacosiarchiae* and that Arrian had loosely called him a *chiliarch*.

113 Arr. *Anab.* 4.30.5, 5.23.7

114 Heckel (*Who's Who, sv. Adaeus*) suggests that Adaeus is 'apparently' a commander of the *hypaspists* and that the term *chiliarch* is anachronistic.

115 Units of 1,000 continued to be used across the Hellenistic period and beyond. A fragment of Poseidonius' book 34, preserved by Athenaeus (4.153b) refers to the Seleucid general Herakleon ordering his troops to eat in their units of 1,000. Similarly, the Bible (1Macc. 6.35) records the Seleucid army at the battle of Beth-Zacharia in 162BC as being arranged into thirty-two units of 1,000 men each.

116 Ptolemy's rank: Arr. *Anab.* 2.8.4; Ptolemy's death: Arr. *Anab.* 2.10.7

117 Arr. *Anab.* 2.10.4-5

118 Arr. *Anab.* 3.9.4

119 Sekunda (*Seleucid Army*, 8) suggests that the standard-bearer and other supernumeraries, equivalent to modern day non-commissioned officers, were attached to the *syntagma*, rather than the *taxis*, as this would have otherwise created an ineffective ratio of staff to combatants. Sekunda also suggests that the attachment of the supernumeraries to the units of the phalanx was a later development in the structure of the pike-phalanx. However, the evidence shows that at least some of these additional men was also part of *the army of Alexander the Great* (see following).

120 For example see: Polyb. 6.24

121 Asclep. *Tact.* 2.8-10; Ael. Tact. *Tact.* 9; Arr. *Tact.* 10; the *Suda* (sv. Σύνταγμα) states 'some people [call this unit] a *parataxis*. Two *taxeis*, 256 men, [constitute it]; and the commander [is called a] *syntagmatarch*. But some people also call this a *xenagia* and the commander a *xenagos*'. This last definition also agrees with the military manuals. Under the entry for the *xenagia* (sv. Ξεναγία) the *Suda* states: 'this is a unit of

1,024 men and usually called a *chilarchie'*.

122 Curt. 3.10.3; Curtius (4.6.19) also says that there were standards in use during Alexander's siege of Gaza.

123 Curt. 5.2.2-3

124 Asclep. *Tact.* 2.8-10; Ael. Tact. *Tact.* 9; Arr. *Tact.* 10

125 Ael. Tact. *Tact.* 9, 16; see also: Asclep. *Tact.* 2.8-10; Arr. *Tact.* 10

126 *Suda* sv. Ἔκτακτοι: Ἔκτακτοι ἢ Ἔκτατοι: τούτους τὸ μὲν παλαιὸν ἡ τάξις εἶχεν, ὡς καὶ τοὔνομα δηλοῖ, διότι τῆς τάξεως ἐξάριθμοι ἦσαν. εἰσὶ δὲ ε': στρατοκῆρυξ, σαλπιγκτής, σημειοφόρος, ὑπηρέτης, οὐραγός. νῦν δὲ καὶ τοῦ συντάγματος λέγονται καὶ τῶν ἄλλων. ἔχειν δὲ δεῖ τούτους τὴν τάξιν ἢ καὶ τὸ σύνταγμα, τὸν μὲν ὅπως τῇ φωνῇ σημαίνῃ τὸ προσταττόμενον, τὸν δὲ ὅπως τῷ σημείῳ, εἰ μὴ ἡ φωνὴ κατακούοιτο διὰ θόρυβον, τὸν δὲ σαλπιγκτήν, ὁπότε μηδὲ σημεῖον βλέποιεν διὰ κονιορτόν, καὶ τὸν ὑπηρέτην, ὥστε τι τῶν εἰς τὴν χρείαν παρακομίσαι: τόν γε μὴν ἕκτατον οὐραγὸν πρὸς τὸ ἐπανάγειν τὸν λειπόμενον ἐπὶ τὴν τάξιν: ὃς τῶν τεσσάρων ἄνω τυπουμένων κατὰ μέτωπον κάτω τάσσεται.

127 Connolly, *Greece and Rome*, 76; G. Wrightson, 'The Nature of Command of the Macedonian *Sarissa* Phalanx' *AHB* 24.3-4 (2010) 73-94. Wrightson also suggests that the unit commander was stationed behind the unit. However, if supernumeraries such as the standard-bearer and trumpeter were part of the commander's means of controlling the unit and issuing orders, and these supernumeraries were positioned at the front of the formation as the ancient sources suggest, then the unit commander hand to be positioned at the front of the line as well – just as is outlined in the narrative histories and military manuals.

128 C. Nylander, 'The Standard of the Great King: A Problem with the Alexander Mosaic' *OpRom* 14.2 (1983) 19-37; interestingly, while Nylander uses many visual and literary descriptions of Persian standards to support his claim, he does not engage with similar literary and visual evidence for the use of banners by the Macedonians.

129 Nylander ('Standard of the Great King', 29-32) suggests that the colours and details of the standard bearer are not clear enough to positively identify him as either a Macedonian or a Persian. However, a close inspection of the mosaic shows the bearer to be wearing the more rounded headwear, which is grey in colour, of the other Macedonian troops depicted elsewhere in the image. Some have suggested that the banner is actually held by the figure to the right (as one views the image) of the somewhat ambiguous Macedonian. This more right-hand figure is clearly identifiable as a Persian due to the higher level of detail of his depiction. However, the placement of the hand that is actually holding the staff of the banner would suggest otherwise unless an extreme level of bodily contortion is assumed. Some of the early line drawings of the Mosaic from when it was first excavated (reproduced in Nylander's article) even omit the ambiguous Macedonian standard bearer entirely – clearly showing that the illustrator believed the standard was held by the more right-hand Persian.

130 J. Serrati, 'The Hellenistic World at War – Stagnation or Development?' in B. Campbell and L.A. Tritle (eds.), *The Oxford Handbook of Warfare in the Classical World* (Oxford University Press, 2013) 183

131 Connolly, *Greece and Rome*, 77; see also, for example, Polyb. 11.23

132 M. Feyel, 'Un nouveau fragment du règlement militaire trouvé à Amphipolis' *Revue archéologique* (1935) 47

133 Sekunda ('Land Forces', 336) likens the *speira* to the early *lochos* which he incorrectly claims is a unit of 512 men.

134 Plut. *Phil.* 9

135 *SEG* 40.524; the inscription also refers to *strategoi* and *tetrarachs*. The configuration of the passage ([στρατηγοὶ] καὶ οἱ σπειράρχαι καὶ τετράρχαι) seems to list these officers in descending order according to the size of respective units. Thus the *strategos*, in command of a *phalangarchia/strategia* of 4,096 men according to the manuals, appears first while the *tetrarch*, commanding sixty four men, appears last. This would suggest that the speira was larger than the *tetrarchia* and smaller than the *phalangarchia*. This would correlate with the speira being synonymous with the *syntagma*.

136 Sekunda (*Seleucid Army*, 7), who examines the use of the word *semaia* (rather than *speira*) by Polybius questions why Asclepiodotus, whom he says was writing about the Seleucid Army of the second century BC in his manual, would use the term *syntagma* rather than *semaia*. He concludes, based upon Polybius' account of Antiochus the Great's Bactrian campaign (10.49.7), in which the cavalry are arranged in *semaiai*, that the term *syntagama* was adopted as a Seleucid term to distinguish infantry formations from those of cavalry.

137 Connolly, *Greece and Rome*, 69

138 App. *Syr.* 32

139 Sekunda, *Seleucid Army*, 10
140 Connolly, *Greece and Rome*, 75-77
141 Sekunda (*Seleucid Army*, 7) suggests that the two file *dilochia* and the four file *tetrarchia* were simply additions made by Asclepiodotus (and possibly by his potential source Poseidonius) 'to provide a complete binary organizational structure to their 'perfect' army'. Sekunda also suggests that the *tetrarchia*, if it did exist, was not a unit of 'four files', but was a quarter of a *syntagma*. This again is an exercise in semantics as a quarter of a *syntagma* (sixteen files) is a unit of four files.
142 Arr. *Anab.* 3.18.5; see also: Curt. 5.2.5 for a Latin variant of the term.
143 Diodorus (17.34.1) states that the officers of each unit in Alexander's army were positioned at the head of their men.
144 Ael. Tact. *Tact.* 10; See also: Asclep. *Tact.* 3.1-6
145 The US Army employs a similar, albeit varied, method for the distribution of command among its armoured platoons. Within an armoured platoon, consisting of four tanks, the platoon commander is located in tank #1, the second in command, the platoon sergeant, is in tank #4, and tanks #2 and #3 are commanded by lower ranks – referred to as 'wingmen' (see: US Army, *FM7-8 Infantry Rifle Platoon and Squad* [US Dept. Of the Army, Washington, 1992] 2.44 Combined Operations with Armored Vehicles). This means that, when the tank platoon is deployed in an extended line in numerical order, the distribution of the officers would be the platoon commander, one wingman, the other wingman, the platoon sergeant. Thus both wings of a modern American tank platoon would be held by the most senior officers of the unit – the same as the Macedonian pike-phalanx of more than two millennia earlier.
146 Ael. Tact. *Tact.* 10
147 Ael. Tact. *Tact.* 10
148 C.B. Wells, 'New Texts from the Chancery of Philip V of Macedonia and the problem of the *Diagramma*', *AJA* 42 (1938) 249
149 This is one of the weaknesses of many modern reconstructions of units such as the *syntagma* and its associated officers, as many illustrations accompanying modern examinations only focus on one unit (usually that on the very right hand side of a larger formation) and do not account for the symmetrical distribution of officers outlined in the manuals. For example see: Connolly, *Greece and Rome*, 76; Cascarino, *Tecnica della Falange*, 57; this also means that the exact identity of the phalangite depicted on the Pergamon Plaque can not accurately be ascertained. Due to the presence of the standard in the image, it can be assumed that the unit being depicted is a *syntagma* at the very least. Yet the combined *pentacosarchia* would also have a standard at its right hand side (depending upon where it was in a larger formation) and the officer standing next to the banner in that instance would be a *pentacosiarch* rather than a *syntagmatarch*. The only thing that can be said with any level of certainty about the plumed officer in the Pergamon Plaque is that the lowest rank that he can be is that of *syntagmatarch*.
150 Cascarino (*Tecnica della Falange*, 53), in a series of confusing calculations, concludes that within a 1,500 man 'taxis' in Alexander's army there were the following officers: 1 x *taxiarch*, 3 x *pentacosiarch*, 6 x *lochargoi*, 12 x *hemilochites*, 24 x *tetrarchs*, 96 x *dekadarkes*, and 1,152 x *lochites* – a total of 1,294 front rank officers (plus 96 *dimoirites* and 192 *dekastateroi*). There are a number of issues with such conclusions. For example, even if the '*taxis*' was three combined *pentacosiarchiae* as Cascarino assumes, in their standard deployment of thirty-two files of sixteen men per *pentacosiarchia*, there should only be a maximum of ninety-six officers across the front of the whole formation (the same number as there is for the half-file leading *dimorites*) not 1,294. A number of factors have contributed to this miscalculation of numbers and misnaming of the ranks of the associated officers. Firstly, it assumes that the *lochos* was a unit of 256 men rather than a file of the phalanx. Secondly, it is based upon the assumption that a 1,500-man '*taxis*' was a standard unit of Alexander's army. Finally, it fails to take into account the symmetrical nature of the positioning of the officers and how the command of a particular unit was taken by a more superior officer when that unit was combined into a larger one. This accounts for why there are far too many officers given in Cascarino's conclusion than there would need to be even if Alexander's army was based upon a 1,500 man 'taxis'. For similar miscalculations of the number of officers in the phalanx see: English, *Army of Alexander*, 111; Milns, 'Hypaspists' 195
151 In my earlier translation of Aelian's *Tactics* (C.A. Matthew, *The Tactics of Aelian* (Pen & Sword, Barnsley, 2012) 146-147), I suggested that there were another four *merarchs* which Aelian had failed to provide details or positions for. However, in this model I failed to consider that the positions where I placed the four lowest grade *merarchs* would actually be occupied by officers holding the ranks of *chiliarch* – and so this model cannot be considered correct. It is only the method of positioning the

officers outlined above which correlates with the descriptions of the placement of officers in the ancient texts and conforms, mathematically, with the correct number of men positioned at the head of the files of the phalanx.

152 Sekunda (*Seleucid Army*, 10) suggests that the division of the phalanx into left and right wings is 'late Hellenistic reality'. While Alexander's pike infantry cannot easily be divided into two wings with symmetrical command qualities, Alexander's army as a whole was divided into wings – with Alexander commanding the right and Parmenion the left (see following). In regards to the pike-phalanx alone, the even division into wings must post-date the campaign of Alexander, but whether it is actually a 'late Hellenistic' development as Sekunda suggests, of whether it occurred under the Successors, is far from certain.

153 Arr. *Anab*. 1.14.2-3; Arrian actually lists the commanders in a somewhat confusing order stating: 'beside [the *hypaspists*] was the unit of Perdiccas...then that of Coenus, then that of Craterus, then that of Amyntas...then the troops under Philip...on the right were the units of Craterus, of Meleager and of Philip up to the centre of the whole line'. It seems that Arrian lists the units from right to left, beginning with that of Perdiccas, up to the unit of Philip in the centre. He then lists them from left to right, beginning with the unit of Craterus, also up to Philip's in the centre which accounts for why Philip is listed twice – although some have argued that there were, in fact seven units, with two of them commanded by different officers with the name of Philip, but this seems unlikely due to the organizational aspects of the pike-phalanx. It is also argued that the first reference to the unit of Craterus, which places it on the right wing, is either an error or interpolation (scribal or otherwise) or indicative that Arrian had based his account of two different sources – most likely those of Ptolemy and Aristobulos (see: D. Campbell, 'Alexander's Great Cavalry Battle: What Really Happened at the River Granicus?' *AncWar* 7.2 (2013) 51)

154 Arr. *Anab*. 2.8.4; There have been various interpretations of Alexander's deployment at Issus. Most scholars agree that the right of the line had Coenus' unit next to the *hypaspists* and then the unit of Perdiccas next to that. The arrangement of the left wing, on the other hand, is a lot more problematic. Fuller (*Generalship of Alexander*, 157-159), for example, says that (from right to left) Alexander's line was the pike-units of Coenus, Perdiccas, Craterus, Meleager, Ptolemaeus and Amyntas. Warry (*Alexander 334-323BC* (Osprey, Oxford, 1991) 31) states 'the infantry line (right to left was:...phalanx units under Coenus, Perdiccas, Meleager, Ptolemy, and Amyntas respectively; left-wing infantry was commanded by Craterus'. Due to the passage, it is uncertain whether Warry suggests that Craterus commanded a unit on the far left, as well as overall command of the left-wing infantry, or whether he was just in command of the left-wing infantry (which would then assume that one of the pike-units was either under the command of someone else (unnamed) or was missing). English (*Field Campaigns*, 87, 104) lists Alexander's deployment as (right to left) Coenus, Perdiccas, Meleager, Ptolemy, Amyntas, Craterus. All of these interpretations seem incorrect. Arrian begins his description of Alexander's deployment by stating that 'from the right' (ἀπὸ τοῦ δεξιοῦ) were the units of Coenus and the Perdiccas. Arrian then states that 'on the left [of these units]' (ἐπὶ δὲ τοῦ εὐωνύμου) came the units of Amyntas, Ptolemaeus, Meleager and Craterus. Thus the entire line, from right to left according to Arrian, was Coenus, Perdiccas, Amyntas, Ptolemaeus, Meleager and Craterus. Warry and English seem to have misinterpreted Arrian's description of the left hand units to mean 'from the left' rather than 'on the left' and so their order for the four units on the left of the line is reversed. Fuller not only also seems to have reversed the order of the right hand units, but also places Craterus on the right side of the left wing rather than at the very left-hand end of the line. This confusion may partially be due to how Curtius' account of the battle (4.13.28) similarly orders the units (from right to left) as Coenus, Perdiccas, Meleager, Ptolemy, Amyntas and Craterus. This is most likely a transcription error, possibly from using a similar source as Arrian but misinterpreting the Greek. What suggests that the order outlined above is correct is that it then makes Alexander's deployment at Issus almost the same as his previous deployment at Granicus, and almost exactly the same for the subsequent battle at Gaugamela (see following).

155 Arr. *Anab*. 1.28.3; this also occurred at the battle of the Hydaspes in 326BC; see: Arr. *Anab*. 5.13.4.

156 English, *Army of Alexander*, 111-112. English cites two passages of Arrian (*Anab*. 1.28.4 and 4.13.4) as support for this claim. However, as will be shown, 1.28.4 only indicates that the position of the commanders on the right of the line swapped at different engagements, and 4.13.4 refers to those in command of the night watch rather than a deployment for battle.

157 Arr. *Anab*. 3.11.10

158 Diod. Sic. 17.53.7; Curt. 4.13.28

159 Arr. *Anab.* 3.11.10; Craterus also held this position of authority at Issus in 333BC see: Arr. *Anab.* 2.8.4

160 For a discussion of some of the changes of command that took place in the pike-phalanx across Alexander's campaign see: Tarn, *Alexander the Great Vol.II*, 142-148

161 Arr. *Anab.* 2.8.4

162 See: Bosworth, *Conquest and Empire*, 60

163 Curt. 3.9.12

164 A. R. Burn, 'The Generalship of Alexander' *Greece and Rome* Second Series 12.2 (1965) 145; see also: Fuller, *Generalship of Alexander*, 157

165 Arr. *Anab.* 2.8.3

166 Polyb. 12.19

167 Arr. *Anab.* 1.6.6

168 Livy, 33.8

169 Cascarino, *Tecnica della Falange*, 149

170 Curt. 3.8.22

171 Curt. 3.8.24

172 Fuller, *Generalship of Alexander*, 157

173 Fuller, *Generalship of Alexander*, 157

174 Asclep. *Tact.* 10.13-16; Ael. Tact. *Tact.* 26-27; Arr. *Tact.* 22-24

175 Aelian (*Tact.* 27) states that this manoeuvre has the effect of appearing as if the formation is fleeing from an enemy that has appeared behind it only to then suddenly about face to engage.

176 Aelian (*Tact.* 27) states that this manoeuvre has the effect of appearing as if the formation is advancing towards an enemy that has appeared behind it.

177 Turning to face the right, or 'to the [side of the] pike/spear' (ἐπὶ δόρυ κλῖνον), and turning to the left, or 'to the [side of the] shield' (ἐπ' ἀσπίδα κλῖνον) are common drill movements outlined in the military manuals. For example see: Asclep. *Tact.* 12.11; Ael. Tact. *Tact.* 53; Arr. *Tact.* 32

178 Polyb. 11.12

179 See: Asclep. *Tact.* 10.2-12, 12.1-8; Ael. Tact. *Tact.* 24, 31; Arr. *Tact.* 21

180 See: Asclep. *Tact.* 11.2-7; Ael. Tact. *Tact.* 35-36; Arr. *Tact.* 28

The Pike-Phalanx in Battle

1 F.E. Adcock, *The Greek and Macedonian Art of War* (University of California Press, Berkeley, 1957) 26

2 J. Pietrykowski, *Great Battles of the Hellenistic World* (Pen & Sword, Barnsley, 2009) 198, 235

3 E.M. Anson, 'The Introduction of the Sarisa in Macedonian Warfare' *AncSoc* 40 (2010) 64-65

4 A.R. Burn, 'The Generalship of Alexander' *Greece & Rome* Second Series 12.2 (1965) 145

5 D. Featherstone, *Warriors and Warfare in Ancient and Medieval Times* (Constable, London, 1988) 58-59

6 J. Warry, *Warfare in the Classical World* (University of Oklahoma Press, Norman, 1980) 73

7 A.B. Bosworth, 'The Argeads and the Phalanx' in E. Carney and D. Ogden (eds.), *Philip II and Alexander the Great: Father and Son, Lives and Afterlives* (Oxford University Press, 2010) 9

8 R. Gabriel, *Philip II of Macedonia – Greater than Alexander* (Potomac, Washington, 2010) 67-68

9 W.W. Tarn, *Hellenistic Military and Naval Developments* (Cambridge University Press, 1930) 28

10 J. Pietrykowski, 'In the School of Alexander – Armies and Tactics in the Age of the Successors' *AncWar* 3.2 (2009) 24; Pietrykowski, *Great Battles*, 169

11 Diod. Sic. 16.86.1

12 Diod. Sic. 17.11.1, 17.12.1-2

13 For a discussion of the idea that some of Alexander's units fought in 'relays' see: R.K. Sinclair, 'Diodorus Siculus and Fighting in Relays' *CQ* 16.2 (1966) 249-255

14 At Miletus: Diod. Sic. 17.22.1; at Halicarnassus: Diod. Sic. 17.24.4, 17.26.4; in Uxianē: Diod. Sic. 17.67.5; at Aornus: Diod. Sic. 17.85.6

15 Arr. *Anab.* 3.13.1; Diod. Sic. 17.57.6

16 Curt. 4.15.1

17 Arr. *Anab.* 3.14.2

18 Arr. *Anab.* 3.14.4

19 Dion. Hal. 20.1; Polyb. 18.28

20 Polyb. 2.66

21 Livy, 37.37; App. *Syr.* 32

22 Polyb. 11.12

23 Polyb. 11.11

24 Polyb. 11.15

25 Diod. Sic. 31.16.2

26 Crescent formations: Ael. Tact. *Tact.* 43, 46; open-ended square: Ael. Tact. *Tact.* 45; wedge formations: Ael. Tact. *Tact.* 47; hollow square formations: Ael. Tact. *Tact.* 48

27 Illyria: Arr. *Anab.* 1.6.3; Gabiene: Diod. Sic. 19.43.5; Magnesia: App. *Syr.* 35

28 Arr. *Anab.* 2.11.1

29 R.E. Dickinson, 'Length Isn't Everything – Use of the Macedonian *Sarissa* in the Time of Alexander the Great' *JBT* 3.3 (2000) 58

30 Anson, 'Introduction of the Sarisa', 65

31 A.M. Snodgrass, *Arms and Armour of the Greeks* (Johns Hopkins University Press, Baltimore, 1999) 121

32 M.M. Markle, 'Use of the *Sarissa* by Philip and Alexander of Macedon' *AJA* 82.4 (1978) 488

33 S. English, *The Field Campaigns of Alexander the Great* (Barnsley, Pen & Sword, 2011) 29; interestingly, Arrian (1.4.2) states that Alexander's men advanced beyond this tilled land to form up for battle. As such, the fighting which ensued did not take place on the flattened grain field which throws English's conclusion into question. It is more likely that the grain was flattened merely to facilitate the advance of the army into position.

34 G.T. Griffith, 'Philip as a General and the Macedonian Army' in M.B. Hatzopoulos and L.D. Loukopoulos (eds.), *Philip of Macedon* (Ekdotike Athenon, Athens, 1980) 59

35 Polyb. 18.31 – καὶ μὴν ὅτι χρείαν ἔχει τόπων ἐπιπέδων καὶ ψιλῶν ἡ φάλαγξ, πρὸς δὲ τούτοις μηδὲν ἐμπόδιον ἐχόντων, λέγω δ' οἷον τάφρους, ἐκρήγματα, συναγκείας, ὀφρῦς, ῥεῖθρα ποταμῶν, ὁμολογούμενόν ἐστι.

36 Livy. 44.37.11 – *quam inutilem vel mediocris iniquitas loci efficeret*

37 Plut. *Aem.* 16 - ὁ δὲ τόπος καὶ πεδίον ἦν τῇ φάλαγγι βάσεως ἐπιπέδου

38 See: J.D. Morgan, 'Sellasia Revisited' *AJA* 85.3 (1981) 330

39 N.G.L. Hammond, *Alexander the Great: King, Commander and Statesman* (Duckworth, London, 1980) 33

40 Plut. *Phoc.* 13

41 Polyaenus, Strat. 4.2.2; for other accounts of the battle see: Polyaenus, *Strat.* 4.2.7; Diod. Sic. 16.86.1-6; Plut. *Pel.* 18; Plut. *Alex.* 9

42 Markle ('Use of the *Sarissa*', 488-489) questions Polyaenus' description of the battle by suggesting that the length of the *sarissa* would make it impossible for the members of the phalanx to about face as described. However, Markle does not seem to have considered the use of the countermarch manoeuvre as a means of reversing the phalanx's facing – which would have been conducted with the pikes raised vertically. If, as the Athenians advanced, Philip had his troops conduct a countermarch using the 'Macedonian Method' (see page 318) not only would this have his troops face to the rear and so be able to march up the slope of the high ground but, according to Aelian (*Tact.* 27) the advantage of use of the Macedonian method of countermarching (as opposed to either the Lacedaemonian or Cretan/Choral/Persian methods) is that the Macedonian countermarch has the appearance of withdrawing from the enemy – the exact thing that Philip would have wanted in order to draw the Athenian left wing forward. Once on the heights, Philip's troops could have then simply conducted another countermarch, employing any method, to face the Athenians, lower their pikes and advance back down the slope to engage.

43 N.G.L. Hammond, 'Training in the Use of a *Sarissa* and its Effects in Battle 359-333 B.C.' *Antichthon* 14 (1980) 62

44 Polyb. 2.65

45 Polyb. 2.65

46 Polyb. 2.66

47 For a description of the topography see: P. McDonnell-Staff, 'Sparta's Last Hurrah – The Battle of Sellasia (222BC)' *AncWar* 2.2 (2008) 27

48 See: Pietrykowski, *Great Battles*, 173

49 Polyb. 2.66

50 Polyb. 2.68

51 Polyb. 2.68

52 Polyb. 2.69

53 Polyb. 2.69

54 Polybius (2.69) says 'for a considerable distance' while Plutarch (*Cleo.* 28) says for a distance of five

stades or just under one kilometre.

55 Plut. *Cleo.* 28; Polybius (2.69), on the other hand, says that Antigonus' beleaguered left wings reformed and counter-attacked the Spartans to gain victory.

56 Plut. *Phil.* 6

57 Plut. *Flam.* 8

58 Polyb. 18.22

59 Pietrykowski, *Great Battles*, 198

60 Morgan, 'Sellasia Revisited', 330

61 For Philip's view of the ground see: Polyb. 18.22; Livy, 33.8.2

62 Polyb. 18.24; Livy, 33.8.7

63 Livy, 33.8.13-14; see also: Polyb. 18.24

64 Polyb. 18.24; Plut, *Flam*, 8

65 Livy, 33.9.7-9; see also: Polyb. 18.26; Plutarch (*Flam.* 8) says that the phalanx was attacked in the flank.

66 Livy, 33.10.3; see also: Polyb. 18.26

67 Plut. *Flam.* 8; see also: Polyb. 18.25

68 Polyb. 18.25; Plut. *Flam.* 8

69 For the whole account of the battle see: Arr. *Anab.* 1.31.1-1.16.7; for the description of the banks see: Arr. *Anab.* 1.13.4-5; for examinations of the site as it is today, coupled with re-examinations of the battle, see: E. Badian, 'The Battle of the Granicus: A New Look' *Ancient Macedonia II* (1977) 271-293; N.G.L. Hammond, 'The Battle of the Granicus River' *JHS* 100 (1980) 73-88; A.M. Devine, 'Demythologizing the Battle of the Granicus' Phoenix 40.3 (1986) 265-278; D. Campbell, 'Alexander's Great Cavalry Battle: What Really Happened at the River Granicus?' *AncWar* 7.2 (2013) 48-53

70 Arr. *Anab.* 1.14.7; Plut. *Alex.* 16; Badian ('Battle of the Granicus', 279-280) suggests that the topographical features described in such passages should be ignored and offers that they are literary exaggerations to highlight the boldness of Alexander's actions.

71 Diod. Sic. 17.19.2; in an irreconcilable passage, Diodorus then states (17.19.3) that the Macedonians crossed the river and formed up in good order before the Persians had a chance to stop them. Diodorus' account of the battle, which conflicts with the other ancient accounts, had caused considerable controversy over the years among those attempting to understand the engagement. Bosworth (*A Historical Commentary on Arrian's History of Alexander Vol I* (Clarendon Books, Oxford, 1998) 114-116; *Conquest and Empire: The Reign of Alexander the Great* (Cambridge University Press, 2008) 40-41), for example, doubts Arrian's account and favours that of Diodorus in which the Macedonians cross the river, the Persian position is further back, and the battle is fought on the plain to the east of the watercourse. Campbell ('Alexander's Great Cavalry Battle' 50), on the other hand, finds nothing illogical in Arrian's account that the Persians had drawn up their cavalry along the river's edge and posted their Greek mercenary infantry to the rear. Hammond (*The Genius of Alexander the Great* (Duckworth, London, 1997) 69) calls Diodorus' account of the battle 'worthless' and Heckel (*The Conquests of Alexander the Great* (Cambridge University Press, 2008) 48) calls it 'implausible'. Campbell supports Bosworth's position by stating that the strength of the Persian cavalry was in their role as missile troops and that their deployment on the bank of the river was to 'ensure that the slow moving Macedonian infantry were deluged in a storm of javelins, as they cross the riverbed'. Such operational tactics would make sense. By acting as skirmishers, the Persian cavalry could disrupt part of the Macedonian phalanx with its missiles as the formation crossed and then, once the phalanx had reached the far back, wheel backwards into the open ground behind them to allow the Greek mercenaries to advance and engage (which for some reason does not happen). As such, the role of the Persian cavalry would have been little different from infantry skirmishers except that they would have been more mobile. For other discussions on the possible course of the battles see: n.69

72 Plut. *Alex.* 16

73 Polyaenus, *Strat.* 4.3.16

74 Arrian (*Anab.* 1.16.5) = 30; Plutarch (*Alex.* 16) and Justin (*Epit.* 11.6.12) = 9;

75 Arr. *Anab.* 2.10.1, 2.10.5; Curtius' account (3.8.24-3.11.20) contains few details of the terrain or the actions of the infantry, while Diodorus' brief account (17.33.1-17.34.9) only mentions that the Macedonian phalanx and the Persian infantry were engaged for a short time.

76 Arr. *Anab.* 2.10.4-5; it was during this fragmentation of the line that Ptolemaeus and his *taxis* of 120 men were killed (Arr. *Anab.* 10.2.7)

77 Arr. *Anab.* 2.11.1

78 Markle, 'Use of the *Sarissa*', 494

79 Markle ('Use of the *Sarissa*', 494) suggests that Alexander's infantry did not use the *sarissa* until their more open field engagement at Gaugamela in 331BC. See also: S. English, *The Army of Alexander the Great* (Pen & Sword, Barnsley, 2009) 19-22.

80 Polyb. 12.22

81 Polyb. 12.20; Markle ('Use of the *Sarissa*' 494) uses these passages as the basis for an argument that the Macedonians did not use the *sarissa* until the battle of Gaugamela in 331 (see also: English, *The Army of Alexander*, 22). However, while commentaries on the description of the terrain and how it may, or may not, have impeded the phalanx, they nowhere state that Alexander's army was not using the *sarissa* because of the terrain and seems to ignore all of the other evidence for pike-phalanxes fighting on difficult ground both before and after the engagement at Issus. Interestingly, in his critique, Polybius (citing Callisthenes) himself says that Alexander's Phalanx was '*sarissa*-armed' (12.20) which would go against what Markle suggests based upon another comment made by the same writer.

82 Plut. *Alex.* 20

83 Polyb. 11.15-16

84 Plut. *Eum.* 14

85 Plut. *Pyrr.* 27

86 Plut. *Pyrr.* 28

87 Plut. *Pyrr.* 27

88 Plut. *Aem.* 16

89 Plut. *Aem.* 16

90 See: N.G.L. Hammond, 'The Battle of Pydna' *JHS* 104 (1984) 31-47

91 Plut. *Aem.* 17

92 Plut. *Aem.* 17

93 Plut. *Aem.* 18

94 Plut. *Aem.* 20

95 Plut. *Aem.* 20

96 Plut. *Aem* 21

97 Strangely Pietrykowski, despite calling the ridge at Cynoscephalae a 'liability' to the functionality of the phalanx, says that the plain of Pydna, crossed by two rivers, 'was almost perfect for the smooth operation of the...phalanx' (*Great Battles*, 225).

98 Livy, 44.41

99 For example see: Polyb. 18.29; Plut. *Aem.* 20; Plut. *Flam.* 8; *Excerpta Polyaeni*, 18.4

100 Polyb. 18.32

101 Plut. *Aem.* 20

102 Plut. *Aem.* 20

103 Arr. *Anab.* 3.9.5

104 G. Cascarino (*Tecnica della Falange* (Il Cerchio, Rome, 2011) 65) says that the benefit of the phalanx was that it was easy to control so long as the line was maintained. However, it was the very lack of control at the small unit level which meant that phalangites had trouble maintaining the line.

105 Arr. *Anab.* 3.14.4

106 M.M. Markle, 'The Macedonian *Sarissa*, Spear and Related Armour' *AJA* 81.3 (1977) 331

107 Plut. *Aem.* 20; Polybius (18.26) and Livy (33.9) also state that it was almost impossible for a phalangite to fight man-to-man.

108 Plut. *Pyrr.* 21

109 Plut. *Pyrr.* 21; for another account of the battle see: Dion. Hal. 20.1-3. Similarly, Polybius (4.64.3-11) describes how, in 219BC, three units of Macedonian *peltasts* (by which he means pikemen) were able to cross the river Achelous against Achaean cavalry holding the far bank. The phalangites were able to maintain such good order of their lines that the opposing cavalry simply abandoned the position and withdrew. What is strange is that Polybius recounts such events which clearly show that units of phalangites could effectively fight across water-courses, yet later in his work suggests that the phalanx could only operate on unbroken terrain.

110 Polyb. 18.28

111 Polyb. 18.31

112 Livy, 36.14; Appian (*Syr.* 17) says the name of the Roman commander was Acilius Manius Glabrio.

113 Livy, 36.18

114 App. *Syr.* 19
115 Livy, 36.18
116 Livy, 36.18
117 Livy, 36.19; App. *Syr.* 20
118 Front. *Strat.* 2.2.1
119 N.G.L. Hammond, 'What May Philip Have Learnt as a Hostage in Thebes?' *GRBS* 38.4 (1997) 368
120 This also assumes that, in urban areas, a unit in the streets was not attacked through windows or doorways as it passed either. The vulnerability of troops in built-up areas was a lesson harshly learnt by Pyrrhus of Epirus who was struck on the base of the neck by a roof-tile, hurled by an angry female defender, when he invaded Argos in 272BC (see: Plut. *Pyrr.* 34). Diodorus also notes (19.35.2) that when the Aetolians occupied the Thermopylae pass against Cassander in 317BC, Cassander recognized the strength of the position due to the terrain and so chose to ferry his troops around it rather than engage. When the two sides met at Thermopylae again the following year, the Macedonians managed to dislodge the Aetolians with great difficulty (Diod. Sic. 19.53.1) and Cassander then chose to ferry his troops from Megara to Epidaurus rather than try and fight his way across the Isthmus of Corinth (Diod. Sic. 19.54.3). This shows that in some cases even commanders of pike-phalanxes would not fight on narrow fronted fields, regardless of how the opposing side was armed, if an easier means of overcoming or circumventing a position was available.
121 Plut, Pyrr. 33
122 Tyre: Arr. *Anab.* 2.24.3; Gaza: Arr. *Anab.* 2.27.5
123 Markle, 'Use of the *Sarissa*', 495
124 J.K. Anderson, 'Shields of Eight Palms Width' *CSCA* 9 (1976) 4
125 Arr. *Anab.* 1.13.2
126 Arr. *Anab.* 1.13.3-7; Badian ('The Battle of the Granicus', 290) suggests that Alexander dismissed Parmenion's advice to wait and attack the next day as, had they done so, the Macedonians would have been advancing into the rising sun while, by attacking immediately in the afternoon, the setting sun would have been shining in the eyes of the Persians on the far bank (see also: Campbell, 'Alexander's Great Cavalry Battle' 48).
127 For the rotating roster of command see: Arr. *Anab.* 1.28.3
128 Curt. 3.8.22, 3.8.24
129 Arr. *Anab.* 2.16.8
130 Polyb. 11.2
131 Arr. *Anab.* 3.9.3-4
132 Hammond, *Genius of Alexander*, 66
133 Alexander issued such orders at Gaugamela. See: Arr. *Anab.* 3.10.1
134 Arr. *Anab.* 3.9.5-8
135 Arr. *Anab.* 3.10.1-2
136 Arr. *Anab.* 5.11.4
137 For examples of the command to 'lower pikes' see: Polyb. 2.67, 5.85, 11.15, 11.16, 18.24.
138 *Suda* sv. Ἔκτακτοι
139 Arr. *Anab.* 3.9.5; for a discussion on the level of discipline in the Macedonian army, in respect to following orders and instructions, in the time of Alexander see: E. Carney, 'Macedonians and Mutiny: Discipline and Indiscipline in the Army of Philip and Alexander' *CP* 91.1 (1996) 19-44
140 Sabin ('Land Battles' in P. Sabin, H. Van Wees and M. Whitby [eds.]), *The Cambridge History of Greek and Roman Warfare Vol. I: Greece the Hellenistic World and the rise of Rome* (Cambridge University Press, 2007) 406-407) suggests that 'battlefield communications were so primitive and ancient armies so unwieldy, most forces could do little more in battle than to put into practice what had been planned and ordered beforehand...'and that 'it is hard to see how complex grand tactical plans...could have been improvised on the spur of the moment'. Such a conclusion clearly ignores all of the evidence for the various methods of relaying information across the battlefield employed by Hellenistic armies, and the descriptions of phalanxes changing their direction or altering their formation during the course of an engagement.
141 Ael. Tact. *Tact.* 52; Arr. *Tact.* 31
142 Arr. *Anab.* 3.9.5
143 App. *Syr.* 33
144 *Suda* sv. Ἔκτακτοι

145 Arr. *Anab.* 5.12.1
146 Arr. *Anab.* 5.14.1
147 Arr. *Anab.* 5.16.3
148 Arr. *Anab.* 3.15.1; Curt. 4.16.1; Diod. Sic. 17.60.7; Plut. Alex. 33; for a discussion of this event see: G.T. Griffith, 'Alexander's Generalship at Gaugamela' *JHS* 67 (1947) 87-88
149 For the standards across the front of Alexander's phalanx at Gaugamela see: Curt. 3.10.3
150 *Suda* sv. Ἔκτακτοι
151 Dion. Hal. 20.2
152 Philip at Athens: Livy, 31.24; Antiochus at Thermopylae: Livy, 36.18
153 Polyb. 2.66; Plutarch (*Phil.* 6) says that the scarlet flag was a cloak suspended from the tip of a spear.
154 Polyb. 2.67
155 *Suda* sv. Ἔκτακτοι
156 Polyb. 2.69
157 Halicarnassus: Diod. Sic. 17.26.5, 17.27.4; Issus: Diod. Sic. 17.33.4; Gaugamela: Diod. Sic. 17.58.1; Hydaspes: Diod. Sic. 17.89.1
158 Hellespont: Diod. Sic. 18.32.2; Megalopolis: Diod. Sic. 18.71.1
159 Diod. Sic. 19.97.2
160 1 Macc. 6.38
161 R. Sheppard (ed.), *Alexander the Great at War* (Osprey, Oxford, 2008) 80
162 Adcock, *Greek and Macedonian Art of War*, 26
163 Pietrykowski, *Great Battles*, 204
164 The drill manual for the Australian Army (*Land Warfare Procedures – General LWP-G 7-7-5 Drill*, (Commonwealth of Australia, Australian Army, 2005) 3A-1) for example defines the quick march as 116 paces per minute with each step being 75cm in length. The US Army Drill Manual (*FM 22-5 Drill and Parades* (US Dept. of the Army, Washington, 1986) 3.6f) defines the 'quick time' march as 120 steps per minute with each pace being 30 inches (75cm) in length from heel to toe. The exact pace and distance of the step is often dictated by the nationality of the military institution, and even the particular regiment, and can also vary based upon the style of marching employed, such as the 'goose-stepping' method of march employed by many former Soviet-bloc and Asian countries.
165 While many modern armies march at the 'quick step' while on parade, soldiers do not generally march at this pace while burdened with their combat equipment except when conducting a 'forced march'. Yet even here the soldiers are not required to remain in step or maintain a very close formation.
166 Polyb. 18.24; Livy, 33.8.7
167 Arr. *Anab.* 3.14.2
168 Arr. *Anab.* 4.26.4
169 Polyaenus, *Strat.* 4.2.2
170 Plut. *Alex.* 33
171 Plut. *Aem.* 18
172 Arr. *Anab.* 2.10.3
173 For an examination of the 'charge' of hoplite formations see: C.A. Matthew. *A Storm of Spears* (Barnsley, Pen & Sword, 2012) 225-230
174 Afric. *Cest.* 1.1.13-15 – σπάνιος τῇ ὁπλίσαι τούτῃ δρόμος, οὐ πολὺς μών, ὀξὺς δὲ καὶ τοσοῦτος ὅσος ἂν γένοιτο τοῦ σπεύδοντος ἐντὸς βέλους γενέσθαι φθάσαι. This is most likely a reference to the Greek charge at the battle of Marathon in 490BC (see: Hdt. 6.112).
175 For Philip's feigned retreat see: Polyaenus, *Strat.* 4.2.2
176 Arr. *Anab.* 4.26.2; Tarn (*Military and Naval Developments*, 13) states that feigned retreats such as those of the pike-phalanx at Chaeronea demonstrate the professionalism and training of the Macedonian army in the time of Philip and Alexander as, had such manoeuvres been attempted with ill-trained troops, a pretend rout would have ended in a real one.
177 App. *Syr.* 35
178 App. *Syr.* 35
179 For example see: Aesch. *Pers.* 393; Thuc. 1.50, 4.43, 4.96, 5.70, 7.44, 7.83; Xen. *Hell.* 2.4.17, 4.2.19; Xen. *An.* 1.8.17, 1.10.20, 4.3.19-31, 4.8.16, 5.2.14, 6.5.27; Plut. *Lyc.* 21-22; Plut. *Mor.* 238a-b
180 For example see: Thuc. 5.70; Plut. *Mor.* 210f; Ath. *Deip.* 14.627D; Pausanias, 3.17.5; Xen. *An.* 6.1.11; Polyaenus, *Strat.* 1.10.1; *Excerpta Polyaeni*, 18.1
181 Dickinson, 'Length Isn't Everything', 58

182 Polyb. 2.69, 5.85, 11.15, 11.16, 18.24

183 Asclep. *Tact.* 12.11; Ael. Tact. *Tact.* 53; Arr. *Tact.* 32; In his experiments with pike-formations, Connolly ('Experiments with the *Sarissa* – The Macedonian Pike and Cavalry Lance – A Functional View' *JRMES* (2000) 111) found it difficult to have the members of the phalanx lower their pikes unless it was conducted one rank at a time beginning with the front rank (Connolly states that starting from the back just caused entanglement). While working one rank at a time from the front does work, if the members of the first five ranks all lowered their weapons at the same time, the same result would be accomplished.

184 Asclep. *Tact.* 5.1; Ael. Tact. *Tact.* 13.3; Arr. *Tact.* 12.3; Polyb. 18.29

185 Polyb. 2.69

186 Livy, 44.41

187 Plut. *Aem.* 19

188 J.F.C. Fuller, *The Generalship of Alexander the Great* (Wordsworth, Hertfordshire, 1998) 106

189 In a diagram accompanying his discussion of the phalanx, Hammond (*Genius of Alexander*, 14) has each file, which he has arranged to a depth of eight, is angled obliquely to the right-rear so that the pikes of the first ranks can be deployed. However, as the whole unit is depicted as rectangular in shape, this angled arrangement of the files has resulted in the two right-hand files containing less than eight men. The second file from the right, for example, contains only six men, while the most right-hand file only contains two men. Such an arrangement seems unlikely for a number of reasons. Firstly, it is unlikely that any file formed up with a depth of two or six when the rest of the file was configured differently. Secondly, due to the angled nature of the whole file, the lateral (i.e. left-to-right) interval that the whole file occupies – that is from the left-hand side of the man in the front rank to the right-hand side of the man in the rear rank – equates to the open-order of 192cm with a depth of only eight men rather than the standard intermediate-order of 96cm which could accommodate a file of sixteen. It is therefore more likely that the first five ranks were arranged in a slightly offset manner, occupying the intermediate-order spacing, while the remaining ranks simply stood in a line directly behind the man in the fifth rank. This would then mean that, no matter that depth of the file, it would always conform with the intermediate-order interval.

190 This idea is outlined by Wheeler ('The Legion as Phalanx in the Late Empire (I)' in Y. le Bohec and C. Wolff (eds.), *L'Armée Romaine de Dioclétien à Valentinien Ier* (Paris, De Boccard, 2004) 329). As Connolly observed in his experiments ('Experiments', 111) there is a natural flex in the length of the *sarissa* which does allow the weapon of the more rearward ranks to arc over those held by the more forward members of the phalanx. However, for weapons to overlap each other in such a way, the members of the phalanx would have to be standing front-on (which is physically impossible while carrying a heavy pike) in order for the file to take up most of the 96cm interval allocated to it. It seems more likely that, even with the flex of the *sarissa* taken into account, the phalangites were deployed in such a manner that both they and their weapons occupied their full interval.

191 Curt. 6.1.10

192 Arr. *Tact.* 11

193 Arr. *Anab.* 2.8.3-4; Campbell ('Alexander's Great Cavalry Battle'51) suggests that Alexander also deployed his phalanx in a half-depth of eight for the battle at the Granicus River in 334BC so that his forces could occupy a frontage equal to that of the opposing Persians. This is not recorded in any of the ancient accounts of the engagement.

194 Polyb. 12.21; Polybius continues to say that 'even if they had, as the poet says, 'laid shield against shield and leaned on each other' [i.e. adopted the close-order formation of 48cm per man] ground twenty *stades* [3.6km] wide would still have been needed while he [i.e. Callisthenes] says the ground was no more than fourteen [*stades* wide]. The terminology used by Polybius, 'even if they had', in reference to the close-order deployment suggests that this was not normally used by pike-wielding phalangites.

195 Polybius, in his examination of Callisthenes' description of the battlefield at Issus (12.19), states '...the distances which must be kept on the march...each man covering six feet' (...ἄνδρας ἐν τοῖς πορευτικοῖς διαστήμασιν...ἑκάστου τῶν ἀνδρῶν ἓξ πόδας ἐπέχοντος). Six Greek feet, or 192cm, is the open-order interval.

196 Ael. Tact. *Tact.* 49; see also: Arr. *Tact.* 29

197 Front. *Strat.* 2.3.2

198 English (*Army of Alexander*, 124), citing a passage of Arrian (*Anab.* 1.2.2), suggests that Alexander marched against the Triballians in 335BC with his army arranged in column to hide their number. Arrian, however, does not actually detail how the phalanx was arranged in this passage.

199 Polyaenus, *Strat.* 4.6.19

200 Arr. *Strat.* 17

201 Arr. *Tact.* 11 – τάσσεται δὲ ἡ φάλαγξ ἐπὶ μῆκος μὲν ὅπου ἀραιοτέρα, εἰ ἥ τε χώρα παρέχοι καὶ ὠφελιμώτερον εἴη, κατὰ βάθος δὲ ὅπου πυκνοτέρα, εἰ αὐτῇ τῇ πυκνότητι καὶ τῇ ῥύμῃ τοὺς πολεμίους ἐξῶσαι δέοι...ἢ αὖ εἰ δέοι τοὺς ἐπελαύνοντας ἀποκρούσασθαι...χρὴ τάσσειν.

202 Polyb. 18.30

203 Ael. Tact. *Tact.* 14; see also: Asclep. *Tact.* 5.2

204 Arr. *Tact.* 12

205 For example see: G. Cawkwell, *Philip of Macedon* (Faber &Faber, London, 1978) 154-155; Adcock, *Art of War*, 4; J. Champion, *Pyrrhus of Epirus* (Pen & Sword, Barnsley, 2009) 68

206 Warry, *Warfare,* 68

207 Asclep. *Tact.* 5.2; Ael. Tact. *Tact.* 14; Arr. *Tact.* 12

208 The exact nature of the *othismos* in Classical Greek warfare has been a long debated topic amongst scholars. However what is clear is that, from a review of the literary evidence, a battle which involved a physical pushing of 'shield against shield' only occurred at three major engagements across the whole Classical Age – Delium (424BC), Coronea (394BC) and at Leuctra (371BC). The remainder of the battles fought during this time seem to have been conducted with both sides 'at spear length' from each other where they could effectively use their weapons. See. C. A. Matthew, *A Storm of Spears: Understanding the Greek Hoplite at War* (Pen & Sword, Barnsley, 2012) 205-217, 222-225

209 The term *othismos* has numerous meanings. Under ὠθέω in the Liddell and Scott *Greek-English Lexicon* the following definitions are given: thrust, push (mostly of human force), throw down, push (with weapons), force out, stuff into, force open, non-human forces (e.g. streams or wind), force back in battle, banish, push matters on (i.e. hurry them along), push off from land, throw (as a horse throws a rider), press forward, to fall violently, a crowd, a throng, to jostle. Under ὠθισμός specifically are the following definitions: thrusting, pushing (as of shield against shield), jostling, struggling (as of combatants in a melee). For the pike-phalanx, the most appropriate definition for any such action is to 'push with weapons'.

210 Connolly, 'Experiments', 111

211 Plut. *Cleom.* 28

212 B. Bennett and M. Roberts, *The Wars of Alexander's Successors 323-281BC Vol.II: Battles and Tactics* (Barnsley, Pen & Sword, 2009) 10

213 This accounts for the numerous descriptions found in the ancient texts which describe the pike-phalanx as 'invincible' or something similar. For example see: Polyaenus, Excepts, 18.4; Livy, 44.41; Plut. *Alex.* 33; Plut. *Aem.* 19; Polyb. 18.29

214 Arr. *Anab.* 1.6.10

215 Gabriel, *Philip*, 63; Cascarino (*Tecnica della Falange*, 65) similarly states that the formation was easy to control so long as it was maintained.

216 Gabriel, *Philip*, 63

217 Asclep. *Tact.* 10.2-12; Ael. Tact. *Tact.* 24; Arr. *Tact.* 21

218 Asclep. *Tact.* 10.1, 11.2-7; Ael. Tact. *Tact.* 23, 35, 36; Arr. *Tact.* 20, 28

219 Basically, using the layout in Fig.41 as an example, *syntagmae* #1 and #4 would advance to a point where their file closing *ouragoi* were just past the front rank of *syntagmae* #2 and #3. The men of *syntagmae* #1 and #4 would then turn to face the left and both formations would march to take up position in front of *syntagmae* #2 and #3. Once in position, the men of *syntagmae* #1 and #4 would then turn to face the front – resulting in a square formation, thirty-two deep, with the officers in their correct locations, and with every man facing forward. Alternatively, if the ground ahead of the line did not permit such a move, the same result could be accomplished by having *syntagmae* #2 and #3 move back and behind *syntagmae* #1 and #4 following a similar, but mirrored, series of steps.

220 Arr. *Anab.* 3.12.1; see also: Curt. 4.13.30; Front. *Strat.* 2.3.19

221 See: P.A. Brunt's note to this passage in the Loeb edition of the text. Other scholars have also followed this line of argument – for example see: Bosworth, *Conquest and Empire*, 81; J. Warry, A*lexander 334-323BC* (Osprey, Oxford, 1998) 59,61; Warry, *Warfare,* 81, 83; Pietrykowski, *Great Battles*, 63; English, *Field Campaigns*, 123; D. Head, *Armies of the Macedonian and Punic Wars 259BC to 146BC* (Wargames Research Group, Sussex, 1982) 65; Hammond, *Genius of Alexander*, 109. Fuller (*Generalship of Alexander*, 169) just calls this line a 'second phalanx...deployed at an unspecified distance behind the front line'. Connolly (*Greece and Rome*, 82) simply calls it 'second line inf[antry]'.

222 Arrian provides no details of which units made up this second line. Diodorus (17.52.5) does not even mention the second line deployment at all, but only states that the wings of the formation were refused to counter any encirclement. The conclusion that Alexander's second line was composed of allies and mercenaries seems to be based upon a passage of Curtius (4.3.31) who says that contingents of Illyrian, mercenary and Thracian light troops were positioned behind Alexander's line. However, according to Arrian, the Paeonians and the troops of Balacrus (who were probably Balkan in origin) were positioned towards the front (*Anab.* 3.12.3), the so-called 'old mercenaries' were positioned on the right wing – refused at an angle from the front (*Anab.* 3.12.2), and the Thracian infantry partially guarded the pack animals and formed part of the refused line on the left (*Anab.* 3.12.4-5). While mainly on the wings, both the Thracian and mercenary units, due to their refused deployment, were still 'behind Alexander's line' as Curtius states. However, this does not necessarily mean that they formed a second line behind the pike-phalanx and their refused deployment would actually suggest otherwise, and would additionally correlate the accounts of Arrian and Diodorus. As Devine ('Grand Tactics at Gaugamela' *Phoenix* 29.4 (1975) 375) points out, the refused wings of the Macedonian formation, regardless of the arrangement of the main line, would have resulted in a roughly trapezoid shape if there was a second line running across the rear. If, however, the second line units were Greek allies and mercenaries, there would not have been enough troops to cover this rearward frontage. This would have resulted in large gaps being present in the rear line which encircling cavalry could exploit – the very thing that Arrian says this second line was guarding against. However, had the second line merely been rearward *chiliarchiae* of a pike-phalanx drawn up in double depth, the overall formation would more closely resemble a flattened wedge, rather than a trapezoid, and the second line units would cover the same frontage as the front-centre of the line. These units, if required, could simply about face, and the refused wings move inward, to form a hollow square if the position was encircled. Additionally, later in the battle, when the right wing of the Macedonian line advanced, gaps formed between this wing and the left. Arrian (*Anab.* 3.14.5-6) tells us that Indian and Persian cavalry were able to penetrate through this gap and attack the Macedonian camp (see also: Diod. Sic. 17.59.5-8; Curt. 4.15.5; Plut. *Alex.* 32). Interestingly, this cavalry never seems to encounter the troops forming the second line of the Macedonian formation. This leaves several possibilities: a) the second line was made up of Greeks and mercenaries and that, due to their smaller number, the Persian cavalry was simply able to avoid them, b) that the units of the second line were positioned just behind the front of the formation and moved with it and, as such, the gaps between the right and left wings fragmented the second line as well. As noted, the first possibility seems unlikely as using small numbers of men to hold the rear of the line would seem tactically unsound. This leaves only the second option for the deployment of the second line. Consequently, the use of a double-depth pike-phalanx to create the second line in Alexander's centre at Gaugamela cannot be discounted.

223 Arr. *Anab.* 1.13.1

224 Arr. *Anab.* 3.12.1

225 Ael. Tact. *Tact.* 37, 38; Arr. *Tact.* 29; in his military manual Arrian, and seemingly in his Alexander narrative as well, actually confuses the *antistomos* formation in which the phalanx is divided laterally with the front half facing forward and the rear half facing backwards, and the *amphistomos* formation where the phalanx is divided down the centre of the line and one side faces left while the other faces to the right. Thus Alexander's double-line at Gaugamela should have been described as being an *antistomos* deployment. Asclepiodotus (*Tact.* 11.3) says that the *antistomos* formation was when an army marched in two parallel columns, but with the file leaders on the inside. This is called the *antistomos diphalangarchia* by Aelian (*Tact.* 39). Aelian seems to be the only one who gets the descriptions of these basic *antistomos* and *amphistomos* formations correct.

226 Ael. Tact. *Tact.* 37

227 Arr. *Anab.* 3.12.1; Curt. 4.13.30

228 Fuller, *Generalship of Alexander*, 48

229 Some scholars merely outline the dominant role of the cavalry, and/or the immobilising function of the pike-phalanx, in Hellenistic warfare. For example see: Gabriel, *Philip*, 65, 69; Pietrykowski, 'In the School of Alexander' 24; W.W. How, 'Arms, Tactics and Strategy in the Persian Wars' *JHS* 43.2 (1923) 119; Warry, *Warfare*, 104; Anson, 'Introduction of the Sarisa' 65; Cawkwell, *Philip of Macedon*, 155, 158; Featherstone, *Warriors and Warfare*, 59; S. Skarmintzos, 'Phalanx versus Legion: Greco-Roman Conflict in the Hellenistic Era' *AncWar* 2.2 (2008) 30; Bosworth, *Conquest and Empire*, 266. Other scholars go as far as to use the terminology of the 'hammer and anvil' analogy in their examinations. Snodgrass (*Arms & Armour*, 115), for example, specifically calls the phalanx an 'anvil', while Cummings (*Alexander the*

Great (Riverside Press, Boston, 1940) 208) says that the result of Alexander's deployment at Gaugamela was to 'catch Darius...between the hammer of the Macedonian cavalry and the anvil of the phalanx'. Tarn (*Military and Naval Developments*, 1, 11) calls the Hellenistic Period the 'Age of Cavalry' and says (p.26) that at Heraclea in 280BC Pyrrhus' phalanx defeated the Romans 'in a battle of sheer hammering'.

230 Fuller, *Generalship of Alexander*, 292-293; see also: C. von Clausewitz, *On War* (Penguin, London, 1982) 1.2

231 English, *Field Campaigns of Alexander*, 140; English, *Army of Alexander*, 22-23, 36. Interestingly, despite such claims English (*Army of Alexander*, 22) still calls the cavalry 'the main strike force' of a Hellenistic army.

232 See: *US Army, FM7-8 Infantry Rifle Platoon and Squad* (US Dept. of the Army, Washington, 1992) 2.12: Movement to Contact, 4.2: Battle Drill 1 – Platoon Attack, Battle Drill 1a – Squad Attack

233 Prior to the offensive to liberate Kuwait in 1991, more than 17,000 US Marines, supported by the battleships Missouri and Wisconsin, conducted near continuous drills in rehearsal for an amphibious landing on the Kuwaiti coast. This ruse, designed to make the occupying Iraqis think that the main offensive was going to come from the sea, forced the Iraqis to deploy considerable resources to their left wing to meet this threat. This finds parallels with Darius sending Bessus and his cavalry to shadow Alexander's oblique movement to the right with his cavalry at the battle of Gaugamela. Then, on 24 February 1991, more than 200,000 French, British, American and Arab forces poured across the desert frontier (Saddam Hussein's so called 'line of death') through vacated 'penetration zones' - similar to the gap that opened in the Persian line at Gaugamela, but on a much larger scale. By moving troops into these gaps, the Iraqis in Kuwait were essentially pinned in place - threatened with amphibious forces from the east, from two Egyptian-led tank divisions crossing into Kuwait from the south, and by the US Marine Second Division and the US Army Eighteenth Airborne Corp who had pushed into Kuwait from the west. Other coalition forces operated further to the west, striking northward into Iraq, along a front extending more than 600km. The strategy was to cut Iraqi supply lines running into Kuwait and to use the flanking forces to encircle the now immobilised Iraqi troops in Kuwait. An anonymous US Defence official summed up the strategy as: 'what we want to do is put them in a bag, tie the top of the bag, seal it, and then punch the bag!' For a summary of this first day of the ground offensive of Operation Desert Storm see: I. Bickerton and M. Pearson, *43 Days - The Gulf War* (The Text Publishing Company, East Melbourne, 1991) 154-157

234 Paraetacene: Diod. Sic. 19.27-32; Gabiene: Diod. Sic. 19.39-43; Gaza: Diod. Sic. 19.80-85; Ipsus: Diod. Sic. 20.113-21.2; Plut. *Dem.* 28-29; Heraclea: Plut. *Pyrr.* 16-17; Asculum: Plut. *Pyrr.* 21; Dion. Hal. 20.1-3; Raphia: Polyb. 5.63-65, 5.79-86; Mantinea: Polyb. 11.11-18; Plut. *Phil.* 10; Magnesia: Livy, 37.39-44; App. *Syr.* 30-36

235 Chaeronea: Diod. Sic. 16.86; Plut. *Pel.* 18; Plut. *Alex.* 9; Polyaenus, *Strat.* 4.2.2; Sellasia: Polyb. 2.65-69; Plut. *Cleom.* 28; Plut. *Phil.* 6

236 Plut. *Phoc.* 13

237 Plut. *Eum.* 5

The Anvil in Action

1 Polyb. 18.28

2 Plut. *Aem.* 19

3 See. C.A. Matthew, *A Storm of Spears: Understanding the Greek Hoplite at War* (Pen & Sword, Barnsley, 2012) 71-92

4 Polyb. 18.30

5 For details of the use of intermediate-order by hoplite armies see: Matthew, *A Storm of Spears*, 179-197

6 Diod. Sic. 16.86.1-6; Plut. Pel. 18; Plut. Alex. 9

7 Plut. *Alex.* 16; there are many similar references to hoplites apparently engaging phalangites effectively from across the Hellenistic Period such as Phocion's troops driving the forces of Philip II out of Boeotia in 348BC (Plut. *Phoc.* 13-14), and the Greeks defeating the Macedonians again in 322BC (Plut. *Phoc.* 25; Diod. Sic. 18.12.4). The elderly Silver Shields, Alexander's former *hypaspists*, are also said to have so effectively engaged the opposing phalangites of Antigonus at the battle of Gabiene that they did not lose a single man while reportedly killing 5,000 of their enemy (Diod. Sic. 19.43.1). The armament of the *hypaspists*/Silver Shields is a strongly contested issue. However, possible literary exaggerations aside, it would seems unlikely that the Silver Shields would have been able to so effectively engage phalangites if they were equipped as hoplites unless it is assumed that Antigonus' line had fragmented – for which

there is no evidence. This suggests that, even if they had been initially armed as hoplites in the time of Alexander, by the Successor period the equipment of the Silver Shields was more akin to that of the phalangite.

8   For example see: M.M. Markle, 'The Macedonian *Sarissa*, Spear and Related Armour' *AJA* 81.3 (1977) 338-339; P.A. Rahe, 'The Annihilation of the Sacred Band at Chaeronea' *AJA* 85.1 (1981) 84-87; J. Warry, *Warfare in the Classical World* (University of Oklahoma Press, Norman, 1995) 68; I. Worthington, *Philip II of Macedonia* (Yale University Press, New Haven, 2008) 147-151

9   Plut. *Pel.* 18

10 N. Sekunda, *The Persian Army 560-330BC* (Osprey, Oxford, 2005) 27

11 For an overview of the depth of the Classical hoplite phalanx see: Matthew, *A Storm of Spears*, 172-179

12 Diodorus (17.38.4) states that some of the Persian dead at Issus were not wearing any armour while others were in a full panoply. Classical Greek writers never seem to have tired of emphasising the superiority of the Greek hoplite against the Persians. For example see: For example see: Hdt. 5.49, 5.97, 7.63-80, 7.211, 9.62; Diod. Sic. 11.7.3; Aesch. *Pers.* 26-51, 85-86, 239-240, 269, 278, 817, 926; Xen. *Hell.* 7.6.1; Pind. *Pyth.* 1.72-80. In his comparison of Greek, Macedonian, Persian and Roman fighting styles, Julius Africanus (*Cest.* 7.1.22-23) flatly states that 'because the [Persian] infantry were lightly armed, they could not endure the onslaught of [the Greeks]'. Livy, in his vitriolic critique of the abilities of Alexander the Great (9.17-19), says that the Persian king Darius whom Alexander faced had 'dragging after him a train of women and eunuchs, wrapped in gold and purple, and encumbered with all the trappings of state', that Alexander was able to defeat the Persians 'without loss, without being called to do anything more daring than to show a just contempt for the idle [Persian] display of power', and that, had Alexander chosen to fight a campaign against Rome, 'he would have been tempted to wish the Persians, Indians and effeminate Asiatics were his enemies [rather than Romans], and would have confessed that his former wars had been waged against women'.

13 Afric. *Cest.* 7.1.43-44

14 Diod. Sic. 17.53.1

15 Arr. *Anab.* 3.14.1; Diodorus (17.58.1) says the troops charged each other with a loud shout. Curtius (4.15.19) says 'then, raising a shout as victors do, the Persians made a ferocious rush at the enemy...'. All of these passages suggest a massed charge.

16 Plut. *Mor.* 180b; see also: Afric. *Cest.* 1.1.40-44

17 Plut. *Mor.* 339b-c

18 Plut. *Aem.* 20

19 Polyaenus, *Strat.* 2.29.2

20 Plut. *Aem.* 18

21 Arr. *Anab.* 3.13.5; Diod. Sic. 17.58.2; Curt. 4.15.14

22 Arrian (*Anab.* 3.13.6) says the chariots passed right through the lines and the drivers were then slain by the *hypaspists*. This seems unlikely as, at 3.11.9 Arrian states that the *hypaspists* were positioned on the right wing between the pike-phalanx and the cavalry. Diodorus (17.58.3-4) says that some of the chariots were scared off by the Macedonians clashing their pikes on their shields, while others continued forward to either be picked off with javelins, to pass right through the Macedonian lines (which had been ordered to open) and flee, or to crash into the phalanx and inflict serious injury.

23 P. Connolly, *Greece and Rome at War* (London, Greenhill Books, 1998) 69

24 Heckel, Willekes and Wrightson ('Scythed Chariots at Gaugamela – A Case Study' in E. Carney and D. Ogden (eds.) *Philip II and Alexander the Great: Father and Son, Lives and Afterlives* (Oxford University Press, 2010) 104-105) outline several practical reasons why the rapid creation of lanes by a pike-phalanx facing the chariot charge at Gaugamela is unlikely. These include: a) the phalangite is not very nimble due to his equipment; b) the phalanx would have begun to compress at this time of the battle; c) densely packed formations have little room to move and cannot react quickly; d) the lowering of the pikes to engage inhibits lateral movement; and e) once lowered, a *sarissa* can only be used to engage directly ahead, rather than against a passing chariot.

25 App. *Syr.* 19 - τὰς σαρίσσας ἐν τάξει πυκνὰς προύβάλοντο

26 Curt. 4.15.14-15

27 Lucian, *Dial. mort.* 27

28 Pietrykowski (*Great Battles of the Hellenistic World* (Pen & Sword, Barnsley, 2009) 81) suggests that Alexander's troops adopted Lucian's kneeling position to counter Porus' elephants at the Hydaspes

River in 326BC. However, there is no reference to such an undertaking in any of the accounts of the engagement.

29 Arr. *Anab.* 3.15.1; Diod. Sic. 17.59.5, 17.60.5-7

30 Hdt. 9.18; although the Macedonians had 2,000 cavalry at the battle of Chaeronea in 338BC (Diod. Sic. 16.85.5), the reluctance of cavalry to frontally charge hoplite formations throws a question mark over any model of the battle which suggests that a young Alexander led a charge of cavalry against the Theban Sacred Band (for example see; Markle, 'Spear', 338-339; Rahe, 'The Annihilation of the Sacred Band' 84-87; Warry, *Warfare,* 68; Worthington, *Philip II of Macedonia*, 147-151).

31 Afric. *Cest.* 7.1.30-31

32 Diod. Sic. 17.60.4, 17.61.1; see also: Curt. 4.15.32

33 Arrian's and Diodorus' accounts of the opening of the Macedonian lines and the failure of the Persian chariot charge at Gaugamela bear striking similarities to Xenophon's account of the battle of Cunaxa in 401BC (see: Xen. *Anab.* 1.8.18-20). It is possible that there was clear knowledge that the charge of the Persian chariots resulted in failure and that this earlier account was drawn upon for inspiration by Arrian and Diodorus (or their sources) to account for it. Lucian (*Zeux.* 8-11) also describes how the phalanx of Antiochus I opened its ranks to allow enemy scythe-bearing chariots to pass through at the so-called 'elephant victory' in 273BC. However, half of Antiochus' army was light-infantry which would have been much more mobile and able to open its lines quickly to avoid the attacking chariots.

34 Heckel, Willekes and Wrightson, 'Scythed Chariots at Gaugamela', 107-109; for the separation of the phalanx see: Arr. *Anab.* 3.14.4

35 Arrian (*Anab.* 3.8.6, 3.11.6), supposedly reporting the disposition of the enemy army from written plans captured after the battle (3.11.3), states that there was a unit of fifteen Indian elephants in the Persian army at the battle of Gaugamela in 331BC. Aristobulos (*FrGrHist* 139 F17), also referring to captured battle plans, states that these elephants were positioned directly in front of the position of Darius. Hammond (*The Genius of Alexander the Great* (Duckworth, London, 1997) 106), despite what Arrian and Aristobulos states, claims that the elephants were left in the Persian camp as only the horses of the Indian cavalry would have been accustomed to them. If these beasts were present, they seem to have played no part in the battle as they are not mentioned again in any source. A contingent of Indian 'cavalry' is reported to have broken through the Macedonian left wing – although this may not be the elephant unit which is said to have been positioned in the centre of the Persian line 'opposite the phalanx' (Arr. *Anab.* 3.11.6). Curtius (4.12.9) says that the Indians (as well as troops from the region of the Red Sea) provided only nominal, as opposed to real, support – an odd claim to make if these troops involved a contingent of pachyderms. As Bosworth (*Conquest and Empire: The Reign of Alexander the Great* (Cambridge University Press, 2008) 128) notes, the Macedonians may have encountered elephants in smaller engagements in the months prior to the battle of the Hydaspes. However, the Hydaspes was the first large set-piece battle in which the Macedonians fought a large contingent of these beasts.

36 Arr. *Anab.* 5.15.2; Curt. 8.14.4, 8.14.19; Arrian (*Anab.* 5.15.4) states that the Indian king Porus selected ground for the deployment of his main force that was 'sandy and level'. Due to this being the sand on an alluvial plain following a period of rainfall, there would have been very little dist kicked up by either side.

37 Arr. *Anab.* 5.15.1-2; Curt. 8.14.2; Plut. Alex. 60

38 Plut. *Alex.* 60

39 Arr. *Anab.* 5.14.2

40 Curt. 8.14.5

41 Arr. *Anab.* 5.14.5-7; Diod. Sic. 17.87.4; Curt. 8.14.9

42 Arr. *Anab.* 5.16.3

43 For the trumpeting of the elephants: Curt. 8.14.23; for Alexander's confidence: Curt. 8.14.16

44 Plut. *Alex.* 60

45 Diod. Sic. 17.88.1

46 Diod. Sic. 17.88.2

47 Arr. *Anab.* 5.17.3; 5.17.6; Diod. Sic. 17.88.3, 17.88.5; Curt. 8.14.24-29

48 Curt. 8.14.18

49 For example see: Ael. Tact. *Tact.* 43, 44, 45, 47

50 Figures taken from A.M. Snodgrass, *Arms and Armour of the Greeks* (Johns Hopkins University Press, Baltimore, 1999) 122

51 Afric. *Cest.* 7.18.1-20

52   Diod. Sic. 18.70.3

53   Plut. *Alex.* 62; Curt. 9.2.1-9.3.19; Diod. Sic. 17.93.2; Arr. *Anab.* 5.25.1-5.27.9

54   Diod. Sic. 18.71.2-6

55   Diod. Sic. 19.83.2

56   Diod. Sic. 19.84.1-5

57   Afric. *Cest.* 7.18.43-46

58   Dion. Hal. 20.1

59   Dion. Hal. 20.2

60   Aelian (*Tact.* 22) even outlines the arrangement of elephant contingents into standardized units and sub-units, each with their own commanding officer, just as he does for all of the other arms of a Hellenistic army (see also: Asclep. *Tact.* 9.1; Arr. *Tact.* 19).

61   Livy (30.26) also states that 4,000 Macedonians served in the army of the Carthaginian general Hannibal which fought at the battle of Zama in 202BC (see also: Front. *Strat.* 2.3.16). Livy states that these troops had been sent by the Macedonian king in support of Carthage while he made inroads into Greece. This would have been mostly likely to bolster the threat to Rome posed by Carthage and to keep Roman troops, who could have been used in a war against Macedon, occupied elsewhere. Livy goes on to say that the Romans sent delegations to the Macedonian court to express their displeasure at this act and that the view of the Romans was that their current treaty with Macedon had been broken. Many scholars, however, see Livy's record of this event as mere Roman propaganda and dispute the presence of any Macedonian troops in Hannibal's army (see: B.T. Carey, *Hannibal's Last Battle: Zama and the Fall of Carthage* (Pen & Sword, Barnsley, 2007) 116). If there were Macedonian troops at Zama, on the other hand, this then would be another instance of phalangites being defeated by the maniples of Rome.

62   Polyb. 18.30

63   Polyb. 18.30; see also Smith ('The Anatomy of Battle – Testing Polybius' Formations' *AncWar* 5.5 (2011) 44-45) who, by examining the amount of room required to accommodate a person in the combative posture to wield the *gladius* and *scutum*, also arrives at a figure of 192cm per man. Wheeler ('The Legion as Phalanx in the Late Empire (I)' in Y. le Bohec and C. Wolff (eds.), *L'Armée Romaine de Dioclétien à Valentinien Ier* (Paris, De Boccard, 2004) 331), on the other hand, rejects this idea and says that Polybius' suggestion that 'each legionary had to face ten *sarissae* offers another example of his exaggeration in an excursus of propagandistic intent'.

64   Plut. *Aem.* 19

65   Plutarch (*Pyrr.* 17) says that the outcome of the battle swung back and forth seven times at Heraclea.

66   Polyb. 18.32; similarly Appian (*Syr.* 37) claims that many people had not expected the Romans to defeat the Macedonians at Magnesia in 190BC as the pike-phalanx was, at that time, renowned for its discipline and valour, and was reported to be invincible. Champion (*Pyrrhus of Epirus* [Pen & Sword, Barnsley, 2009] 68) suggests that references to the pike-phalanx being unbeatable are simply literary motifs to make victories over them seem more impressive. However, such a conclusion fails to consider the narratives of battles like Heraclea, Asculum, Cynoscephalae, Magnesia and Pydna which show that the pike-phalanx could effectively engage the legions of Rome, nor consider the motive behind Polybius' discussion of the Roman and Macedonian ways of fighting.

67   Polyb. 18.32

68   Afric. *Cest.* 7.1.52-57

69   Polyb. 18.31-32

70   Livy 9.19

71   Afric. *Cest.* 7.1.49-50

72   Livy 9.19

73   Plut. *Aem.* 20

74   Afric. *Cest.* 7.1.53-55

75   Livy (31.34) states that the Greeks 'had seen wounds caused by spears, arrows and, occasionally, lances since they were accustomed to fighting against other Greeks and Illyrians. But now they saw bodies dismembered with the "Spanish sword" [i.e. the *gladius*], arms cut off with the shoulder attached, or heads severed from bodies – with the necks completely cut through, internal organs exposed, and other horrible wounds, and a general sense of panic ensued when they discovered the kind of weapons, and the kind of men, they would have to contend with.'

76   Polyb. 18.32

77   Afric. *Cest.* 7.1.68-70

78 Polyb. 6.23
79 Livy 9.19
80 Polyb. 18.30
81 Caes. *BG* 1.25
82 S. Skarmintzos, 'Phalanx versus Legion: Greco-Roman Conflict in the Hellenistic Era' *AncWar* 2.2 (2008) 34
83 W.W. Tarn, *Hellenistic Military and Naval Developments* (Cambridge University Press, 1930) 28

The Legacy of the Hellenistic Pike-Phalanx

1 Diod. Sic. 21.2.2 - Μακεδόνων τῶν τὴν Ἀσιαν καὶ τὴν Εὐρῶπην πεποιημένων δορίκτητον. Didodorus makes this statement in the section of his work for the year 299BC. This is not long after the battle of Ipsus in 301BC – one of the last major battles between the Successors of Alexander the Great and one of the last times two opposing pike-phalanxes would face off against each other.
2 Plut. *Sull.* 18-19
3 Plut. *Sull.* 18; for other accounts of the battle see: App. *Mith.* 42-43; Front. *Strat.* 2.3.17
4 For example see: J. Pietrykowski, *Great Battles of the Hellenistic World* (Pen & Sword, Barnsley, 2009) 235
5 Suet. *Ner.* 19
6 B. Bennett and M. Roberts, *The Wars of Alexander's Successors 323-281BC Vol.II: Battles and Tactics* (Pen & Sword, Barnsley, 2009) 3
7 Seut. *Ner.* 19
8 Webster (*The Roman Imperial Army of the First and Second Centuries A.D.* (University of Oklahoma Press, Norman, 1988) 45, 106) claims that the creation of 'Alexander's Phalanx' was 'the product of a disordered mind' and part of Nero's 'strange romantic delusions'.
9 Cass. Dio 78.7.1-2; for Caracalla and Alexander see also: S.H.A. *M. Ant.* 2.1
10 Herodian 4.8.2
11 R. Cowan, 'The Battle of Nisibis, AD217 – The Last Battle of the Parthian Wars' *AncWar* 3.5 (2009) 34; for a discussion of the description of Caracalla's phalangites see also: D. Karunanithy, 'Of Ox-Hide Helmets and Three-Ply Armour: The Equipment of Macedonian Phalangites as Described through a Roman Source' *Slingshot* 213 (2001) 33-40; G. Sumner, *Roman Military Clothing (2) AD200-400* (Osprey, Oxford, 2003) 39,45, pl.C3; R. Cowan, *Imperial Roman Legionary AD161-284* (Osprey, Oxford, 2003) 27-28
12 Curt. 9.7.19; Diod. Sic. 17.100.2
13 Ael. Tact. *Tact.* 9; see also: Asclep. *Tact.* 2.8-10; Arr. *Tact.* 10
14 Cass. Dio 78.18.1
15 Cowan, 'The Battle of Nisibis' 34
16 For example see: Arr. *Alan.* 5-6, 15
17 Vitr. *De arch.* 10.3.7
18 Plut. *Mar.* 13; Front. *Str.* 4.1.7; Festus, *De Verborum Significatione* s.v. 'muli Mariani'; Josephus (BJ 3.87) describes troops carrying equipment in the manner of the Marian Mules in the first century AD and the frieze on Trajan's Column depicts Marian Mules crossing the Danube into Dacia in the second.
19 Cowan, 'The Battle of Nisibis' 34; for the *triarii* of the later empire see: Veg. Mil. 3.14; see also: E.L. Wheeler, 'The Legion as Phalanx in the Late Empire (II)' *RÉMA* 1 (2004) 170-173
20 Herodian 4.15.2; Afric. *Cest.* 1.1.55-56
21 S.H.A. *Alex. Sev.* 50.4-5
22 For a discussion of other early editions of Aelian's *Tactica* and other military manuals see: C. A. Matthew, *The Tactics of Aelian* (Pen & Sword, Barnsley, 2012) xiv-xv

Plates

1 Author's photos
2 Image courtesy of the Great North Museum, Newcastle upon Tyne.
3 Images taken from: K.F. Kinch, 'Le tombeau de Naiusta. Tombeau Macédonien' *Videnskabernes Selskabs Skrifter – Historisk og Filosofisk Afdeling* 7.4 (1917-1937) pl.2-3. Sadly, this unique tomb painting is now lost and only the images made for the original report on the find, shown above, remain.
4 Image courtesy of Kate Matthew
5 Images courtesy of Kate Matthew

6   Image courtesy of the Allard Pierson Museum, Amsterdam

7   Image taken from A. Conze, *Stadt und Landschaft – Altertümer von Pergamon* 1.2 (Verlag von Georg Reimer, Berlin, 1913) 250-251; see also: Liampi, *Schild*, 81-82; in personal correspondence (10-Sept-2012), Martin Maischberger of the Staatliche Museen zu Berlin has confirmed that this plaque was inventoried into the museum's collection early in the twentieth century with an inventory number of P50. However, the plaque has been missing since the fall of Berlin at the end of World War II and recent investigations with the Pushkin Museum, Moscow, have not located it among the 'Russian war booty' that was taken out of Germany in the last months of the war. The plaque, the only visual reference of the phalangite in action of its kind, is now regarded as 'completely lost'. Fortunately, we still have the line drawing from the work of Conze (reproduced here) which is the only visual representation of this unique and valuable missing artefact.

8   Image courtesy of the 16th Ephorate of Prehistoric and Classical Antiquities, Greece

9   Author's photos

10 Author's photo

11 Author's photo – coin in the author's personal collection

12 Author's photos

13 Author's photos

14 Author's photo

15 Authors' photos

16 Authors' photos

17 Authors' photos

18 Author's photo

19 Author's photo

20 Author's photo

Bibliography

ANCIENT TEXTS

Aelian, *Historical Miscellany* (trans. N.G. Wilson), (Harvard University Press – Loeb Classical Library, Cambridge, 1997)

Aelian (Tacticus), *The Tactics of Aelian* (trans. C.A. Matthew), (Pen & Sword, Barnsley, 2012)

Aeneas Tacticus/ Asclepiodotus/ Onasander (trans. Illinois Greek Club), (Harvard University Press – Loeb Classical Library, Cambridge, 2001)

Aeschines, *Speeches* (trans. C.D. Adams), (Harvard University Press – Loeb Classical Library, Cambridge, 1958)

Aeschylus, *Vol.I – Suppliant Maidens/ Persians/ Prometheus/ Seven Against Thebes* (trans. H.W. Smyth), (Harvard University Press – Loeb Classical Library, Cambridge, 1973)

Appian, *Roman History Vol.II* (trans. H. White), (Harvard University Press – Loeb Classical Library, Cambridge, 2005)

Aristophanes, *Vol. II – The Peace/ The Birds/ The Frogs* (trans. B.B. Rogers), (Harvard University Press – Loeb Classical Library, Cambridge, 1979)

Aristotle, *Nicomachean Ethics* (trans. H. Rackham), (Harvard University Press – Loeb Classical Library, Cambridge, 1926)

Aristotle, *Politics* (trans. H. Rackham), (Harvard University Press – Loeb Classical Library, Cambridge, 1967)

Aristotle, *The Athenian Constitution* (trans. H. Rackham), (Cambridge, Harvard University Press – Loeb Classical Library, 1952)

Arrian, *Anabasis of Alexander Vol.I* (trans. P.A. Brunt), (Harvard University Press – Loeb Classical Library, Cambridge, 1976)

Arrian, *Anabasis of Alexander Vol.II* (trans. P.A. Brunt), (Harvard University Press – Loeb Classical Library, Cambridge, 1983)

Arrian, *Tactical Handbook / Expedition Against the Alans* (trans. J.G. DeVoto), (Ares Publishers, Chicago, 1993)

Arrian's History of Alexander's Expedition Vol.I (trans. J. Rooke), (Allen and Co., London, 1814)

Athenaeus, *The Learned Banqueters Vol.I* (trans. S.D. Olson), (Harvard University Press – Loeb Classical Library, Cambridge, 2007)

Athenaeus, *The Learned Banqueters Vol.II* (trans. S.D. Olson), (Harvard University Press – Loeb Classical Library, Cambridge, 2007)

Athenaeus, *The Learned Banqueters Vol.V* (trans. S.D. Olson), (Harvard University Press – Loeb Classical Library, Cambridge, 2009)

Athenaeus, *The Learned Banqueters Vol.VII* (trans. S.D. Olson), (Harvard University Press – Loeb Classical Library, Cambridge, 2011)

Caesar, *Gallic War* (trans. H.J. Edwards), (Harvard University Press – Loeb Classical Library, Cambridge, 1963)

Cicero, *Letters to Friends Vol.I* (trans. D.R. Shackleton Bailey), (Harvard University Press – Loeb Classical Library, Cambridge,2001)

Cicero, *On the nature of the Gods/ Academics* (trans. H. Rackham), (Harvard University Press – Loeb Classical Library, Cambridge, 1933)

Cicero, *Tusculan Disputations* (trans. J.E. King), (Harvard University Press – Loeb Classical Library, Cambridge, 1927)

Curtius (Quintus Curtius Rufus), *History of Alexander Vol.I* (trans. J.C. Rolfe), (Harvard University Press – Loeb Classical Library, Cambridge, 1971)

Curtius (Quintus Curtius Rufus), *History of Alexander Vol.II* (trans. J.C. Rolfe), (Harvard University Press – Loeb Classical Library, Cambridge, 1962)

Curtius (Quintus Curtius Rufus), *The History of Alexander* (trans. J. Yardley), (Penguin, London, 2001)

Demosthenes, *Vol.I – Olynthiacs/ Philippics/ Minor Public Speeches/ Speech Against Leptines*, (trans. J.H. Vince), (Harvard University Press – Loeb Classical Library, Cambridge, 1962)

Didymus, *On Demosthenes* (trans. P. Harding), (Oxford University Press, 2006)

Dio Cassius, *Roman History Vol.IX* (trans. E. Cary and H.B. Foster), (Cambridge, Harvard University Press – Loeb Classical Library, 1927)

Dio Chrysostom, *Vol.I Discourses 1-11* (trans. J.W. Cohoon), (Harvard University Press – Loeb Classical Library, Cambridge, 1932)

Diodorus Siculus, *Library of History Vol.VII* (trans. C.L. Sherman), (Harvard University Press – Loeb Classical Library, Cambridge, 1952)

Diodorus Siculus, *Library of History Vol.VIII* (trans. C.B. Welles), (Harvard University Press – Loeb Classical Library, Cambridge, 1963)

Diodorus Siculus, *Library of History Vol.IX* (trans. R.M. Geer), (Cambridge, Harvard University Press – Loeb Classical Library, 1947)

Diodorus Siculus, *Library of History Vol.X* (trans. R.M. Geer), (Harvard University Press – Loeb Classical Library, Cambridge, 2006)

Diodorus Siculus, *Library of History Vol.XI* (trans. F.R. Walton), (Harvard University Press – Loeb Classical Library, Cambridge, 1957)

Diogenes Laertius, *Lives of Eminent Philosophers Vol.I* (trans. R.D. Hicks), (Harvard University Press – Loeb Classical Library, Cambridge, 1966)

Dionysius of Halicarnassus, *Critical Essays Vol.I: Ancient Orators/ Lysias/ Isocrates/ Isaeus/ Demosthenes/ Thucydides* (trans. S. Usher), (Harvard University Press – Loeb Classical Library, Cambridge, 1974)

Dionysius of Halicarnassus, *Roman Antiquities Vol.VII* (trans. E. Cary), (Harvard University Press – Loeb Classical Library, Cambridge, 1950)

Empedocles, *The Extant Fragments* (trans. M.R. Wright), (Yale University Press, New Haven, 1981)

Etymologicum Magnum (ed. F. Sylburg), (Apud J.A.G. Weigel, Lipsiae, 1816)

Euripides, *Vol.I – Iphigeneia in Aulus/ Rhesus/ Hecuba/ Daughters of Troy/ Helen* (trans. A.S. Way), (Harvard University Press – Loeb Classical Library, Cambridge, 1978)

Euripides, *Vol.III – Bacchanals/ Madness of Hercules/ Children of Hercules/ Phoenician Maidens/ Suppliants* (trans. A.S. Way), (Harvard University Press – Loeb Classical Library, Cambridge, 1962)

Eustathius, *Commentarii Ad Homeri Iliadem* (ed. J.G. Stallbaum), (Cambridge University Press, Cambridge, 2010)

Festus, *De Verborum Significatione*, (Georg Olms Verlagsbuchhandlung, Hildesheim, 1913)

Frontinus, *Stratagems* (trans. M.B. McElwain), (Harvard University Press – Loeb Classical Library, Cambridge, 1950)

Greek Anthology Vol.I (trans. W.R. Paton), (Harvard University Press – Loeb Classical Library, Cambridge, 1969)

Greek Anthology Vol.IV (trans. W.R. Paton), (Harvard University Press – Loeb Classical Library, Cambridge, 1969)

Greek Bucolic Poets (trans. J.M. Edmonds), (Cambridge, Harvard University Press – Loeb Classical Library, 1960)

Greek Elegiac Poetry: From the Seventh to the Fifth Centuries BC (trans. D.E. Gerber), (Harvard University Press – Loeb Classical Library, Cambridge, 1999)

Greek Lyric Vol.II - Anacreon, Anacreontea, Choral Lyric from Olympus to Alcman (trans. D.A. Campbell), (Harvard University Press – Loeb Classical Library, Cambridge, 1988)

Harpocration, *Lexicon*, (C.H.F. Hartmann, Lipsiae, 1824)

Hellenica Oxyrhynchia (trans. P.R. McKechnie and S.J. Kern), (Aris and Phillips, Wiltshire, 1993)

Herodian, *History of the Empire Vol.I* (trans. C.R. Whittaker), (Harvard University Press – Loeb Classical Library, Cambridge, 1969)

Herodotus, *Histories Vol.I* (trans. A.D. Godley), (Harvard University Press – Loeb Classical Library, Cambridge, 1971)

Herodotus, *Histories Vol.II* (trans. A.D. Godley), (Harvard University Press – Loeb Classical Library,

Cambridge, 1921)

Herodotus, *Histories Vol.III* (trans. A.D. Godley), (Harvard University Press – Loeb Classical Library, Cambridge, 1971)

Herodotus, *Histories Vol.IV* (trans. A.D. Godley), (Harvard University Press – Loeb Classical Library, Cambridge, 1971)

Heron, *Dioptra*, (ed. H. Schöne), (De Gruyter, Berlin, 1899)

Hesiod, *Theogony/ Works and Days/ Testimonia* (trans. G.W. Most), (Harvard University Press – Loeb Classical Library, Cambridge, 2007)

Hesiod, *The Shield/ Catalogue of Women/ Other Fragments* (trans. G.W. Most), (Harvard University Press – Loeb Classical Library, Cambridge, 2007)

Historia Augusta Vol.II (trans. D. Magie), (Harvard University Press – Loeb Classical Library, Cambridge, 1924)

Holy Bible (New Revised Standard Version), (Oxford University Press, 1989)

Homer, *Iliad Vol.I* (trans. A.T. Murray), (Harvard University Press – Loeb Classical Library, Cambridge, 1978)

Homer, *Iliad Vol.II* (trans. A.T. Murray), (Harvard University Press – Loeb Classical Library, Cambridge, 1976)

Inscriptiones Graecae IV-2: Inscriptiones Aeginae insulae (ed. K. Hallof), (de Gruyter, Berlin, 2007)

Josephus, *The Jewish War Vol.II* (trans. H. St.J. Thackeray), (Harvard University Press – Loeb Classical Library, Cambridge, 1927)

Julius Africanus, *Cesti* (trans. W. Adler) (de Gruyter, Berlin, 2012)

Justin, *Epitome of the Philippic History of Pompeius Trogus* (trans. J.C. Yardley), (Scholars Press, Atlanta, 1994)

Livy, *History of Rome Vol.IV* (trans. B.O. Foster), (Harvard University Press – Loeb Classical Library, Cambridge, 1926)

Livy, *History of Rome Vol.IX* (trans. E.T. Sage), (Harvard University Press – Loeb Classical Library, Cambridge, 1935)

Livy, *History of Rome Vol.X* (trans. E.T. Sage), (Harvard University Press – Loeb Classical Library, Cambridge, 1935)

Livy, *History of Rome Vol. XIII* (trans. A.C. Schlesinger), (Harvard University Press – Loeb Classical Library, Cambridge, 1951)

Lucian, *Vol.VI* (trans. K. Kilburn), (Harvard University Press – Loeb Classical Library, Cambridge, 1959)

Lucian, *Vol.VIII* (trans. M.D. MacLeod), (Harvard University Press – Loeb Classical Library, Cambridge, 1961)

Menander, *Vol.II – Heros/ Theophoroumene/ Karchedonios/ Kitharistes/ Kolax/ Koneiazomenai/ Leukadia/ Misoumenos/Perikeiromene/ Perinthia* (trans. W.G. Arnott), (Harvard University Press – Loeb Classical Library, Cambridge, 1997)

Minor Latin Poets Vol.I - Publilius Syrus/ Elegies on Maecenas/ Grattius/ Calpurnius Siculus/ Laus Pisonis/ Einsiedeln Eclogues/ Aetna (trans. J.W. Duff and A.M. Duff), (Harvard University Press – Loeb Classical Library, Cambridge, 1934)

Nepos (Cornelius), *On Great Generals* (trans. J.C. Rolfe), (Harvard University Press – Loeb Classical Library, Cambridge, 1966)

Pausanias, *Description of Greece Vol.II* (trans. W.H.S. Jones and H.A. Ormerod), (Harvard University Press – Loeb Classical Library, Cambridge, 1926)

Pausanias, *Description of Greece Vol.III* (trans. W.H.S. Jones), (Harvard University Press – Loeb Classical Library, Cambridge, 1933)

Pausanias, *Description of Greece Vol.IV* (trans. W.H.S. Jones), (Harvard University Press – Loeb Classical Library, Cambridge, 1935)

Pausanias, *Guide to Greece Vol.II: Southern Greece* (trans. P. Levi), (Penguin, London, 1985)

Philo of Byzantium, *Mechanicae Syntaxis* (ed. R. Schöne), (G. Reimeri, Berlin, 1893)

Philostratus, *Life of Apollonius of Tyana Vol.I* (trans. C.P. Jones), (Harvard University Press – Loeb Classical Library, Cambridge, 2005)

Photius, *Photii Patriarchae Lexicon Vol II* (ed. C. Theodoridis), (Walter de Gruyter, Berlin, 1998)

Pindar, *Nemean Odes/ Isthmian Odes/ Fragments* (trans. W.H. Race), (Harvard University Press – Loeb Classical Library, Cambridge, 1997)

Pindar, *Olympian Odes/ Pythian Odes* (trans. W.H. Race), (Harvard University Press – Loeb Classical Library, Cambridge, 1997)

Pliny (the Elder), *Natural History Vol.II* (trans. H. Rackham), (Harvard University Press – Loeb Classical Library, Cambridge, 1942)

Pliny (the Elder), *Natural History Vol.IV* (trans. H. Rackham), (Harvard University Press – Loeb Classical Library, Cambridge, 1945)

Pliny (the Elder), *Natural History Vol.IX* (trans. H. Rackham), (Harvard University Press – Loeb Classical Library, Cambridge, 1952)

Plutarch, *Lives Vol.I – Theseus and Romulus/ Lycurgus and Numa/ Solon and Publicola* (trans. B. Perrin), (Harvard University Press – Loeb Classical Library, Cambridge, 1967)

Plutarch, *Lives Vol.III – Pericles and Fabius Maximus/ Nicias and Crassus* (trans. B. Perrin), (Harvard University Press – Loeb Classical Library, Cambridge, 1967)

Plutarch, *Lives Vol. IV – Alcibiades and Coriolanus/ Lysander and Sulla* (trans. B. Perrin), (Harvard University Press – Loeb Classical Library, Cambridge, 1968)

Plutarch, *Lives Vol.V – Agesilaus and Pompey/ Pelopidas and Marcellus* (trans. B. Perrin), (Harvard University Press – Loeb Classical Library, Cambridge, 1968)

Plutarch, *Lives Vol.VI – Dion and Brutus/ Timoleon and Aemilius Paulus* (trans. B. Perrin), (Harvard University Press – Loeb Classical Library, Cambridge, 1961)

Plutarch, *Lives Vol.VII – Demosthenes and Cicero/ Alexander and Caesar* (trans. B. Perrin), (Harvard University Press – Loeb Classical Library, Cambridge, 1967)

Plutarch, *Lives Vol.VIII – Sertorius and Eumenes/ Phocion and Cato the Younger* (trans. B. Perrin), (Harvard University Press – Loeb Classical Library, Cambridge, 1969)

Plutarch, *Lives Vol.IX – Demetrius and Antony/ Pyrrhus and Caius Marius* (trans. B. Perrin), (Harvard University Press – Loeb Classical Library, Cambridge, 1968)

Plutarch, *Lives Vol.X – Agis and Cleomenes/ Tiberius and Caius Gracchus/ Philopoemen and Flamininus*, (trans. B. Perrin), (Harvard University Press – Loeb Classical Library, Cambridge, 1968)

Plutarch, *Moralia Vol.III* (trans. F.C. Babbitt), (Harvard University Press – Loeb Classical Library, Cambridge, 1968)

Plutarch, *Moralia Vol.IV* (trans. F.C. Babbitt), (Harvard University Press – Loeb Classical Library, Cambridge, 1936)

Polyaenus, *Stratagems of War Vol.I* (trans. P. Krentz and E.L. Wheeler), (Ares Publishers, Chicago, 1994)

Polyaenus, *Stratagems of War Vol.II* (trans. P. Krentz and E.L. Wheeler), (Ares Publishers, Chicago, 1994)

Polybius, *The Histories Vol.I* (trans. W.R. Paton), (Harvard University Press – Loeb Classical Library, Cambridge, 2010)

Polybius, *The Histories Vol.II* (trans. W.R. Paton), (Harvard University Press – Loeb Classical Library, Cambridge, 2010)

Polybius, *The Histories Vol.III* (trans. W.R. Paton), (Harvard University Press – Loeb Classical Library, Cambridge, 2011)

Polybius, *The Histories Vol.IV* (trans. W.R. Paton), (Harvard University Press – Loeb Classical Library, Cambridge, 2011)

Polybius, *The Histories Vol.V* (trans. W.R. Paton), (Harvard University Press – Loeb Classical Library, Cambridge, 2012)

Polybius, *The Histories Vol.VI* (trans. W.R. Paton and S.D. Olson), (Harvard University Press – Loeb Classical Library, Cambridge, 2012)

Remains of Old Latin Vol.I – Ennius/ Caecilius (trans. E.H. Warmington), (Harvard University Press – Loeb Classical Library, Cambridge, 1935)

Sophocles, *Vol.II – Ajax/ Electra/ Trachiniae/ Philoctetes* (trans. F. Storr), (Harvard University Press – Loeb Classical Library, Cambridge, 1961)

Statius, *Thebaid Vol.I* (trans. D.R. Shackleton Bailey), (Harvard University Press – Loeb Classical Library, Cambridge, 2004)

Strabo, *The Geography of Strabo Vol.II* (trans. H.L. Jones), (Harvard University Press – Loeb Classical Library, Cambridge, 1960)

Strabo, *The Geography of Strabo Vol.V* (trans. H.L. Jones), (Harvard University Press – Loeb Classical Library, Cambridge, 1961)
Strabo, *The Geography of Strabo Vol.VII* (trans. H.L. Jones), (Harvard University Press – Loeb Classical Library, Cambridge, 1961)
Suetonius, *Lives of the Caesars Vol.II* (trans. J.C. Rolfe), (Harvard University Press – Loeb Classical Library, Cambridge, 1914)
Suda Vol.I (ed. L. Kusterus), (Typis Academicis, Cambridge, 1705)
Suda Vol.II (ed. L. Kusterus), (Typis Academicis, Cambridge, 1705)
Suda Vol.III (ed. L. Kusterus), (Typis Academicis, Cambridge, 1705)

Theophrastus, *Enquiry into Plants Vol.I* (trans. A. Hort), (Harvard University Press – Loeb Classical Library, Cambridge, 1968)
Theophrastus, *On Stones* (E.R. Caley and J.F.C. Richards eds.) (Ohio State University Press, Columbus, 1956)
Thucydides, *History of the Peloponnesian War Vol.I* (trans. C.F. Smith), (Harvard University Press – Loeb Classical Library, Cambridge, 1969)
Thucydides, *History of the Peloponnesian War Vol.II* (trans. C.F. Smith), (Cambridge, Harvard University Press – Loeb Classical Library, 1965)
Thucydides, *History of the Peloponnesian War Vol.III* (trans. C.F. Smith), (Harvard University Press – Loeb Classical Library, Cambridge, 1966)
Thucydides, *History of the Peloponnesian War Vol.IV* (trans. C.F. Smith), (Harvard University Press – Loeb Classical Library, Cambridge, 1965)

Vegetius, *Epitome of Military Science* (trans. N.P. Milner), (Liverpool University Press, 2001)
Vitruvius, *On Architecture Vol.II* (trans. F. Granger), (Harvard University Press – Loeb Classical Library, Cambridge, 1934)

Xenophon, *Anabasis* (trans. C.L. Brownson), (Harvard University Press – Loeb Classical Library, Cambridge, 1968)
Xenophon, *Cyropaedia Vol.I* (trans. W. Miller), (Harvard University Press – Loeb Classical Library, Cambridge, 1968)
Xenophon, *Cyropaedia Vol.II* (trans. W. Miller), (Harvard University Press – Loeb Classical Library, Cambridge, 1968)
Xenophon, *Hellenica Vol.I* (trans. C.L. Brownson), (Harvard University Press – Loeb Classical Library, Cambridge, 1978)
Xenophon, *Hellenica Vol.II* (trans. C.L. Brownson), (Harvard University Press – Loeb Classical Library, Cambridge, 1968)
Xenophon, *Memorabilia* (trans. E.C. Marchant), (Harvard University Press – Loeb Classical Library, Cambridge, 1968)
Xenophon, Scripta Minora (trans. E.C. Marchant), (Harvard University Press – Loeb Classical Library, Cambridge, 2000)
Xenophon, *The Persian Expedition* (trans. R. Warner), (Penguin Books, London, 1972)

MODERN TEXTS

Adam-Veleni, P., 'Χάλκινη ασπίδα από τή Βεγόρα τής Φλωρίνας' *Ancient Macedonia 5.1 – Papers Read at the Fifth International Symposium, Thessaloniki* (1993) 17-28
Adcock, F.E., *The Greek and Macedonian Art of War* (University of California Press, Berkeley, 1957)
Aldrete, G.S., Bartell, S. and Aldrete, A., *Reconstructing Ancient Linen Armour – Unraveling the Linothorax Debate* (The Johns Hopkins University Press, Baltimore, 2013)
Anderson, J.K., 'Philopoemen's Reform of the Achaean Army' *CP* 62.2 (1967) 104-106
Anderson, J.K., *Military Theory and Practice in the Age of Xenophon* (University of California Press, Berkeley, 1970)
Anderson, J.K., 'Shields of Eight Palms' Width' *CSCA* 9 (1976) 1-6
Andronicos, M., 'Sarissa' *BCH* 94.1 (1970) 91-107
Andronicos, M., *Olympia* (Ekdotike Athenon, Athens, 1999)
Andronikos, M., 'Vergina: The Royal Graves in the Great Tumulus' *Athens Annals of Archaeology* 10

(1977) 26-27

Anonymous, *The Exercise of the English, in the Militia of the Kingdome of England* (London, 1642)

Anson, E.M., 'Alexander's Hypaspists and the Argyraspids' *Historia* 30.1 (1981) 117-120

Anson, E.M., 'The Hypaspists: Macedonia's Professional Citizen-Soldiers' *Historia* (1985) 246-248

Anson, E.M., 'The Introduction of the Sarisa in Macedonian Warfare' *AncSoc* 40 (2010) 51-68

Australian Army, *Land Warfare Procedures – General LWP-G 7-7-5 Drill*, (Australian Army, 2005)

Aymard, A., 'Philippe de Macédoine, otage á Thèbes' *REA* 56 (1954) 15-26

Badian, E., 'The Battle of the Granicus: A New Look' *Ancient Macedonia* II (1977) 271-293

Baitinger, H., *Die Angriffswaffen aus Olympia* (de Gruyter, Berlin, 2000)

Bar-Sharrar, B. and Borza, E.N. (eds.), *Macedonia and Greece in Late Classical and Early Hellenistic Times* (National Gallery of Art, Washington, 1982)

Bardunias, P., 'Don't Stick to Glued Linen 486– The *Linothorax* Debate' *AncWar* 4.3 (2011) 48-53

Bardunias, P., 'The *Linothorax* Debate: A Response' *Ancient Warfare* 5.1 (2011) 4-5

Bennett, B. and Roberts, M., *The Wars of Alexander's Successors 323-281BC Vol.II: Battles and Tactics* (Barnsley, Pen & Sword, 2009)

Bertosa, B., 'The Supply of Hoplite Equipment by the Athenian State Down to the Lamian War' *JMH* 67.2 (2003) 361-379

Berve, H., *Das Alexanderreich auf prosopographischer Grundlage Vol.II* (Beck, München, 1926)

Best, J.G.P., *Thracian Peltasts and their Influence on Greek Warfare* (Wolters-Noordhoff, Groningen, 1969)

Bickerton, I. and Pearson, M., *43 Days - The Gulf War* (The Text Publishing Company, East Melbourne, 1991)

Bieber, M., 'The Portraits of Alexander the Great' *PAPS* 93.5 (1949) 373-421 + 423-427

Blumberg, A., 'Inspired by the Bard: Philip II, Alexander the Great and the Homeric Ethos' *AncWar* 3.3 (2009) 18-22

Blyth, P.H., *The Effectiveness of Greek Armour against Arrows in the Persian Wars (490-479B.C.): An Interdisciplinary Approach* (British Library Lending Division [unpublished thesis – University of Reading], London, 1977)

Bol, P.C., *Argivische Schilde* (de Gruyter, Berlin, 1989)

Borza, E.N., 'Royal Macedonian Tombs and the Paraphernalia of Alexander the Great' *Phoenix* 41.2 (1987) 105-121

Borza, E.N., *In the Shadow of Olympus: The Emergence of Macedonia* (Princeton University Press, 1990)

Bosworth, A.B., 'ΑΣΘΕΤΑΙΡΟΙ' *CQ* New Series 23.2 (1973) 245-253

Bosworth, A.B., *A Historical Commentary on Arrian's History of Alexander Vol I and II* (Clarendon Books, Oxford, 1998)

Bosworth, A.B., *Conquest and Empire: The Reign of Alexander the Great* (Cambridge University Press, 2008)

Broneer, O., *Isthmia Vol.I: Temple of Poseidon* (American School of Classical Studies at Athens, Princeton, 1971)

Brunt, P.A., 'Anaximenes and King Alexander I of Macedon' *JHS* 96 (1976) 151-153

Burn, A.R., 'The Generalship of Alexander' *Greece & Rome* Second Series 12.2 (1965) 140-154

Bury, J.B., Cook, S.A. and Adcock, F.E. (eds.), *The Cambridge Ancient History Vol.6 – Macedon 401–301BC* (Cambridge University Press, London, 1933)

Callaghan, P.J., 'Macedonian Shields, 'Shield Bowls' and Corinth: A Fixed Point in Hellenistic Ceramic Chronology?' *AAA* 11 (1978) 53-60

Callaghan, P.J., 'On the Date of the Great Altar of Zeus at Pergamon' *BCIS* 28.1 (1981) 115-121

Campbell, B. and Tritle, L.A. (eds.) *The Oxford Handbook of Warfare in the Classical World* (Oxford University Press, 2013)

Campbell. D., 'Alexander's Great Cavalry Battle: What Really Happened at the River Granicus?' *AncWar* 7.2 (2013) 48-53

Carey, B.T., *Hannibal's Last Battle: Zama and the Fall of Carthage* (Pen & Sword, Barnsley, 2007)

Carney, E., 'Macedonians and Mutiny: Discipline and Indiscipline in the Army of Philip and Alexander' *CP* 91.1 (1996) 19-44

Carney, E. and Ogden, D. (eds.), *Philip II and Alexander the Great: Father and Son, Lives and Afterlives* (Oxford University Press, 2010)

Cartledge, P., 'Hoplites and Heroes: Sparta's Contribution to the Technique of Ancient Warfare' *JHS* 97 (1977) 11-27

Carey, M.G., *Operational Art in Classical Warfare: The Campaigns of Alexander the Great* (US Army Command and General Staff College, Fort Leavenworth 1997)

Cascarino, G., *Tecnica della Falange* (Il Cerchio, Rome, 2011)

Cawkwell, G., *Philip of Macedon* (Faber & Faber, London, 1978)

Champion, J., *Pyrrhus of Epirus* (Pen & Sword, Barnsley, 2009)

Chananie, J., 'The Physics of Karate Strikes' *Journal of How Things Work* 1 (1999) 1-4

Chase, G.H., *The Shield Devices of the Greeks in Art and Literature* (Ares, Chicago, 1979)

Choremis, A., 'Metallimos Oplismos apo tou tapho sto Prodromi tes Thesprotias' *Athens Annals of Archaeology* 13 (1980) 10-12

Collins, A.W., 'The Office of Chiliarch under Alexander and the Successors' *Phoenix* 55.3-4 (2001) 259-283

Connolly, P., *Greece and Rome at War* (Greenhill Books, London, 1998)

Connolly, P., 'Experiments with the Sarissa – the Macedonian Pike and Cavalry Lance – a Functional View' *JRMES* 11 (2000) 103-112

Conze, A., *Stadt und Landschaft – Altertümer von Pergamon* 1.2 (Verlag von Georg Reimer, Berlin, 1913) 250-251

Cooke, E., *The Character of Warre* (Tho. Purfoot, London, 1626)

Cotterrell, B. and Kamminga, J., *Mechanics of Pre-Industrial Technology: An Introduction to the Mechanics of Ancient and Traditional Material* (Cambridge University Press, Melbourne, 1990)

Cowan, R., *Imperial Roman Legionary AD161-284* (Osprey, Oxford, 2003)

Cowan, R., 'The Battle of Nisibis, AD217 – The Last Battle of the Parthian Wars' *AncWar* 3.5 (2009) 29-35

Cummings, L.V., *Alexander the Great* (Riverside Press, Boston, 1940)

Daniel, T., 'The Taxeis of Alexander and the change to Chiliarch, the Companion Cavalry and the change to Hipparchies: A Brief Assessment' *AncW* 23.2 (1992) 43-57

Davey, P.R., Thorpe, R.D. and Williams, C., 'Fatigue Decreases Tennis Performance' *Journal of Sports Sciences* 20.4 (2002) 311-318

Dekoulakou-Sideris, I., 'A Metrological Relief from Salamis' *AJA* 94:3 (1990) 445-451

Delbrück, H., *Geschichte der Kriegskunst im Rahmen der politischen Geschichte I* (G. Stilke, Berlin, 1900) 404

Descamps-Lequime, S. (ed.), *Peinture et couleur dans le monde grec antique* (Musée du Louvre, Paris, 2007)

Develin, R., 'Anaximenes ("F Gr Hist" 72) F 4' *Historia* 34.4 (1985) 493-496

Devine, A.M., 'Grand Tactics at Gaugamela' *Phoenix* 29.4 (1975) 374-385

Devine, A.M., 'Demythologizing the Battle of the Granicus' *Phoenix* 40.3 (1986) 265-278

Devine, A.M., 'The Short Sarissa: Tactical reality or Scribal Error?' *AHB* 8:4 (1994) 132

Dickinson, R.E., 'Length Isn't Everything – Use of the Macedonian Sarissa in the Time of Alexander the Great' *JBT* 3.3 (2000) 51-62

Didbin, T.F., *An Introduction to the Knowledge of Rare and Valuable Editions of the Greek and Latin Classics* (London, W. Dwyer, 1804)

Dinsmoor, W.B., 'The Basis of Greek Temple Design in Asia Minor, Greece and Italy' *Atti VII Congresso Internazionale di Archologia Classica I* (L'Erma di Bretschneider, Rome, 1961) 358-361

Dintsis, P., *Hellenistische Helme* (Giorgio Bretschneider Editore, Rome, 1986)

Donlan, W. and Thompson, J., 'The Charge at Marathon: Herodotus 6.112' *Classical Journal* 71.4 (1976) 339-343

Donvito, F., 'Clad in Gold and Silver' *Ancient Warfare* V.6 (2012) 8-10

Dörpfeld, W., 'Beiträge zur antiken Metrologie 1: Das solonisch-attische System' *Ath. Mitt. VII* (Athens, 1882)

Ducrey, P., *Warfare in Ancient Greece* (Schocken Books, New York, 1986)

Edelstein, E.J. and Edelstein, L., *Asclepius – A Collection and Interpretation of the Testimonies* (The Johns Hopkins Press, Baltimore, 1945)

Ellis, J.R., 'Alexander's Hypapsits Again' *Historia* 24.4 (1975) 617-618

Ellis, J.R., *Philip II and Macedonian Imperialism* (Thames & Hudson, London, 1976)

Engels, D.W., *Alexander the Great and the Logistics of the Macedonian Army* (University of California Press, Berkeley, 1978)

English, S., 'Hoplite or Peltast? – Macedonian 'Heavy' Infantry', *Ancient Warfare* 2:1 (2008) 35

English, S., *The Army of Alexander the Great* (Pen & Sword, Barnsley, 2009)

English, S., *The Field Campaigns of Alexander the Great* (Pen & Sword, Barnsley, 2011)

Erskine, A., 'The πεζέταιροι of Philip II and Alexander III' *Historia* 38.4 (1989) 385-394

Everson, T., *Warfare in Ancient Greece* (Sutton Publishing, Stroud, 2004)

Featherstone, D., *Warriors and Warfare in Ancient and Medieval Times* (Constable, London, 1997)

Feyel, M., 'Un nouveau fragment du règlement militaire trouvé à Amphipolis' *Revue archéologique* (1935) 29-68

Fine, J.V.A., *The Ancient Greeks: A Critical History* (Cambridge University Press, 1983)

Forestier, N. and Nougier, V., 'The Effects of Muscular Fatigue on the Co-Ordination of Multijoint Movement in Humans' *Neuroscience Letters* 252.3 (1998) 187-190

Foulon, E., 'Hypaspistes, peltastes, chrysaspides, argyraspides, chalcaspides' *REA* 98 (1996) 53-62

Foulon, E., 'La garde à pied, corps d'élite de la phalange hellénistique' *BAGB* 1 (1996) 17-31

Fuller, J.F.C., *The Generalship of Alexander the Great* (Wordsworth, Hertfordshire, 1998)

Gabriel, R. (ed.), *The Battle Atlas of Ancient Military History* (Canadian Defence Academy Press, Ontario, 2008)

Gabriel, R., *Philip II of Macedonia – Greater than Alexander* (Potomac, Washington, 2010)

Gabriel, R. and Metz, K., *From Sumer to Rome* – The Military Capabilities of Ancient Armies (Greenwood Press, Connecticut, 1991)

Geyer, F., *Makedonien bis zur Thronbesteigung Philipps II* (R. Oldenbourg, Munich, 1939)

Griffith, G.T., *Mercenaries of the Hellenistic World* (Cambridge University Press, 1935)

Griffith, G.T., 'Alexander's Generalship at Gaugamela' *JHS* 67 (1947) 77-89

Griffith, G.T., 'MAKEDONIKA: The Macedonians Under Philip and Alexander' *Proceedings of the Cambridge Philological Society (1956-1957)* 3-10

Griffiths, W.B., 'Re-enactment as Research: Towards a Set of Guidelines for Re-enactors and Academics' *JRMES* 11 (2000) 135-139

Grossman, D., *On Killing* (Back Bay Books, New York, 1996)

Guimier-Sorbets, A.M., Hatzopoulos, M.B. and Morizot, Y. (eds.), *Rois, cites, necropoles: Institutions, rites et monuments en Macedoine* (Meletemata 45) (Research Centre for Greek and Roman Antiquity, Athens, 2006)

Hammond, N.G.L., 'Alexander's Campaign in Illyria' *JHS* 94 (1974) 66-87

Hammond, N.G.L., *Alexander the Great: King, Commander and Statesman* (Duckworth, London, 1980)

Hammond, N.G.L., 'The Battle of the Granicus River' *JHS* 100 (1980) 73-88

Hammond, N.G.L., 'Training in the Use of a Sarissa and its Effects in Battle 359-333 B.C.' *Antichthon* 14 (1980) 53-63

Hammond, N.G.L., 'Arms of the King: The Insignia of Alexander the Great' *Phoenix* 43.3 (1989) 217-224

Hammond, N.G.L., 'Casualties and Reinforcements of Citizen Soldiers in Greece and Macedonia' *JHS* 109 (1989) 56-68

Hammond, N.G.L., 'The Battle between Philip and Bardylis' *Antichthon* 23 (1989) 1-9

Hammond, N.G.L., *The Macedonian State: Origins, Institutions and History* (Clarendon Press, Oxford, 1989)

Hammond, N.G.L., 'Royal Pages, Personal Pages, and Boys Trained in the Macedonian Manner during the Period of the Temenid Monarchy' *Historia* 39.3 (1990) 261-290

Hammond, N.G.L., 'The Various Guards of Philip II and Alexander III' *Historia* 40.4 (1991) 396-418

Hammond, N.G.L., 'A Macedonian Shield and Macedonian Measures' *ABSA* 91 (1996) 365-367

Hammond, N.G.L., *The Genius of Alexander the Great* (Duckworth, London, 1997)

Hammond, N.G.L., 'What May Philip Have Learnt as a Hostage in Thebes?' *GRBS* 38.4 (1997) 355-372

Hammond, N.G.L., 'The Continuity of Macedonian Institutions and the Macedonian Kingdoms of the Hellenistic Era' *Historia* 49.2 (2000) 141-160

Hammond, N.G.L. and Griffith, G.T., *A History of Macedonia Vol.II* (Oxford University Press, 1979)

Hammond, N.G.L. and Walbank, F.W., *A History of Macedonia III 336-167 BC* (Oxford University Press, 1988)

Hatzopoulos, M.B., 'La Béotie et la Macédoine à l'époque de l'hégémonie Thébaine' in *La Béotie antique: Lyon-Saint-Étienne – 16-20 Mai 1983* (Editions du Centre national de la recherche scientifique, Paris, 1985) 247-257

Hatzopoulos, M.B., *L'organization de l'armee macédonienne sous les Antigonides* (Kentron Hellēnikēs kai Rōmaïkēs Archaiotētos, Ethnikon Hidryma Ereunōn, Athens, 2001)

Hatzopoulos, M.B., 'The Burial of the Dead (at Vergina) or The Unending Controversy on the Identity of the Occupants of Tomb II' *Tekmeria* 9 (2008) 91-118

Hatzopoulos, M.B. and Juhel, P., 'Four Hellenic Funerary Stelae from Graphyra, Macedonia' *AJA* 113.3 (2009) 423-437

Hatzopoulos, M.B. and Loukopoulos, L.D. (eds.), *Philip of Macedon* (Ekdotike Athenon, Athens, 1980)

Head, D., *Armies of the Macedonian and Punic Wars 359 BC to 146 BC* (Wargames Research Group, Sussex, 1982)

Head, D., 'The Thracian Sarissa' *Slingshot* 214 (2001) 10-13

Heckel, W., *The Wars of Alexander the Great 336-323BC* (Osprey, Oxford, 2002)

Heckel, W., *The Conquests of Alexander the Great* (Cambridge University Press, 2008)

Heckel, W. and Jones, R., *Macedonian Warrior: Alexander's Elite Infantryman* (Osprey, Oxford, 2006)

Heckel, W., *Who's Who in the Age of Alexander the Great* (Wiley-Blackwell, West Sussex, 2009)

Heisserer, A.J., *Alexander the Great and the Greeks: the Epigraphic Evidence* (University of Oaklahoma Press, Norman, 1980)

Hockey, R. (ed.), *Stress and Fatigue in Human Performance* (Wiley and Sons, Chichester, 1983)

Holt, F.L., 'Imperium Macedonicum and the East: The Problem of Logistics' *Ancient Macedonia V* (1989) 585-592

Homolle, M.T., *Fouilles de Delphes – Tome V* (Ancienne Librairie Thorin et Fils, Paris, 1908)

How, W.W., Arms, Tactics and Strategy in the Persian Wars *JHS* 43.2 (1923) 117-132

Hultsch, F., *Griechische und Römische Metrologie – 2nd Ed.* (Weidmann, Berlin, 1882)

Jacoby, F., *Die Fragmente der Griechischen Historiker I-III* (Brill, Lieden, 1954)

Jacoby, F., *Die Fragmente der Griechischen Historiker IV* (Brill, Lieden, 1957)

Jaeckel, P., 'Pergamenische Waffenreliefs' *Jarbuch Waffen und Kostumkunde* (Dt. Kunstverl, Munich, 1965)

Jarva, E., *Archaiologia on Archaic Greek Body Armour* (Studia Archaeologica Septentrionalia 3), (: Pohjois-Suomen Historiallinen Yhdistys, Societas Historica Finlandiae Septentrionalis, Rovaniemi, 1995)

Juhel, P., 'Fragments de "boucliers macédoniens" au nom de Roi Démétrios trouvés à Staro Bonce (République de Macédoine)' *Zeitschrift für Papyrologie und Epigrafik 162* (2007) 165-180

Kagan, D. and Viggiano, G.F., *Men of Bronze – Hoplite Warfare in Ancient Greece* (Princeton University Press, 2013)

Karunanithy, D., 'Of Ox-Hide Helmets and Three-Ply Armour: The Equipment of Macedonian Phalangites as Described through a Roman Source' *Slingshot* 213 (2001) 33-40

Kastorchis, E., 'ΠΕΡΙ ΤΟΥ ΕΝ ΧΑΙΡΩΝΕΙΑ ΛΕΟΝΤΟΣ' *Athenaion* 8 (1879) 486-491

Kinch, K.F., 'Le tombeau de Naiusta. Tombeau Macédonien' *Videnskabernes Selskabs Skrifter – Historisk og Filosofisk Afdeling* 7.4 (1917-1937) 281-288

Kingsley, B.M., 'The Cap that Survived Alexander' *AJA* 85 (1981) 39-46

Koepfer, C., 'The Sarissa', *AncWar* 3.2 (2009) 37

Krueger, G.P., 'Sustained Work, Fatigue, Sleep Loss and Performance: A Review of the Issues' *Work and Stress* 3:2 (1989) 129-141

Kromayer, J. and Veith, G., *Heerwesen und Kriegführung und Geichen und Römer* (C.H. Beck, Munich, 1928)

Kunze, E., *Bericht Über die Ausgrabungen in Olympia – Vol.V* (Verlag Walter de Gruyter and Co., Berlin, 1956)

Kunze, E., *Bericht Über die Ausgrabungen in Olympia – Vol.VI* (Verlag Walter de Gruyter and Co., Berlin, 1958)

Lane Fox, R., *Alexander the Great* (Penguin, London, 1973)

Lane Fox, R. (ed.), *Brill's Companion to Ancient Macedon: Studies in the Archaeology and History of Macedon, 650 BC-300 AD* (Brill, Leiden, 2011)

Lammert, F., 'Sarisse' *RE Second Series IA* (Stuttgart 1920)

Launey, M., *Recherches sur les armées hellénistiques: Bibliothèque des écoles françaises d'Athènes et de Rome* (de Boccard, Paris, 1949)

Lawrence, D.R., *The Complete Soldier – Military Books and Military Culture in Early Stuart England, 1603-1645* (Brill, Leiden, 2009)

le Bohec, Y. and Wolff, C. (eds.), *L'Armée Romaine de Dioclétien à Valentinien Ier* (de Boccard, Paris, 2004)

Liampi, K., 'Der makedonische Schild als propagandistiches Mittel' *Meletemata* 10 (1990) 157-171

Liampi, K., *Der makedonische Schild* (Rudolf Habelt, Bonn, 1998)

Lock, R.A., 'The Origins of the Argyraspids' *Historia* 26.3 (1977) 373-378

Lorimer, H.L., *Homer and the Monuments* (Macmillan, London, 1950)

Lumpkin, H., 'The Weapons and Armour of the Macedonian Phalanx' *JAAS* 58.3 (1975) 193-208

Lush, D, 'Body Armour in the Phalanx of Alexander's Army' *AncW* 38.1 (2007) 15-37

McDonnell-Staff, P., 'Sparta's Last Hurrah – The Battle of Sellasia (222BC)' *AncWar* 2.2 (2008) 23-29

McDonnell-Staff, P., 'Hypaspists to Peltasts: The Elite Guard Infantry of the Antigonid Macedonian Army' *Ancient Warfare* 5.6 (2012) 20-25

McKechnie, P. 'Greek Mercenary Troops and their Equipment' *Historia* 43.3 (1994) 297-305

Manti, P., 'The Sarissa of the Macedonian Infantry' *AncW* 23.2 (1992) 31-42

Manti, P., 'The Macedonian Sarissa, Again' *AncW* 25.2 (1994) 77-91

Markle, M.M., 'The Macedonian Sarissa, Spear and Related Armour' *AJA* 81.3 (1977) 323-339

Markle, M.M., 'Use of the Sarissa by Philip and Alexander of Macedon' *AJA* 82.4 (1978) 483-497

Markle, M.M., 'A Shield Monument from Veria and the Chronology of Macedonian Shield Types' *Hesperia* 68.2 (1999) 219-254

Marsden, E.W., *Greek and Roman Artillery Vol.II – Technical Treatises* (Clarendon Press, Oxford, 1971)

Mashiro, N., *Black Medicine: The Dark Art of Death – The Vital Points of the Human Body in Close Combat* (Paladin Press, Boulder, 1978)

Matthew, C.A., 'When Push comes to Shove: What was the Othismos of Hoplite Combat?' *Historia* 58.4 (2009) 395-415

Matthew, C.A., 'For Valour: The 'Shield Coins' of Alexander and the Successors' *JNAA 20* (2009/2010) 15-35

Matthew, C.A., *On the Wings of Eagles: The Reforms of Gaius Marius and the Creation of Rome's First Professional Soldiers* (Cambridge Scholars, Newcastle upon Tyne, 2010)

Matthew, C.A., 'Testing Herodotus – Using Re-creation to Understand the Battle of Marathon' *AncWar* 5.4 (2011) 41-46

Matthew, C.A., *A Storm of Spears: Understanding the Greek Hoplite at War* (Pen & Sword, Barnsley, 2012)

Matthews, R., *The Battle of Thermopylae – A Campaign in Context* (History Press, Stroud, 2008)

Mattingly, H.B., 'The Athenian Coinage Decree' *Historia* 10 (1961) 148-188

Mattingly, H.B., 'Epigraphy and the Athenian Empire' *Historia* 41 (1992) 129-138

Mattingly, H.B., 'New Light on the Athenian Standards Decree' *Klio* 75 (1993) 99-102

Meiggs, R., *Trees and Timber in the Ancient Mediterranean World* (Clarendon Press, Oxford, 1982)

Meiggs, R. and Lewis, D., *A Selection of Greek Historical Inscriptionsto the End of the Fifth Century BC* (Oxford University Press, Oxford, 2004)

Melville Jones, J.R., *Testimonia Numaria Vol.II* (Spink and Sons, London, 2007)

Michaelis, A., 'The Metrological Relief at Oxford' *JHS* 4 (1883) 335-350

Milns, R.D., 'Alexander's Pursuit of Darius through Iran' *Historia* 15 (1966) 256

Milns, R.D., 'Philip II and the Hypaspists' *Historia* 16.4 (1967) 509-510

Milns, R.D., 'The Hypaspists of Alexander III – Some Problems' *Historia* 20 (1971) 187-188

Milns, R.D., 'Arrian's Accuracy in Troop Details: A Note' *Historia* 27:2 (1975) 375

Mitchiner, M., *Indo-Greek and Indo-Scythian Coinage Vol.I – The Early Indo-Greeks and their Antecedents* (Hawkins Publications, Sanderstead, 1975)

Mixter, J.R., 'The Length of the Macedonian Sarissa During the Reigns of Philip II and Alexander the Great' *AncW* 23.2 (1992) 21-29

Moormann, E.M., *Ancient Sculpture in the Allard Pierson Museum Amsterdam* (Allard Pierson Series, Amsterdam, 2000)

Morgan, J.D., 'Sellasia Revisited' *AJA* 85.3 (1981) 328-330

Neumann, C., 'A Note on Alexander's March-Rates' *Historia* 20 (1971) 196-198

Noguera, B.A., 'L'évolution de la phalange macédonienne: le cas de la sarisse' *Ancient Macedonia* 6.2 (1999) 839-850

Nylander, C., 'The Standard of the Great King: A Problem with the Alexander Mosaic' *OpRom* 14.2 (1983) 19-37

Pandermalis, D., 'Basile[ōs Dēmētr]iou' Myrtos: Mnēme Ioulias Vokotopoulou (Thessaloniki, 2000) xviii-xxii

Park, M., 'The Fight for Asia – The Battle of Gabiene' *AncWar* 3.2 (2009) 33-35

Parke, H.W., *Greek Mercenary Soldiers* (Ares, Chicago, 1981)

Peltz, U., 'Der Makedonische Schild aus Pergamon der Antikensammlung Berlin', *Jahrbuch der Berliner Museen*, Bd. 43 (2001) 331-343

Petsas, P.M., Ἀνασκοφὴ ἀρχαίου νεκροταφείου Βεργίνης [1960-1961]' Archaeologikon Deltion 17A (1961-1962) 218-288

Phytalis, L., 'ΕΡΕΥΝΑΙ ΕΝ ΤΩ ΠΟΛΥΑΝΔΡΙΩ ΧΑΙΡΩΝΕΙΑΣ' *Athenaion* 9 (1880) 347-352

Pietrykowski, J., *Great Battles of the Hellenistic World* (Pen & Sword, Barnsley, 2009)

Pietrykowski, J., 'In the School of Alexander – Armies and Tactics in the Age of the Successors' *AncWar* 3.2 (2009) 21-28

Plant, R., *Greek Coins Types and their Identification* (Seaby, London, 1979)

Pleket, H.W. and Stroud, R.S. (eds), *Supplementum Epigraphicum Graecum Vol.XL* (Brill, Lieden, 1990)

Pleket, H.W. and Stroud, R.S. (eds), *Supplementum Epigraphicum Graecum Vol.XXXVI* (Brill, Lieden, 1986)

Post, R., 'Alexandria's Colourful Tombstones – Ptolemaic Soldiers Reconstructed' *AncWar* 1.1 (2007) 38-43

Price, M.J., *The Coinage in the name of Alexander the Great and Philip Arrhidaeus Vol.I: Introduction and Catalogue* (British Museum Press, London, 1991)

Price, M.J., *The Coinage in the name of Alexander the Great and Philip Arrhidaeus Vol.II: Concordances, Indexes and Plates* (British Museum Press, London, 1991)

Pritchett, W.K., *Ancient Greek Military Practices: Part I* (University of California Press, Berkeley, 1971)

Pritchett, W.K., *The Greek State at War – Part II* (University of California Press, Berkeley, 1974)

Pritchett, W.K., *The Greek State at War – Part IV* (University of California Press, Berkeley, 1985)

Rahe, P.A., 'The Annihilation of the Sacred Band at Chaeronea' *AJA* 85.1 (1981) 84-87

Reinach, A.J., 'La frise du monument de Paul-Émile à Delphes' *BCH* 34 (1910) 444-446

Richardson, W.F., *Numbering and Measuring in the Classical World* (Bristol Phoenix Press, Bristol, 2004)

Roesch, P., 'Un Décret inédit de la Ligue Thébaine et la flotte d'Epaminindas' *REG* 97 (1984) 45-60

Roy, I. (ed.), *Military Memoirs: Blaise de Monluc: The Valois-Habsburg Wars and the French Wars of Religion* (Longman, London, 1971)

Royal, K., Farrow, D., Mujika, I., Hanson, S., Pyne, D. and Abernethy, B., 'The Effects of Fatigue on Decision Making and Shooting Skill in Water Polo Players' *Journal of Sports Sciences* 24.8 (2006) 807-815

Saatsoglou-Paliadeli, C., 'Aspects of Ancient Macedonian Costume' *JHS* 113 (1993) 122-147

Sabin, P., *Lost Battles* (Continuum, London, 2009)

Sabin, P., van Wees, H., and Whitby, M. (eds.), *The Cambridge History of Greek and Roman Warfare Vol. I: Greece, the Hellenistic World and the Rise of Rome* (Cambridge University Press, 2007)

Salazar, C.F., *The Treatment of War Wounds in Greco-Roman Antiquity* (Brill, Leiden, 2000)

Sekunda, N., *The Seleucid Army* (Montvert, Stockport, 1994)

Sekunda, N., *The Army of Alexander the Great* (Osprey, Oxford, 1999)

Sekunda, N., *Greek Hoplite 480-323 BC* (Osprey, Oxford, 2004)

Sekunda, N., *The Persian Army 560-330BC* (Osprey, Oxford, 2005)

Sekunda, N., *Macedonian Armies after Alexander 323-168BC* (Osprey, Oxford, 2012)

Seymour, T.D., *Life in the Homeric Age* (Biblo & Tannen, New York, 1963)

Sheppard, R. (ed.), *Alexander the Great at War* (Osprey, Oxford, 2008)

Sinclair, R.K., 'Diodorus Siculus and Fighting in Relays' *CQ* 16.2 (1966) 249-255

Skarmintzos, S., 'Phalanx versus Legion: Greco-Roman Conflict in the Hellenistic Era' *AncWar* 2.2 (2008) 30-34

Smith, A., 'The Anatomy of Battle – Testing Polybius' Formations' *AncWar* 5.5 (2011) 41-45

Smythe, J., *Instructions, Observations and Orders Mylitarie*, (R. Johnes, London, 1595)

Snodgrass, A.M., *Early Greek Armour and Weapons* (Edinburgh University Press, Edinburgh, 1964)

Snodgrass, A.M., *Arms and Armour of the Greeks* (The John Hopkins University Press, Baltimore, 1999)

Sotiriades, G., 'Das Schlachtfeld von Chäronea' *AthMitt* 28 (1903) 301-330

Spence, I.G., *The Cavalry of Classical Greece* (Clarendon Press, Oxford, 1993)

Stamatopoulou, M. and Yeroulanou, M. (eds.), *Excavating Classical Greek Culture: Recent Archaeological Discoveries in Greece* (Beazley Archive and Archaeopress, Oxford, 2002)

Stein, A., *On Alexander's Track to the Indus* (Asian Publications, Lucknow, 1985)

Stroszeck, J., 'Lakonisch-rotfigurige Keramik aus den Lakedaimoniergräbern am Kerameikos von Athen' *AA* 2 (2006) 101-120

Sumner, G., *Roman Military Clothing (2) AD200-400* (Osprey, Oxford, 2003)

Tarn, W.W., *Hellenistic Military and Naval Developments* (Cambridge University Press, 1930)

Tarn, W.W., *Alexander the Great Vol.II* (Ares, Chicago, 1981)

Thomas, C.G., 'What you seek is Here: Alexander the Great' *Journal of the Historical Society* 7.1 (2007) 61-83

Thompson, M., *Granicus 334BC: Alexander's First Persian Victory* (Osprey, Oxford, 2007)

Tsimbidou-Avloniti, M., *Makedonikoi Taphoi ston Phoinika kai ston Aghio Athanasio Thessalonikes* (Ekdose tou Tameiou Archaiologikon Poron kai Apallotrioseon, Athens, 2005)

Tudela, F., 'The Hoplite's Second Shield – Defensive Roles of Greek Cavalry', *AncWar* 2:4 (2008) 36-39

Ueda-Sarson, L., 'The Evolution of Hellenistic Infantry Part 1: The Reforms of Iphicrates' *Slingshot* 222 (2002) 30-36

Ueda-Sarson, L., 'The Evolution of Hellenistic Infantry Part 2: Infantry of the Successors' *Slingshot* 223 (2002) 23-28

US Army, *FM7-8 Infantry Rifle Platoon and Squad* (US Dept. of the Army, Washington, 1992)

US Army, *FM22-5 Drill and Parades* (US Dept. of the Army, Washington, 1986)

van Hook, L., 'On the Lacedaemonians Buried in the Kerameikos' *AJA* 36.3 (1932) 290-292

van Wees, H., *Greek Warfare – Myths and Realities* (Duckworth, London, 2004)

von Clausewitz, C., *On War* (Penguin, London, 1982)

Walbank, F.W., *Philip V of Macedon* (Cambridge University Press, Cambridge, 1940)

Warry, J., *Alexander 334-323BC* (Osprey, Oxford, 1991)

Warry, J., *Warfare in the Classical World* (University of Oklahoma Press, Norman, 1995)

Webster, G., *The Roman Imperial Army of the First and Second Centuries A.D.* (University of Oklahoma Press, Norman, 1988)

Webster, J., 'In Search of the Iphikratean Peltast/Hoplite' *Slingshot* 287 (2013) 5-6

Wells, C.B., 'New Texts from the Chancery of Philip V of Macedonia and the problem of the Diagramma', *AJA* 42 (1938) 245-260

Wheatley, P. and Hannah, R. (eds.), *Alexander and His Successors – Essays from the Antipodes* (Regina Books, Claremont, 2009)

Wheeler, E.L., 'The Legion as Phalanx in the Late Empire (II)' *RÉMA* 1 (2004) 147-175

Williams, A., *The Knight and the Blast Furnace – A History of the Metallurgy of Armour in the Middle Ages and the Early Modern Period* (Brill, Leiden, 2003)

Williams, M.F., 'Philopoemen's Special Forces: *Peltasts* and a New Kind of Greek Light-Armed Warfare (Livy 35.27)' *Historia* 53.3 (2004) 257-277

Wood, M., *In the Footsteps of Alexander the Great* (BBC Worldwide, London, 2001)

Worthington, I., *Philip II of Macedonia* (Yale University Press, New Haven, 2008)

Worthington, I. (ed.), *Alexander the Great – A Reader* (Routledge, London, 2012)

Wrightson, G., 'The Nature of Command of the Macedonian Sarissa Phalanx' *AHB* 24.3-4 (2010) 73-94

REFERENCE TEXTS
Liddell, H.G. and Scott, R. (eds.), *A Greek-English Lexicon* (Clarendon Press, Oxford, 1968)

Index

spear, 89–90, 101, 376
    defensive use of, 187, 200,
205–206, 215, 223
drill movements with *see*
  Phalangite(s)-drill movements
  of
grip used to carry, 83–8, 133,
  376
    effect on intervals of the
    phalanx, 83–7, 143–8,
    152–3, 363
Macedonian word for, 18
possible Balkan origins of, 12
possible Egyptian origins of, 13
possible Homeric origins of, 32
possible Macedonian origins
  of, 32
possible term for a single-piece
  weapon, 59
possible Thracian origins of,
  12, 25, 31
reach of, 3, 31, 79, 167, 187–8,
  336–7
    combat (battle) range of,
    168–70, 227, 247, 253, 382
    effective range of, 168–83,
    201, 247, 253, 375, 383, 399
    tests to determine, 171–82
    trajectory of attacks, 178–81
repair of, 56–7, 64–5, 89
similarity to Iphicratean pike,
  14–18
transportation of, 56
unsuitability for close combat,
  127
unsuitability for hunting, 53
use as an improvised weapon,
  57, 99, 164
weight of, 91, 130–1, 165, 237,

350
sarissophoroi ('*sarissa* bearers'),
  xxvi
Sarpedon, 193
Scythia/Scythians,
  use of cavalry wedge by, 45
    adoption by Philip II, 45
Seleucus,
  at the battle of the Hydaspes
    (327BC), 347
Seleucids, 14
  at battle of Beth Zachariah
    (162BC), 102, 350
  coins of, 110
Sellasia, battle of (222BC), 159,
  165, 329, 373, 388–9
  casualties at, 197
  description of terrain, 329,
    331–2, 364
  Macedonian actions at, 332–3
  Macedonian deployment at,
    332, 360, 364–5
  pre-battle meeting at, 349
  signals used at, 349–50
  Spartan actions at, 333, 364
  Spartan deployment at, 331–2
Septimius Severus, 403
Servius Tullius, 8
Severus Alexander, 404
Shield Coins, 103–104, 111
Shields, *see aspis, peltē*
Sicily/Sicilians,
  system of measurements used
    by, 71
Silver Shields (*argyrasapides*),
  212
  see also hypaspists
  at the battle of Magnesia
    (190BC), 102